Leif B.G. Andersen and Vladimir V. Piterbarg

Interest Rate Modeling

Volume III: Products and Risk Management

Atlantic Financial Press

London New York

Leif B.G. Andersen
Bank of America Merrill Lynch
One Bryant Park
New York, NY 10036
USA
leif.andersen@baml.com

Vladimir V. Piterbarg
Barclays
5 The North Colonnade
London, E14 4BB
UK
vladimir.piterbarg@barclays.com

Mathematics Subject Classification (2010): 60H10, 60H35, 62P05, 65C05, 65C20, 91G20, 91G30, 91G60, 91G80

JEL Classification: G12, G13, E43

LSI Subjects: BUS036540 (Business & Economics : Investments & Securities - Options), BUS091000 (Business & Economics : Business Mathematics), MAT003000 (Mathematics : Applied)

Library of Congress Control Number: 2010905508

Suggested reference:

Interest Rate Modeling, Volume III: Products and Risk Management, by Leif B.G. Andersen and Vladimir V. Piterbarg, 1st edition, Atlantic Financial Press, 2010

The present volume is the original first hardcover edition, third printing.

Errata is available at www.andersen-piterbarg-book.com

Submit typos and corrections at www.andersen-piterbarg-book.com

ISBN-10: 0-9844221-2-9
ISBN-13: 978-0-9844221-2-8

Printed by Lightning Source Ltd.

ref v3.hc.003

To our families

— L.B.G.A, V.V.P

Preface to Volume III

The first half of Volume III of this book contains a detailed study of several classes of fixed income securities, ranging from simple vanilla options to highly exotic cancelable and path-dependent trades. The analysis is done in product-specific fashion, covering, among other subjects, risk characterization, calibration strategies, and valuation methods. In its second half, Volume III studies the general topic of derivative portfolio risk management, with a particular emphasis on the challenging problem of computing smooth price sensitivities to market input perturbations.

London, New York,
June 2004 — August 2010

Leif B.G Andersen
Vladimir V. Piterbarg

Table of Contents for All Volumes

VOLUME I Foundations and Vanilla Models

Part I Foundations

1   Introduction to Arbitrage Pricing Theory   3
   1.1   The Setup   3
   1.2   Trading Gains and Arbitrage   7
   1.3   Equivalent Martingale Measures and Arbitrage   8
   1.4   Derivative Security Pricing and Complete Markets   11
   1.5   Girsanov's Theorem   12
   1.6   Stochastic Differential Equations   14
   1.7   Explicit Trading Strategies and PDEs   17
   1.8   Kolmogorov's Equations and the Feynman-Kac Theorem .   18
   1.9   Black-Scholes and Extensions   21
       1.9.1   Basics   22
       1.9.2   Alternative Derivation   25
       1.9.3   Extensions   27
           1.9.3.1   Deterministic Parameters and Dividends   27
           1.9.3.2   Stochastic Interest Rates   28
   1.10 Options with Early Exercise Rights   30
       1.10.1   The Markovian Case   32

1.10.2 Some General Bounds 34

1.10.3 Early Exercise Premia 36

2 Finite Difference Methods 43

2.1 1-Dimensional PDEs: Problem Formulation 43

2.2 Finite Difference Discretization 45

    2.2.1 Discretization in x-Direction. Dirichlet Boundary Conditions 45

    2.2.2 Other Boundary Conditions 48

    2.2.3 Time-Discretization 49

    2.2.4 Finite Difference Scheme 50

2.3 Stability 52

    2.3.1 Matrix Methods 52

    2.3.2 Von Neumann Analysis 53

2.4 Non-Equidistant Discretization 56

2.5 Smoothing and Continuity Correction 58

    2.5.1 Crank-Nicolson Oscillation Remedies 58

    2.5.2 Continuity Correction 59

    2.5.3 Grid Shifting 59

2.6 Convection-Dominated PDEs 61

    2.6.1 Upwinding 62

    2.6.2 Other Techniques 63

2.7 Option Examples 63

    2.7.1 Continuous Barrier Options 64

    2.7.2 Discrete Barrier Options 66

    2.7.3 Coupon-Paying Securities and Dividends 67

    2.7.4 Securities with Early Exercise 68

    2.7.5 Path-Dependent Options 69

    2.7.6 Multiple Exercise Rights 71

2.8 Special Issues 72

    2.8.1 Mesh Refinements for Multiple Events 72

    2.8.2 Analytics at the Last Time Step 76

    2.8.3 Analytics at the First Time Step 78

2.9 Multi-Dimensional PDEs: Problem Formulation 79

2.10 Two-Dimensional PDE with No Mixed Derivatives 80

    2.10.1 Theta Method 81

    2.10.2 The Alternating Direction Implicit (ADI) Method 82

    2.10.3 Boundary Conditions and Other Issues 85

2.11 Two-Dimensional PDE with Mixed Derivatives 85

    2.11.1 Orthogonalization of the PDE 86

    2.11.2 Predictor-Corrector Scheme 89

2.12 PDEs of Arbitrary Order 91

3   Monte Carlo Methods   95
    3.1   Fundamentals   95
        3.1.1   Generation of Random Samples   97
            3.1.1.1   Inverse Transform Method   98
            3.1.1.2   Acceptance-Rejection Method   99
            3.1.1.3   Composition   101
        3.1.2   Correlated Gaussian Samples   103
            3.1.2.1   Cholesky Decomposition   103
            3.1.2.2   Eigenvalue Decomposition   104
        3.1.3   Principal Components Analysis (PCA)   105
    3.2   Generation of Sample Paths   106
        3.2.1   Example: Asian Basket Options in Black-Scholes
                 Economy   106
        3.2.2   Discretization Schemes, Convergence, and Stability   108
        3.2.3   The Euler Scheme   110
            3.2.3.1   Linear-Drift SDEs   112
            3.2.3.2   Log-Euler Scheme   112
        3.2.4   The Implicit Euler Scheme....................   113
            3.2.4.1   Implicit Diffusion Term   114
        3.2.5   Predictor-Corrector Schemes   115
        3.2.6   Ito-Taylor Expansions and Higher-Order Schemes.   116
            3.2.6.1   Ordinary Taylor Expansion of ODEs ...   117
            3.2.6.2   Ito-Taylor Expansions   118
            3.2.6.3   Milstein Second-Order Discretization
                       Scheme   119
        3.2.7   Other Second-Order Schemes   121
        3.2.8   Bias vs. Monte Carlo Error   122
        3.2.9   Sampling of Continuous Process Extremes   124
        3.2.10   PCA and Bridge Construction of Brownian
                  Motion Paths   128
            3.2.10.1   Brownian Bridge and Quasi-Random
                        Sequences   128
            3.2.10.2   PC Construction   130
    3.3   Sensitivity Computations   131
        3.3.1   Finite Difference Estimates   132
            3.3.1.1   Black-Scholes Delta.................   132
            3.3.1.2   General Case   133
        3.3.2   Pathwise Estimate   135
            3.3.2.1   Black-Scholes Delta.................   135
            3.3.2.2   General Case   136
            3.3.2.3   Sensitivity Path Generation..........   138
        3.3.3   Likelihood Ratio Method   139
            3.3.3.1   Black-Scholes Delta.................   139
            3.3.3.2   General Case   140
            3.3.3.3   Euler Schemes   140

                3.3.3.4   Some Remarks 142
    3.4 Variance Reduction Techniques........................ 142
         3.4.1   Variance Reduction and Efficiency 143
         3.4.2   Antithetic Variates 144
                3.4.2.1   The Gaussian Case 144
                3.4.2.2   General Case 145
         3.4.3   Control Variates.............................. 145
                3.4.3.1   Basic Idea 145
                3.4.3.2   Non-Linear Controls 147
         3.4.4   Importance Sampling 149
                3.4.4.1   Basic Idea 149
                3.4.4.2   Density Formulation 149
                3.4.4.3   Importance Sampling and SDEs 151
                3.4.4.4   More on SDE Path Simulation 152
                3.4.4.5   Rare Event Simulation and Linearization 154
    3.5 Some Notes on Bermudan Security Pricing.............. 158
         3.5.1   Basic Idea 158
         3.5.2   Parametric Lower Bound Methods 159
         3.5.3   Parametric Lower Bound: An Example 160
         3.5.4   Regression-Based Lower Bound 161
         3.5.5   Upper Bound Methods 163
         3.5.6   Confidence Intervals 164
         3.5.7   Other Methods 164
    3.A Appendix: Constants for Φ^{-1} Algorithm 165

4   Fundamentals of Interest Rate Modeling............... 167
    4.1 Fixed Income Notations 167
         4.1.1   Bonds and Forward Rates 167
         4.1.2   Futures Rates 169
         4.1.3   Annuity Factors and Par Rates 170
    4.2 Fixed Income Probability Measures 171
         4.2.1   Risk Neutral Measure 172
         4.2.2 T-Forward Measure 174
         4.2.3   Spot Measure 175
         4.2.4   Terminal and Hybrid Measures 176
         4.2.5   Swap Measures 178
    4.3 Multi-Currency Markets............................. 178
         4.3.1   Notations and FX Forwards................... 178
         4.3.2   Risk Neutral Measures 179
         4.3.3   Other Measures 180
    4.4 The HJM Analysis................................. 181
         4.4.1   Bond Price Dynamics 181
         4.4.2   Forward Rate Dynamics 182
         4.4.3   Short Rate Process 183
    4.5 Examples of HJM Models 184

4.5.1   The Gaussian Model 184
4.5.2   Gaussian HJM Models with Markovian Short Rate   187
4.5.3   Log-Normal HJM Models 189

5   Fixed Income Instruments 191
5.1   Fixed Income Markets and Participants 191
5.2   Certificates of Deposit and Libor Rates................. 194
5.3   Forward Rate Agreements (FRA)...................... 195
5.4   Eurodollar Futures 196
5.5   Fixed-for-Floating Swaps 197
5.6   Libor-in-Arrears Swaps............................. 200
5.7   Averaging Swaps 201
5.8   Caps and Floors................................... 201
5.9   Digital Caps and Floors 203
5.10 European Swaptions 203
      5.10.1 Cash-Settled Swaptions 205
5.11 CMS Swaps, Caps and Floors 206
5.12 Bermudan Swaptions 207
5.13 Exotic Swaps and Structured Notes 208
      5.13.1 Libor-Based Exotic Swaps 209
      5.13.2 CMS-Based Exotic Swaps 210
      5.13.3 Multi-Rate Exotic Swaps 210
      5.13.4 Range Accruals 211
      5.13.5 Path-Dependent Swaps 212
5.14 Callable Libor Exotics 213
      5.14.1 Definitions................................. 213
      5.14.2 Pricing Callable Libor Exotics................ 215
      5.14.3 Types of Callable Libor Exotics 216
      5.14.4 Callable Snowballs.......................... 216
      5.14.5 CLEs Accreting at Coupon Rate 216
      5.14.6 Multi-Tranches 217
5.15 TARNs and Other Trade-Level Features 217
      5.15.1 Knock-out Swaps 218
      5.15.2 TARNs...................................... 218
      5.15.3 Global Cap 219
      5.15.4 Global Floor 219
      5.15.5 Pricing and Trade Representation Challenges 220
5.16 Volatility Derivatives.............................. 220
      5.16.1 Volatility Swaps 220
      5.16.2 Volatility Swaps with a Shout 221
      5.16.3 Min-Max Volatility Swaps 222
      5.16.4 Forward Starting Options and Other Forward
              Volatility Contracts 222
5.A   Appendix: Day Counting Rules and Other Trivia 224
      5.A.1 Libor Rate Definitions 224
      5.A.2 Swap Payments 225

Part II Vanilla Models

6   **Yield Curve Construction and Risk Management** 229
  6.1 Notations and Problem Definition 230
      6.1.1 Discount Curves 230
      6.1.2 Matrix Formulation 232
      6.1.3 Construction Principles and Yield Curves........ 232
  6.2 Yield Curve Fitting with N-Knot Splines 234
      6.2.1 C^0 Yield Curves: Bootstrapping 234
          6.2.1.1 Piecewise Linear Yields 235
          6.2.1.2 Piecewise Flat Forward Rates 236
      6.2.2 C^1 Yield Curves: Hermite Splines............... 238
      6.2.3 C^2 Yield Curves: Twice Differentiable Cubic Splines 240
      6.2.4 C^2 Yield Curves: Twice Differentiable Tension
          Splines 243
  6.3 Non-Parametric Optimal Yield Curve Fitting 245
      6.3.1 Norm Specification and Optimization 245
      6.3.2 Choice of λ 248
      6.3.3 Example 249
  6.4 Managing Yield Curve Risk........................... 250
      6.4.1 Par-Point Approach........................... 251
      6.4.2 Forward Rate Approach 252
      6.4.3 From Risks to Hedging: The Jacobian Approach .. 254
      6.4.4 Cumulative Shifts and other Common Tricks 256
  6.5 Various Topics in Discount Curve Construction.......... 258
      6.5.1 Curve Overlays and Turn-of-Year Effects 258
      6.5.2 Cross-Currency Curve Construction 259
          6.5.2.1 Basic Problem 259
          6.5.2.2 Separation of Discount and Forward
                Rate Curves 260
          6.5.2.3 Cross-Currency Basis Swaps 262
          6.5.2.4 Modified Curve Construction Algorithm 263
      6.5.3 Tenor Basis and Multi-Index Curve Group
          Construction 265
  6.A Appendix: Spline Theory 270
      6.A.1 Hermite Spline Theory 270
      6.A.2 C^2 Cubic Splines 273
      6.A.3 C^2 Exponential Tension Splines 274

7   **Vanilla Models with Local Volatility** 277
  7.1 General Framework 278
      7.1.1 Model Dynamics 278
      7.1.2 Volatility Smile and Implied Density 278
      7.1.3 Choice of φ 279

7.2   CEV Model.. 280
     7.2.1   Basic Properties............................. 280
     7.2.2   Call Option Pricing 282
     7.2.3   Regularization 284
     7.2.4   Displaced Diffusion Models 285
7.3   Quadratic Volatility Model 287
     7.3.1   Case 1: Two Real Roots to the Left of $S(0)$ 287
     7.3.2   Case 2: One Real Root to the Left of $S(0)$ 291
     7.3.3   Extensions and Other Root Configurations....... 291
7.4   Finite Difference Solutions for General φ 292
     7.4.1   Multiple λ and T 293
     7.4.2   Forward Equation for Call Options 293
7.5   Asymptotic Expansions for General φ 295
     7.5.1   Expansion around Displaced Log-Normal Process . 296
     7.5.2   Expansion around Gaussian Process 298
7.6   Extensions to Time-Dependent φ 299
     7.6.1   Separable Case 300
     7.6.2   Skew Averaging 301
          7.6.2.1   Examples 304
          7.6.2.2   A Caveat About the Process Domain .. 306
     7.6.3   Skew and Convexity Averaging by Small-Noise
          Expansion 307
     7.6.4   Numerical Example 311

8   **Vanilla Models with Stochastic Volatility I**............. 315
8.1   Model Definition 315
8.2   Model Parameters 317
8.3   Basic Properties................................. 319
8.4   Fourier Integration 324
     8.4.1   General Theory 324
     8.4.2   Applications to SV Model 327
     8.4.3   Numerical Implementation.................... 330
     8.4.4   Refinements of Numerical Implementation 332
     8.4.5   Fourier Integration for Arbitrary European Payoffs 336
8.5   Integration in Variance Domain 339
8.6   CEV-Type Stochastic Volatility Models and SABR 343
8.7   Numerical Examples: Volatility Smile Statics............ 345
8.8   Numerical Examples: Volatility Smile Dynamics 348
8.9   Hedging in Stochastic Volatility Models 353
     8.9.1   Hedge Construction, Delta and Vega 353
     8.9.2   Minimum Variance Delta Hedging 355
     8.9.3   Minimum Variance Hedging: an Example 357
8.A   Appendix: General Volatility Processes 359

9   Vanilla Models with Stochastic Volatility II 363
  9.1   Fourier Integration with Time-Dependent Parameters 363
  9.2   Asymptotic Expansion with Time-Dependent Volatility .. 366
  9.3   Averaging Methods 371
    9.3.1   Volatility Averaging 371
    9.3.2   Skew Averaging 373
    9.3.3   Volatility of Variance Averaging 374
    9.3.4   Calibration by Parameter Averaging 377
  9.4   PDE Method 381
    9.4.1   PDE Formulation............................. 382
    9.4.2   Range for Stochastic Variance 382
    9.4.3   Discretizing Stochastic Variance 383
    9.4.4   Boundary Conditions for Stochastic Variance..... 385
    9.4.5   Range for Underlying 386
    9.4.6   Discretizing the Underlying 387
  9.5   Monte Carlo Method............................... 387
    9.5.1   Exact Simulation of Variance Process 388
    9.5.2   Biased Taylor-Type Schemes for Variance Process 389
      9.5.2.1   Euler Schemes 389
      9.5.2.2   Higher-Order Schemes 390
    9.5.3   Moment Matching Schemes for Variance Process.. 390
      9.5.3.1   Log-normal Approximation 390
      9.5.3.2   Truncated Gaussian 391
      9.5.3.3   Quadratic-Exponential 392
      9.5.3.4   Summary of QE Algorithm 394
    9.5.4   Broadie-Kaya Scheme for the Underlying 395
    9.5.5   Other Schemes for the Underlying 396
      9.5.5.1   Taylor-Type Schemes 396
      9.5.5.2   Simplified Broadie-Kaya............. 396
      9.5.5.3   Martingale Correction 398
  9.A   Appendix: Proof of Proposition 9.3.4................. 398
  9.B   Appendix: Coefficients for Asymptotic Expansion........ 402

VOLUME II Term Structure Models

Part III Term Structure Models

10 One-Factor Short Rate Models I 407
   10.1 The One-Factor Gaussian Short Rate Model 407
      10.1.1 The Ho-Lee Model 408
         10.1.1.1 Notations and First Steps 408
         10.1.1.2 Fitting the Term Structure of Discount
                   Bonds 409
         10.1.1.3 Analysis and Comparison with HJM
                   Approach........................... 411
      10.1.2 The Mean-Reverting GSR Model 413
         10.1.2.1 The Vasicek Model 413
         10.1.2.2 The General One-Factor GSR Model... 415
         10.1.2.3 Time-Stationarity and Caplet Hump ... 418
      10.1.3 European Option Pricing 420
         10.1.3.1 The Jamshidian Decomposition 420
         10.1.3.2 Gaussian Swap Rate Approximation ... 422
      10.1.4 Swaption Calibration 423
      10.1.5 Finite Difference Methods 424
         10.1.5.1 PDE and Spatial Boundary Conditions. 425
         10.1.5.2 Determining Spatial Boundary
                   Conditions from PDE 426
         10.1.5.3 Upwinding 427
      10.1.6 Monte Carlo Simulation 427
         10.1.6.1 Exact Discretization 427
         10.1.6.2 Approximate Discretization 429
         10.1.6.3 Using other Measures for Simulation ... 430
   10.2 The Affine One-Factor Model 431
      10.2.1 Basic Definitions 431
         10.2.1.1 SDE 431
         10.2.1.2 Regularity Issues 432
         10.2.1.3 Volatility Skew 432
         10.2.1.4 Time-Dependent Parameters.......... 433
      10.2.2 Discount Bond Pricing and Extended Transform.. 433
         10.2.2.1 Constant Parameters 434
         10.2.2.2 Piecewise Constant Parameters 436
      10.2.3 Discount Bond Calibration.................... 437
         10.2.3.1 Change of Variables 437
         10.2.3.2 Algorithm for $\omega(t)$..................... 438
      10.2.4 European Option Pricing 439

        10.2.5 Swaption Calibration 441
               10.2.5.1 Basic Problem 441
               10.2.5.2 Calibration Algorithm 442
        10.2.6 Quadratic One-Factor Model.................... 443
        10.2.7 Numerical Methods for the Affine Short Rate Model 444

11 One-Factor Short Rate Models II 445
  11.1 Log-Normal Short Rate Models 445
        11.1.1 The Black-Derman-Toy Model 445
        11.1.2 Black-Karasinski Model 447
        11.1.3 Issues in Log-Normal Models.................... 447
        11.1.4 Sandmann-Sondermann Transformation 448
  11.2 Other Short Rate Models 451
        11.2.1 Power-Type Models and Empirical Model
              Estimation.................................... 451
        11.2.2 The Black Shadow Rate Model 452
        11.2.3 Spanned and Unspanned Stochastic Volatility:
              the Fong and Vasicek Model 454
  11.3 Numerical Methods for General One-Factor Short Rate
     Models 455
        11.3.1 Finite Difference Methods 456
        11.3.2 Calibration to Initial Yield Curve............... 457
               11.3.2.1 Forward Induction................... 458
               11.3.2.2 Forward-from-Backward Induction 460
               11.3.2.3 Yield Curve and Volatility Calibration . 461
               11.3.2.4 The Dybvig Parameterization 463
               11.3.2.5 Link to HJM Models................. 464
               11.3.2.6 The Hagan and Woodward
                        Parameterization 465
        11.3.3 Monte Carlo Simulation 469
               11.3.3.1 SDE Discretization 469
               11.3.3.2 Practical Issues with Monte Carlo
                        Methods 470
  11.A Appendix: Markov-Functional Models 472
        11.A.1 State Process and Numeraire Mapping 472
        11.A.2 Libor MF Parameterization 473
        11.A.3 Swap MF Parameterization 475
        11.A.4 Non-Parametric Calibration.................... 476
        11.A.5 Numerical Implementation..................... 477
        11.A.6 Comments and Comparisons 478

12 Multi-Factor Short Rate Models 479
  12.1 The Gaussian Model 480
        12.1.1 Development from Separability Condition........ 480
               12.1.1.1 Mean-Reverting State Variables 481

12.1.1.2   Further Changes of Variables 485
12.1.2   Classical Development 487
12.1.2.1   Diagonalization of Mean Reversion
Matrix 488
12.1.3   Correlation Structure 490
12.1.4   The Two-Factor Gaussian Model 491
12.1.4.1   Some Basics 491
12.1.4.2   Variance and Correlation Structure 492
12.1.4.3   Volatility Hump 494
12.1.4.4   Another Formulation of the
Two-Factor Model 494
12.1.5   Multi-Factor Statistical Gaussian Model 497
12.1.6   Swaption Pricing 501
12.1.6.1   Jamshidian Decomposition 502
12.1.6.2   Gaussian Swap Rate Approximation ... 506
12.1.7   Calibration via Benchmark Rates 507
12.1.8   Monte Carlo Simulation 510
12.1.9   Finite Difference Methods 511
12.2 The Affine Model 512
12.2.1   Introduction 512
12.2.2   Basic Model 512
12.2.3   Regularity Issues 514
12.2.4   Discount Bond Prices 515
12.2.5   Some Concrete Models 517
12.2.5.1   Fong-Vasicek Model 517
12.2.5.2   Longstaff-Schwartz Model 517
12.2.5.3   Multi-Factor CIR Models 519
12.2.6   Brief Notes on Option Pricing 520
12.3 The Quadratic Gaussian Model 520
12.3.1   Quadratic Gaussian Models are Affine 521
12.3.2   The Basics 522
12.3.3   Parameterization 524
12.3.3.1   Smile Generation 524
12.3.3.2   Quadratic Term 525
12.3.3.3   Linear Term 527
12.3.4   Swaption Pricing 528
12.3.4.1   State Vector Distribution Under the
Annuity Measure 528
12.3.4.2   Exact Pricing of European Swaptions .. 529
12.3.4.3   Approximations for European Swaptions 530
12.3.5   Calibration 533
12.3.6   Spanned Stochastic Volatility 534
12.3.7   Numerical Methods 534
12.A Appendix: Quadratic Forms of Gaussian Vectors 535

13 The Quasi-Gaussian Model 539
13.1 One-Factor Quasi-Gaussian Model 539
    13.1.1 Definition 539
    13.1.2 Local Volatility.............................. 541
    13.1.3 Swap Rate Dynamics 542
    13.1.4 Approximate Local Volatility Dynamics for Swap
              Rate 543
          13.1.4.1   Simple Approximation 544
          13.1.4.2   Advanced Approximation............. 544
    13.1.5 Linear Local Volatility 547
    13.1.6 Linear Local Volatility for a Swaption Strip 549
    13.1.7 Volatility Calibration 550
    13.1.8 Mean Reversion Calibration.................... 552
          13.1.8.1   Effects of Mean Reversion 552
          13.1.8.2   Calibrating Mean Reversion to
                    Volatility Ratios..................... 554
          13.1.8.3   Calibrating Mean Reversion to
                    Inter-Temporal Correlations 557
          13.1.8.4   Final Comments on Mean Reversion
                    Calibration 559
    13.1.9 Numerical Methods 560
          13.1.9.1   Direct Integration 560
          13.1.9.2   Finite Difference Methods 562
          13.1.9.3   Monte Carlo Simulation 565
          13.1.9.4   Single-State Approximations 565
13.2 One-Factor Quasi-Gaussian Model with Stochastic Volatility 569
    13.2.1 Definition 569
    13.2.2 Swap Rate Dynamics 571
    13.2.3 Volatility Calibration 572
    13.2.4 Mean Reversion Calibration.................... 573
    13.2.5 Non-Zero Correlation 574
    13.2.6 PDE and Monte Carlo Methods 574
13.3 Multi-Factor Quasi-Gaussian Model..................... 575
    13.3.1 General Multi-Factor Model 575
    13.3.2 Local and Stochastic Volatility Parameterization . 576
    13.3.3 Swap Rate Dynamics and Approximations 579
    13.3.4 Volatility Calibration 584
    13.3.5 Mean Reversions, Correlations, and Numerical
          Schemes 584
13.A Appendix: Density Approximation 585
    13.A.1 Simplified Forward Measure Dynamics 585
    13.A.2 Effective Volatility 587
    13.A.3 The Forward Equation for Call Options 587
    13.A.4 Asymptotic Expansion 588
    13.A.5 Proof of Theorem 13.1.14 590

14 The Libor Market Model I 591
  14.1 Introduction and Setup 592
    14.1.1 Motivation and Historical Notes 592
    14.1.2 Tenor Structure 593
  14.2 LM Dynamics and Measures 593
    14.2.1 Setting 593
    14.2.2 Probability Measures 594
    14.2.3 Link to HJM Analysis 597
    14.2.4 Separable Deterministic Volatility Function 598
    14.2.5 Stochastic Volatility 600
    14.2.6 Time-Dependence in Model Parameters 603
  14.3 Correlation 603
    14.3.1 Empirical Principal Components Analysis 604
      14.3.1.1 Example: USD Forward Rates 605
    14.3.2 Correlation Estimation and Smoothing 606
      14.3.2.1 Example: Fit to USD Data 609
    14.3.3 Negative Eigenvalues 610
    14.3.4 Correlation PCA 611
      14.3.4.1 Example: USD Data 613
      14.3.4.2 Poor Man's Correlation PCA 614
  14.4 Pricing of European Options 615
    14.4.1 Caplets 615
    14.4.2 Swaptions 616
    14.4.3 Spread Options 619
      14.4.3.1 Term Correlation 620
      14.4.3.2 Spread Option Pricing 621
  14.5 Calibration 622
    14.5.1 Basic Principles 622
    14.5.2 Parameterization of $\|\lambda_k(t)\|$ 622
    14.5.3 Interpolation on the Whole Grid 624
    14.5.4 Construction of $\lambda_k(t)$ from $\|\lambda_k(t)\|$ 625
      14.5.4.1 Covariance PCA 626
      14.5.4.2 Correlation PCA 626
      14.5.4.3 Discussion and Recommendation 627
    14.5.5 Choice of Calibration Instruments 627
    14.5.6 Calibration Objective Function 630
    14.5.7 Sample Calibration Algorithm 633
    14.5.8 Speed-Up Through Sub-Problem Splitting 633
    14.5.9 Correlation Calibration to Spread Options 635
    14.5.10 Volatility Skew Calibration 637
  14.6 Monte Carlo Simulation 637
    14.6.1 Euler-Type Schemes 638
      14.6.1.1 Analysis of Computational Effort 639
      14.6.1.2 Long Time Steps 641
      14.6.1.3 Notes on the Choice of Numeraire 642

14.6.2 Other Simulation Schemes   643
        14.6.2.1 Special-Purpose Schemes with Drift
                  Predictor-Corrector   643
        14.6.2.2 Euler Scheme with Predictor-Corrector .   644
        14.6.2.3 Lagging Predictor-Corrector Scheme . . .   644
        14.6.2.4 Further Refinements of Drift Estimation   646
        14.6.2.5 Brownian-Bridge Schemes and Other
                  Ideas   647
        14.6.2.6 High-Order Schemes   650
14.6.3 Martingale Discretization   650
        14.6.3.1 Deflated Bond Price Discretization   651
        14.6.3.2 Comments and Alternatives   652
14.6.4 Variance Reduction   653
        14.6.4.1 Antithetic Sampling   653
        14.6.4.2 Control Variates   654
        14.6.4.3 Importance Sampling   655

15 The Libor Market Model II   657
   15.1 Interpolation   657
        15.1.1 Back Stub, Simple Interpolation   658
        15.1.2 Back Stub, Arbitrage-Free Interpolation   659
        15.1.3 Back Stub, Gaussian Model   661
        15.1.4 Front Stub, Zero Volatility   662
        15.1.5 Front Stub, Exogenous Volatility   663
        15.1.6 Front Stub, Simple Interpolation   666
        15.1.7 Front Stub, Gaussian Model   666
   15.2 Advanced Swaption Pricing via Markovian Projection   667
        15.2.1 Advanced Formula for Swap Rate Volatility   670
        15.2.2 Advanced Formula for Swap Rate Skew   672
        15.2.3 Skew and Smile Calibration in LM Models   674
   15.3 Near-Markov LM Models   676
   15.4 Swap Market Models   676
   15.5 Evolving Separate Discount and Forward Rate Curves   678
        15.5.1 Basic Ideas   679
        15.5.2 HJM Extension   680
        15.5.3 Applications to LM Models   683
        15.5.4 Deterministic Spread   687
   15.6 SV Models with Non-Zero Correlation   687
   15.7 Multi-Stochastic Volatility Extensions   690
        15.7.1 Introduction   690
        15.7.2 Setup   690
        15.7.3 Pricing Caplets and Swaptions   691
        15.7.4 Spread Options   692
        15.7.5 Another Use of Multi-Dimensional Stochastic
               Volatility   693

VOLUME III Products and Risk Management

Part IV Products

16 Single-Rate Vanilla Derivatives 697
16.1 European Swaptions 697
    16.1.1 Smile Dynamics 698
    16.1.2 Adjustable Backbone........................ 699
    16.1.3 Stochastic Volatility Swaption Grid 702
    16.1.4 Calibrating Stochastic Volatility Model to
            Swaptions 703
    16.1.5 Some Other Interpolation Rules 705
16.2 Caps and Floors.................................... 706
    16.2.1 Basic Problem 706
    16.2.2 Setup and Norms 707
    16.2.3 Calibration Procedure....................... 708
16.3 Terminal Swap Rate Models 709
    16.3.1 TSR Basics 709
    16.3.2 Linear TSR Model 711
    16.3.3 Exponential TSR Model 714
    16.3.4 Swap-Yield TSR Model 715
16.4 Libor-in-Arrears 716
16.5 Libor-with-Delay 719
    16.5.1 Swap-Yield TSR Model 720
    16.5.2 Other Terminal Swap Rate Models 721
    16.5.3 Approximations Inspired by Term Structure Models 721
    16.5.4 Applications to Averaging Swaps 722
16.6 CMS and CMS-Linked Cash Flows 723
    16.6.1 The Replication Method for CMS.............. 724
    16.6.2 Annuity Mapping Function as a Conditional
            Expected Value 726
    16.6.3 Swap-Yield TSR Model 727
    16.6.4 Linear and Other TSR Models 728
    16.6.5 The Quasi-Gaussian Model 730
    16.6.6 The Libor Market Model 731
    16.6.7 Correcting Non-Arbitrage-Free Methods 733
    16.6.8 Impact of Annuity Mapping Function and Mean
            Reversion................................... 735
    16.6.9 CDF and PDF of CMS Rate in Forward Measure . 737
    16.6.10 SV Model for CMS Rate 739
    16.6.11 Dynamics of CMS Rate in Forward Measure 741

16.6.12 Cash-Settled Swaptions 744
16.7 Quanto CMS 745
    16.7.1 Overview 746
    16.7.2 Modeling the Joint Distribution of Swap Rate
            and Forward Exchange Rate 748
    16.7.3 Normalizing Constant and Final Formula 749
16.8 Eurodollar Futures 750
    16.8.1 Fundamental Results on Futures............... 750
    16.8.2 Motivations and Plan 752
    16.8.3 Preliminaries................................ 753
    16.8.4 Expansion Around the Futures Value 754
    16.8.5 Forward Rate Variances 757
    16.8.6 Forward Rate Correlations.................... 758
    16.8.7 The Formula 759
16.9 Convexity and Moment Explosions 760

17 Multi-Rate Vanilla Derivatives......................... 765
17.1 Introduction to Multi-Rate Vanilla Derivatives 765
17.2 Marginal Distributions and Reference Measure 767
17.3 Dependence Structure via Copulas.................... 768
    17.3.1 Introduction to Gaussian Copula Method 768
    17.3.2 General Copulas............................. 770
    17.3.3 Archimedean Copulas 772
    17.3.4 Making Copulas from Other Copulas........... 773
17.4 Copula Methods for CMS Spread Options 776
    17.4.1 Normal Model for the Spread 776
    17.4.2 Gaussian Copula for Spread Options 777
    17.4.3 Spread Volatility Smile Modeling with the Power
            Gaussian Copula 780
    17.4.4 Copula Implied From Spread Options 781
17.5 Rates Observed at Different Times.................... 784
17.6 Numerical Methods for Copulas 785
    17.6.1 Numerical Integration Methods................ 786
    17.6.2 Dimensionality Reduction for CMS Spread Options 789
    17.6.3 Dimensionality Reduction for Other Multi-Rate
            Derivatives 791
    17.6.4 Dimensionality Reduction by Conditioning....... 793
    17.6.5 Dimensionality Reduction by Measure Change ... 797
    17.6.6 Monte Carlo Methods 799
17.7 Limitations of the Copula Method 801
17.8 Stochastic Volatility Modeling for Multi-Rate Options.... 803
    17.8.1 Measure Change by Drift Adjustment 803
    17.8.2 Measure Change by CMS Caplet Calibration..... 804
    17.8.3 Impact of Correlations on the Spread Smile 805
    17.8.4 Connection to Term Structure Models.......... 806

17.9 CMS Spread Options in Term Structure Models   808
 17.9.1 Libor Market Model   808
 17.9.2 Quadratic Gaussian Model   810
17.A Appendix: Implied Correlation in Displaced Log-Normal
 Models   811
 17.A.1 Preliminaries   811
 17.A.2 Implied Log-Normal Correlation   812
 17.A.3 A Few Numerical Results   813

18 **Callable Libor Exotics**   817
18.1 Model Calibration for Callable Libor Exotics   817
 18.1.1 Risk Factors for CLEs   818
 18.1.2 Model Choice and Calibration   821
18.2 Valuation Theory   822
 18.2.1 Preliminaries   822
 18.2.2 Recursion for Callable Libor Exotics   823
 18.2.3 Marginal Exercise Value Decomposition   824
18.3 Monte Carlo Valuation   825
 18.3.1 Regression-Based Valuation of CLEs, Basic Scheme   825
 18.3.2 Regression for Underlying   827
 18.3.3 Valuing CLE as a Cancelable Note   829
 18.3.4 Using Regressed Variables for Decision Only   830
 18.3.5 Regression Valuation with Boundary Optimization   832
 18.3.6 Lower Bound via Regression Scheme   833
 18.3.7 Iterative Improvement of Lower Bound   835
 18.3.8 Upper Bound   838
  18.3.8.1 Basic Ideas   838
  18.3.8.2 Nested Simulation (NS) Algorithm   839
  18.3.8.3 Bias and Computational Cost of NS
   Algorithm   842
  18.3.8.4 Confidence Intervals and Practical Usage   844
  18.3.8.5 Non-Analytic Exercise Values   845
  18.3.8.6 Improvements to NS Algorithm   847
  18.3.8.7 Other Upper Bound Algorithms   849
 18.3.9 Regression Variable Choice   850
  18.3.9.1 State Variables Approach   850
  18.3.9.2 Explanatory Variables   851
  18.3.9.3 Explanatory Variables with Convexity .   854
 18.3.10 Regression Implementation   856
  18.3.10.1 Automated Explanatory Variable
   Selection   856
  18.3.10.2 Suboptimal Point Exclusion   858
  18.3.10.3 Two Step Regression   859
  18.3.10.4 Robust Implementation of Regression
   Algorithm   860

18.4 Valuation with Low–Dimensional Models 864
    18.4.1   Single-Rate Callable Libor Exotics 864
    18.4.2   Calibration Targets for the Local Projection
             Method 864
    18.4.3   Review of Suitable Local Models 866
    18.4.4   Defining a Suitable Analog for Core Swap Rates .. 867
    18.4.5   PDE Methods for Path-Dependent CLEs 870
             18.4.5.1   CLEs Accreting at Coupon Rate 870
             18.4.5.2   Snowballs 872

19 **Bermudan Swaptions** 875
19.1 Definitions.. 875
19.2 Local Projection Method 876
19.3 Smile Calibration.................................... 878
19.4 Amortizing, Accreting, Other Non-Standard Swaptions ... 880
    19.4.1   Relationship Between Non-Standard and
             Standard Swap Rates 882
    19.4.2   Same-Tenor Approach......................... 883
    19.4.3   Representative Swaption Approach 884
    19.4.4   Basket Approach 887
    19.4.5   Super-Replication for Non-Standard Bermudan
             Swaptions 890
    19.4.6   Zero-Coupon Bermudan Swaptions 894
    19.4.7   American Swaptions 895
             19.4.7.1   American Swaptions vs. High-
                        Frequency Bermudan Swaptions 896
             19.4.7.2   The Proxy Libor Rate Method 897
             19.4.7.3   The Libor-as-Extra-State Method 898
    19.4.8   Mid-Coupon Exercise 899
19.5 Flexi-Swaps... 900
    19.5.1   Purely Global Bounds......................... 901
    19.5.2   Purely Local Bounds......................... 901
    19.5.3   Marginal Exercise Value Decomposition 903
    19.5.4   Narrow Band Limit 904
19.6 Monte Carlo Valuation 905
    19.6.1   Regression Methods.......................... 905
    19.6.2   Parametric Boundary Methods 906
             19.6.2.1   Sample Exercise Strategies for
                        Bermudan Swaptions 906
             19.6.2.2   Some Numerical Tests 909
             19.6.2.3   Additional Comments............... 911
19.7 Other Topics.. 912
    19.7.1   Robust Bermudan Swaption Hedging with
             European Swaptions 912
    19.7.2   Carry and Exercise 914

19.7.3   Fast Pricing via Exercise Premia Representation .   916
19.A Appendix: Forward Volatility and Correlation   919
19.B Appendix: A Primer on Moment Matching.   920
19.B.1   Basics.   920
19.B.2   Example 1: Asian Option in BSM Model   922
19.B.3   Example 2: Basket Option in BSM Model   924

20   TARNs, Volatility Swaps, and Other Derivatives   925
20.1 TARNs   925
20.1.1   Definitions and Examples.   925
20.1.2   Valuation and Risk with Globally Calibrated
Models   927
20.1.3   Local Projection Method   928
20.1.4   Volatility Smile Effects   929
20.1.5   PDE for TARNs.   931
20.2 Volatility Swaps   933
20.2.1   Local Projection Method   934
20.2.2   Shout Options   935
20.2.3   Min-Max Volatility Swaps   938
20.2.4   Impact of Volatility Dynamics on Volatility Swaps   940
20.3 Forward Swaption Straddles   945

21   Out-of-Model Adjustments   951
21.1 Adjusting the Model   952
21.1.1   Calibration to Coupons   952
21.1.2   Adjusters.   954
21.1.3   Path Re-Weighting   956
21.1.4   Proxy Model Method   961
21.1.5   Asset-Based Adjustments.   963
21.1.6   Mapping Function Adjustments   965
21.2 Adjusting the Market   965
21.3 Adjusting the Trade   966
21.3.1   Fee Adjustments   967
21.3.2   Fee Adjustment Impact on Exotic Derivatives   968
21.3.3   Strike Adjustment   969

Part V Risk Management

22   Introduction to Risk Management   975
22.1 Risk Management and Sensitivity Computations   976
22.1.1   Basic Information Flow.   976
22.1.2   Risk: Theory and Practice   978
22.1.3   Example: the Black-Scholes Model   980

22.1.4 Example: Black-Scholes Model with
Time-Dependent Parameters 983
22.1.5 Actual Risk Computations 985
22.1.6 What about Θ_{prm} and Θ_{num}? 986
22.1.7 A Note on Trading P&L and the Computation of
Implied Volatility 987
22.2 P&L Analysis 990
22.2.1 P&L Predict 991
22.2.2 P&L Explain 993
22.2.2.1 Waterfall Explain 993
22.2.2.2 Bump-and-Reset Explain 994
22.3 Value-at-Risk 995
22.A Appendix: Alternative Proof of Lemma 22.1.1 998

23 Payoff Smoothing and Related Methods 1001
23.1 Issues with Discretization Schemes 1001
23.1.1 Problems with Grid Dimensioning 1002
23.1.2 Grid Shifts Relative to Payout 1002
23.1.3 Additional Comments 1005
23.2 Basic Techniques 1006
23.2.1 Adaptive Integration 1006
23.2.2 Adding Singularities to the Grid 1007
23.2.3 Singularity Removal 1009
23.2.4 Partial Analytical Integration 1010
23.3 Payoff Smoothing For Numerical Integration and PDEs . . 1012
23.3.1 Introduction to Payoff Smoothing 1012
23.3.2 Payoff Smoothing in One Dimension 1014
23.3.2.1 Box Smoothing 1015
23.3.2.2 Other Smoothing Methods 1018
23.3.3 Payoff Smoothing in Multiple Dimensions 1019
23.4 Payoff Smoothing for Monte Carlo 1022
23.4.1 Tube Monte Carlo for Digital Options 1022
23.4.2 Tube Monte Carlo for Barrier Options 1024
23.4.3 Tube Monte Carlo for Callable Libor Exotics 1029
23.4.4 Tube Monte Carlo for TARNs 1029
23.A Appendix: Delta Continuity of Singularity-Enlarged Grid
Method 1030
23.B Appendix: Conditional Independence for Tube Monte Carlo 1032

24 Pathwise Differentiation 1035
24.1 Pathwise Differentiation: Foundations 1035
24.1.1 Callable Libor Exotics 1035
24.1.1.1 CLE Greeks 1036
24.1.1.2 Keeping the Exercise Time Constant . . . 1038
24.1.1.3 Noise in CLE Greeks 1040

24.1.2 Barrier Options 1041
24.2 Pathwise Differentiation for PDE Based Models 1044
   24.2.1 Model and Setup 1044
   24.2.2 Bucketed Deltas 1045
   24.2.3 Survival Density 1048
24.3 Pathwise Differentiation for Monte Carlo Based Models .. 1051
   24.3.1 Pathwise Derivatives of Forward Libor Rates 1051
   24.3.2 Pathwise Deltas of European Options 1054
       24.3.2.1 Pathwise Deltas of the Numeraire 1054
       24.3.2.2 Pathwise Deltas of the Payoff 1055
   24.3.3 Adjoint Method For Greeks Calculation 1056
   24.3.4 Pathwise Delta Approximation for Callable Libor
           Exotics.................................... 1058
24.4 Notes on Likelihood Ratio and Hybrid Methods 1060

25 **Importance Sampling and Control Variates** 1063
25.1 Importance Sampling In Short Rate Models............. 1063
25.2 Payoff Smoothing by Importance Sampling 1065
   25.2.1 Binary Options.............................. 1065
   25.2.2 TARNs...................................... 1068
   25.2.3 Removing the First Digital 1068
   25.2.4 Smoothing All Digitals by One-Step Survival
           Conditioning 1069
   25.2.5 Simulating Under the Survival Measure Using
           Conditional Gaussian Draws 1072
   25.2.6 Generalized Trigger Products in Multi-Factor LM
           Models 1074
25.3 Model-Based Control Variates........................ 1077
   25.3.1 Low-Dimensional Markov Approximation for LM
           models 1078
   25.3.2 Two-Dimensional Extension.................... 1081
   25.3.3 Approximating Volatility Structure 1082
   25.3.4 Markov Approximation as a Control Variate 1084
25.4 Instrument-Based Control Variates 1086
25.5 Dynamic Control Variates 1090
25.6 Control Variates and Risk Stability 1093

26 **Vegas in Libor Market Models**........................ 1095
26.1 Basic Problem of Vega Computations 1095
26.2 Review of Calibration 1097
26.3 Vega Calculation Methods 1098
   26.3.1 Direct Vega Calculations 1098
       26.3.1.1 Definition and Analysis 1098
       26.3.1.2 Numerical Example.................. 1101
   26.3.2 What is a Good Vega? 1102

26.3.3 Indirect Vega Calculations 1105
    26.3.3.1   Definition and Analysis 1105
    26.3.3.2   Numerical Example and Performance
               Analysis........................... 1107
26.3.4 Hybrid Vega Calculations..................... 1111
    26.3.4.1   Definition and Analysis 1111
    26.3.4.2   Numerical Example.................. 1113
26.4 Skew and Smile Vegas................................ 1114
26.5 Vegas and Correlations............................... 1115
26.5.1 Term Correlation Effects 1115
26.5.2 What Correlations should be Kept Constant? 1116
26.5.3 Vegas with Fixed Term Correlations 1118
26.5.4 Numerical Example 1119
26.6 Deltas with Backbone................................ 1120
26.7 Vega Projections 1122
26.8 Some Notes on Computing Model Vegas................ 1124

Appendix

A   Markovian Projection 1129
A.1 Marginal Distributions of Ito Processes................. 1129
A.2 Approximations for Conditional Expected Values 1134
A.2.1 Gaussian Approximation 1134
A.2.2 Least-Squares Projection 1136
A.3 Applications to Local Stochastic Volatility Models 1137
A.3.1 Markovian Projection onto an SV Model 1137
A.3.2 Fitting the Market with an LSV Model.......... 1139
A.3.3 On Calculating Proxy Local Volatility 1143
A.4 Basket Options in Local Volatility Models 1145
A.5 Basket Options in Stochastic Volatility Models 1149
A.A Appendix: $E(\sqrt{z_n(t)z_m(t)})$ and $E(\sqrt{z_n(t)})$.............. 1152
A.A.1 Proof of Proposition A.A.1 1153
    A.A.1.1   Step 1. Reduction to Covariance....... 1153
    A.A.1.2   Step 2. Linear Approximation 1154
    A.A.1.3   Step 3. Coefficients 1154
    A.A.1.4   Step 4. Order of Approximation 1155
A.A.2 Proof of Lemma A.A.2 1155

References... i

Index... xxi

Commonly Used Notations

Probability Notations

- $(\Omega, \mathcal{F}, \mathrm{P})$: probability space.
- $\mathcal{F}_t$, $\mathcal{B}_t$: filtrations of σ-algebras.
- P: probability measure.
- Q: risk-neutral measure.
- Q^B: spot Libor measure.
- Q^T, Q^n: forward measure for time T or T_n (given tenor structure).
- $\mathrm{Q}^{n,m}$: swap measure for swap rate $S_{n,m}$ (given tenor structure).
- Q^N: measure for numeraire N.
- E, E^P, E^Q, E^T, E^n, $\mathrm{E}^{n,m}$, E^A, ...: expectations under various measures.
- E_t, E_t^P, E_t^Q, E_t^T, E_t^n, $\mathrm{E}_t^{n,m}$, E_t^A, ...: expectations conditional on $\mathcal{F}_t$ under various measures.
- $Z(t)$, $W(t)$, $W^T(t)$, $W^n(t)$, $W^{n,m}(t)$, $W^A(t)$, ...: Brownian motions under various probability measures.
- $\mathrm{Var}(X)$: variance of X.
- $\mathrm{Stdev}(X)$: standard deviation of X.
- $\mathrm{Cov}(X,Y)$: covariance of X, Y.
- $\mathrm{Corr}(X,Y)$: correlation of X, Y.
- $\mathcal{N}(\mu, \Sigma)$: Gaussian distribution with mean μ and variance-covariance matrix Σ.
- $\mathcal{LN}(\mu, \sigma^2)$: log-normal distribution with mean μ and variance σ^2.
- $\mathcal{U}(a,b)$: uniform distribution on an interval $[a,b]$.
- $\Phi(z)$: standard Gaussian CDF, $\phi(z)$: standard Gaussian PDF.
- $\Gamma(a,x)$: the (upper) incomplete Gamma function, $\Gamma(a,x) = \int_x^\infty u^{a-1} e^{-u}\, du$.
- $\Gamma(a)$: the Gamma function, $\Gamma(a) = \Gamma(a,0)$.
- $\mathcal{E}(X(t))$: Doléans exponential martingale for the process $X(t)$.
- $\langle X(t) \rangle$, $\langle X(t), Y(t) \rangle$: quadratic variation and covariation.

Finance Notations

- $T_0 < T_1 < \ldots < T_N$: tenor structure.
- τ_n: year fraction between T_n and T_{n+1}.
- $\beta(t)$: continuously compounded money market account.
- $B(t)$: discretely compounded money market account.
- $P(t, T)$: zero-coupon (or discount) bond price at time t for maturity T.
- $P(t, T, S)$: forward bond price at time t, for delivery of S-maturity discount bond at time T, $T \leq S$.
- $y(t, T, S)$: continuously compounded forward yield at time t for the period $[T, S]$.
- $f(t, T)$: instantaneous forward rate at t for maturity T.
- $r(t)$: short rate at time t, $r(t) = f(t, t)$.
- $L(t, T, S)$: forward Libor rate at time t for the period $[T, S]$.
- $L_n(t)$: forward Libor rate at t for the period $[T_n, T_{n+1}]$, given a tenor structure, $L_n(t) = L(t, T_n, T_{n+1})$.
- $S_{n,m}(t)$: forward swap rate at time t, starting at T_n and with the final payment date at T_{n+m} (given a tenor structure).
- $A_{n,m}(t)$: annuity at time t, with the first payment date T_{n+1} and the final payment date T_{n+m} (given a tenor structure).
- $U_n(t)$: the n-th exercise ("underlying") value of a Bermudan swaption or a callable Libor exotic.
- $H_n(t)$: the n-th hold value of a Bermudan swaption or a callable Libor exotic.
- $\sigma_B(t, S; T, K)$: an implied Black volatility smile, parameterized by the time t spot S, strike K and expiry T.
- $c_B(t, S; T, K)$, $c_B(t, S; T, K, \sigma)$: price of a call option in the Black model with time t spot S, strike K, expiry T and Black volatility σ.
- $c_N(t, S; T, K)$, $c_N(t, S; T, K, \sigma)$: price of a call option in the Gaussian (Normal, or Bachelier) model with time t spot S, strike K, expiry T and Normal volatility σ.

Miscellaneous Notations

- $\operatorname{Re}(z)$, $\operatorname{Im}(z)$: real and imaginary part of a complex number z.
- $O(\cdot)$, $o(\cdot)$: "Big O" and "Little o" order symbols.
- $1_{\{A\}}$: indicator of A.
- L^1 and L^2: spaces of integrable and square-integrable random variables, vectors, or functions.
- C^n: space of functions with the n-th continuous derivative, i.e $C = C^0$ are continuous functions, C^1 are differentiable functions with continuous derivative, C^2 are twice-differentiable functions with continuous second-order derivative, etc.

- $\mathcal{L}$, $\mathcal{J}$: differential operators, e.g. $a\,\partial/\partial x + b\,\partial^2/\partial x^2$ or $\partial/\partial t + a\,\partial/\partial x + b\,\partial^2/\partial x^2$.
- $(\mathcal{F}f)(\omega), (\mathcal{F}^{-1}\varphi)(x)$: direct and inverse Fourier transforms.
- $\triangleq$: "is defined as", e.g. $f(x) \triangleq x^2$.
- x^+, x^-: maximum and minimum of x and 0, i.e. $x^+ = \max(x, 0)$, $x^- = \min(x, 0)$.
- $\lfloor x \rfloor$: integer part of real number x.
- $A^\top$: transpose of matrix A.
- $\det(A)$: the determinant of a square matrix A.
- $\operatorname{tr}(A)$: the trace of a square matrix A.
- $\operatorname{diag}(a)$: a square matrix with the vector a on the diagonal and zeros elsewhere.

Part IV

Products

16

Single-Rate Vanilla Derivatives

We start our analysis of interest rate securities with the class of *single-rate vanilla derivatives*. This class contains derivative securities whose payoffs can, either exactly or approximately, be decomposed into a linear combination of functions of individual rate observations. The class is very broad and includes all of the most liquid volatility-dependent interest rate derivatives, including caps, swaptions, Eurodollar futures, CMS swaps, CMS caps, and so on. Basic definitions of these products were given in Chapter 5; in this chapter we proceed to analyze carefully a range of issues in pricing, calibration and risk management.

16.1 European Swaptions

We have observed on a number of occasions that a European swaption[1] can be conveniently written as a European option on a swap rate that happens to be a martingale in the corresponding annuity measure. For outright pricing of swaptions, we can therefore use the Black or Bachelier models, or any of the "vanilla" diffusion models developed in Chapters 7, 8, or 9.

For a given expiry and swap maturity, liquid swaption quotes are typically available for more than a single strike; as discussed earlier, these quotes can be inverted into an implied volatility smile that is normally downward-sloping in strikes (but may eventually, for high strikes, flatten or increase). In itself, the presence of volatility smile does not necessarily imply the need to use a sophisticated model (e.g., a stochastic volatility model) for European swaption pricing. For instance, for pricing purposes it suffices to simply keep track of implied volatilities for a number of key strikes; prices for swaptions with strikes outside of the set of key strikes can be found by interpolation[2].

---

[1]We treat caplets as one-period European swaptions and do not specifically mention them in this section.

[2]The collection of implied swaption volatilities at a range of different expiries, swap tenors, and strikes is known as a *swaption volatility cube*.

Nevertheless, it is important to realize that different models, while possibly producing identical swaption prices, may imply different — sometimes *very* different — hedging strategies and, ultimately, P&L (*Profit-And-Loss*) of a vanilla options desk. We elaborate on this in the next two sections, and then turn to a few relevant topics associated with practicalities of swaption model calibration.

16.1.1 Smile Dynamics

Delta hedging in vanilla models is intimately linked to the model-implied *volatility smile dynamics*, i.e. how the smile moves with the underlying. To expand on this, let us focus on a T-maturity, K-strike European call option $c(t) = c(t, S(t); T, K)$, and consider the computation of its time t delta

$$\Delta(t) = \frac{\partial c(t)}{\partial S}.$$

Recall (see (7.6)) the notion of implied volatility $\sigma_B(t, S(t); T, K)$,

$$c(t) = c_B\left(t, S(t); T, K; \sigma_B\left(t, S(t); T, K\right)\right),$$

where $c_B(t, S; T, K; \sigma)$ represents the usual Black formula at a volatility level σ (see Remark 7.2.8). By the chain rule, it follows that

$$\Delta(t) = \frac{\partial c_B}{\partial S} + \frac{\partial c_B}{\partial \sigma_B}\frac{\partial \sigma_B}{\partial S} = \Delta_B(t) + \Upsilon_B(t)\frac{\partial \sigma_B}{\partial S}, \tag{16.1}$$

where $\Delta_B(t)$ and $\Upsilon_B(t)$ are, respectively, the delta and vega[3] computed in a Black model at a volatility of σ_B. From the Black formula (7.6), it follows that

$$\Delta_B(t) = \Phi\left(d_+\right), \quad \Upsilon_B(t) = S(t)\sqrt{T-t}\phi\left(d_+\right),$$

with $d_\pm$ defined by (7.6). According to (16.1), for models more sophisticated than the Black model, one can expect that a certain (possibly negative) amount of Black vega will "leak" into the Black delta, to produce a proper model-consistent delta. The amount of this leakage is controlled by the smile dynamics of the model, as given by the term $\partial \sigma_B / \partial S$. This term is often called the *backbone* of the model.

For arguments sake, suppose we use the Black model with strike-specific volatility for risk-managing options, i.e. we use the pricing formula

$$c(t, S(t); T, K) = c_B\left(t, S(t); T, K; \psi\left(T, K\right)\right) \tag{16.2}$$

where $\psi(T, K)$ is calibrated, at time $t = 0$, to market for each T and K. According to (16.1), the delta calculated in our model will be exactly $\Delta_B(t)$, i.e. the ordinary Black delta. On the other hand, this would not be the

---

[3]Vega is the volatility sensitivity, see Remark 8.9.3.

case in, say, the Heston model, which we recall (from Section 8.8) to have "sticky delta" dynamics when the variance variable $z(t)$ is kept fixed. For the common case of a downward-sloping smile (i.e. a negative correlation parameter in the Heston model), it follows that $\Delta_{\text{Heston}}(t) > \Delta_{\text{B}}(t)$, since $\Upsilon_{\text{B}}(t) > 0$ and — according to Figure 8.8 in Section 8.8 — $\partial \sigma_{\text{B}}/\partial S > 0$. For a downward sloping smile in a local volatility model, on the other hand, $\partial \sigma_{\text{B}}/\partial S < 0$ (see Figure 8.7) and the model delta would be *less* than the Black delta.

In general, the usefulness of a given swaption model to a trading desk lies perhaps not so much in its ability to fit the market — an easy feat if model parameters are allowed to depend on strike — but by how closely the model can match the realized dynamics of the volatility smile. Indeed, to the extent that volatility smile moves predicted by the model differ markedly from observations, the hedging strategies prescribed by the model will not be successful in practice, and traders will not be able to predict the P&L implications of a move in market variables. Due to the importance of the backbone, it is not uncommon for a trading desk to exogenously supply an ad-hoc rule for smile moves that overrides the model-computed value of $\partial \sigma_{\text{B}}/\partial S$; this practice is sometimes known as *shadow delta hedging*. Common rules include the earlier mentioned sticky delta rule, as well as the *sticky strike* rule (16.2) which assumes that the smile remains fixed as a function of K when S moves[4]. While applications of ad-hoc rules when computing deltas will compromise the theoretical integrity of the underlying model, in practice the efficacy and stability of the hedging strategy may nevertheless improve.

16.1.2 Adjustable Backbone

As shadow delta hedging is a very common practice, let us proceed to elaborate on the basic idea, by describing a possible approach for exogenously controlling the backbone in a more nuanced way than through simple sticky delta/strike rules. We present our ideas in the context of a displaced log-normal model; adding stochastic volatility to the model would follow standard procedures and will not affect the backbone. Recalling the model (7.21),

$$dS(t) = \lambda \left(bS(t) + (1-b) L \right) dW(t), \tag{16.3}$$

we first focus on calculating its backbone. To find approximately the implied Black volatility for the model, we apply the expansion method of Proposition 7.5.1 with $\beta = 0$, $\zeta = 1$. Keeping only the first term, we obtain

$$\sigma_{\text{B}}(0, S(0); T, K) \approx \lambda \frac{\ln(S(0)/K)}{\int_K^{S(0)} \frac{du}{bu + (1-b)L}} = \lambda b \frac{\ln(S(0)/K)}{\ln\left(\frac{bS(0)+(1-b)L}{bK+(1-b)L}\right)}.$$

---

[4]It can be shown that the sticky strike rule is arbitrageable.

The backbone for the at-the-money strike is given by

$$\frac{\partial \sigma_B \left(0, S(0); T, K\right)}{\partial S(0)}\bigg|_{K=S(0)} \approx -\frac{1-b}{2}\frac{L}{S(0)^2}\lambda$$

and with $L \approx S(0)$, as is typically the case, we get

$$\frac{\partial \sigma_B \left(0, S(0); T, K\right)}{\partial S(0)}\bigg|_{K=S(0)} \approx -\frac{1-b}{2}\frac{\lambda}{S(0)}.$$

Clearly, the backbone is controlled by b. When $b = 1$, we obtain the Black backbone ($\partial \sigma_B / \partial S = 0$) and, when $b = 0$, we obtain what we call the *Gaussian* backbone, as it is the backbone that is consistent with Gaussian dynamics — see Remark 7.2.9. The model (16.3), however, does not allow for an independent control over the backbone as b is not a free parameter, but is determined by the slope of the market-observed volatility smile.

Let us see what would happen if we had specified the model (16.3) somewhat differently:

$$dS(t) = \lambda \left(bS(t) + (1-b) S(0)\right) dW(t). \tag{16.4}$$

Then, following the same steps as above, we would get

$$\sigma_B \left(0, S(0); T, K\right) \approx \lambda b\frac{\ln \left(S(0)/K\right)}{\ln \left(\frac{S(0)}{bK+(1-b)S(0)}\right)},$$

$$\frac{\partial \sigma_B \left(0, S(0); T, K\right)}{\partial S(0)}\bigg|_{K=S(0)} \approx \frac{1-b}{2}\frac{\lambda}{S(0)}.$$

We see that the backbone is now different — *positive* (for $b \in [0, 1)$) rather than *negative*, as in the model (16.3). This difference, of course, originates with the fact that a perturbation to $S(0)$ now affects the local volatility function in (16.4): a shock of size $\delta S(0)$ to $S(0)$ increases the local volatility function by $\lambda(1 - b)\delta S(0)$, and the impact propagates into the implied volatility itself. On the other hand, the volatility smile generated by the model (16.4) has the same slope as the smile generated by the model (16.3), so we have arrived at two models with the same (static) smile but different smile dynamics. Clearly, by "mixing" the two, we can get a model where the backbone is controlled independently of the smile.

Here is how we proceed. Introducing a new parameter, "mixing" m, we specify

$$dS(t) = \lambda \left(bS(t) + (m - b) S(0) + (1 - m) L\right) dW(t). \tag{16.5}$$

For L close to $S(0)$, λ still has the meaning of relative (log-normal) volatility, and the slope of the smile is still controlled by b. On the other hand, simple calculations yield

$$\sigma_{\mathrm{B}}\left(0, S(0); T, K\right) \approx \lambda b \frac{\ln\left(S(0)/K\right)}{\ln\left(\frac{mS(0)+(1-m)L}{bK+(m-b)S(0)+(1-m)L}\right)},$$

$$\left.\frac{\partial\sigma_{\mathrm{B}}\left(0, S(0); T, K\right)}{\partial S(0)}\right|_{K=S(0)} \approx \frac{1}{2}\frac{(m-b)S(0)-(1-m)L}{S(0)}\frac{\lambda}{S(0)}$$

and, with $L \approx S(0)$,

$$\left.\frac{\partial\sigma_{\mathrm{B}}\left(0, S(0); T, K\right)}{\partial S(0)}\right|_{K=S(0)} \approx \left(m-\frac{1+b}{2}\right)\frac{\lambda}{S(0)}.$$

Clearly, for any b, we can adjust the backbone by choosing a suitable m. For example, we can always obtain the Black backbone by setting $m = (1+b)/2$, or the Gaussian backbone by setting $m = b/2$.

To check that m does not, indeed, have an impact on the (static) smile, let us calculate the slope of the implied volatility smile in the model (16.5). We have

$$\left.\frac{\partial\sigma_{\mathrm{B}}\left(0, S(0); T, K\right)}{\partial K}\right|_{K=S(0)} \approx \frac{1}{2}\frac{-mS(0)+Lm+bS(0)-L}{S(0)}\frac{\lambda}{S(0)}$$

and, with $L \approx S(0)$,

$$\left.\frac{\partial\sigma_{\mathrm{B}}\left(0, S(0); T, K\right)}{\partial K}\right|_{K=S(0)} \approx -\frac{1-b}{2}\frac{\lambda}{S(0)}. \qquad (16.6)$$

Thus, as claimed, the slope of the smile is independent of m and, in particular, is the same for the models (16.3), (16.4) and (16.5).

We should note that sometimes a different definition of the backbone is used, a definition that we call the *ATM backbone*:

$$\left.\frac{\partial\sigma_{\mathrm{B}}\left(0, S; T, S\right)}{\partial S}\right|_{S=S(0)}.$$

This quantity specifies how the at-the-money volatility $\sigma_{\mathrm{B}}(0, S(0); T, S(0))$ changes with the underlying $S(0)$. Simple calculus yields

$$\left.\frac{\partial\sigma_{\mathrm{B}}\left(0, S; T, S\right)}{\partial S}\right|_{S=S(0)} = \left.\frac{\partial\sigma_{\mathrm{B}}\left(0, S(0); T, K\right)}{\partial S(0)}\right|_{K=S(0)}$$

$$+ \left.\frac{\partial\sigma_{\mathrm{B}}\left(0, S(0); T, K\right)}{\partial K}\right|_{K=S(0)}$$

and we obtain, for the model (16.5) assuming $L \approx S(0)$,

$$\left.\frac{\partial\sigma_{\mathrm{B}}\left(0, S; T, S\right)}{\partial S}\right|_{S=S(0)} \approx -\left(1-m\right)\frac{\lambda}{S(0)}. \qquad (16.7)$$

The ATM backbone in the model (16.5) is independent of the skew b. When $m = b$ (model (16.3)) the ATM backbone is twice the slope of the volatility smile (see (16.6)), and when $m = 1$ (model (16.4)), the ATM backbone is zero, i.e. at-the-money implied volatility does not change as the underlying moves. Other regimes are easy to simulate. For example, a trader may believe that the at-the-money implied volatility should "slide along" the smile, i.e. exhibit (a weaker form of) the "sticky strike" behavior. Mathematically, this is expressed as

$$\frac{\partial \sigma_B (0, S; T, S)}{\partial S}\bigg|_{S=S(0)} = \frac{\partial \sigma_B (0, S(0); T, K)}{\partial K}\bigg|_{K=S(0)}$$

which, using (16.6) and (16.7), gives us the following condition on the mixing parameter:

$$m = \frac{b+1}{2}.$$

16.1.3 Stochastic Volatility Swaption Grid

With backbone issues out of the way, let us now discuss a typical setup for vanilla options modeling. As mentioned before, a stochastic volatility model offers a good compromise between tractability and the ability to represent typical shapes of volatility smiles. To describe a typical setup, let us introduce a tenor structure

$$0 < T_0 < T_1 < T_2 < \ldots < T_N, \quad \tau_n = T_{n+1} - T_n,$$

and a collection of forward swap rates of different expiries/tenors, as in Section 5.10, see (5.13)–(5.14). For each swap rate $S_{n,m}(t)$, we specify the following SV-style dynamics (see Chapter 8) in the corresponding annuity measure

$$dS_{n,m}(t) = \lambda_{n,m} \left(b_{n,m} S_{n,m}(t) + (1 - b_{n,m}) S_{n,m}(0) \right) \sqrt{z_{n,m}(t)} \, dW^{n,m}(t),$$

$$\tag{16.8}$$

$$dz_{n,m}(t) = \theta \left(1 - z_{n,m}(t) \right) dt + \eta_{n,m} \sqrt{z_{n,m}(t)} \, dZ^{n,m}(t), \tag{16.9}$$

with $\langle dZ^{n,m}(t), dW^{n,m}(t) \rangle = 0$. Alternative local volatility parameterizations as in (16.3) or even (16.5) could of course be used, but we abstain from doing so to simplify notations. Further, we assume that the mean reversion of variance parameter θ is *global*, i.e. the same for all swaptions. This does not in any way restrict the range of available smiles for each individual swap rate as explained in Section 8.2, yet allows for a measure of consistency in term structure models (e.g., as in Section 13.2) that we, eventually, calibrate to the vanilla market. The rest of the parameters

$$\{(\lambda_{n,m}, b_{n,m}, \eta_{n,m})\}, \quad n = 0, \dots, N-1, \quad m = 1, \dots, N-n,$$

form the so-called *SV swaption grid*, with the meanings of various parameters explained in Section 8.2. Relative to a full swaption volatility cube (see footnote 2), an SV swaption grid typically requires storage of fewer parameters, as the SV model produces a fairly parsimonious (and guaranteed arbitrage-free) interpolation rule in the strike dimension, eliminating the need for outright storage of implied Black or Gaussian volatilities on a strike grid. Of course, multiple other — possibly heuristic — interpolation rules can be used instead. We return to this briefly in Section 16.1.5 below.

16.1.4 Calibrating Stochastic Volatility Model to Swaptions

SV parameters for swaptions are usually obtained by individually calibrating swaptions for each expiry/maturity grid point, with an exception of the mean reversion parameter θ. Recall (Section 8.2) that the parameter θ controls the speed at which the volatility smile flattens with time to expiry, so it is possible to choose a single θ for the whole grid, in such a way as to minimize the variability of $\eta_{n,m}$ across different n's (expiries).

Selection of θ can be done manually by choosing a particular θ, calibrating SV parameters to each swaption grid point, and then assessing how constant $\eta_{n,m}$ for different n, m are. If not sufficiently constant, a different θ can be selected, and the procedure iterated until a sufficiently good choice is found. In general, if $\eta_{n,m}$'s increase with n, we need a smaller θ to prevent volatility smiles from flattening out too fast with expiry. Conversely, if $\eta_{n,m}$'s decrease in n, a larger θ is needed.

Apart from θ, swaption calibration is performed individually for each grid point. Let us sketch the algorithm. First, we fix a particular swaption maturity and a swap tenor, as represented by indices n, m. Suppressing these indices for the moment, suppose a collection of strikes $K_1, \dots, K_J$ is given, along with corresponding market prices of swaptions $\widehat{V}_1, \dots, \widehat{V}_J$. Given λ, b, η let

$$V(K_j; \lambda, b, \eta), \quad j = 1, \dots, J,$$

be the model prices of swaptions in the model (16.8)–(16.9) with parameters λ, b, η. Our goal is to find λ, b, η to match as closely as possible the market prices $\widehat{V}_1, \dots, \widehat{V}_J$, where nearly always $J \geq 3$. This type of problem is most conveniently solved by non-linear optimization methods. Defining the objective function

$$\mathcal{I}_1(\lambda, b, \eta) = \sum_{j=1}^{J} w_j \left(V(K_j; \lambda, b, \eta) - \widehat{V}_j \right)^2, \tag{16.10}$$

where $w_1, \dots, w_J$ are user-specified weights, we obtain the calibrated parameters by solving the problem

$$(\lambda^*, b^*, \eta^*) = \underset{\{\lambda, b, \eta\}}{\operatorname{argmin}} \mathcal{I}_1 (\lambda, b, \eta)$$

with a specialized algorithm such as the Fletcher-Reeves or the Levenberg-Marquardt method (see Press et al. [1992]). As the optimization problem is solved numerically, the solution typically involves multiple calculations of option prices in the SV model, and having an efficient valuation algorithm such as the one developed in Section 8.4 is important for performance.

The weights $w_1, \ldots, w_J$ serve two purposes. One is to express the view on which swaptions should be matched more accurately: the higher the weight w_j is, the more closely the algorithm will try to match the price of the swaption with strike K_j. As we typically have more confidence in the at-the-money swaption prices, we would often set the weights higher for at-the-money strikes and lower for strikes away from the ATM. The other important purpose of the weights is to normalize the magnitude of different terms in the sum in (16.10), as different scales of different terms (i.e. some $\widehat{V}_j$'s are bigger than others) will influence which terms are matched closer. As we often seek to ensure a good fit in terms of implied volatilities rather than absolute option values, a commonly-used scaling involves vegas of the options in the optimization problem; each weight w_j then represents a product of a user-specified importance weight and a scaling weight equal to the inverse of the swaption vega. To simplify user interface, the vega scaling can be internalized, with (16.10) replaced by

$$\mathcal{I}_2 (\lambda, b, \eta) = \sum_{j=1}^{J} w_j \left(\frac{V (K_j; \lambda, b, \eta) - \widehat{V}_j}{\widehat{\Upsilon}_j} \right)^2, \qquad (16.11)$$

where $\widehat{\Upsilon}_j$'s are vegas of corresponding options. For efficiency reasons, the vegas should not be calculated inside the calibration loop; a common shortcut is to just use vegas obtained in the Black model. The resulting objective function is a (numerically efficient) approximation to an objective function expressed in terms of implied volatilities:

$$\mathcal{I}_3 (\lambda, b, \eta) = \sum_{j=1}^{J} w_j (\sigma_{\mathrm{B}} (K_j; \lambda, b, \eta) - \widehat{\sigma}_j)^2, \qquad (16.12)$$

where $\sigma_{\mathrm{B}}(K_j; \lambda, b, \eta)$ is the Black volatility implied by the model for the option with strike K_j, and $\widehat{\sigma}_j$ is its market-implied volatility. While (16.12) could be used directly, the expense of calculating implied volatilities inside the calibration loop typically makes it less attractive than (16.11).

Finally, let us remind the reader that optimization of a precision norm (e.g. either $\mathcal{I}_1, \mathcal{I}_2, \mathcal{I}_3$) must be undertaken for each pair of swaption expiries and swap tenors in the SV swaption grid, a total of $N(N + 1)/2$ separate optimization problems.

16.1.5 Some Other Interpolation Rules

Usage of the SV model for swaption calibration is particularly convenient if one ultimately needs to use swaption market data for calibration of term structure models such as the quasi-Gaussian model of Section 13.2 or the Libor market model of Section 14.2.5. See, in particular, the discussion in Section 15.2. It is, however, certainly possible to represent the strike-dependence of implied swaption volatilities by different means. A particularly popular choice is to calibrate a SABR model (Section 8.6) to the smile, using the principles outlined above. The existence of a reasonably accurate expansion for implied volatilities in this model makes optimization of the norm (16.12) particularly convenient.

To improve the fitting capability of the model, we note that it is not uncommon for practitioners to "improve" the SABR model with heuristic modifications, such as making the power c or correlation ρ a smooth, bounded function of swaption strikes. Such alterations of the original model make little sense dynamically speaking, but may still represent a valid representation of the marginal distribution of forward swap rates. In a sense, the original SABR model has been used to produce a particular parametric interpolation rule for implied volatilities, where some of the parameters happen to have a convenient intuitive interpretation. Of course, it may then be tempting to skip the entire concept of a dynamic model and simply jump straight to the specification of a smooth parametric form for implied volatilities as a function of strike. There are numerous such forms in circulation; one representative example is the SVI ("stochastic volatility inspired") 5-parameter form proposed in Gatheral [2004]:

$$\sigma_{\mathrm{B}}\left(0, S(0); T, K\right) = a + b\left(\rho(k - h) + \sqrt{(k - h)^2 + s^2}\right), \qquad (16.13)$$

where $k \triangleq \ln(K/S(0))$. More details about the valid range and intuition for the parameters a, b, h, ρ, s can be found in Gatheral [2004] and Gatheral and Jacquier [2010], and shall not be repeated here. Let us just note that one drawback of parametric forms is that they can produce arbitrages, in the sense that there typically are parameter combinations that will imply negative marginal densities[5] for swap rates. This issue must be taken into consideration, e.g. by imposing constraints on the parameter space when calibrating the parametric form against market prices.

We note that similar issues with violation of arbitrage can arise if crude interpolation schemes are used for strike interpolation in a swaption cube — for instance, linear interpolation should never be used, since the second derivative of implied volatilities will not exist everywhere. A better choice for an interpolation scheme would be a twice differentiable spline, such as those described at length in Chapter 6.

---

[5]These can be computed by differentiating swaption prices twice with respect to the strike, see Section 7.1.2.

16.2 Caps and Floors

While pricing caps and floors (see Section 5.8) is typically no more complicated than pricing swaptions, calibrating a model to quoted cap or floor
prices is more involved than calibrating it to swaptions. This is due to
the fact that market prices of individual caplets are not directly available,
since only prices of full caps — i.e. collections of caplets — are quoted. For
example, in the short- to medium-term market in the US, caps of maturities
1, 2, 3, 5 and 10 years are traded. With each caplet covering 3 months, a
total of $10 \times 4 = 40$ different caplet maturities are involved, each requiring
its own rule for strike interpolation. The scarcity of actual quotes, and the
fact that these quotes represent sums of option prices, can potentially lead
to overfitting unless extra constraints, either implicit or explicit, are imposed
during calibration.

16.2.1 Basic Problem

Assume for a moment that market prices of both 2 year and 2 year 3 months
caps are known at multiple strikes. By simple subtraction, we could then
recover the prices of individual caplets fixing in two years time, and would
be able to fit model parameters to prices of those caplets across strikes, in
a manner identical to that employed for swaptions. In reality, there is no
market in 2 year 3 months caps, but we could always attempt to obtain
the required prices by interpolating between the known prices of 2 year and
3 year caps. Alas, this idea is hampered by the fact that 2 year caps are
typically quoted in a range of strikes that is *different* from the strikes for a
3 year cap: for a cap with a given maturity, the quoted strikes are typically
fixed offsets from the forward swap rate of the corresponding maturity, i.e.
ATM±100 basis points, ATM±200 basis points, and so on. As a consequence,
we would need to perform interpolation between 2 year and 3 year cap prices
across both expiries *and strikes*. Ensuring that such an interpolation scheme
is both free of arbitrage and will give rise to reasonable (i.e. smooth over
time) model parameters is not an easy task; our advice is to avoid it.

A more reasonable approach to cap calibration is to employ interpolation
directly in model parameters, borrowing the ideas from yield curve construction theory (see Chapter 6). Specifically, we can formalize the cap calibration
problem as finding model parameter curves indexed by expiry, such that a
price precision norm is minimized subject to penalties for non-smooth model
parameters. Encouraging smooth model parameters across expiry makes any
subsequent parameter interpolation across time (as required for seasoned
trades with fixing schedules deviating from that used in calibration) both
more stable and more believable. Moreover, imposing smoothness constraints
promotes the stability of calibration through time, a property important for
consistency of risk management.

16.2.2 Setup and Norms

To formalize our approach, we specify the tenor structure

$$0 = T_0 < T_1 < T_2 < \ldots < T_N, \quad \tau_n = T_{n+1} - T_n,$$

such that $[T_n, T_{n+1}]$ is a caplet tenor (3 months in the US). For concreteness, we use an SV model to define our volatility interpolation scheme in strike space; let the SV parameters to be used for a caplet that fixes at T_n and pays at T_{n+1} be denoted (λ_n, b_n, η_n), $n = 1, \ldots, N-1$. We denote the price of the n-th caplet with strike K in the SV model with parameters λ, b, η by $V_n(K; \lambda, b, \eta)$. Let n_i, $n_1 < \ldots < n_I$, be the number of caplets in the i-th standard market cap. Furthermore, let us suppose that the i-th standard cap is available with strikes

$$K_{i,1}, \ldots, K_{i,J},$$

where we for simplicity have assumed that caps of different tenors are quoted for the same *number* of strikes J (but we allow for different *values* of those strikes). Finally, let us denote by $\widehat{V}_{i,j}$, $i = 1, \ldots, I$, $j = 1, \ldots, J$, the market price of the i-th standard cap (with n_i caplets) at strike $K_{i,j}$.

Let us first consider the introduction of a precision norm that quantifies the amount of mispricing associated with a given set of SV parameters. For instance, we could use a standard weighted least-squares norm

$$\mathcal{I}_1 = \sum_{i=1}^{I} \sum_{j=1}^{J} w_{i,j} \left(\sum_{n=1}^{n_i} V_n\left(K_{i,j}; \lambda_n, b_n, \eta_n\right) - \widehat{V}_{i,j} \right)^2,$$

where $w_{i,j}$ is the weight associated with the i-th cap of strike $K_{i,j}$.

In principle, we can treat parameter triples for all n, $n = 1, \ldots, N-1$, as independent variables to be recovered in the solution of an optimization problem. However, significant performance improvements can be realized if we reduce the number of free parameters by allowing only the parameters that correspond to the *expiries of market caps* to be free inputs, while interpolating the rest. Linear interpolation seems to perform well, although more sophisticated interpolation schemes borrowed from Chapter 6 could result in further improvements. In any case, if we denote by $\mathcal{X}$ the collection of $\{(\lambda_{n_i}, b_{n_i}, \eta_{n_i})\}$ for $i = 1, \ldots, I$, then we can rewrite the objective function as

$$\mathcal{I}_2\left(\mathcal{X}\right) = \sum_{i=1}^{I} \sum_{j=1}^{J} w_{i,j} \left(\sum_{n=1}^{n_i} V_n\left(K_{i,j}; \lambda_n\left(\mathcal{X}\right), b_n\left(\mathcal{X}\right), \eta_n\left(\mathcal{X}\right)\right) - \widehat{V}_{i,j} \right)^2,$$

where, as explained, $\lambda_n(\mathcal{X})$, $b_n(\mathcal{X})$, $\eta_n(\mathcal{X})$ are obtained from the elements in $\mathcal{X}$ by suitable interpolation.

Various types of penalties for lack of smoothness are possible, with the discussion of the similar issues in LM calibration in Section 14.5.6 imminently applicable here; a reasonable choice would minimize the discrete equivalent of the integral of the square of the first-order derivative. In particular, we define a norm that, in essence, penalizes deviations of SV parameters from being constant over time,

$$
\mathcal{I}_{\text{smooth}}\left(\mathcal{X}\right) = w^\lambda \sum_{n=2}^{N-1} \left(\lambda_n\left(\mathcal{X}\right) - \lambda_{n-1}\left(\mathcal{X}\right)\right)^2
$$

$$
+ w^b \sum_{n=2}^{N-1} \left(b_n\left(\mathcal{X}\right) - b_{n-1}\left(\mathcal{X}\right)\right)^2 + w^\eta \sum_{n=2}^{N-1} \left(\eta_n\left(\mathcal{X}\right) - \eta_{n-1}\left(\mathcal{X}\right)\right)^2. \quad (16.14)
$$

Here weights w^λ, w^b, w^η determine the relative importance of smoothing different model parameters. While the representation (16.14) is quite transparent, we can improve performance somewhat by reformulating — exactly or approximately, depending on the interpolation used — the objective function solely in terms of the components of $\mathcal{X}$, i.e. by imposing smoothness directly on the "free" parameters $\{(\lambda_{n_i}, b_{n_i}, \eta_{n_i})\}$ for $i = 1, \ldots, I$.

16.2.3 Calibration Procedure

Having introduced precision and smoothing norms above, the SV cap calibration problem can be cast as a minimization of the following objective function,

$$
\mathcal{X} = \operatorname{argmin}\left\{w_{\text{precision}}\mathcal{I}_2\left(\mathcal{X}\right) + w_{\text{smooth}}\mathcal{I}_{\text{smooth}}\left(\mathcal{X}\right)\right\},
$$

over the allowed domain for $\mathcal{X}$. The weights $w_{\text{precision}}$ and w_{smooth} determine relative importance of achieving smoothness over a good fit to market prices. This optimization problem may be solved by numerical methods (see Press et al. [1992]), just as many other calibration problems we discussed previously. To improve efficiency of the algorithm, one could here attempt to split volatility level calibration from smile slope calibration, as dividing optimization problems into smaller ones and tackling them separately often gives us better performance (see relevant discussion in Section 14.5.8). Specifically, significant gains can often be found by iterating over a split scheme where we first calibrate the volatility parameter to at-the-money cap prices only and then calibrate the other model parameters to out-of-the-money cap prices. The success of such a "relaxation" scheme lies with the relative independence of the impacts of volatility parameter and the other parameters: by successfully optimizing in two relatively orthogonal dimensions, we reach a joint minimum faster than in the full calibration. We omit straightforward details.

16.3 Terminal Swap Rate Models

The relative simplicity with which European swaptions (and caps) can be priced comes from the fact that valuation here requires only knowledge of the terminal distribution of a single swap rate, in the appropriate annuity measure. This holds true for all securities whose payoffs can be expressed as deterministic functions of the swap rate $S(T)$ in the annuity measure, as should be clear from the replication argument of Proposition 8.4.13. Unfortunately, such payoffs are relatively rare. Much more common are relatively simple payoffs that appear to depend on the rate $S(T)$ only but, in fact, require the knowledge of certain additional discount bonds, often observed on the same date. As multiple discount bonds are involved and the knowledge of the distribution of a swap rate is not sufficient for valuation, it would appear that a full term structure model is needed to price such derivatives. This, of course, is an option that is always available. However, if the dependence on additional discount bonds is sufficiently mild, we often can avoid computational cost of a full-blown term structure model through certain approximations that aim at functionally linking the values of discount bonds on date T to the "driving" rate $S(T)$, i.e. the rate that primarily determines the payoff. The basic modeling idea, which we denote the *Terminal Swap Rate* (TSR) approach, is extremely useful in handling a range of actively traded European derivatives that are not, strictly speaking, functions of a single rate, but can still be approximated accurately as such. We use the (somewhat loose, admittedly) term *approximately single-rate* for this class of securities; several common securities in the class will be presented in subsequent sections, after the TSR method has been described in detail.

16.3.1 TSR Basics

As briefly outlined in the previous section, the TSR approach treats the swap rate $S(T)$ as the single fundamental state variable for the yield curve at time T. To define the method formally, we continue with the notations of $A(t)$ being the annuity corresponding to the swap rate $S(t)$; for concreteness, we assume that (see (5.4)–(5.5))

$$A(t) \triangleq A_{0,N}(t) = \sum_{n=0}^{N-1} \tau_n P\left(t, T_{n+1}\right), \tag{16.15}$$

$$S(t) \triangleq S_{0,N}(t) = \frac{P\left(t, T\right) - P\left(t, T_N\right)}{A(t)}, \tag{16.16}$$

where

$$0 < T = T_0 < T_1 < \ldots < T_N, \quad \tau_n = T_{n+1} - T_n,$$

is a tenor structure of dates. We continue denoting by Q^A the annuity measure, i.e. the measure for which $A(t)$ is the numeraire. We recall that the market-implied distribution of $S(T)$ in $\mathbb{Q}^A$ can be found from calibrating a vanilla model to N-period swaptions with expiry T, across multiple strikes.

The no-arbitrage valuation formula (1.15) states that the value of a derivative with an $\mathcal{F}_T$-measurable payoff X is given by

$$V(0) = A(0)\mathrm{E}^A\left(\frac{X}{A(T)}\right). \tag{16.17}$$

Let $\{P(T, M)\}_{M \geq T}$ be the discount bonds of various maturities, all observed at time T. A TSR model specifies a map

$$P(T, M) = \pi(S(T), M), \quad M \geq T, \tag{16.18}$$

where $\{\pi(\cdot, M)\}_{M \geq T}$ is a collection of exogenously specified maturity-indexed functions. In other words, each discount factor is assumed to be a deterministic, known function of the swap rate.

In a proper term structure model, the relationship between the market rate $S(T)$ and the discount factors $\{P(T, M)\}_{M \geq T}$ emerges from the model itself, and is ultimately derived from no-arbitrage conditions. While now we seek to impose the functional relationships (16.18) exogenously, consideration of no-arbitrage must also play a role. Indeed, a first condition to be imposed on the functions $\{\pi(\cdot, M)\}_{M \geq T}$ will be the *no-arbitrage condition*: the valuation formula (16.17), when applied to the specification (16.18), must reproduce initial discount bond prices, i.e. the following must hold for any[6] $M \geq T$,

$$P(0, M) = A(0)\mathrm{E}^A\left(\frac{\pi(S(T), M)}{\sum_{n=0}^{N-1} \tau_n \pi(S(T), T_{n+1})}\right), \quad M \geq T. \tag{16.19}$$

The no-arbitrage condition by itself is not sufficient to obtain a workable model. Another restriction on the mapping functions is obtained by observing that the swap rate $S(T)$ itself is a function of discount factors, as evidenced by (16.16). This suggests the introduction of a *consistency condition*, i.e. the requirement that the following holds for all x,

$$x = \frac{1 - \pi(x, T_N)}{\sum_{n=0}^{N-1} \tau_n \pi(x, T_{n+1})}. \tag{16.20}$$

The final condition that we impose on a TSR model is that the set of functions $\{\pi(\cdot, M)\}_{M \geq T}$ should be *reasonable*. While somewhat harder to quantify than the other conditions, we shall mostly impose the following restrictions:

---

[6]In some applications of TSR models, it may suffice that this expression holds only for a single value of M (namely the payment date of the security in question). In such cases, certain simplifications may be possible, as demonstrated in Section 16.6.4.

- For each x and $M \geq T$, $\pi(x, M)$ is between 0 and 1,

$$0 < \pi(x, M) \leq 1.$$

- For each x, $\pi(x, \cdot)$ is monotonic in M,

$$M_1 < M_2 \implies \pi(x, M_1) \geq \pi(x, M_2).$$

- The function $\pi(x, M)$ is continuous in (x, M).

Some of these conditions are more important than others. For example, one may choose to tolerate negative interest rates, i.e. having $\pi(x, M) > 1$ for some x, M, but not negative prices of bonds, i.e. having $\pi(x, M) < 0$ for some x, M.

The conditions listed above do not define the functions $\{\pi(\cdot, M)\}_{M \geq T}$ uniquely; however, they do, as a rule, specify the functions uniquely within a particular parametric class. A concrete model is then obtained by postulating a particular parametric class for the functions $\{\pi(\cdot, M)\}_{M \geq T}$ first, and then choosing functions within the class uniquely from the no-arbitrage and consistency conditions. Let us consider a few representative examples.

16.3.2 Linear TSR Model

The linear TSR model is obtained by specifying

$$\frac{\pi(x, M)}{\sum_{n=0}^{N-1} \tau_n \pi(x, T_{n+1})} = a(M)x + b(M), \quad M \geq T, \tag{16.21}$$

for deterministic functions $a(\cdot)$ and $b(\cdot)$. The no-arbitrage condition requires

$$P(0, M) = A(0)\mathrm{E}^A (a(M)S(T) + b(M)),$$

implying a condition on the free coefficient $b(\cdot)$,

$$b(M) = \frac{P(0, M)}{A(0)} - a(M)S(0). \tag{16.22}$$

The consistency condition requires that

$$x = \frac{1 - \pi(x, T_N)}{\sum_{n=0}^{N-1} \tau_n \pi(x, T_{n+1})} = (a(T_0)x + b(T_0)) - (a(T_N)x + b(T_N)),$$

so that

$$b(T_0) = b(T_N), \tag{16.23}$$
$$a(T_0) = 1 + a(T_N). \tag{16.24}$$

It follows from (16.22) that if (16.23) is satisfied, then (16.24) is satisfied as well, and vice versa.

The definition (16.21) imposes new restrictions on $a(\cdot)$, $b(\cdot)$ that go beyond those considered in the previous section. In particular, the following must now hold,

$$\sum_{n=0}^{N-1} \tau_n \left(a\left(T_{n+1}\right) x + b\left(T_{n+1}\right)\right) \equiv 1,$$

implying

$$\sum_{n=0}^{N-1} \tau_n a\left(T_{n+1}\right) = 0, \qquad (16.25)$$

$$\sum_{n=0}^{N-1} \tau_n b\left(T_{n+1}\right) = 1. \qquad (16.26)$$

It is enough to ensure that one of these two conditions is satisfied; the other will follow automatically by (16.22).

To complete the model specification, we may proceed as follows. First, choose coefficients $\{a(T_1), \ldots, a(T_N)\}$, subject to the condition (16.25); a scheme for this will be discussed shortly. Then, calculate $a(T) = a(T_0)$ from (16.24), and the rest of $a(M)$'s by, for example, linear interpolation of $\{a(T), a(T_1), \ldots, a(T_N)\}$. Finally, calculate all $b(M)$'s via (16.22).

The specification of a TSR model above enjoys a fair amount of numerical tractability, owning to the simple linear relationship between the market rate and annuity-discounted bonds. For that reason it is rather popular in applications. However, the linear relationship imposed by the model is not wholly realistic, as bond prices may become negative in certain states of the world. Whether this is a problem for a particular application should be decided on a case-by-case basis.

The linear TSR model (16.21) is rather flexible, as the coefficients $\{a(T_1), \ldots, a(T_N)\}$ can be selected essentially independently, subject to (16.25) only. Setting these coefficients individually is not particularly convenient, however, as financial implications of various choices are not transparent. As promised above, let us therefore look for a more meaningful way of parameterizing $a(\cdot)$. For this, let us first observe that the coefficients $a(\cdot)$ essentially define the shape of the yield curve at time T for different levels of the "state variable" $S(T)$. Of course, we have seen previously that the same role is played by the mean reversion parameter in the context of a Gaussian term structure model, see Section 10.1.2. This suggests connecting the coefficients of the TSR model to a mean reversion parameter, which would not only reduce the number of parameters we need to specify, but also parametrize the model with a single parameter that has strong financial interpretation and that, in principle, could be derived from prices of traded derivatives (see Section 13.1.8).

To connect $a(\cdot)$ to mean reversion, we interpret the equality (16.21) as defining $a(M)$ via

$$a(M) = \frac{\partial}{\partial S(T)} \frac{P(T,M)}{\sum_{n=0}^{N-1} \tau_n P(T,T_{n+1})},$$

which we rewrite, in the context of a Gaussian one-factor model, as

$$a(M) = \frac{\partial}{\partial x} \frac{P(T,M,x)}{\sum_{n=0}^{N-1} \tau_n P(T,T_{n+1},x)}\bigg|_{S(T,x)=S(0)}$$

$$\times \left(\frac{\partial S(T,x)}{\partial x}\bigg|_{S(T,x)=S(0)}\right)^{-1},$$

where x is now the short rate state in the Gaussian model on which all discount bonds and swap rates depend. We denote by $A(T,x)$ the annuity as the function of the short rate state x,

$$A(T,x) = \sum_{n=0}^{N-1} \tau_n P(T,T_{n+1},x),$$

so that

$$S(T,x) = (1 - P(T,T_N,x))/A(T,x)$$

and

$$\frac{\partial}{\partial x} \frac{P(T,M,x)}{A(T,x)} = -\frac{P(T,M,x)G(T,M)}{A(T,x)} - \frac{P(T,M,x)}{A(T,x)^2}\frac{\partial A(T,x)}{\partial x},$$

$$\frac{\partial}{\partial x} S(T,x) = \frac{P(T,T_N,x)G(T,T_N)}{A(T,x)} - \frac{S(T,x)}{A(T,x)}\frac{\partial A(T,x)}{\partial x},$$

where $G(\cdot,\cdot)$ is a function of mean reversion (see (10.18)). By using the approximation

$$P(T,t,x)|_{S(T,x)=S(0)} \approx P(0,t)/P(0,T)$$

for all $t \geq T$, we obtain

$$a(M) = \frac{P(0,M)(\gamma - G(T,M))}{P(0,T_N)G(T,T_N) + A(0)S(0)\gamma}, \qquad (16.27)$$

where

$$\gamma \triangleq -\frac{1}{A(T,x)}\frac{\partial A(T,x)}{\partial x}\bigg|_{P(T,t,x)=P(0,t),\ \forall t \geq T}$$

$$= \frac{\sum \tau_n P(0,T_{n+1})G(T,T_{n+1})}{\sum \tau_n P(0,T_{n+1})}. \qquad (16.28)$$

As explained before, the coefficients $b(\cdot)$ are obtained from $a(\cdot)$ by (16.22).

With this parameterization, instead of a collection $\{a(T_1), \ldots, a(T_N)\}$, only one parameter — the mean reversion $\varkappa$ — needs to be specified. As we shall see later in Section 16.6.8, the choice of mean reversion has a mild but non-vanishing impact on values of many approximately single-rate derivatives.

Linking $a(\cdot)$ to mean reversion leads to a more intuitive parameterization of the model, and also facilitates better risk management. This is so because the mean reversion parameter can in principle be hedged by European swaptions of the same expiry (here T) and different tenors, as we discussed in Section 13.1.8.2. For truly precise vega hedging, however, this somewhat indirect linkage of $a(\cdot)$ to swaption volatilities is less than ideal. In fact, it is not difficult to link $a(\cdot)$ to swap rate volatilities (expiring on the same date T but of different tenors) directly, using an approach similar to what we have developed in this section. We leave the details of this idea for the reader to fill.

16.3.3 Exponential TSR Model

The linear TSR model is a convenient, but by no means unique, representative of the TSR approach. The *exponential TSR model* belongs to the same class, but uses exponential functions to connect a swap rate to discount bonds; this idea originates with the exponential relationship between a discount bond and a corresponding continuously compounded spot yield.

We start the development of the exponential specification by postulating that

$$\pi(x, M) \propto \exp(-l(M)x), \quad M \geq T. \tag{16.29}$$

The intuitive meaning of the loading $l(M)$ is best understood by recalling that a continuously compounded spot yield for the period $[T, M]$, observed at T, is given by

$$y(T, M) = -\frac{1}{M-T} \ln P(T, M) = -\frac{1}{M-T} \ln \pi(S(T), M).$$

Coupled with the specification (16.29), this tells us

$$y(T, M) \propto \frac{l(M)}{M-T} S(T), \quad M \geq T,$$

so that the curve $l(M)/(M-T)$, $M \geq T$, defines the shape of the shock to the yield curve, expressed in terms of yields $\{y(T, M)\}_{M \geq T}$, for a perturbation of the market rate $S(T)$. Recycling the idea of connecting this shape to something similar in a term structure model, we use a one-factor Gaussian model and specify

$$l(M) = \frac{1 - e^{-\varkappa(M-T)}}{\varkappa}. \tag{16.30}$$

If more precision is desired, we can alternatively write

$$l(M) = \frac{1 - e^{-\varkappa(M-T)}}{\varkappa} \left(\left. \frac{\partial S(T,x)}{\partial x} \right|_{S(T,x)=S(0)} \right)^{-1},$$

where the partial derivative is computed in a one-factor Gaussian model, with x being the short rate state and $S(T,x)$ the swap rate as function of the short rate state in the Gaussian model (see Section 10.1.2).

Unfortunately, the expression on the right-hand side of (16.29) cannot be used directly in a TSR model, since it lacks flexibility to satisfy the consistency requirement. This, however, is easily rectified by modifying the functional form slightly and replacing $x \rightarrow \psi(x)$ in (16.29),

$$\pi(x, M) = \exp(-l(M)\psi(x) + b(M)), \quad M \geq T. \tag{16.31}$$

The maturity-dependent function $b(\cdot)$ is obtained from the no-arbitrage conditions (16.19), and the function $\psi(\cdot)$ is defined implicitly by the consistency condition: for any x, $\psi(x)$ is set to be the solution z^* of the equation

$$x = \frac{1 - \exp(-l(T_N)z + b(T_N))}{\sum_{n=0}^{N-1} \tau_n \exp(-l(T_{n+1})z + b(T_{n+1}))}.$$

This equation can easily be solved with just a couple of iterations of a numerical root search algorithm; not surprisingly, it turns out that $\psi(x) \approx x$ to high precision.

16.3.4 Swap-Yield TSR Model

Another example of a TSR model is inspired by the coupon bond yield formula (see Burghardt [2005]). The mapping functions for the *swap-yield TSR model* are defined by

$$\pi(x, M) = \left(\prod_{i=0}^{n-1} \frac{1}{1 + \tau_i x} \right) \times \left(\frac{1}{1 + \tau_n x} \right)^{(M-T_n)/\tau_n}, \quad M \geq T, \tag{16.32}$$

where $n = q(M) - 1$ with the index function $q(M)$, $M \geq T$, defined by (14.2), such that

$$M \in [T_{q(M)-1}, T_{q(M)}), \tag{16.33}$$

with the assumption that $T_{N+1} = +\infty$.

The consistency condition (16.20) is satisfied automatically as the following identity holds,

$$\frac{1 - \prod_{i=0}^{N-1} (1 + \tau_i x)^{-1}}{\sum_{n=0}^{N-1} \tau_n \prod_{i=0}^{n} (1 + \tau_i x)^{-1}} \equiv x. \tag{16.34}$$

The formula (16.32) essentially tells us to discount all cash flows after T at the same rate, namely a rate given by the realized swap rate $S(T)$. As mentioned, this is motivated by traditional definitions of a coupon bond yield or by the payoff of a cash-settled swaption, see Section 5.10.1. The swap-yield TSR specification is motivated by real financial constructs and it can be said to be "reasonable" in the sense of Section 16.3.1; not surprisingly the model has enjoyed widespread popularity in the financial industry. Despite this, the model is not arbitrage-free, as (16.19) is *not* satisfied. Empirically, the extent of violation of no-arbitrage conditions is fairly small, but not always negligible. Another issue with the model, at least in its basic form, is its lack of explicit control over the shape of the yield curve at time T, something we managed to introduce into the linear and exponential specifications by imposing a link between parameters of these models to a mean reversion parameter.

In defense of the swap-yield model, it should be said that, in principle, the model could be improved by few modifications. Violations of no-arbitrage conditions could be addressed by introducing a scaling parameter $b(M)$ as in (16.32).The consistency condition that would no longer hold could be satisfied by introducing a function $\psi(x)$ in the place of x in (16.32), similar to (16.31); this function would need to be calculated numerically. Even ideas about mean reversion could be applied. We invite the reader to attempt these improvements, although we generally feel that the resulting model would offer few, if any, advantages over the linear or exponential TSR models.

16.4 Libor-in-Arrears

We start our study of approximately single-rate derivatives with a closer look at Libor-in-arrears, or LIA (see Section 5.6), cash flows, probably the simplest single-rate derivatives apart from European swaptions and caps. Interestingly, it turns out that LIA cash flow valuation does not require the machinery we developed in Section 16.3, as in fact a LIA cash flow can be stated as a true single-rate product. Our discussion nevertheless shall allow us to give a convenient introduction to issues that are relevant for more complicated products to be covered in later sections.

We recall (Section 5.6) that the defining characteristic of an LIA cash flow is that it pays the Libor rate on the date when the rate fixes, rather than on the date it matures (i.e. the payment date). While most often a whole strip of such cash flows is used as a leg in a Libor-in-arrears swap, we focus our attention on a single cash flow; the valuation of a full strip follows by additivity. Let $T > 0$ be the start date, and M the end date of the period covered by a Libor rate. The forward Libor rate is given, for t such that $0 \leq t \leq T$, by

$$L(t, T, M) = \frac{P(t, T) - P(t, M)}{\tau P(t, M)}, \quad \tau = M - T;$$

we use simplified notation $L(t) = L(t, T, M)$ when there is no chance of confusion. The value, at time 0, of a Libor-in-arrears cash flow is then given by

$$V_{\text{LIA}}(0) = \beta(0)\text{E}\left(\beta(T)^{-1}L(T)\right),$$

where $\beta(t)$ is the continuously compounded money market account, and the expected value is taken under the risk-neutral measure Q. The standard approach to valuing payoffs that pay at time T would involve a switch to the T-forward measure, as then the expression under the expected value operator simplifies accordingly (see Section 4.2.2). Unfortunately, this is not convenient for LIA cash flows as traded caplets provide information about the distribution of the Libor rate in the M-forward measure, *not* the T-forward measure. We shall apply the M-forward measure in a moment; but for now, using the T-forward measure, we obtain

$$V_{\text{LIA}}(0) = P\left(0, T\right)\text{E}^T\left(L(T)\right).$$

While the expression looks rather simple, our progress along this route is hampered by the fact that $L(t) = L(t, T, M)$ is a martingale in the M-forward measure, not the T-forward measure. Thus,

$$\text{E}^T\left(L(T)\right) \neq L(0).$$

To characterize this situation, let us define the concept of a *Libor-in-arrears convexity adjustment*, defined by the difference

$$D_{\text{LIA}}(0) \triangleq \text{E}^T\left(L(T)\right) - L(0).$$

This adjustment arises from the mismatch between the measure appropriate for the given payment date and the measure in which the market rate is a martingale. We shall encounter many similar examples later in the chapter; the difference of valuations under different measures is often described generically as *convexity*.

Returning to the issue of valuing an LIA cash flow, we now write the valuation formula in the M-forward measure to obtain that

$$V_{\text{LIA}}(0) = P\left(0, M\right)\text{E}^M\left(\frac{1}{P\left(T, M\right)}L(T)\right).$$

Fortunately, the factor $\frac{1}{P(T,M)}$ can be rewritten in terms of the Libor rate,

$$\frac{1}{P\left(T, M\right)} = 1 + \tau L(T),$$

so that

$$V_{\text{LIA}}(0) = P\left(0, M\right)\text{E}^M\left((1 + \tau L(T))\, L(T)\right). \tag{16.35}$$

The rate $L(t)$ is a martingale in the M-forward measure, i.e. it has no drift,

$$E^M\left(L(T)\right) = L(0).$$

In particular,

$$V_{\text{LIA}}(0) = P\left(0, M\right)\left(L(0) + \tau E^M\left(L(T)^2\right)\right). \qquad (16.36)$$

The full distribution of $L(T)$ in this measure is encoded in prices of caplets on $L(T)$ with different strikes, see Section 7.1.2. So, to compute

$$E^M\left(L(T)^2\right), \qquad (16.37)$$

we merely need to integrate the function x^2 against the probability density of $L(T)$ in measure Q^M. If one has fitted a particular vanilla caplet model to the market, the density could, in principle, be extracted from this model; in some cases (e.g. the Black or Bachelier models), the density integral can be computed in closed form. In general, however, it is preferable to establish the density directly from observed market prices of T-maturity caplets[7] on $L(T)$, and to use the replication method of Proposition 8.4.13. Applying the proposition to the problem (16.37), we obtain

$$E^M\left(L(T)^2\right) = L(0)^2$$
$$+ 2 \int_{-\infty}^{L(0)} p\left(0, L(0); T, K\right) dK + 2 \int_{L(0)}^{\infty} c\left(0, L(0); T, K\right) dK, \qquad (16.38)$$

where $p(t, L; T, K)$ $(c(t, L; T, K))$ are undiscounted values of put (call) options on the rate $L(T)$ with strike K, i.e. simple undiscounted floorlets (caplets). The values of such options are available from the market, and the value of the Libor-in-arrears cash flow is computed by integrating them up.

The power of the replication method goes beyond a mere calculation of the convexity value. As should be clear from the formula above, it also provides a way to hedge the Libor-in-arrears cash flow with standard puts and calls in a *model-independent* way. In particular, to hedge a contract with the payoff

$$(1 + \tau L(T))\, L(T),$$

we would

- Enter a short FRA (forward rate agreement, see Section 5.3) on $L(T)$.
- Put $\tau P(0, M)L(0)^2$ dollars into a money market account.
- Sell $2\tau \cdot (dK)$ K-strike puts for all $K \in (-\infty, L(0)]$.
- Sell $2\tau \cdot (dK)$ K-strike calls for all $K \in [L(0), \infty)$.

The hedge is static, i.e. it never needs adjustment throughout the life of the trade. And, as mentioned earlier, the hedge is model-independent, as it does not rely on any modeling assumptions.

---

[7]Of course establishing caplet prices may itself require some work, as only prices of full caps are quoted. See Section 16.2 for more on this.

To account for the fact that in reality one does not have an infinite number of options on the rate to construct the integrals in (16.38), the integrals can be discretized, and the following formula may be used instead,

$$\mathrm{E}^M \left(L(T)^2 \right) \approx L(0)^2 + \sum_i w_i^p p\left(0, L(0); T, K_i \right) + \sum_i w_i^c c\left(0, L(0); T, K_i \right),$$

$$(16.39)$$

for a collection of strikes $\{K_i\}$, with weights w_i^p and w_i^c chosen so that the sums in (16.39) approximate the integrals in (16.38) at fixing time T. In particular, for a given range $x \in (-x_{\min}, x_{\max})$, the weights can be chosen to *super-replicate* the actual payoff for all values $x \in (-x_{\min}, x_{\max})$ of the realized Libor rate $x = L(T)$:

$$\sum_i \overline{w}_i^p p\left(T, x; T, K_i \right) + \sum_i \overline{w}_i^c c\left(T, x; T, K_i \right)$$

$$\geq 2 \int_{-\infty}^{L(0)} p\left(T, x; T, K \right) dK + 2 \int_{L(0)}^{\infty} c\left(T, x; T, K \right) dK = x^2 - L(0)^2.$$

Similarly, we can choose the weights to *sub-replicate*:

$$\sum_i \underline{w}_i^p p\left(T, x; T, K_i \right) + \sum_i \underline{w}_i^c c\left(T, x; T, K_i \right)$$

$$\leq 2 \int_{-\infty}^{L(0)} p\left(T, x; T, K \right) dK + 2 \int_{L(0)}^{\infty} c\left(T, x; T, K \right) dK = x^2 - L(0)^2.$$

The minimum value over all super-replicating (maximum over all sub-replicating) portfolios can be regarded as the upper (lower) arbitrage bound on the value of the long (short) LIA cash flow. A value of the LIA cash flow outside of these bounds is arbitrageable with options available in the market in a static, model-independent way.

16.5 Libor-with-Delay

Having considered Libor-in-arrears, we now move on to a more interesting case of Libor cash flows with an arbitrary payment delay. For this product, we can apply the lessons learned in Section 16.4, and also start using the techniques of Section 16.3 for the first time.

Continuing with the notation $L(t) = L(t, T, M)$ for the forward Libor rate covering the period $[T, M]$, we consider a cash flow that pays this rate at some arbitrary payment time T_p, $T_p \geq T$. Switching to the M-forward measure, the measure in which the market-implied distribution of $L(T)$ is known, it follows that the value of the *Libor-with-delay* cash flow is given by

$$V_{\mathrm{LD}}(0) = P\left(0, M\right) \mathrm{E}^M \left(\frac{P\left(T, T_p \right)}{P\left(T, M \right)} L(T) \right).$$

$$(16.40)$$

The presence of the term $P(T, T_p)$ inside the expected value operator now generally prevents us from representing the payoff as a function of the rate $L(T)$ only, making this payoff a simple example of what we defined as an approximately single-rate derivative in Section 16.3.

Valuing Libor-with-delay cash flows presents no theoretical difficulties if one uses a full term structure model, such as the quasi-Gaussian model or a version of the Libor market model. While possible, such a brute-force approach is generally not recommended here. For instance, we note the value of a Libor-in-arrears cash flow and, by extension, of a Libor-with-delay cash flow will depend on values of options of all strikes on a given rate, suggesting that the underlying model should ideally match the entire volatility smile for the underlying Libor rate, something that will be a stretch for a full-blown term structure model. On top of this, there are obvious computational issues in employing a full term structure model for something as vanilla as Libor-with-delay cash flows.

A more sensible approach to the pricing of a Libor-with-delay cash flow utilizes the replication method from Section 16.4 above, along with a method to represent the payoff in (16.40) as a function of the single rate $(L(T))$ only. For the latter, we may use the methods in Section 16.3; we proceed to show an example.

16.5.1 Swap-Yield TSR Model

The simplest and probably most popular method for establishing a functional relationship between the payoff in (16.40) and the rate $L(T)$ is an application of the swap-yield terminal swap rate model of Section 16.3.4. While this method is not fully arbitrage-free (as mentioned), the degree to which no-arbitrage is violated is typically immaterial for Libor-with-delay cash flows. To apply the model to the problem at hand, the rate $L(T) = L(T, T, M)$ is specified to be the market rate, and $P(T, T_p)$ is linked to this rate via the relationship

$$P(T, T_p) = \left(\frac{1}{1 + \tau L(T)} \right)^{(T_p - T)/\tau}.$$

The formula (16.40) then becomes

$$V_{\mathrm{LD}}(0) = P(0, M) \, \mathrm{E}^M \left((1 + \tau L(T))^{1 - (T_p - T)/\tau} L(T) \right),$$

allowing us to apply the replication method outlined in Section 16.4 to obtain the value, and the model-independent hedge, of the cash flow. While the resulting formula would not be perfectly arbitrage-free, it will handle correctly the two special cases $T_p = T$ (the formula (16.35) is recovered) and $T_p = M$ (zero convexity adjustment for Libor paid at the end date of the Libor period). Of course, as with any approximation, one should be mindful of pushing the formula beyond its limits; in particular it should not be used for $T_p \gg M$.

16.5.2 Other Terminal Swap Rate Models

As an alternative to the approach in Section 16.5.1, we may apply any of the arbitrage-free terminal swap rate models of Sections 16.3.2 and 16.3.3. As these two TSR models are specifically designed to relate discount factors observed at a particular date T to a market rate observed on the same date, the task of linking $P(T, T_p)$ to $L(T)$ for the Libor-with-delay contract is easily accomplished. We trust the reader can see how to proceed; if not, a more general case is considered in Section 16.6.4 below.

16.5.3 Approximations Inspired by Term Structure Models

While we do not recommend using a full term structure model to value Libor-with-delay cash flows, it is possible to use such models to derive the relationship between the discount factor $P(T, T_p)$ and the Libor rate $L(T)$. For instance, Andreasen [2002] suggests the one-factor quasi-Gaussian model for the task. While a purely Gaussian model (which has been our standard for these types of approximations in earlier chapters) would give very similar results, we use the qG model here for variety. To develop the approximation, we recall (13.5),

$$P(T, M) = P(T, M, x(T), y(T)), \quad M \geq T,$$

$$P(T, M, x, y) = \frac{P(0, M)}{P(0, T)} \exp\left(-G(T, M) x - \frac{1}{2} G(T, M)^2 y\right),$$

where $x(T)$, $y(T)$ are the state variables in the model and $G(T, M)$ is a function of mean reversion, see (13.3).

Writing the forward Libor rate $L(t) = L(t, T, M)$ as $L(t, x, y)$ to emphasize its dependence on state variables, we note that $P(T, T_p, x, y)$ and $L(T, x, y)$ are monotonic in x, for any fixed value of the state variable $y(T)$. We recall from Chapter 13 that the state variable $y(T)$ is a locally deterministic auxiliary variable whose role is to keep the model arbitrage-free. As done many times in Chapter 13, for the purposes of deriving an approximation let us fix its time T value at some deterministic level $\overline{y}(T)$. Then the discount bond can be expressed in terms of the Libor rate directly,

$$P(T, T_p) = P(T, T_p, X(T, L(T)), \overline{y}(T)), \tag{16.41}$$

where $X(T, l)$ is an inverse (in x) function to $L(T, x, \overline{y}(T))$.

To derive a suitable expression for $\overline{y}(T)$, we recall (Proposition 13.1.4, equation (10.41)) that in the quasi-Gaussian model

$$\mathrm{E}(y(T)) \approx \mathrm{Var}(x(T)).$$

Then, assuming a nearly linear relationship between the Libor rate and the state variable x, we write

$$\text{Var}\left(L(T)\right) \approx \left(\left.\frac{\partial L\left(T, x, y\right)}{\partial x}\right|_{x=y=0}\right)^2 \text{Var}\left(x(T)\right).$$

This expressions suggests the following approximation for $\bar{y}(T)$,

$$\bar{y}(T) = \left(\left.\frac{\partial L\left(T, x, y\right)}{\partial x}\right|_{x=y=0}\right)^{-2} \text{Var}\left(L(T)\right),$$

where we approximate $\text{Var}(L(T))$ with the variance in the corresponding forward measure, $\text{Var}(L(T)) \approx \text{Var}^M(L(T))$, and compute the latter either in the vanilla model calibrated to the volatility smile of options on $L(T)$, or directly by the replication method. With $\bar{y}(T)$ set this way, we can solve (16.41) numerically, thereby establishing the required relationship between $P(T, T_p)$ and $L(T)$ and allowing the replication method to be applied. The model thus constructed will violate the no-arbitrage condition (16.19), but only mildly so; moreover we can fix this violation with the application of the scaling idea from Section 16.6.7.

Other term structure models can be used to link $P(T, T_p)$ to the Libor rate. For example, later in the chapter (in Section 16.6.6) we develop approximations that are inspired by Libor market models.

16.5.4 Applications to Averaging Swaps

Libor-with-delay cash flows do not, as a rule, trade individually but instead serve as building blocks for other derivatives; common among them are the so-called *averaging swaps*, i.e. swaps that are composed of *averaging cash flows*. An averaging cash flow (recall Section 5.7) pays, at time T_p, an average Libor rate $\bar{L}$ over the period,

$$\bar{L} = \sum_{i=1}^{k} w_i L\left(t_i^f, t_i^s, t_i^e\right),$$

where we use the notation of Section 5.7. The value of the averaging cash flow at time 0 is given by

$$V_{\text{avg}}(0) = \beta(0)\text{E}\left(\beta(T_p)^{-1}\bar{L}\right)$$

which, using linearity of the payoff and applying appropriate measure changes, is given by

$$V_{\text{avg}}(0) = \sum_{i=1}^{k} w_i \beta(0)\text{E}\left(\beta(T_p)^{-1} L\left(t_i^f, t_i^s, t_i^e\right)\right)$$

$$= \sum_{i=1}^{k} w_i P\left(0, t_i^e\right)\text{E}^{t_i^e}\left(\frac{P\left(t_i^f, T_p\right)}{P\left(t_i^f, t_i^e\right)} L\left(t_i^f, t_i^s, t_i^e\right)\right).$$

Each term in the sum is the value of a Libor-with-delay cash flow, and can be evaluated by one of the methods developed above.

16.6 CMS and CMS-Linked Cash Flows

The discussion of issues around valuation of Libor-in-arrears and Libor-with-delay cash flows provides us with a useful blueprint for modeling the larger, and more important, class of CMS and CMS-linked cash flows. Using the notations (16.15)–(16.16), we recall from Section 5.11 that a CMS cash flow pays the swap rate $S(T)$ at time T_p, $T_p \geq T$, typically with[8] $T_p \leq T_1$. More generally, a CMS-linked cash flow pays some function of the swap rate $S(T)$.

In a direct analogy to the LIA case, the market-implied distribution of $S(T)$ in the swap measure Q^A is known from market values of European swaptions; yet, a CMS cash flow is more naturally valued in the T_p-forward measure, the measure linked to a discount bond maturing on the CMS payment date. Not surprisingly, this gives rise to a convexity adjustment. More precisely, note that the value of a CMS cash flow is given by the following expectation in the annuity measure,

$$V_{\text{CMS}}(0) = A(0)\text{E}^A \left(\frac{P(T,T_p)}{A(T)} S(T) \right), \tag{16.42}$$

whereby we can define the *CMS convexity adjustment* to be

$$D_{\text{CMS}}(0) \triangleq \text{E}^{T_p}(S(T)) - S(0) \tag{16.43}$$

$$= \frac{A(0)}{P(0,T_p)}\text{E}^A \left(\frac{P(T,T_p)}{A(T)} S(T) \right) - S(0).$$

At a high level, our discussion of Libor-with-delay valuation issues (Section 16.5) readily extends to the case of CMS cash flows. As always, the expected value in (16.42) can, in principle, be computed with the help of a term structure model, but this is typically too inaccurate and too slow[9], so the replication method is typically a better choice. Of course, in order to apply it, the payoff of (16.42) and, in particular, the multiplier $P(T,T_p)/A(T)$, needs to be represented as a function of $S(T)$ only. We shall discuss this in a moment, but first we briefly review the replication method for CMS cash flows. We have already discussed replication for Libor-linked cash flows, but we find it worthwhile to examine the method again in a CMS-specific setting.

---

[8] Recall that T_1 is the first payment date of the swap underlying the swap rate S.

[9] It is nevertheless often useful to be able to calculate CMS convexity adjustments in a given term structure model in closed form, e.g. for assessing the loss of precision of the model when pricing exotics linked to CMS rates. We return to this task later in the chapter.

16.6.1 The Replication Method for CMS

In close analogy to the LIA case, when the replication method of Proposition 8.4.13 is applied to the CMS payoff in (16.42), it decomposes the CMS payoff into a portfolio of standard European options on the swap rate, i.e. swaptions; from this representation the market value of the payoff may be obtained by simply summing up swaption values. These swaption values can be taken directly from the market or, if our goal is to compute a model price, from a given model.

We shall discuss ways of linking the term $P(T, T_p)/A(T)$ to the swap rate $S(T)$ momentarily; for now let us simply assume that an *annuity mapping function* $\alpha(s)$ has been found such that

$$\mathrm{E}^A \left(\frac{P(T, T_p)}{A(T)} S(T) \right) = \mathrm{E}^A \left(\alpha(S(T)) S(T) \right). \tag{16.44}$$

Proposition 8.4.13 then stipulates that

$$\mathrm{E}^A \left(\alpha(S(T)) S(T) \right) = S(0) \alpha(S(0))$$

$$+ \int_{-\infty}^{S(0)} w(K) p\left(0, S(0); T, K\right) \, dK + \int_{S(0)}^{\infty} w(K) c\left(0, S(0); T, K\right) \, dK, \tag{16.45}$$

where the hedge weights $w(s)$ are given by

$$w(s) = \frac{d^2}{ds^2} (\alpha(s) s),$$

and $p(t, S; T, K)$ $(c(t, S; T, K))$ are put (call) options on the rate $S(T)$ with strike K, forward S and fixing at T, as observed at t. Combining (16.42), (16.44), and (16.45), we obtain

$$V_{\mathrm{CMS}}(0) = A(0) S(0) \alpha(S(0))$$

$$+ \int_{-\infty}^{S(0)} w(K) V_{\mathrm{rec}}\left(0, K\right) \, dK + \int_{S(0)}^{\infty} w(K) V_{\mathrm{pay}}\left(0, K\right) \, dK, \tag{16.46}$$

where $V_{\mathrm{rec}}(0, K)$ $(V_{\mathrm{pay}}(0, K))$ are the values, at time 0, of receiver (payer) European swaptions, respectively:

$$V_{\mathrm{rec}}(0, K) = A(0) \mathrm{E}^A \left((K - S(T))^+ \right),$$

$$V_{\mathrm{pay}}(0, K) = A(0) \mathrm{E}^A \left((S(T) - K)^+ \right).$$

As mentioned above, the swaption values can either be computed in a model of choice or directly observed in the market. We emphasize again that not only does the replication method calculate the value of a CMS cash flow consistently with the market in swaptions for all strikes, but it also provides a static, model-independent (up to the choice of $\alpha(s)$) hedging portfolio of payer and receiver swaptions.

In the basic expression (16.46) we can impose various restrictions on swaption hedge positions to, say, incorporate liquidity constraints into the price of the CMS cash flow. For example, swaptions of very low or very high strikes may not be easily tradeable. Then, supposing that the lowest available strike is $K_{\min}$, and the highest one is $K_{\max}$, one can choose to pay no more than

$$A(0)S(0)\alpha(S(0)) + \int_{K_{\min}}^{S(0)} w(K)V_{\mathrm{rec}}(0, K) \, dK + \int_{S(0)}^{K_{\max}} w(K)V_{\mathrm{pay}}(0, K) \, dK,$$

on the grounds that this is the value that can be "locked in" by hedging with available vanillas. Adjustment for the fact that only a finite number of strikes are traded can proceed along the lines of the discussion in Section 16.4.

The replication approach extends virtually unchanged to cash flows that pay an arbitrary — but reasonably smooth — function of the swap rate, say $g(S(T))$, paid at time $T_p \geq T$. Notable examples of such cash flows include CMS caplets and floorlets (see Section 5.11),

$$g(s) = g_{\mathrm{caplet}}(s) = (s - K)^+, \quad g(s) = g_{\mathrm{floorlet}}(s) = (K - s)^+. \quad (16.47)$$

The value of a CMS-linked cash flow is then, naturally, equal to

$$V_{\mathrm{gCMS}}(0) = A(0)\mathrm{E}^A \left(\frac{P(T, T_p)}{A(T)} g(S(T)) \right) = A(0)\mathrm{E}^A \left(\alpha(S(T))g(S(T)) \right), \tag{16.48}$$

and the replication method, as given by (16.46), applies with the weights calculated by

$$w(s) = \frac{d^2}{ds^2} \left(\alpha(s)g(s) \right). \tag{16.49}$$

For many payoffs of interest the second derivative here will not be defined in a conventional sense at all points, and may, in particular, contain Dirac delta functions. For example, for the CMS caplets and floorlets with strike K, the second derivative in (16.49) would include a delta function centered at K. This is, however, not a cause for concern, as delta functions are easy to handle in the integrals in (16.46): a delta function centered at some point s_0 would just contribute a term $V_{\mathrm{rec}}(0, s_0)$ (or $V_{\mathrm{pay}}(0, s_0)$, depending on the relationship between s_0 and K) to the integrals in the replication method.

We observe in passing that the replication method requires calculation of values for a collection of swaptions of different strikes. For some models, such calculations can be optimized, see e.g. the discussion of Section 8.4.5.

16.6.2 Annuity Mapping Function as a Conditional Expected Value

The previous section introduced the annuity mapping function $\alpha(s)$ as a critical ingredient of CMS valuation, but stopped short of developing a method to determine it. From examples presented earlier in this chapter, the reader might reasonably expect that terminal swap rate models and/or approximations inspired by term structure models could be used for that purpose. This is indeed the case, as we shall show momentarily. First, however, we find it useful to step back a little to determine the actual theoretical meaning of annuity mapping functions; this analysis is illuminating in its own right and is also helpful in developing a systematic approach to finding good approximations.

Let us start with the main valuation formula (16.42). We obtain

$$
\begin{aligned}
V_{\mathrm{CMS}}(0) &= A(0)\mathrm{E}^A\left(\frac{P\left(T,T_p\right)}{A(T)}S(T)\right)\\
&= A(0)\mathrm{E}^A\left(\mathrm{E}^A\left(\frac{P\left(T,T_p\right)}{A(T)}S(T)\bigg|\,S(T)\right)\right)\\
&= A(0)\mathrm{E}^A\left(S(T)\mathrm{E}^A\left(\frac{P\left(T,T_p\right)}{A(T)}\bigg|\,S(T)\right)\right).
\end{aligned}
$$

Now, if we compare this formula to (16.44), we obtain the following useful result.

Proposition 16.6.1. *The annuity mapping function $\alpha(s)$ in (16.44) or, more generally, (16.48) may be written as the conditional expectation*

$$
\alpha(s) = \mathrm{E}^A\left(\frac{P\left(T,T_p\right)}{A(T)}\bigg|\,S(T) = s\right). \tag{16.50}
$$

This result is model-independent.

The proposition clarifies the role of various methods of linking discount bond values to rates that we introduced previously in order to value approximately single-rate derivatives. These methods, in fact, can be seen as approximations to the true annuity mapping function defined by the conditional expected value in (16.50). We shall return to this interpretation and explore it in more detail as we discuss various methods individually below. For now, we note that the problem of calculating the conditional expected value in (16.50) could be attacked directly, as we demonstrate later in Section 16.6.6 for the LM model, or by projection methods. To elaborate briefly, the expected value of random variable X conditional on some other random variable Y can be interpreted as a projection of X on the space of all (suitably regular) functions of Y. Let us denote such a space by $\mathcal{B}$; then

$$E\left(X|Y\right) = f^*(Y), \quad \text{where } f^* = \operatorname{argmin}\left\{E\left((X - f(Y))^2\right), \quad f \in \mathcal{B}\right\}.$$

Following Antonov and Misirpashaev [2009a], we can then obtain a tractable approximation to the true value of the conditional expected value by restricting the subspace of functions of Y to project on. For a given subspace $\widetilde{\mathcal{B}} \subset \mathcal{B}$, an approximation is then defined as the closest, in the least-squares sense, element of the subspace to X,

$$E\left(X|Y\right) \approx f^*(Y), \quad \text{where } f^* = \operatorname{argmin}\left\{E\left((X - f(Y))^2\right), \quad f \in \widetilde{\mathcal{B}}\right\}.$$

If the subspace $\widetilde{\mathcal{B}}$ is defined by a parametric functional form,

$$\widetilde{\mathcal{B}} = \{f\left(y; \theta\right), \ \theta \in \Theta\}$$

for some parametric set $\Theta \subset \mathbb{R}^d$, then the necessary condition for $f^*(y) \triangleq f(y; \theta^*)$ to be optimal is given by the equations

$$\frac{\partial}{\partial \theta_i} E\left((X - f\left(Y; \theta\right))^2\right) = 0, \quad i = 1, \ldots, d.$$

For later use, let us formalize this result as a proposition.

Proposition 16.6.2. *Given two random variables X and Y and a parametric set of functions $\{f(y; \theta)\}$, $\theta \in \Theta \subset \mathbb{R}^d$, an approximation to $E(X|Y)$ is given by*

$$E\left(X|Y\right) \approx f\left(Y; \theta^*\right),$$

where θ^* is a solution to the set of equations

$$E\left(X \frac{\partial f}{\partial \theta_i}\left(Y; \theta\right)\right) = E\left(f\left(Y; \theta\right) \frac{\partial f}{\partial \theta_i}\left(Y; \theta\right)\right), \quad i = 1, \ldots, d. \qquad (16.51)$$

16.6.3 Swap-Yield TSR Model

As our first concrete model, let us consider the swap-yield model of Section 16.3.4, a model that has long been a de-facto standard for linking the annuity to the swap rate. Recalling the index function $q(M)$ defined by (16.33), we link discount factors $P(T, M)$ to the swap rate by the formula

$$P\left(T, M\right) = \left(\prod_{i=0}^{q(M)-1} \frac{1}{1 + \tau_i S(T)}\right) \times \left(\frac{1}{1 + \tau_{q(M)} S(T)}\right)^{\left(M - T_{q(M)}\right)/\tau_{q(M)}},$$

with $M \geq T$. As (16.34) holds, we have

$$A(T) = \sum_{n=0}^{N-1} \tau_n P\left(T, T_{n+1}\right)$$

$$= \sum_{n=0}^{N-1} \tau_n \prod_{i=0}^{n} \frac{1}{1 + \tau_i S(T)} = \frac{1}{S(T)}\left(1 - \prod_{i=0}^{N-1} \frac{1}{1 + \tau_i S(T)}\right).$$

Also, assuming $T_p \in [T_0, T_1]$ (with obvious modifications for the general case),

$$P\left(T, T_p\right) = \left(\frac{1}{1 + \tau_0 S(T)}\right)^{(T_p - T)/\tau_0}.$$

Then

$$\alpha(s) = s \frac{\left(\frac{1}{1+\tau_0 s}\right)^{(T_p - T)/\tau_0}}{1 - \prod_{i=0}^{N-1} \frac{1}{1+\tau_i s}}$$

defines the function $\alpha(s)$ to be used in (16.44). We note, as before, that the model will violate basic arbitrage restrictions, in the sense that

$$E^A\left(\frac{P(T, T_p)}{A(T)}\right) \neq \frac{P(0, T_p)}{A(0)}.$$

A method to correct for this is shown in Section 16.6.7.

16.6.4 Linear and Other TSR Models

As all terminal swap rate models are specifically designed to relate discount bonds of various maturities on a particular date T to a market rate $S(T)$, it is easy to extend the discussion in Section 16.6.3 to the general TSR model class. In the TSR class, the linear TSR model (see Section 16.3.2) is arguably the simplest and probably the most popular, so let us use this model as our second concrete example. Applied to the problem of CMS cash flow valuation, the model postulates a linear relationship between the inverse annuity and the swap rate,

$$\alpha(s) = \alpha_1 s + \alpha_2 \tag{16.52}$$

(in the notation of Section 16.3.2 we have $\alpha_1 = a(T_p)$, $\alpha_2 = b(T_p)$). The parameter α_1 can be considered an exogenous input, and α_2 determined by the no-arbitrage requirement that

$$\frac{P(0, T_p)}{A(0)} = E^A\left(\frac{P(T, T_p)}{A(T)}\right) = E^A\left(\alpha_1 S(T) + \alpha_2\right) = \alpha_1 S(0) + \alpha_2,$$

implying

$$\alpha_2 = \frac{P(0, T_p)}{A(0)} - \alpha_1 S(0). \tag{16.53}$$

With this specification,

$$
\begin{aligned}
V_{\text{CMS}}(0) &= A(0)\text{E}^A\left((\alpha_1 S(T)+\alpha_2)\,S(T)\right) \\
&= \alpha_2 A(0)S(0) + \alpha_1 A(0)\text{E}^A\left(S(T)^2\right) \\
&= P\left(0,T_p\right)S(0) - \alpha_1 A(0)S(0)^2 + \alpha_1 A(0)\text{E}^A\left(S(T)^2\right) \\
&= P\left(0,T_p\right)S(0) + \alpha_1 A(0)\text{Var}^A\left(S(T)\right),
\end{aligned}
\tag{16.54}
$$

and the convexity adjustment is then given simply by

$$
D_{\text{CMS}}(0) = \alpha_1 \frac{A(0)}{P\left(0,T_p\right)}\text{Var}^A\left(S(T)\right).
\tag{16.55}
$$

The variance of $S(T)$ is computed either directly in a model of choice, or by integrating the function $(s - S(0))^2$ against the market-implied probability density of the swap rate obtained from swaption prices or, equivalently[10], by the replication method. The elegant formula (16.55), reminiscent of the Libor-in-arrears formula (16.36), is commonly used in practice, despite the fact that discount factors can become negative in certain states of the simulated world under the specification (16.52). The parameter α_1 can be linked to mean reversion as discussed in Section 16.3.2; we touch on this in more detail in Section 16.6.8. More crudely, we can estimate α_1 from a boundary argument, where we simply observe that for scenarios where the time T yield curve is very low (and $S(T)$ therefore is close to zero), we must have

$$
\frac{P(T,T_p)}{A(T)} \approx \frac{1}{\sum_{n=0}^{N-1} \tau_n}.
$$

As we may write $\alpha_2 = \text{E}^A(\alpha_1 S(T)+\alpha_2 | S(T) = 0)$, this suggests setting

$$
\alpha_2 = \frac{1}{\sum_{n=0}^{N-1} \tau_n}, \qquad \alpha_1 = \frac{1}{S(0)}\left(\frac{P\left(0,T_p\right)}{A(0)} - \alpha_2\right),
\tag{16.56}
$$

where the equation for α_1 follows from (16.53). While not exactly state-of-the-art, this simplified approach often yields decent precision. See Figure 16.1 on p. 736 for some representative test results.

To wrap up the discussion of the applications of linear TSR models to CMS pricing, let us briefly touch upon its relationship to the result of Proposition 16.6.1. Setting $X = P(T,T_p)/A(T)$, $Y = S(T)$, and $f(y;\theta) = \theta_1 + \theta_2 x$, $(\theta_1,\theta_2)^\top \in \mathbb{R}^2$, we obtain from Proposition 16.6.2 that the coefficients of the best linear approximation to the annuity mapping function $\alpha(s)$,

$$
\alpha(s) = \theta_1^* s + \theta_2^*,
$$

are given by the solution to the equations

---

[10]In theory; numerical implementation could lead to slight differences.

$$\mathrm{E}^A \left(\frac{P(T,T_p)}{A(T)} \right) = \mathrm{E}^A \left(\theta_1 S(T) + \theta_2 \right),$$

$$\mathrm{E}^A \left(\frac{P(T,T_p)}{A(T)} S(T) \right) = \mathrm{E}^A \left((\theta_1 S(T) + \theta_2) S(T) \right).$$

Solving the equations, we obtain the optimal coefficients

$$\theta_1^* = \frac{P(0,T_p)}{A(0)} \frac{D_{\mathrm{CMS}}(0)}{\mathrm{Var}^A(S(T))}, \quad \theta_2^* = \frac{P(0,T_p)}{A(0)} - \theta_1^* S(0). \tag{16.57}$$

The same result could have been obtained by backing out α_1 and α_2 from (16.53) and (16.55), a fact that is not surprising given that both calculations started with a linear approximation to $\alpha(s)$.

The result (16.57), even if somewhat trivially obtainable from (16.53) and (16.55), emphasizes the point that a known magnitude of the CMS adjustment will often uniquely identify the annuity mapping function within a parametric class. Importantly, this can be used to calibrate the annuity mapping function (within a given parametric class) to liquidly traded CMS swaps, the market values of which reveal the size of the convexity adjustment. The calibrated annuity mapping function can then be used to value more complicated, and less liquid, CMS-linked derivatives such as CMS caps or CMS range accruals.

Let us finally note that while the linear TSR model gives us the most amount of analytic tractability, the ideas behind Proposition 16.6.2 could be applied to other types of TSR models as well. For example, if we were to choose functions for discount bonds from the exponential class and apply Proposition 16.6.2, we would obtain a model that is quite similar to the exponential TSR model.

16.6.5 The Quasi-Gaussian Model

As was the case for Libor-with-delay cash flows (see Section 16.5.3), the quasi-Gaussian (qG) model can be used as a source of inspiration for the functional relationship between the annuity and the swap rate. We recall the bond reconstruction formula in the quasi-Gaussian model (see (13.5))

$$P(T,M) = P(T,M,x(T),y(T)), \quad M \geq T,$$

$$P(T,M,x,y) = \frac{P(0,M)}{P(0,T)} \exp \left(-G(T,M)x - \frac{1}{2}G(T,M)^2 y \right),$$

with $x(T)$, $y(T)$ being the state variables of the model and $G(T,M)$ a deterministic function of mean reversion, and define $A(T,x,y)$, $S(T,x,y)$ accordingly. Motivated by Section 16.5.3, we set

$$\bar{y}(T) = \left(\frac{\partial S(T,x,y)}{\partial x} \bigg|_{x=y=0} \right)^{-2} \mathrm{Var}^A(S(T)),$$

where $\text{Var}^A(S(T))$, the variance of the swap rate $S(T)$ in the annuity measure, is computed consistently with the model used in the replication method, and define $X(T, s)$ to be the inverse function, in x, of $S(T, x, \overline{y}(T))$. Then we can define the mapping function $\alpha(s)$ by

$$\alpha(s) = \frac{P(T, T_p, X(T, s), \overline{y}(T))}{A(T, X(T, s), \overline{y}(T))}, \tag{16.58}$$

and calculate $V_{\text{CMS}}(0)$ via (16.46). For calculating market-consistent CMS values, the values of swaptions in the replication algorithm should be either taken directly from the market or calculated using a market-calibrated vanilla model. If, on the other hand, our objective is to calculate an analytical approximation to the value of a CMS cash flow in the quasi-Gaussian model (a value that could be used to assess and adjust the valuation of more exotic payoffs linked to CMS rates in the model), then we should value swaptions in the qG model directly, perhaps using an approximation such as Proposition 13.1.10.

16.6.6 The Libor Market Model

A Libor market model can also be used to specify the form of dependency of the annuity on the forward swap rate. While perhaps too complicated for valuing CMS cash flows *per se*, establishing the dependency explicitly would be useful for applications of Libor market models to exotic derivatives that are linked to CMS rates, such as callable CMS range accruals, CMS spread TARNs, and the like (see e.g. Sections 5.13.2 and 5.14). When valuing CMS-linked exotic derivatives, it is often desirable to confirm that the values of CMS convexity adjustments in the Libor market model agree with the "market" convexity adjustments or, at the very least, to quantify, and potentially correct for, any observed differences (see Chapter 21). Of course, one can always use Monte Carlo simulation to calculate CMS adjustments in a Libor market model, but the usual performance considerations favor an analytical or semi-analytical approach.

The subject of calculating CMS adjustments in Libor market models has received some attention in the literature — see e.g. Gatarek [2003] for a representative approach — but most published methods generally boil down to using "freezing" techniques to approximate the drift of the swap rate in the forward measure, a method that is not particularly accurate. A notable exception to this trend is the recent work by Antonov and Arneguy [2009] who calculate the expected value

$$\frac{P(0, T_p)}{A(0)} \text{E}^{T_p}(S(T)) = \text{E}^A\left(\left(\frac{P(T, T_p)}{A(T)}\right) S(T)\right)$$

by deriving an approximate SDE for $P(t, T_p)/A(t)$, and then obtain a linear annuity mapping function via (16.57). Test results given in the paper suggest that the approach is reasonably accurate; however, we believe that it is important to capture the non-linearity of the annuity mapping function in LMM in order to obtain a truly precise approximation. Our preferred alternative is developed below.

First, we recall (16.50) which states that, independently of the underlying model, $\alpha(s)$ can be interpreted as the expected value of $P(T, T_p)/A(T)$ conditioned on $S(T) = s$. We proceed to derive an approximation to this conditional expected value, consistent with the Libor market model. As we did in Chapter 14, we denote the spanning Libor rates by

$$L_n(t) = L(t, T_n, T_{n+1}), \quad n = 0, \ldots, N - 1.$$

Assuming $T_p = T$ for notational simplicity[11], we observe that the argument of the conditional expected value, the inverse numeraire $1/A(T)$, is a deterministic function of the vector of Libor rates $\mathbf{L}(T) = (L_0(T), \ldots, L_{N-1}(T))^\top$,

$$\frac{1}{A(T)} = \rho(\mathbf{L}(T)), \quad \rho(x) = \left(\sum_{n=0}^{N-1} \tau_n \prod_{i=0}^{n} (1 + \tau_i x_i)^{-1} \right)^{-1}.$$

Approximating

$$\alpha(s) = \mathrm{E}^A \left(\rho(\mathbf{L}(T)) \mid S(T_n) = s \right) \approx \rho \left(\mathrm{E}^A \left(\mathbf{L}(T) \mid S(T) = s \right) \right),$$

we reduce the problem to that of computing $\mathrm{E}^A(\mathbf{L}(T)|S(T) = s)$ in a Libor market model, a problem we can tackle in the usual fashion, through an application of a Gaussian approximation. For concreteness, let us consider the following form of the Libor market model (see Section 14.2.5),

$$dL_n(t) = \sqrt{z(t)} \varphi(L_n(t)) \lambda_n(t)^\top dW^{T_{n+1}}(t), \tag{16.59}$$

$$dz(t) = \theta(z_0 - z(t)) \, dt + \eta(t) \sqrt{z(t)} \, dZ(t), \quad z(0) = z_0 = 1,$$

with $\langle dW^{T_{n+1}}(t), dZ(t) \rangle = 0$ for $n = 0, \ldots, N - 1$. Here $\lambda_n(t)$ is an m-dimensional deterministic volatility function and $W^{T_{n+1}}(t)$ is an m-dimensional $Q^{T_{n+1}}$-Brownian motion. To compute $\mathrm{E}^A(\mathbf{L}(T)|S(T) = s)$, we use the following Gaussian approximation to the Q^A-dynamics of Libor and swap rates,

$$L_n(t) \approx \widehat{L}_n(t), \quad S(t) \approx \widehat{S}(t),$$

where

$$d\widehat{L}_n(t) = \varphi(L_n(0))\lambda_n(t)^\top dW^A(t), \quad \widehat{L}_n(0) = L_n(0), \quad n = 0, \ldots, N - 1,$$

$$d\widehat{S}(t) = \varphi(S(0)) \left(\sum_{i=0}^{N-1} w_i \lambda_i(t)^\top \right) dW^A(t), \quad \widehat{S}(0) = S(0),$$

---

[11] For $T \neq T_p \leq T_1$ the function ρ to be defined momentarily would be multiplied by a term that expresses $P(T, T_p)$ as an (approximate) function of $x_0 = L_0(T)$ as in, e.g., Section 16.5.1.

with

$$w_i = \frac{\varphi(L_i(0))}{\varphi(S(0))} \frac{\partial S(0)}{\partial L_i(0)}, \quad i = 0, \ldots, N-1, \quad (16.60)$$

(see Section 14.4.2 for details on approximating swap rate dynamics in Libor market models). The required expected value is then computed by

$$\mathrm{E}^A \left(L_n(T) | S(T) = s \right)$$

$$\approx \mathrm{E}^A \left(\hat{L}_n(T) \Big| \hat{S}(T) = s \right) = L_n(0) \left(1 + c_n \frac{s - S(0)}{S(0)} \right),$$

where

$$c_n = \frac{\varphi(L_n(0)) S(0) \int_0^T \lambda_n(t)^\top \left(\sum_{i=0}^{N-1} w_i \lambda_i(t) \right) dt}{\varphi(S(0)) L_n(0) \int_0^T \left\| \sum_{i=0}^{N-1} w_i \lambda_i(t) \right\|^2 dt}. \quad (16.61)$$

Putting all steps together, we obtain the following proposition.

Proposition 16.6.3. *The mapping function $\alpha(s)$ defined by (16.50) in the Libor market model (16.59) is approximately given by*

$$\alpha(s) = \mathrm{E}^A \left(\frac{1}{A(T)} \Big| S(T) = s \right) \approx \left(\sum_{n=0}^{N-1} \tau_n \prod_{i=0}^{n} (1 + \tau_i l_i(s))^{-1} \right)^{-1},$$

where

$$l_n(s) = L_n(0) \left(1 + c_n \frac{s - S(0)}{S(0)} \right),$$

for $n = 0, \ldots, N-1$, with coefficients c_n given by (16.61) and weights w_n given by (16.60).

16.6.7 Correcting Non-Arbitrage-Free Methods

Several of the annuity mapping methods developed in previous sections (e.g. in Sections 16.6.3, 16.6.5 and 16.6.6) are not arbitrage-free by construction. Others, such as the linear TSR model, may theoretically be arbitrage-free, but numerical methods may induce slight errors. In this section we introduce a simple adjustment to all methods that will remedy the main arbitrage issues, be they theoretical or numerical.

We recall that the principal valuation formula for CMS cash flows specifies

$$V_{\mathrm{CMS}}(0) = A(0) \mathrm{E}^A \left(\frac{P(T, T_p)}{A(T)} S(T) \right) = A(0) \mathrm{E}^A \left(\alpha(S(T)) S(T) \right),$$

where $\alpha(s)$ is obtained by one of the methods discussed above. The quantity $P(T, T_p)/A(T)$, being a ratio of a tradeable asset and the numeraire, is a martingale in the annuity measure. Hence, in any arbitrage-free model the following should hold,

$$E^A \left(\frac{P(T, T_p)}{A(T)} \right) = \frac{P(0, T_p)}{A(0)}.$$

That is, we should have

$$E^A \left(\alpha(S(T)) \right) = \frac{P(0, T_p)}{A(0)}. \tag{16.62}$$

If, however, the function $\alpha(s)$ is obtained by one of the methods that does not satisfy the no-arbitrage condition, we would see that

$$\bar{\alpha} \triangleq E^A \left(\alpha(S(T)) \right) \neq \frac{P(0, T_p)}{A(0)}. \tag{16.63}$$

For purposes of CMS product valuations, the inequality (16.63) is, by far, the most important manifestation of the arbitrage in the model. Pragmatically, we can compensate by rescaling the original function $\alpha(s)$ to force (16.62) to be satisfied. In particular, defining

$$\tilde{\alpha}(s) = \frac{P(0, T_p)}{A(0)} \frac{\alpha(s)}{\bar{\alpha}}, \tag{16.64}$$

we obtain the "improved" CMS valuation formula,

$$V_{\text{CMS}}(0) = A(0) E^A \left(S(T) \tilde{\alpha}(S(T)) \right) = P(0, T_p) \frac{E^A \left(S(T) \alpha(S(T)) \right)}{E^A \left(\alpha(S(T)) \right)}.$$

In fact, the correction (16.64) is useful even for arbitrage-free models; while the no-arbitrage property (16.62) holds in theory, in practice it can be violated in the numerical scheme used.

Apart from the fundamental "test" (16.62) that any annuity mapping method must pass, there are other checks that are useful to keep in mind while looking at any particular method for CMS product valuation. One such test is obtained from the trivial identity

$$\sum_{n=0}^{N-1} \frac{\tau_n P(T, T_{n+1})}{A(T)} = 1,$$

which implies the following relationship between annuity mapping functions $\alpha(s, T_n)$ that correspond to different payment dates T_n, $n = 1, \ldots, N$ (note how we enriched the notation for the annuity mapping function with the payment date for the moment),

$$\sum_{n=0}^{N-1} \tau_n E^A \left(\alpha(S(T), T_{n+1}) S(T) \right) = S(0).$$

This identity, in effect, states that the sum of CMS convexity adjustments with payment dates running over all tenor dates of the swap rate should be equal to zero.

Another useful check is obtained if we recall that

$$\frac{P(T,T_0)}{A(T)} - \frac{P(T,T_N)}{A(T)} = S(T)$$

(here of course $T_0 = T$, but we write it as such to highlight the symmetry in the expression). Multiplying both sides by $S(T)$, applying the expected value operator, and using the extended notation $\alpha(s, M)$ for the M-payment-date annuity mapping function, we obtain another identity that should be satisfied,

$$\mathrm{E}^A\left(\alpha(S(T),T_0)S(T)\right) - \mathrm{E}^A\left(\alpha(S(T),T_N)S(T)\right) = \mathrm{E}^A(S(T)^2).$$

The right-hand side here can be obtained from the Q^A-distribution of the swap rate $S(T)$ by replication and, as such, is independent of any annuity mapping function. Therefore, for any annuity mapping method, this identity — i.e. the requirement that the difference of CMS payments paid on the swap rate fixing date and on the last payment date be annuity-mapping-independent — represents another constraint that should be satisfied by the method.

16.6.8 Impact of Annuity Mapping Function and Mean Reversion

The importance of capturing volatility smile in CMS valuation, typically through the replication method (16.46), is widely acknowledged. On the other hand, the impact of other components entering into CMS valuation, especially the annuity mapping function $\alpha(s)$, is sometimes overlooked. One does not need to look any further than at the formula (16.54) for the CMS value in a linear TSR model to understand potential issues: if the parameter α_1 in (16.54) is allowed to vary freely, the CMS convexity adjustment can be made arbitrarily small or large.

Of course not all values of α_1 are compatible with financial reality, but choosing a reasonable range for α_1 is not entirely trivial. Relating the parameter to mean reversion as we did in Section 16.3.2 is useful, since we understand the role of mean reversion and its impact on model dynamics reasonably well. Moreover, mean reversion can be directly linked to market prices of traded securities, as shown in Section 13.1.8. It turns out that CMS convexity adjustment can vary by 10%-20% (in relative terms) when using different but reasonable levels of mean reversions.

To demonstrate the effect of mean reversion on the CMS convexity adjustment, consider the concrete problem of estimating the time 0 forward rate for a 10 year swap rate with semi-annual fixings. We use the linear

TSR method and assume that interest rates are flat at 5% (continuously compounded), and that the par swap rate is log-normally distributed with a constant volatility of $\sigma_S = 17\%$ for all fixing dates (which is hardly consistent with the presence of mean reversion, but good enough for a numerical example). Under this assumption,

$$\text{Var}^A\left(S(T)\right) = S(0)^2 \left(e^{\sigma_S^2 T} - 1\right),$$

allowing us to compute the convexity adjustment (16.55) in closed form for a given maturity T. We estimate the coefficient α_1 in (16.55) in two ways: by the simplified approach (16.56), or from the more elaborate formula (16.27) that takes mean reversion $\varkappa$ as a parameter. Results are shown in Figure 16.1, at multiple values of T and $\varkappa$.

Fig. 16.1. CMS Convexity Adjustment (Basis Points)

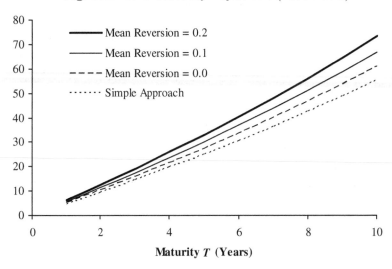

Notes: CMS convexity adjustment (16.55) in basis points for the linear TSR model, as computed by formula (16.27) (at three different mean reversion levels) and by the simplified formula (16.56). The rate and volatility settings are described in the text.

Notice that the convexity adjustment increases with mean reversion, a consequence of the fact that the volatility of the annuity factor effectively increases when the mean reversion goes up. Consistent with this, the slope parameter α_1 increases in mean reversion. The simple approach in formula (16.56) results in lower convexity adjustments than the mean reversion based approach.

16.6.9 CDF and PDF of CMS Rate in Forward Measure

The replication method is very useful for pricing CMS-linked cash flows, but it is not always convenient to apply. In particular, if the payoff is discontinuous, then the calculation of weights in, e.g., (16.49) will require special care. Let us attempt to develop a suitable alternative. We start by noting that the problem of pricing a cash flow that pays $g(S(T))$ at time T_p, $T_p \geq T$, (as in Section 16.6.1) can be seen as a problem of determining a density of $S(T)$ in the T_p-forward measure, $\psi^{T_p}(s)$, as we can always write

$$\mathrm{E}^{T_p}\left(g\left(S(T)\right)\right) = \int_{-\infty}^{\infty} g(s)\psi^{T_p}(s)\, ds. \tag{16.65}$$

This density is not directly available; however, the cumulative distribution function (CDF) $\Psi^A(\cdot)$ and the probability density function (PDF) $\psi^A(\cdot)$ of a swap rate in the *annuity* measure are available in either closed form for a particular vanilla model calibrated to market, or non-parametrically from the market prices of swaptions of all strikes (see Section 7.1.2) via

$$\Psi^A(K) = 1 + \frac{\partial}{\partial K}c(K), \tag{16.66}$$

$$\psi^A(K) = \frac{\partial^2}{\partial K^2}c(K), \tag{16.67}$$

$$c(K) = \mathrm{E}^A\left((S(T) - K)^+\right). \tag{16.68}$$

The following proposition allows us to obtain the CDF and PDF of the swap rate in the forward measure from its distributional characteristics in the annuity measure.

Proposition 16.6.4. *Given an annuity mapping function $\alpha(s)$ defined by (16.50), the PDF $\psi^{T_p}(s)$ and the CDF $\Psi^{T_p}(s)$ of the swap rate in the T_p-forward measure are linked to the PDF $\psi^A(s)$ and the CDF $\Psi^A(s)$ of the swap rate in the annuity measure by*

$$\psi^{T_p}(s) = \frac{A(0)}{P(0, T_p)}\alpha(s)\psi^A(s), \tag{16.69}$$

$$\Psi^{T_p}(s) = \frac{A(0)}{P(0, T_p)}\int_{-\infty}^{s} \alpha(u)\psi^A(u)\, du. \tag{16.70}$$

Proof. Proceeding somewhat informally, we observe that the value of the density $\psi^{T_p}(K)$ at point K is equal to the (undiscounted) value of the security with the delta-function payoff, $\delta\left(S(T) - K\right)$,

$$\psi^{T_p}(K) = \mathrm{E}^{T_p}\left(\delta\left(S(T) - K\right)\right).$$

By switching to the annuity measure, using the law of iterated conditional expectations, and the definition (16.50) of $\alpha(s)$, we obtain

$$\psi^{T_p}(K) = \frac{A(0)}{P(0,T_p)} E^A \left(\frac{P(T,T_p)}{A(T)} \delta(S(T) - K) \right)$$

$$= \frac{A(0)}{P(0,T_p)} E^A \left(\alpha(S(T)) \delta(S(T) - K) \right)$$

$$= \frac{A(0)}{P(0,T_p)} \alpha(K) E^A \left(\delta(S(T) - K) \right)$$

$$= \frac{A(0)}{P(0,T_p)} \alpha(K) \psi^A(K).$$

The statement (16.70) follows trivially. □

In practice, we would use one of the approximations to $\alpha(s)$ as derived in, for example, Sections 16.6.3, 16.6.4, 16.6.5, or 16.6.6. The density integration method (16.65) with the density $\psi^{T_p}(s)$ given by (16.69) is theoretically equivalent to the replication method of Section 16.6.1, but can, as hinted at earlier, have better numerical properties for non-smooth or discontinuous payoffs such as digital options or range accruals on a CMS rate. Indeed, unlike (16.49), the density integration method does not involve payoff differentiation. Another important application of the method arises when valuing cash flows linked to *multiple* CMS rates as we discuss in Chapter 17.

The expression (16.70) for the CDF has a particularly simple form when the function $\alpha(s)$ is linear as in, for example, the linear TSR model (Section 16.3.2).

Corollary 16.6.5. *In the linear TSR model (16.52), the CDF $\Psi^{T_p}(s)$ of the swap rate in the T_p-forward measure is given by*

$$\Psi^{T_p}(s) = \frac{A(0)}{P(0,T_p)} \left(\alpha_1 \left(S(0) - s - c(s) \right) + \alpha(s) \Psi^A(s) \right), \qquad (16.71)$$

where the CDF in the annuity measure $\Psi^A(s)$ is given by (16.66), and $c(s)$ is the option price with strike s in (16.68).

Proof. We have,

$$\frac{P(0,T_p)}{A(0)} \Psi^{T_p}(s) = \int_{-\infty}^{s} (\alpha_1 u + \alpha_2) \psi^A(u) \, du$$

$$= \alpha_1 \left(\int_{-\infty}^{\infty} u \psi^A(u) \, du - \int_{s}^{\infty} (u - s) \psi^A(u) \, du - s \int_{s}^{\infty} \psi^A(u) \, du \right)$$

$$+ \alpha_2 \int_{-\infty}^{s} \psi^A(u) \, du.$$

We note that

$$\int_{-\infty}^{\infty} u \psi^A(u) \, du = S(0), \qquad \int_{s}^{\infty} (u - s) \psi^A(u) \, du = c(s),$$

$$\int_{s}^{\infty} \psi^A(u) \, du = 1 - \Psi^A(s), \qquad \int_{-\infty}^{s} \psi^A(u) \, du = \Psi^A(s),$$

and the result follows.   □

Corollary 16.6.6. *In the linear TSR model (16.52), the PDF $\psi^{T_p}(s)$ of the swap rate in the T_p-forward measure is given by*

$$\psi^{T_p}(s) = \frac{A(0)}{P(0,T_p)} (\alpha_1 s + \alpha_2) \psi^A(s), \qquad (16.72)$$

where its PDF in the annuity measure $\psi^A(s)$ is given by (16.67).

Proof. Either directly from (16.69) or by differentiating (16.71) and using (16.66).   □

At this point the reader may wonder if the PDF $\psi^{T_p}(s)$ is directly linked to prices of traded derivatives. To answer this, we should recall the definition of a CMS caplet payoff in (16.47). Clearly,

$$V_{\text{cmscaplet}}(0,K) = P(0,T_p) \mathrm{E}^{T_p}\left((S(T) - K)^+ \right), \qquad (16.73)$$

and we therefore have the following result (compare with (16.67)):

Lemma 16.6.7. *The market-implied PDF $\psi^{T_p}(s)$ of the swap rate in the T_p-forward measure can be directly obtained from values of CMS caplets by*

$$\psi^{T_p}(K) = \frac{1}{P(0,T_p)} \frac{\partial^2}{\partial K^2} V_{\text{cmscaplet}}(0,K). \qquad (16.74)$$

16.6.10 SV Model for CMS Rate

A forward measure associated with the payment date of a cash flow is often the most convenient measure to use when dealing with vanilla derivatives linked to *multiple* market rates, as we shall find out in the next chapter. So, it would be useful to have PDFs and CDFs of market rates in the forward measure to be of tractable form, i.e. come from some common parameterizations such as the SV model of Chapter 8. Alas, this is not the case if we start with the SV model in the *annuity* measure for the swap rate, which is of course a common procedure. In this section we discuss these issues and proceed to derive useful approximations.

The formulas in Proposition 16.6.4 and Corollary 16.6.6 specify how PDFs of swap rates change under a measure change, from the annuity measure to the T_p-forward measure. These transformations are independent of the actual models (PDFs) used and are, of course, exact (to the extent that $\alpha(s)$ represents a true conditional expected value). As we indicated above it is useful to have an approximation to the PDF $\psi^{T_p}(s)$ that is from the same family as the PDF in the annuity measure $\psi^A(s)$. To elaborate, assume the swap rate follows

$$dS(t) = \lambda \left(bS(t) + (1-b)\, S(0)\right) \sqrt{z(t)}\, dW^A(t), \tag{16.75}$$

$$dz(t) = \theta \left(1 - z(t)\right) dt + \eta \sqrt{z(t)}\, dZ^A(t), \quad z(0) = 1, \tag{16.76}$$

where $\langle dZ^A(t), dW^A(t) \rangle = 0$ and $Z^A(t)$, $W^A(t)$ are Brownian motions in Q^A, the annuity measure. This SDE system defines the distribution, and in particular the PDF $\psi^A(\cdot)$ of $S(T)$, in measure Q^A. Let us now define an adjusted process $\widetilde{S}(t)$ given by the following SV dynamics in the T_p-forward measure,

$$d\widetilde{S}(t) = \widetilde{\lambda} \left(\widetilde{b}\widetilde{S}(t) + \left(1 - \widetilde{b}\right) \widetilde{S}(0)\right) \sqrt{\widetilde{z}(t)}\, dW^{T_p}(t),$$

$$d\widetilde{z}(t) = \theta \left(1 - \widetilde{z}(t)\right) dt + \widetilde{\eta} \sqrt{\widetilde{z}(t)}\, dZ^{T_p}(t), \quad \widetilde{z}(0) = 1, \tag{16.77}$$

where $Z^{T_p}(t)$, $W^{T_p}(t)$ are Q^{T_p}-Brownian motions satisfying $\langle dZ^{T_p}(t), dW^{T_p}(t) \rangle = 0$, and where we align the mean of $\widetilde{S}(T)$ to equal the CMS-adjusted value of $S(T)$, i.e.

$$\widetilde{S}(0) = E^{T_p}\left(S(T)\right).$$

Finally, we define $\psi^{T_p}(\cdot)$ to be the PDF of $\widetilde{S}(T)$, and aim to set the adjusted parameters $\widetilde{\lambda}$, $\widetilde{b}$, $\widetilde{\eta}$ such that the distribution of $\widetilde{S}(T)$ is as close as possible to the distribution of $S(T)$ in the T_p-forward measure.

As measure transformations affect the drift of an SDE, it is clear that the equality (in distribution) of $S(T)$ and $\widetilde{S}(T)$ under Q^{T_p} cannot, in general, be achieved exactly, as we here attempt to represent the measure transform as a parameter change solely affecting the diffusion term in the SDE for $S(t)$. Still, as we said in the beginning of the section, such a representation, even if approximate, is often useful for the multi-rate derivatives that we consider in Chapter 17.

The calculation of $(\widetilde{S}(0), \widetilde{\lambda}, \widetilde{b}, \widetilde{\eta})$ from $(S(0), \lambda, b, \eta)$ is not trivial and best done numerically. The convexity-adjusted CMS rate $\widetilde{S}(0)$ is calculated by the replication method of Section 16.6.1. As we can rewrite (16.73) in the form

$$V_{\text{cmscaplet}}(0, K) = P(0, T_p)\, E^{T_p}\left(\left(\widetilde{S}(T) - K\right)^+\right),$$

we note that CMS caplets are nothing more than European call options on $\widetilde{S}(T)$. Hence, we can obtain $(\widetilde{\lambda}, \widetilde{b}, \widetilde{\eta})$ by direct calibration of the SV model (16.77) (which is in T_p-forward measure) to prices of CMS caplets *that are computed in the original SV model* for $S(T)$ (with dynamics specified in the annuity measure). These CMS caplet prices in the SV model for $S(t)$ are best obtained by the replication algorithm (16.46) with weights (16.49). As we need CMS caplet prices across a range of strikes for $(\widetilde{\lambda}, \widetilde{b}, \widetilde{\eta})$-calibration, we can reuse much of the calculations in (16.46), as only the weights (but not the swaption prices used in replication) change in (16.46) for CMS caplets of different strikes.

At this point one may question whether the SV parameters (λ, b, η) perhaps do not change when we switch from the annuity to the forward measure, i.e. that all measure-related changes can be embedded in the change of the forward value, $S(0) \to \widetilde{S}(0) = \mathrm{E}^{T_p}(S(T))$. The answer is, of course, a clear no: the relationship between the densities as given by, for example, (16.71) is *not* just a shift of the mean of $S(T)$. In fact we see that the measure change affects the whole distribution, in particular re-distributing the probability mass from the region of lower values of the swap rate to the region of the higher values of the rate (as $\alpha_1 > 0$ typically). Hence, we would expect, at the very least, a change in the skew parameter b; other parameters will also be affected. This highlights an important point: CMS caps/floors *should not* be valued by simply convexity-adjusting the forward swap rate and then otherwise using the same model with the same parameters as for European swaptions. Despite the fact that it often produces sizable errors, this type of computation nevertheless appears quite common in practice.

16.6.11 Dynamics of CMS Rate in Forward Measure

While in the previous section we absorbed the measure change into the SV diffusion parameters, in reality measure changes affect only (instantaneous) drifts of the SDEs defining the dynamics. Let us explore how this will work for the SV model. While not a particularly useful consideration for single-rate derivatives, this becomes important when we consider Monte Carlo pricing of derivatives linked to multiple rates; see Section 17.8.1 for such applications.

We continue looking at a single swap rate $S(t)$ associated with the annuity $A(t)$, and assume that the following stochastic volatility model is specified in the annuity measure,

$$dS(t) = \lambda \varphi\left(S(t)\right) \sqrt{z(t)}\, dW^A(t), \tag{16.78}$$

$$dz(t) = \theta\left(1 - z(t)\right) dt + \eta \psi\left(z(t)\right) dZ^A(t), \quad z(0) = 1, \tag{16.79}$$

where $\langle dZ^A(t), dW^{A(t)} \rangle = 0$ and $Z^A(t)$, $W^A(t)$ are Brownian motions in Q^A, the annuity measure.

We are interested in bringing the two-dimensional SDE (16.78)–(16.79) into the T_p-forward measure, where T_p is the payment time of the CMS contract. However, the dynamics (16.78)–(16.79) in the annuity measure *do not* uniquely define a T_p-forward measure — for that we would need a full term structure model (or, at least, an additional specification for the density process $P(t, T_p)/A(t)$). On the other hand, if we assume that we have knowledge of the conditional expectation (16.50) we can nevertheless construct a probability measure that would resemble a forward measure for European-type payoffs fixing at T and paying at T_p (but not for any other payoffs such as e.g. European derivatives fixing at time other than T). More precisely, we are interested in constructing a measure $\widetilde{\mathrm{Q}}^{T_p}$ such that, for any function $f(\cdot)$,

$$\mathrm{E}^A \left(\frac{P(T, T_p)}{A(T)} f(S(T)) \right) = \frac{P(0, T_p)}{A(0)} \widetilde{\mathrm{E}}^{T_p} \left(f(S(T)) \right),$$

where $\widetilde{\mathrm{E}}^{T_p}$ denotes expectation in measure $\widetilde{\mathrm{Q}}^{T_p}$.

Before stating our result, we recall the definition of the function $\alpha(s)$ from (16.50). It is more convenient to deal with a rescaled function, so let us define

$$\widehat{\alpha}(s) = \frac{A(0)}{P(0, T_p)} \alpha(s) = \frac{A(0)}{P(0, T_p)} \mathrm{E}^A \left(\frac{P(T, T_p)}{A(T)} \bigg| S(T) = s \right).$$

Proposition 16.6.8. *Define the measure $\widetilde{\mathrm{Q}}^{T_p}$ by the condition that the process $(z(t), S(t))$ satisfies the following SDE in $\widetilde{\mathrm{Q}}^{T_p}$,*

$$dS(t) = \lambda \varphi \left(S(t) \right) \sqrt{z(t)} \left(dW^{T_p}(t) + v^S(t) \, dt \right), \tag{16.80}$$
$$dz(t) = \theta \left(1 - z(t) \right) dt + \eta \psi \left(z(t) \right) \left(dZ^{T_p}(t) + v^z(t) \, dt \right), \quad z(0) = 1,$$

where $Z^{T_p}(t)$ and $W^{T_p}(t)$ are (uncorrelated) driftless Brownian motions in $\widetilde{\mathrm{Q}}^{T_p}$, the drift adjustments are given by

$$v^z(t) = \eta \psi \left(z(t) \right) \frac{\partial}{\partial z} \ln \left(\Lambda \left(t, z(t), S(t) \right) \right), \tag{16.81}$$
$$v^S(t) = \lambda \varphi \left(S(t) \right) \sqrt{z(t)} \frac{\partial}{\partial S} \ln \left(\Lambda \left(t, z(t), S(t) \right) \right),$$

and the function $\Lambda(t, z, s)$ satisfies the following PDE,

$$\frac{\partial}{\partial t} \Lambda \left(t, z, s \right) + \theta \left(1 - z \right) \frac{\partial}{\partial z} \Lambda \left(t, z, s \right) + \frac{\eta^2}{2} \psi(z)^2 \frac{\partial^2}{\partial z^2} \Lambda \left(t, z, s \right)$$
$$+ \frac{\lambda^2}{2} \varphi(s)^2 z \frac{\partial^2}{\partial s^2} \Lambda \left(t, z, s \right) = 0, \quad t \in [0, T], \tag{16.82}$$

with the terminal condition at $t = T$

$$\Lambda \left(T, z, s \right) = \widehat{\alpha}(s). \tag{16.83}$$

Then for any function $f(\cdot)$ we have

$$\frac{A(0)}{P(0, T_p)} \mathrm{E}^A \left(\frac{P(T, T_p)}{A(T)} f(S(T)) \right) = \widetilde{\mathrm{E}}^{T_p} \left(f(S(T)) \right). \tag{16.84}$$

Proof. Clearly, for any function $f(\cdot)$,

$$\frac{A(0)}{P(0, T_p)} \mathrm{E}^A \left(\frac{P(T, T_p)}{A(T)} f(S(T)) \right) = \mathrm{E}^A \left(\widehat{\alpha}(S(T)) f(S(T)) \right).$$

The condition (16.84) would therefore be satisfied if we locate a measure that satisfies

$$E^A\left(\widehat{\alpha}(S(T))f(S(T))\right) = \widetilde{E}^{T_p}\left(f(S(T))\right)$$

for any $f(\cdot)$. Recalling the results in Section 1.4, we must find a density process, i.e. a positive Q^A-martingale, that equals $\widehat{\alpha}(S(T))$ at time T. Such a martingale is easy to construct,

$$\Lambda(t) = E_t^A\left(\widehat{\alpha}(S(T))\right),$$

and this allows us to specify the measure $\widetilde{Q}^{T_p}$ as the measure for which $\Lambda(t)$ is the Radon-Nikodym derivative (with respect to Q^A).

By Girsanov's theorem, moving from $\widetilde{Q}^A$ to $\widetilde{Q}^{T_p}$ is associated with certain changes in the drift terms of (16.78)–(16.79). Specifically, we recall from Section 1.5 that

$$dZ^A(t) = dZ^{T_p}(t) + \nu^z(t)\,dt,$$
$$dW^A(t) = dW^{T_p}(t) + \nu^S(t)\,dt,$$

where

$$d\Lambda(t)/\Lambda(t) = \nu^z(t)\,dZ^A(t) + \nu^S(t)\,dW^A(t).$$

By Ito's lemma and the fact that $\Lambda(t)$ is a Q^A-martingale,

$$d\Lambda\left(t, z(t), S(t)\right) = \Lambda_z\left(t, z(t), S(t)\right)\eta\psi\left(z(t)\right)\,dZ^A(t)$$
$$+ \Lambda_S\left(t, z(t), S(t)\right)\lambda\varphi\left(S(t)\right)\sqrt{z(t)}\,dW^A(t).$$

Then the expressions for $\nu^z(t)$, $\nu^S(t)$ follow by matching the dZ^A, dW^A terms in the last two equations.

Finally, to find the expression for $\Lambda(t)$ we recall that since the process $(S(t), z(t))$ is Markovian, we have, with slight abuse of notations,

$$\Lambda(t) = \Lambda\left(t, z(t), S(t)\right),$$
$$\Lambda(t, z, s) = E_t^A\left(\widehat{\alpha}\left(S(T)\right)\middle|\, z(t) = z, S(t) = s\right).$$

If follows from the Feynman-Kac theorem that the function $\Lambda(t, z, s)$ satisfies the PDE (16.82)–(16.83) . $\square$

Proposition 16.6.8 establishes a numerical scheme for simulating $(z(t), S(t))$ in $\widetilde{Q}^{T_p}$, for purposes of pricing European-style derivatives fixing at a given time T and paying at T_p. In general, we would determine the function $\Lambda(t, z, s)$ by numerically solving the PDE (16.82)–(16.83) on a grid of (t, z, s), and then perform the Monte Carlo simulation for (16.80), with the drift adjustments $\nu^z(t)$, $\nu^S(t)$ computed for each path using (16.81). In some important cases, however, no finite difference scheme is required, as shown in the following corollary for the often-used case where the function $\alpha(s)$ is linear.

Corollary 16.6.9. *Assume that*

$$\widehat{\alpha}(s) = \widehat{\alpha}_1 s + \widehat{\alpha}_2.$$

Then

$$v^z(t) = 0, \quad v^S(t) = \lambda \varphi\left(S(t)\right) \sqrt{z(t)} \frac{\widehat{\alpha}_1}{\widehat{\alpha}_1 S(t) + \widehat{\alpha}_2}. \tag{16.85}$$

Proof. The swap rate $S(t)$ is a Q^A-martingale, hence

$$\Lambda\left(t, z, s\right) = \mathrm{E}^A\left(\widehat{\alpha}\left(S(T)\right)\middle|\, z(t) = z, S(t) = s\right) = \widehat{\alpha}_1 s + \widehat{\alpha}_2.$$

□

16.6.12 Cash-Settled Swaptions

After the mainly theoretical considerations of Section 16.6.11, let us return to applications and consider the important topic of pricing of cash-settled European swaptions. As explained in Section 5.10.1, cash-settled swaptions are the most common type of vanilla options in European markets, especially for derivatives quoted in EUR and GBP. Cash-settled swaptions are closely linked to the swap-settled European swaptions that are standard in the US, but rather than exercising into a physical swap contract, a cash-settled swaption instead uses a particular "swap-like" formula to determine a cash amount to be paid upon option exercise. As it turns out, the replication methods we developed earlier in this chapter allow us to link a value of a cash-settled swaption to those of swap-settled swaptions across a range of strikes. While the two kinds of swaptions are rarely traded in the same market, this connection is nevertheless important as it allows us to continue treating *swap-settled* swaptions as the fundamental type of vanilla options to which we calibrate all our models, irrespective of market conventions. That is, we would maintain a swaption grid of vanilla model parameters (see Section 16.1.3) that represents values of swap-settled swaptions, even if these are not directly traded. The actual model parameter values for each grid point would be calculated by calibration to the values of the most prevalent type of swaptions in the market; for the case of cash-settled swaptions, this would require usage of the valuation formula (16.86) developed below.

As we recall, the payoff of a cash-settled swaption is given by a deterministic function $g(\cdot)$ applied to the swap rate $S(T)$ and paid at T, where the function $g(\cdot)$ is given by

$$g(s) = \left(\sum_{n=0}^{N-1} \tau_n \prod_{i=0}^{n} (1 + \tau_i s)^{-1}\right) (s - K)^+$$

for a payer swaption. Given the annuity mapping function $\alpha(s)$, the value is then given by

$$V_{\text{CSS}}(0) = A(0)\text{E}^A\left(\alpha(S(T))g(S(T))\right), \tag{16.86}$$

and is easily calculated by the replication method applied to prices for swap-settled swaptions.

In application of (16.86), we would need to fix a choice for the the annuity mapping function α; for this, we typically recommend using the linear TSR model of Section 16.3.2. Interestingly, if we were alternatively to use the swap-yield model of Section 16.6.3, we would get

$$\alpha(s)g(s) = (s - K)^+,$$

as the swap-yield annuity mapping function exactly cancels the annuity discounting term in the payoff. We would therefore get

$$V_{\text{CSS}}(0) = A(0)\text{E}^A\left((S(T) - K)^+\right),$$

which is the value of a swap-settled swaption. In reality the values of swap- and cash-settled swaptions should, of course, be different; the inability of the swap-yield model to distinguish them is a symptom of the fact that the swap-yield model is not a truly arbitrage-free model.

As cash-settled swaptions differ from swap-settled swaptions, they also do not obey the "standard" call-put parity,

$$V_{\text{CSS,pay}}(0) - V_{\text{CSS,rec}}(0)$$
$$= A(0)\text{E}^A\left(\alpha(S(T))\left(\sum_{n=0}^{N-1}\tau_n\prod_{i=0}^{n}(1+\tau_iS(T))^{-1}\right)(S(T) - K)\right)$$
$$\neq A(0)\left(S(0) - K\right).$$

Instead, a combined long-short position in a cash-settled payer swaption and a cash-settled receiver position is equivalent to a "cash-settled swap", i.e. a (typically non-traded) derivative with the payoff

$$\left(\sum_{n=0}^{N-1}\tau_n\prod_{i=0}^{n}(1+\tau_iS(T))^{-1}\right)(S(T) - K).$$

16.7 Quanto CMS

All securities discussed so far in this chapter produce payments in the same currency as the currency of the underlying rates used to calculate the payoff. It is, however, not uncommon to use a different currency for payment, a modification that leads to the creation of so-called "quanto" cash flows, see Section 4.3. While we generally limit the scope of this book to single-currency derivatives only, quanto extensions of CMS-linked derivatives are sufficiently common to warrant a discussion of their valuation.

We have already arrived at some preliminary results on multi-currency markets in Section 4.3, a section that the reader is advised to review before proceeding; the notations of this section will be adopted in what follows.

16.7.1 Overview

Let the swap rate $S(T)$ of Section 16.6 be computed from a *domestic* currency yield curve. A quanto CMS cash flow pays $g(S(T))$ at time T_p in some other *foreign* currency; the value of the cash flow is therefore equal to

$$V_{\text{QuantoCMS}}(0) = \beta_f(0)\mathrm{E}^f\left(\beta_f(T_p)^{-1}g(S(T))\right)$$

in foreign currency units, where $\beta_f(t)$ is the foreign money market account and E^f is the expected value operator for the foreign risk-neutral measure. Since $S(t)$ is defined by the domestic interest rate curve, its distribution in the domestic (annuity) measure is available from the swaption market. By Lemma 4.3.1, the density process relating the foreign and domestic risk-neutral measures is given by

$$\mathrm{E}^d_t\left(\frac{d\mathbb{Q}^f}{d\mathbb{Q}^d}\right) = \frac{\beta_d(0)\beta_f(t)X(t)}{\beta_d(t)\beta_f(0)X(0)}, \quad t \geq 0,$$

where $X(t)$ is the spot FX rate measured in domestic currency per foreign currency units. The value of the contract in the domestic risk-neutral measure may therefore be written (in foreign currency units, naturally) as

$$V_{\text{QuantoCMS}}(0) = \frac{\beta_d(0)}{X(0)}\mathrm{E}^d\left(\beta_d(T_p)^{-1}g(S(T))X(T_p)\right). \tag{16.87}$$

Of course, the same formula can be derived by observing that since $g(S(T))$ is paid in a foreign currency, we can convert the proceeds into the domestic currency at time T_p to create a domestic asset with the payoff $g(S(T))X(T_p)$.

Conditioning in (16.87) on $\mathcal{F}_T$, we get

$$V_{\text{QuantoCMS}}(0) = \frac{\beta_d(0)}{X(0)}\mathrm{E}^d\left(\beta_d(T)^{-1}g(S(T))\left[\beta_d(T)\mathrm{E}^d_T\left(\beta_d(T_p)^{-1}X(T_p)\right)\right]\right)$$

$$= \frac{\beta_d(0)}{X(0)}\mathrm{E}^d\left(\beta_d(T)^{-1}g(S(T))\left[P_d(T,T_p)X_{T_p}(T)\right]\right),$$

where we have used the notation from Section 4.3,

$$X_{T_p}(T) = \frac{P_f(T,T_p)}{P_d(T,T_p)}X(T),$$

to denote the T_p-forward FX rate seen at time T.

As we have seen previously for CMS-linked cash flows, the (domestic) annuity measure provides the most direct market information about the distribution of the underlying swap rate. Switching to this measure, we have

$$V_{\text{QuantoCMS}}(0) = \frac{A(0)}{X(0)} E^{A,d} \left(g(S(T)) \frac{P_d(T,T_p)}{A(T)} X_{T_p}(T) \right).$$

Drawing on the results of Section 16.6, we see that if the payment currency of the cash flow were domestic, then the appropriate valuation formula instead would read

$$V(0) = A(0) E^{A,d} \left(g(S(T)) \frac{P_d(T,T_p)}{A(T)} \right).$$

The *quanto adjustment* is defined to be the ratio,

$$D_{\text{Quanto}}(0) = \frac{E^{A,d} \left(g(S(T)) \frac{P_d(T,T_p)}{A(T)} X_{T_p}(T) \right)}{X(0) E^{A,d} \left(g(S(T)) \frac{P_d(T,T_p)}{A(T)} \right)}.$$

For quanto CMS valuation or, equivalently, calculation of quanto adjustments, it is natural to search for a suitable extension of the methods developed previously for (single-currency) CMS-linked cash flows. In particular, as quanto CMS structures are quite vanilla-like, we would ideally like to avoid the usage of full term structure (multi-currency) interest rate models.

By the arguments similar to those of Section 16.6.2, we have

$$V_{\text{QuantoCMS}}(0) = \frac{A(0)}{X(0)} E^{A,d} \left(g(S(T)) v(S(T)) \right), \tag{16.88}$$

where

$$v(s) \triangleq E^{A,d} \left(\frac{P_d(T,T_p)}{A(T)} X_{T_p}(T) \middle| S(T) = s \right).$$

Let us recall the definition (16.50) of $\alpha(s)$ and also define

$$\chi(s) \triangleq E^{A,d} \left(X_{T_p}(T) \middle| S(T) = s \right). \tag{16.89}$$

Then, approximately[12],

$$v(s) \approx \alpha(s)\chi(s),$$

so that (also approximately)

$$V_{\text{QuantoCMS}}(0) = \frac{A(0)}{X(0)} E^{A,d} \left(g(S(T)) \alpha(S(T)) \chi(S(T)) \right). \tag{16.90}$$

Once the value is represented in the form (16.90) it can be computed by the replication method, as in Section 16.6.1. To complete the valuation, it only remains to determine the function $\chi(s)$ in (16.89).

---

[12] Here we essentially assume that the slope of the yield curve is independent of the "pure" FX component of the forward FX rate.

16.7.2 Modeling the Joint Distribution of Swap Rate and Forward Exchange Rate

To compute the function $\chi(s)$ in the previous section, a joint distribution of the swap rate $S(T)$ and the forward FX rate $X_{T_p}(T)$ needs to be specified. The marginal one-dimensional distribution of $S(T)$ in $Q^{A,d}$ is given by the swaption model; we denote the cumulative distribution function (CDF) of $S(T)$ by $\Psi^A(s)$, see (16.66). The payoff of the quanto CMS cash flow in (16.88) depends on $X_{T_p}(T)$ linearly, indicating that the value of the derivative has rather limited dependence on the particular form of the distribution of the FX rate, a fact well-supported by numerical tests[13]. In the interest of analytical tractability, we simply model $X_{T_p}(T)$ as being log-normal in the domestic annuity measure $Q^{A,d}$, i.e. we assume that there exists a standard Gaussian random variable ξ_1, a volatility σ_X, and a scaling constant m_X such that

$$X_{T_p}(T) = X(0)e^{\sigma_X \sqrt{T} \xi_1 + m_X T}. \tag{16.91}$$

The volatility σ_X is obtained by calibrating (16.91) to T-expiry ATM options on the FX rate[14], whereas the choice of the constant m_X is clarified below.

With marginal distributions of $S(T)$ and $X_{T_p}(T)$ specified, we impose the dependence structure with a simple application of the so-called *copula method*. Chapter 17 contains a full review of copula methods and their applications to multi-rate vanilla derivatives, but for our needs here it suffices to note that if ξ_2 is a standard Gaussian random variable, then clearly

$$S(T) \stackrel{d}{=} \left(\Psi^A\right)^{-1} \left(\Phi\left(\xi_2\right)\right),$$

where $\Phi(\cdot)$ is the standard Gaussian CDF, and the equality is in terms of distribution. The dependence structure between $S(T)$ and $X_{T_p}(T)$ may now be imposed by correlating the two standard Gaussian random variables, ξ_1 and ξ_2, with a correlation ρ_{XS}, leading to the following specification of the joint distribution

$$X_{T_p}(T) = X(0)e^{\sigma_X \sqrt{T} \xi_1 + m_X T}, \quad S(T) = \left(\Psi^A\right)^{-1} \left(\Phi\left(\xi_2\right)\right),$$
$$\text{Corr}\left(\xi_1, \xi_2\right) = \rho_{XS}.$$

The function $\chi(s)$ in (16.89) can now easily be computed,

---

[13] For very long-dated quanto contracts, the FX volatility smile may start to matter. The method we develop here can be extended to incorporate the FX smile, by techniques discussed in Chapter 17.

[14] This is not exact as the T-expiry FX option is written on $X_T(T)$, not $X_{T_p}(T)$. The difference is rarely material as $T_p - T$ is often small, but more elaborate schemes are not difficult to design, if desired.

$$\chi(s) = \mathrm{E}^{A,d}\left(X_{T_p}(T)\big|\,S(T) = s\right)$$
$$= X(0)e^{m_X T}\mathrm{E}^{A,d}\left(e^{\sigma_X \sqrt{T}\xi_1}\big|\,\xi_2 = \Phi^{-1}\left(\Psi^A(s)\right)\right)$$
$$= X(0)e^{m_X T}\widetilde{\chi}(s), \tag{16.92}$$

$$\widetilde{\chi}(s) = \exp\left(\rho_{XS}\sigma_X \sqrt{T}\Phi^{-1}\left(\Psi^A(s)\right) + \frac{\sigma_X^2 T}{2}\left(1 - \rho_{XS}^2\right)\right).$$

16.7.3 Normalizing Constant and Final Formula

To complete the development of a pricing formula for quanto CMS cash flows, we now only need to establish the constant m_X in (16.91) (and (16.92)). For this, we note that $X_{T_p}(\cdot)$ is a martingale in the domestic T_p-forward measure $Q^{T_p,d}$ (see Section 4.3), and in particular,

$$X_{T_p}(0) = \mathrm{E}^{T_p,d}\left(X_{T_p}(T)\right).$$

Changing to the domestic annuity measure $Q^{A,d}$, the following holds,

$$X_{T_p}(0) = \frac{A(0)}{P_d(0,T_p)}\mathrm{E}^{A,d}\left(\frac{P_d(T,T_p)}{A(T)}X_{T_p}(T)\right).$$

Recalling the definition of $\alpha(s)$ and $\chi(s)$ we finally write

$$X_{T_p}(0) = \frac{A(0)}{P_d(0,T_p)}\mathrm{E}^{A,d}\left(\alpha(S(T))\chi(S(T))\right)$$
$$= X(0)e^{m_X T}\frac{A(0)}{P_d(0,T_p)}\mathrm{E}^{A,d}\left(\alpha(S(T))\widetilde{\chi}(S(T))\right),$$

and hence

$$e^{-m_X T} = \frac{X(0)}{X_{T_p}(0)}\frac{A(0)}{P_d(0,T_p)}\mathrm{E}^{A,d}\left(\alpha(S(T))\widetilde{\chi}(S(T))\right).$$

Combining all previous results, we have arrived at the following pricing formula.

Proposition 16.7.1. *Let the forward FX rate $X_{T_p}(T)$ be log-normal with volatility σ_X, and let the co-dependence between $X_{T_p}(T)$ and the swap rate $S(T)$ be characterized by a Gaussian copula with correlation ρ_{XS}. The value of a quanto CMS cash flow $g(S(T))$ paid in a foreign currency at time $T_p \geq T$ is then approximated by*

$$V_{\mathrm{QuantoCMS}}(0) \approx P_f(0,T_p)\frac{\mathrm{E}^{A,d}\left(g(S(T))\alpha(S(T))\widetilde{\chi}(S(T))\right)}{\mathrm{E}^{A,d}\left(\alpha(S(T))\widetilde{\chi}(S(T))\right)}, \tag{16.93}$$

where the annuity mapping function $\alpha(\cdot)$ is defined by (16.50), and

$$\widetilde{\chi}(s) = \exp\left(\rho_{XS}\sigma_X \sqrt{T}\Phi^{-1}\left(\Psi^A(s)\right) + \frac{\sigma_X^2 T}{2}\left(1 - \rho_{XS}^2\right)\right),$$

with $\Psi^A(s) \triangleq \mathrm{P}^{A,d}(S(T) < s)$.

Remark 16.7.2. The expected values in the denominator and the numerator of the right-hand side of (16.93) can be computed by the replication method from Section 16.6.1.

16.8 Eurodollar Futures

Eurodollar (ED) futures are exchange-traded futures contracts on Libor rates (see Section 5.4) and serve as fundamental inputs in the construction of the yield curve, as explained in Chapter 6. As we explained in Section 4.1.2, daily mark-to-market causes the value of the futures contract on a Libor rate to differ from the value of a forward contract on a Libor rate. Only the latter is an input into an interest rate curve construction, while only the former is liquidly quoted in the market. The difference between the two is called the *ED future convexity adjustment*, a quantity that we shall analyze and quantify in the following sections.

16.8.1 Fundamental Results on Futures

As in Section 4.1.2, we let $F(t, T, M)$ denote the futures rate at time t covering the period $[T, M]$, with $0 \leq t \leq T < M$. The forward (Libor) rate for the same period is, as always, denoted by $L(t, T, M)$. The next lemma establishes the relationship between the two.

Lemma 16.8.1. *Let* Q *be the risk-neutral measure and* Q^M *be the M-forward measure, with* E *and* E^M *being the corresponding expected value operators. Then*

$$L(t, T, M) = E_t^M \left(L(T, T, M) \right), \quad F(t, T, M) = E_t \left(L(T, T, M) \right).$$

Proof. The first result is from Lemma 4.2.3 and the second is from Lemma 4.2.2. □

Lemma 16.8.1 holds for futures contracts that are marked-to-market continuously. For calculation purposes it is often more convenient to assume discrete mark-to-market; it has been established (see Flesaker [1993]) that prices with monthly or even quarterly resettlement frequency differ little from the prices with continuous (or daily) resettlement. To work out a version of Lemma 16.8.1 that covers discrete mark-to-market, let us introduce a standard tenor structure

$$0 = T_0 < T_1 < \ldots < T_N, \quad \tau_n = T_{n+1} - T_n,$$

and define Libor rates as before,

$$L_n(t) = L(t, T_n, T_{n+1}) = \frac{P(t, T_n) - P(t, T_{n+1})}{\tau_n P(t, T_{n+1})}, \quad n = 0, \ldots, N-1.$$

The discretely compounded money market account $B(t)$ is defined by (4.24), and the corresponding measure, the spot Libor measure, is defined in Section 4.2.3; we denote it by Q^B and the corresponding expected value operator by E^B. We abbreviate the notation for the expected value in the T_n-forward measure to $E^n \triangleq E^{T_n}$ (and the same for variance) and, finally, in line with the definition of spanning Libor rates, define spanning futures rates

$$F_n(t) = F(t, T_n, T_{n+1}), \quad n = 0, \ldots, N - 1.$$

We are now ready for the valuation formula of discretely-resettled futures rates.

Proposition 16.8.2. *The futures rate that is marked-to-market only on the dates $T_0 < T_1 < \ldots < T_n = T < M$ is given by*

$$F(t, T, M) = E_t^B (L(T, T, M)).$$

In particular,

$$E^B (L_n(T_n)) = F_n(0), \quad n = 0, \ldots, N - 1.$$

Proof. At time $T_n = T$, the cash flow associated with the futures contract is

$$F(T_n, T, M) - F(T_{n-1}, T, M) = L(T, T, M) - F(T_{n-1}, T, M).$$

At time T_{n-1}, the present value of this cash flow is

$$V_{\text{fut}}(T_{n-1}) = B(T_{n-1}) E_{T_{n-1}}^B \left(\frac{L(T, T, M) - F(T_{n-1}, T, M)}{B(T_n)} \right).$$

By the definition of the rolling spot numeraire $B(t)$, the quantity

$$B(T_{n-1})/B(T_n) = P(T_{n-1}, T_n)$$

is non-random at time T_{n-1}, and so is $F(T_{n-1}, T_n, M)$, whereby

$$V_{\text{fut}}(T_{n-1}) = P(T_{n-1}, T_n) \left(E_{T_{n-1}}^B (L(T, T, M)) - F(T_{n-1}, T, M) \right).$$

As futures contracts are always entered into at a price of 0, it follows that $V_{\text{fut}}(T_{n-1}) = 0$ by definition, and therefore

$$F(T_{n-1}, T, M) = E_{T_{n-1}}^B (L(T, T, M)).$$

At time T_{n-2}, we may write

$$V_{\text{fut}}(T_{n-2}) = B(T_{n-2}) E_{T_{n-2}}^B \left(\frac{F(T_{n-1}, T, M) - F(T_{n-2}, T, M)}{B(T_{n-1})} \right)$$

$$= B(T_{n-2}) E_{T_{n-2}}^B \left(\frac{E_{T_{n-1}}^B (L(T, T, M)) - F(T_{n-2}, T, M)}{B(T_{n-1})} \right)$$

$$= P(T_{n-2}, T_{n-1}) \left(E_{T_{n-2}}^B (L(T, T, M)) - F(T_{n-2}, T, M) \right),$$

whereby (as $V_{\text{fut}}(T_{n-2}) = 0$)

$$F(T_{n-2}, T, M) = \mathrm{E}^B_{T_{n-2}} \left(L(T, T, M) \right).$$

Proceeding inductively, the result follows.   □

The ED future convexity adjustment is given by the difference

$$F(t, T, M) - L(t, T, M) = \mathrm{E}_t \left(L(T, T, M) \right) - \mathrm{E}^M_t \left(L(T, T, M) \right)$$

or, for discrete settlement,

$$F(t, T, M) - L(t, T, M) = \mathrm{E}^B_t \left(L(T, T, M) \right) - \mathrm{E}^M_t \left(L(T, T, M) \right).$$

Proposition 4.5.3 derived the convexity adjustment in a general Gaussian multi-factor HJM model in closed-form. However, having demonstrated the importance of incorporating volatility smile in calculations for other types of convexity (Libor-in-arrears, CMS), we can legitimately ask whether the smile has a significant impact on ED convexity as well. This question cannot be answered within the constraints of a Gaussian model as it does not allow for smile control; we instead follow the ideas in Piterbarg and Renedo [2006] and develop a smile-enabled pricing approach in the following.

16.8.2 Motivations and Plan

Performance requirements for valuing ED futures are even more stringent than for other types of derivatives, due to their high trading volumes and, in particular, their use in yield curve construction. This rules out Monte Carlo methods, or even PDE-based schemes, necessitating the development of analytic approximations that incorporate the volatility smile yet allow for efficient numerical algorithms. In addition, we look for the formula for convexity adjustments that depends on observable market inputs in the most direct way possible, with lengthy model and curve calibrations reduced to a minimum or eliminated altogether. To achieve this, we separate model parameters into two categories: those that change often, and those that do not. The former category here covers volatility parameters, and are taken directly from the prices of options on ED contracts across different expiries and strikes. These parameters can be updated in real time as we build yield curves intra-day. The latter category of (slow-moving) parameters are essentially correlation parameters, and originate from calibrating a model with a rich volatility structure to caps and swaptions. Due to computational constraints, these parameters cannot be updated often — but they do not need to be, as they typically do not change much over time.

We use the following road map to derive the ED futures valuation formula:

1. First, an expansion technique is applied to derive a model-independent relationship that expresses a *forward rate* as a functional of a *collection of futures* rates with expiries on or before the expiry of the forward rate.

2. The variance terms that appear in the formula are separated into slow- and fast-moving parameters, as described earlier.

3. Fast-moving volatility parameters are represented in several different ways, both in model-independent fashion as functions of prices of options on ED futures across strikes[15], and as closed-form expressions involving volatility smile parameters.

4. Finally, slow-moving correlation parameters are expressed in terms of the parameters of a Libor market model, properly calibrated to relevant market instruments or, more pragmatically, through the simplified formulas (16.117) or (16.118).

We emphasize that the first step of the algorithm differs from traditional methods which typically express the value of futures rates in terms of forward rates, an approach diametrically opposite of ours. We consider our approach superior, as it eliminates the need to invert equations to obtain forward rates from market-observed futures quotes (a fundamental requirement of curve building algorithms).

16.8.3 Preliminaries

As established previously, the following relations hold,

$$F_n(0) = E^B\left(L_n(T_n)\right) = E^B\left(F_n(T_n)\right), \qquad (16.94)$$
$$L_n(0) = E^{n+1}\left(L_n(T_n)\right) = E^{n+1}\left(F_n(T_n)\right), \qquad (16.95)$$

for all $n = 0, \ldots, N-1$. We assume that all $F_n(0)$, $n = 0, \ldots, N-1$, are known; our goal is to derive formulas that express forward rates $\{L_n(0)\}$ in terms of futures $\{F_n(0)\}$ and, potentially, other market-observed quantities. The following result is straightforward.

Lemma 16.8.3. *For each n, $n = 0, \ldots, N-1$,*

$$L_n(0) = E^B\left(\left[\prod_{i=0}^{n} \frac{1 + \tau_i L_i(0)}{1 + \tau_i L_i(T_i)}\right] L_n(T_n)\right). \qquad (16.96)$$

Proof. Follows by a measure change. □

Lemma 16.8.3 expresses the forward rate $L_n(0)$ as an expectation of a certain payoff in the spot Libor measure (not forward measure as in (16.95)), the measure that is used in defining futures in (16.94). This result is used as a starting point for deriving convexity adjustments.

---

[15]This is similar to the replication method for computing CMS and Libor-in-arrears convexity adjustments.

16.8.4 Expansion Around the Futures Value

To express the expected value in (16.96) in terms of market-observed quantities, we derive a Taylor series expansion of (16.96) in powers of a small parameter that measures the deviation of each of $L_n(T_n)$ from its mean in the spot Libor measure, $F_n(0) = E^B(L_n(T_n))$.

Fix $\epsilon > 0$, and define L_n^ϵ's by

$$L_n^\epsilon(t) = \epsilon\left(L_n(t) - F_n(0)\right) + F_n(0).$$

Note that for any n,

$$L_n^1(t) = L_n(t), \tag{16.97}$$

$$L_n^0(t) = F_n(0), \tag{16.98}$$

$$\frac{\partial L_n^\epsilon(t)}{\partial \epsilon} = L_n(t) - F_n(0), \tag{16.99}$$

$$L_n^\epsilon(T_n) = \epsilon\left(L_n(T_n) - E^B\left(L_n(T_n)\right)\right) + E^B\left(L_n(T_n)\right). \tag{16.100}$$

Define

$$V(\epsilon) = \left[\prod_{i=0}^n \frac{1 + \tau_i L_i(0)}{1 + \tau_i L_i^\epsilon(T_i)}\right] L_n^\epsilon(T_n). \tag{16.101}$$

It should be clear that $V(1)$ is the value inside the expectation on the right-hand side of (16.96),

$$L_n(0) = E^B\left(V(1)\right). \tag{16.102}$$

Expanding $V(\epsilon)$ into a Taylor series in ϵ yields,

$$V(\epsilon) = V(0) + E^B\left(\frac{dV}{d\epsilon}(0)\right) \times \epsilon + \frac{1}{2}E^B\left(\frac{d^2V}{d\epsilon^2}(0)\right) \times \epsilon^2 + O\left(\epsilon^3\right). \tag{16.103}$$

The values of the derivatives of $V(\epsilon)$ are computed in the following lemma.

Lemma 16.8.4. *For any n, $n = 0, \ldots, N-1$,*

$$V(0) = \left[\prod_{i=0}^n \frac{1 + \tau_i L_i(0)}{1 + \tau_i F_i(0)}\right] F_n(0), \tag{16.104}$$

$$E^B\left(\frac{dV}{d\epsilon}(0)\right) = 0, \tag{16.105}$$

$$E^B\left(\frac{d^2V}{d\epsilon^2}(0)\right) = V(0) \sum_{j,m=0}^n D_{j,m} \mathrm{Cov}^B\left(L_j(T_j), L_m(T_m)\right), \tag{16.106}$$

where the coefficients $D_{j,m}$ are given by

$$D_{j,m} = \left(-\frac{\tau_j}{1 + \tau_j F_j(0)} + \frac{1_{\{j=n\}}}{F_n(0)}\right)\left(-\frac{\tau_m}{1 + \tau_m F_m(0)} + \frac{1_{\{m=n\}}}{F_n(0)}\right)$$

$$+ 1_{\{j=m\}}\left(\frac{\tau_j^2}{(1 + \tau_j F_j(0))^2} - \frac{1_{\{j=n\}}}{F_n(0)^2}\right), \quad (16.107)$$

and, by definition,

$$\mathrm{Cov}^B\left(L_j(T_j), L_m(T_m)\right)$$
$$= \mathrm{E}^B\left[(L_j(T_j) - \mathrm{E}^B\left(L_j(T_j)\right))\left(L_m(T_m) - \mathrm{E}^B(L_m(T_m))\right)\right]$$
$$= \mathrm{E}^B\left[(F_j(T_j) - F_j(0))(F_m(T_m) - F_m(0))\right].$$

Proof. It follows from (16.101) that

$$V(\epsilon) = \left(\prod_{i=0}^{n}(1 + \tau_i L_i(0))\right)p\left(L_0^\epsilon(T_0), \ldots, L_n^\epsilon(T_n)\right),$$

where we defined

$$p\left(y_0, \ldots, y_n\right) = \left[\prod_{i=0}^{n}\frac{1}{1 + \tau_i y_i}\right]y_n. \quad (16.108)$$

Obviously, (16.104) follows from (16.98). Moreover, with the help of (16.99),

$$\frac{dV(\epsilon)}{d\epsilon} = \left(\prod_{i=0}^{n}(1 + \tau_i L_i(0))\right)$$
$$\times \sum_{j=0}^{n}\frac{\partial}{\partial y_j}p\left(L_0^\epsilon(T_0), \ldots, L_n^\epsilon(T_n)\right)\left(L_j(T_j) - F_j(0)\right), \quad (16.109)$$

Since

$$\mathrm{E}^B\left(L_j(T_j) - F_j(0)\right) = 0,$$

the statement (16.105) is proved.

Differentiating (16.109) with respect to ϵ again, we obtain

$$\frac{d^2V(0)}{d\epsilon^2} = \left(\prod_{i=0}^{n}(1 + \tau_i L_i(0))\right)p\left(F_0(0), \ldots, F_n(0)\right)$$
$$\times \sum_{j,m}D_{j,m}\left(L_j(T_j) - F_j(0)\right)\left(L_m(T_m) - F_m(0)\right),$$

where we have denoted

$$D_{j,m} = \frac{\partial^2 p(F_0(0), \ldots, F_n(0))/\partial y_j\,\partial y_m}{p(F_0(0), \ldots, F_n(0))}.$$

The expression (16.107) for $D_{j,m}$'s follows by calculating $\partial^2 p/\partial y_j \partial y_m$ from (16.108). Simplifying, we obtain

$$\frac{d^2 V(0)}{d\epsilon^2} = V(0) \sum_{j,m} D_{j,m} \left(L_j(T_j) - F_j(0) \right) \left(L_m(T_m) - F_m(0) \right).$$

Taking the expected value of both sides leads to (16.105). Full details of the proof are in Piterbarg and Renedo [2006]. $\square$

Lemma 16.8.4 expresses quantities $V(0)$, $\mathrm{E}^B(\partial^2 V(0)/\partial\epsilon^2)$ in the series expansion (16.103) in terms of quantities that are either directly observable, such as futures rates, or computable, such as covariances of forward rates. Applying the results of Lemma 16.8.4 to the representation (16.102) and expansion (16.103), we obtain the following result.

Theorem 16.8.5. *For any n, $n = 0, \ldots, N-1$, an approximation to forward rate $L_n(0)$ is obtained from the futures $\{F_i(0)\}_{i=0}^n$ and forward rates for previous periods $\{L_i(0)\}_{i=0}^{n-1}$ by solving the following equation,*

$$L_n(0) = V(0) \left(1 + \frac{1}{2} \sum_{j,m=0}^n D_{j,m} \mathrm{Cov}^B \left(L_j(T_j), L_m(T_m) \right) \right), \qquad (16.110)$$

with $V(0)$ and $D_{j,m}$ given in Lemma 16.8.4.

Remark 16.8.6. Theorem 16.8.5 specifies an algorithm for solving for forward rates $L_n(0)$ sequentially for all n, using futures prices $\{F_j(0)\}$ as inputs.

Remark 16.8.7. The expression on the right-hand side of (16.110) will be simplified in the sections that follow. In many cases the rate to be determined from the expression, $L_n(0)$, will appear on the right-hand side of (16.110) as well. In this case, (16.110) should be treated not as an identity, but as an equation on $L_n(0)$. While this may seem to complicate the problem of finding $L_n(0)$, in reality the dependence of the right-hand side of (16.110) on $L_n(0)$ is typically mild, and the equation can be solved iteratively in just a few steps.

The formula (16.110) depends on covariances between various forward rates. By the definition of the covariance,

$$\mathrm{Cov}^B \left(L_j(T_j), L_m(T_m) \right)$$
$$= \left(\mathrm{Var}^B \left(L_j(T_j) \right) \mathrm{Var}^B \left(L_m(T_m) \right) \right)^{1/2} \mathrm{Corr}^B \left(L_j(T_j), L_m(T_m) \right), \quad (16.111)$$

where the variances and the correlation are computed in the spot measure. We proceed to discuss how to estimate the terms on the right-hand side of (16.111) from market observations.

16.8.5 Forward Rate Variances

The variance of $L_n(T_n)$ in the formula (16.111) is to be computed in the spot measure. As an approximation, we compute the variance instead in the measure in which $L_n(t)$ is a martingale,

$$\mathrm{Var}^B\left(L_n(T_n)\right) \approx \mathrm{Var}^{n+1}\left(L_n(T_n)\right), \quad n = 1, \ldots, N-1, \qquad (16.112)$$

where by definition

$$\mathrm{Var}^{n+1}\left(L_n(T_n)\right) \triangleq \mathrm{E}^{n+1}\left(L_n(T_n) - \mathrm{E}^{n+1}L_n(T_n)\right)^2.$$

The error of this approximation is typically small, and allows us to rewrite the formula for computing forward rates from futures rates as

$$L_n(0) \approx V(0)\left(1 + \frac{1}{2}\sum_{j,m=0}^{n} D_{j,m}\mathrm{Var}^{j+1}\left(L_j(T_j)\right)^{1/2}\right.$$

$$\left. \times\ \mathrm{Var}^{m+1}\left(L_m(T_m)\right)^{1/2}\mathrm{Corr}^B\left(L_j(T_j), L_m(T_m)\right)\right). \quad (16.113)$$

The market in options on ED futures contracts is very liquid — perhaps the most liquid market of options on interest rates. Applying the familiar replication method allows estimation of the variance of a forward rate in a model-independent way from observable prices of ED futures options (compare to (16.38)):

$$\mathrm{Var}^{n+1}\left(L_n(T_n)\right) = \mathrm{E}^{n+1}\left(\left(L_n(T_n) - \mathrm{E}^{n+1}L_n(T_n)\right)^2\right) \qquad (16.114)$$

$$= 2\int_{-\infty}^{L_n(0)} \mathrm{E}^{n+1}\left((K - L_n(T_n))^+\right) dK$$

$$+ 2\int_{L_n(0)}^{\infty} \mathrm{E}^{n+1}\left((L_n(T_n) - K)^+\right) dK.$$

In the formula (16.113), observable option prices are used directly for variance calculations and the forward rate $L_n(0)$, the rate to solve for, enters the right-hand side of the equation only as an integration limit in (16.114), so the comments of Remark 16.8.7 still apply. Equation (16.114) demonstrates explicitly that the ED convexity adjustment depends on prices of ED futures options at all strikes, i.e. on the volatility smile.

Equation (16.114) may not be easy to use in practice as only a discrete set of strikes is typically traded, and not all of them are very liquid. For these reasons, we may wish to capture the smile by a low-parametric vanilla model — or perhaps just a functional form, as in Section 16.1.5 — calibrated to liquid strikes. For instance, we could use a standard stochastic volatility model[16] for the rate $L_n(t)$,

---

[16]We use σ instead of our customary λ to avoid notational conflict with LM model volatilities used later on.

$$dL_n(t) = \sigma_n \left(b_n L_n(t) + (1 - b_n) L_n(0) \right) \sqrt{z(t)} \, dW(t), \qquad (16.115)$$
$$dz(t) = \theta \left(1 - z(t) \right) dt + \eta_n \sqrt{z(t)} \, dZ(t),$$

with correlation $\langle dW(t), dZ(t) \rangle = 0$. These SDEs are understood to be in the T_{n+1}-forward measure. The set of parameters (σ_n, b_n, η_n) defines the volatility smile in options on the rate $L_n(T_n)$ and is calibrated to market as described in Section 16.1.4. The variance of the Libor rate in the model (16.115) can easily be calculated:

Proposition 16.8.8. *Recall the definition (8.11) of $\Psi_{\bar{z}}(v, u; t)$. Then*

$$\mathrm{Var}^{n+1} \left(L_n(T_n) \right) = \frac{L_n(0)^2}{b_n^2} \left[\Psi_{\bar{z}} \left((\sigma_n b_n)^2, 0; T_n \right) - 1 \right].$$

Proof. Direct calculations. □

We again comment that the expression for the variance $\mathrm{Var}^{n+1}(L_n(T_n))$ involves $L_n(0)$, which here makes (16.113) a quadratic equation in $L_n(0)$. In addition, it should be noted that the implied values of parameters (σ_n, b_n, η_n) also, in principle, depend on the parameter $L_n(0)$ through the calibration process. However, the loss of accuracy is negligible if (σ_n, b_n, η_n) are calibrated with the "previous" value of the forward rate $L_n(0)$, i.e. the value before the update of the convexity adjustment.

16.8.6 Forward Rate Correlations

Once forward rate variances have been computed, the computation of (16.113) can be completed provided that we can establish the correlations $\mathrm{Corr}^B(L_j(T_j), L_m(T_m))$. There are many ways this can be done, but some type of model assumption will generally be required. For instance, if we have a calibrated LM model (16.59) lying around, we may calculate correlations in the Libor market model from the formula (14.35). Specifically, assuming $T_j \leq T_m$, we have for the model (16.59),

$$\mathrm{Corr}^B \left(L_j(T_j), L_m(T_m) \right) = \frac{\int_0^{T_j} \lambda_j(s)^\top \lambda_m(s) \, ds}{\left(\int_0^{T_j} \| \lambda_j(s) \|^2 \, ds \right)^{1/2} \left(\int_0^{T_m} \| \lambda_m(s) \|^2 \, ds \right)^{1/2}}.$$
$$(16.116)$$

Extraction of the model parameters $\{\lambda_j(\cdot)\}$ from the market is described in Chapter 14. Since correlations do not change often, this calibration can be performed "off-line", i.e. on an infrequent basis with the results reused as needed.

The approach above assumes that a full LM model has been implemented and calibrated. This may not always be practical, so let us consider a simplified method that retains the general spirit of a full LM model. First, we assume that the dynamics of Libor rates originate from a time-stationary Gaussian process of the mean-reverting type,

$$dL_i(t) = O(dt) + \sigma_0 e^{-\varkappa(T_i - t)} dW_i(t), \quad i = j, m,$$

where $W_j(t)$ and $W_m(t)$ are scalar Q^B-Brownian motions with correlation

$$\langle dW_j(t), dW_m(t) \rangle = q\,(T_j - t, T_m - t)\ dt,$$

for some function $q : \mathbb{R}^2 \to [-1, 1]$. Representative examples of the correlation function q are listed in Section 14.3.2. Ignoring drift terms and assuming $T_j \leq T_m$, we have

$$\mathrm{Cov}^B\,(L_j(T_j), L_m(T_m)) = \sigma_0^2 \int_0^{T_j} e^{-\varkappa(T_j - t)} e^{-\varkappa(T_m - t)} q\,(T_j - t, T_m - t)\ dt,$$

$$\mathrm{Var}^B\,(L_i(T_i)) = \frac{\sigma_0^2}{2\varkappa}\left(1 - e^{-2\varkappa T_i}\right), \quad i = j, m.$$

Hence

$$\mathrm{Corr}^B\,(L_j(T_j), L_m(T_m)) = 2\varkappa \frac{\int_0^{T_j} e^{-\varkappa(T_j - t)} e^{-\varkappa(T_m - t)} q\,(T_j - t, T_m - t)\ dt}{\sqrt{\left(1 - e^{-2\varkappa T_j}\right)\left(1 - e^{-2\varkappa T_m}\right)}}.$$
$$\tag{16.117}$$

A special case arises when $q(T_j - t, T_m - t) = \rho_{j,m}$ and does not depend on t, in which case (16.117) reduces to

$$\mathrm{Corr}^B\,(L_j(T_j), L_m(T_m))$$

$$= \rho_{j,m} \frac{\left(e^{2\varkappa T_j} - 1\right) e^{-\varkappa(T_j + T_m)}}{\sqrt{\left(1 - e^{-2\varkappa T_j}\right)\left(1 - e^{-2\varkappa T_m}\right)}} = \rho_{j,m}\sqrt{\frac{e^{2\varkappa T_j} - 1}{e^{2\varkappa T_m} - 1}}. \tag{16.118}$$

Formulas (16.117) and (16.118) do not require a full calibration of an LM model, only the estimation of basic forward rate correlations and a single mean reversion. We note that the role of the latter quantity is to govern the amount of de-correlation caused by the fact that L_j and L_m fix at different times. The mean reversion can potentially be estimated from market data (see Section 13.1.8), or could be a direct trader input.

16.8.7 The Formula

For convenience, let us now pull all previous results together into a single, easily referenced formula. First, let us summarize the notations. By $\{L_n(0)\}_{n=1}^{N-1}$ we denote the (unknown) sequence of forward rates for the tenor structure $\{T_n\}_{n=0}^{N}$, and by $\{F_n(0)\}_{n=1}^{N-1}$ we denote the (known) sequence of futures rates. For each n, $n = 1, \ldots, N - 1$, let the triple (σ_n, b_n, η_n) be the set of parameters of the model (16.115) implied from market prices of options on the rate $L_n(T_n)$ of different strikes.

Theorem 16.8.9. *For each* n, $n = 0, \ldots, N - 1$, *the forward rate* $L_n(0)$ *is obtained from the futures rates* $\{F_i(0)\}_{i=0}^n$ *and forward rates for previous periods* $\{L_i(0)\}_{i=0}^{n-1}$ *by solving the following equation,*

$$
L_n(0) = V(0) \left(1 + \frac{1}{2} \sum_{j,m=0}^n D_{j,m} \frac{L_j(0)L_m(0)}{b_j b_m} \right.
$$

$$
\left. \times \left(\Psi_{\bar{z}} \left((\sigma_j b_j)^2, 0; T_j \right) - 1 \right)^{1/2} \left(\Psi_{\bar{z}} \left((\sigma_m b_m)^2, 0; T_m \right) - 1 \right)^{1/2} c_{j,m} \right),
$$

(16.119)

with

$$
V(0) = \left[\prod_{i=0}^n \frac{1 + \tau_i L_i(0)}{1 + \tau_i F_i(0)} \right] F_n(0),
$$

$$
D_{j,m} = \left(-\frac{\tau_j}{1 + \tau_j F_j(0)} + \frac{1_{\{j=n\}}}{F_n(0)} \right) \left(-\frac{\tau_m}{1 + \tau_m F_m(0)} + \frac{1_{\{m=n\}}}{F_n(0)} \right)
$$

$$
+ 1_{\{j=m\}} \left(\frac{\tau_j^2}{(1 + \tau_j F_j(0))^2} - \frac{1_{\{j=n\}}}{F_n(0)^2} \right),
$$

and

$$
c_{j,m} = \mathrm{Corr}^B (L_j(T_j), L_m(T_m))
$$

as given by, for example, (16.116), (16.117) or (16.118). The function $\Psi_{\bar{z}}(v, u; t)$ *is defined by (8.11) and is available in closed form per Proposition 8.3.8.*

16.9 Convexity and Moment Explosions

When dealing with convexity products — Libor-in-arrears, CMS, ED futures, and so forth — we find (equations (16.36), (16.54) and (16.110)) that their values depend on the variance of some underlying rate, i.e. a *second moment* of the appropriate terminal distribution. Some care must be taken in the model setup to ensure that this second moment is well-behaved. For instance, if we have elected to work in a stochastic volatility setup, Proposition 8.3.10 and the discussion around it tell us that the second moment of the underlying in a stochastic volatility model may fail to exist, even for fairly reasonable values of model parameters. Should that occur, the theoretical convexity value will become infinite.

Intuitively, convexity value depends on prices of options at a continuum of strikes, from 0 to $+\infty$. In the market however, only prices of options over a finite range of strikes are observed, and infinite prices arise solely from the model-based extrapolation of the volatility smile beyond the observable range. We can control the smile extrapolation by altering the model specification in the manner discussed after Proposition 8.3.10. Alternatively, at least for pricing vanilla derivatives, we can control smile extrapolation explicitly, e.g. by restricting the domain of integration in the replication method, as we have already discussed in Sections 16.4 and 16.6.1. In particular, when evaluating the variance of some rate $S(t)$ we would replace

$$\text{Var}\left(S(T)\right) = 2 \int_{-\infty}^{S(0)} \text{E}\left(K - S(T)\right)^+ dK + 2 \int_{S(0)}^{\infty} \text{E}\left(S(T) - K\right)^+ dK$$

with

$$\text{Var}\left(S(T)\right) \approx 2 \int_{K_{\min}}^{S(0)} \text{E}\left(K - S(T)\right)^+ dK + 2 \int_{S(0)}^{K_{\max}} \text{E}\left(S(T) - K\right)^+ dK$$

for some $-\infty < K_{\min} \leq S(0) \leq K_{\max} < \infty$, as justified by the fact that only options with market-observable strikes can be used in hedging. The same idea can be applied to all convexity products evaluated by the replication method. The extra parameters, $K_{\min}$ and $K_{\max}$, could even be used to calibrate CMS (and other) convexity values to market, if these market values are available.

An alternative to outright cropping of the integration domain would be to institute an ad-hoc modification of the model-implied density of $S(T)$ for small/large values of the swap rate. This can be accomplished in many different ways, e.g. by artificially flattening out the implied volatility smile for large strikes[17]. The particulars of this scheme are left to the reader to explore, but let us note that any scheme to flatten out the implied volatility smile should, of course, be smooth as a function of strikes, to avoid generation of negative densities.

If we wish to control smile wing behavior in such a way that we are always certain that a valid density arises, we can also contemplate ad-hoc measures to modify model densities directly to prevent moment explosion (or to otherwise regularize the model). Assume that we have implemented a model where the density for $S(T)$ in its annuity measure Q^A is $\psi(s)$. A call payout (as needed for a payer swaption) is therefore valued as

$$\text{E}^A\left((S(T) - K)^+\right) = \int_K^{\infty} (s - K)\,\psi(s)\,ds.$$

Let us introduce some user-specified strikes $K_{\min}$ and $K_{\max}$, and rewrite the expectation as (assuming $K_{\min} < K_{\max}$)

---

[17] We note in passing that this technique can also be used to control errors in volatility expansion formulas, such as the one used in SABR, which may yield negative densities in the wings unless some kind of regularization is performed.

$$\mathrm{E}^A\left((S(T) - K)^+\right)$$

$$= \int_{K_{\min}}^{K_{\max}} (s - K)^+ \psi(s)\,ds$$

$$+ \int_{-\infty}^{K_{\min}} (s - K)^+ \psi(s)\,ds + \int_{K_{\max}}^{\infty} (s - K)^+ \psi(s)\,ds$$

$$= \int_{K_{\min}}^{K_{\max}} (s - K)^+ \psi(s)\,ds$$

$$+ Q^A\left(S(T) \le K_{\min}\right) \int_{-\infty}^{K_{\min}} (s - K)^+ \psi\left(s\,|\,S(T) \le K_{\min}\right)\,ds$$

$$+ Q^A\left(S(T) > K_{\max}\right) \int_{K_{\max}}^{\infty} (s - K)^+ \psi\left(s\,|\,S(T) > K_{\max}\right)\,ds.$$

Here we have introduced conditional densities

$$\psi\left(s\,|\,S(T) \le K_{\min}\right)\,ds = Q^A\left(S(T) \in [s, s+ds]\,|\,S(T) \le K_{\min}\right),$$
$$\psi\left(s\,|\,S(T) > K_{\max}\right)\,ds = Q^A\left(S(T) \in [s, s+ds]\,|\,S(T) > K_{\max}\right).$$

Suppose now that we wish to replace the density ψ with another density in the tails, i.e. for values of $S(T)$ outside of the range $[K_{\min}, K_{\max}]$. To do this, let us write

$$\mathrm{E}^A\left((S(T) - K)^+\right)$$

$$= \int_{K_{\min}}^{K_{\max}} (s - K)^+ \psi(s)\,ds$$

$$+ Q^A\left(S(T) \le K_{\min}\right) \int_{-\infty}^{K_{\min}} (s - K)^+ \psi_{\min}\left(s\,|\,S(T) \le K_{\min}\right)\,ds$$

$$+ Q^A\left(S(T) > K_{\max}\right) \int_{K_{\max}}^{\infty} (s - K)^+ \psi_{\max}\left(s\,|\,S(T) > K_{\max}\right)\,ds,$$

where we have introduced two new conditional densities $\psi_{\min}$ and $\psi_{\max}$. Consider now some $K \in [K_{\min}, K_{\max}]$ and let us require that $\mathrm{E}^A((S(T) - K)^+)$ is unchanged by the introduction of $\psi_{\min}$ and $\psi_{\max}$; i.e., we insist that the smile in the strike region $[K_{\min}, K_{\max}]$ is preserved after manipulation of the tail densities. Additionally, we of course should demand that $\mathrm{E}^A(S(T)) = S(0)$. The first of these restrictions requires that

$$Q^A\left(S(T) > K_{\max}\right) \int_{K_{\max}}^{\infty} (s - K)\,\psi_{\max}\left(s\,|\,S(T) > K_{\max}\right)\,ds$$

$$= \int_{K_{\max}}^{\infty} (s - K)\psi(s)\,ds, \quad K \in [K_{\min}, K_{\max}].$$

As

$$\int_{K_{max}}^{\infty} \psi_{max}\left(s|\, S(T) > K_{max}\right) ds = 1,$$

it follows that

$$Q^A\left(S(T) > K_{max}\right)\left(\int_{K_{max}}^{\infty} s\psi_{max}\left(s|\, S(T) > K_{max}\right) ds - K\right)$$

$$= \int_{K_{max}}^{\infty} s\psi(s)\, ds - KQ^A\left(S(T) > K_{max}\right)$$

or

$$\int_{K_{max}}^{\infty} s\psi_{max}\left(s|\, S(T) > K_{max}\right) ds = \frac{\int_{K_{max}}^{\infty} s\psi(s)\, ds}{Q^A\left(S(T) > K_{max}\right)}. \qquad (16.120)$$

Insisting also that $E^A(S(T))$ is unchanged by the introduction of ψ_{min} and ψ_{max} then leads to

$$\int_{-\infty}^{K_{min}} s\psi_{min}\left(s|\, S(T) \le K_{min}\right) ds = \frac{\int_{-\infty}^{K_{min}} s\psi(s)\, ds}{Q^A\left(S(T) \le K_{min}\right)}. \qquad (16.121)$$

The right-hand sides of (16.120)–(16.121) can be computed from the given model, yielding two consistency requirements any density modification in the tails must satisfy. If we, say, postulate that the conditional densities ψ_{min} and ψ_{max} originate from Black models with unknown constant volatilities σ_{min} and σ_{max}, respectively, the consistency requirements will allow us to back out σ_{min} and σ_{max}. Indeed, notice that in such a setup

$$\psi_{min}\left(s|\, S(T) \le K_{min}\right)$$

$$= \frac{1}{S_0 \Phi\left(d(K_{min})\right)\sigma_{min}\sqrt{2\pi T}} \exp\left(-\frac{1}{2}d(s)^2\right), \quad s \le K_{min},$$

where

$$d(x) = \frac{\ln\left(x/S(0)\right) + \frac{1}{2}\sigma_{min}^2 T}{\sigma_{min}\sqrt{T}},$$

which can be used to show that

$$\int_{-\infty}^{K_{min}} s\psi_{min}\left(s|\, S(T) \le K_{min}\right) ds = S(0)\frac{\Phi\left(d(K_{min}) - \sigma_{min}\sqrt{T}\right)}{\Phi\left(d(K_{min})\right)}.$$

Similarly,

$$\int_{K_{max}}^{\infty} s\psi_{max}\left(s|\, S(T) > K_{max}\right) ds = S(0)\frac{\Phi\left(-d(K_{max}) + \sigma_{max}\sqrt{T}\right)}{\Phi\left(-d(K_{max})\right)}.$$

which allows us to uncover σ_{min} and σ_{max} from (16.120)–(16.121). We point out that the scheme above can easily be modified to handle more complicated density tails, e.g. the ones from CEV or displaced diffusion models.

17

Multi-Rate Vanilla Derivatives

After our analysis of single-rate vanilla derivatives in the previous chapter, we now proceed to consider European-type payoffs that are linked to more than one swap or Libor rate. The most important member of this class is the CMS spread, but securities such as floating range accruals and floating range accruals on a CMS spread are also popular in the market.

Valuation of these *multi-rate vanilla derivatives* in a dynamic term structure model presents no conceptual difficulties. However, given the (relatively) high traded volume in derivatives of this type, application of a full term structure model may not be accurate or fast enough (or both) in practice. Hence, in this chapter we look for extensions of the vanilla models of Chapter 16 that allow us to price derivatives linked to multiple rates in an efficient manner. As in Chapter 16, convexity adjustments are an inherent part of valuation, but require only straightforward extensions of the methods developed for the single-rate case. New challenges do arise in the multi-rate setting, however, as we now face the need to specify and control the dependence structure between the rates involved. We spend most of the chapter discussing this issue in detail.

17.1 Introduction to Multi-Rate Vanilla Derivatives

We define a multi-rate vanilla derivative to be a derivative security with a European-type payoff linked to more than one market rate. Given a payment date T_p, a collection of swap or Libor rates $S_1(\cdot), \ldots, S_d(\cdot)$, a collection of fixing dates $t_1, \ldots, t_d$, and a d-argument function $f(s_1, \ldots, s_d)$, a multi-rate derivative is defined by its payoff

$$V(T_p) = f(S_1(t_1), \ldots, S_d(t_d)) \text{ paid at time } T_p. \qquad (17.1)$$

Of course, we require $t_i \leq T_p$ for all $i = 1, \ldots, d$. Our main focus in this chapter is on the case when all t_i's are the same: $t_i = T$ for some $T \leq T_p$

and all $i = 1, \ldots, d$. That said, we do develop methods to deal with fixing dates that are not all exactly equal — a case important for floating range accruals, for example — but we still insist here on the fixing dates being "not too far" from each other. If fixing dates are wide apart, valuation is typically best handled in a dynamic term structure model (e.g., the LM model).

As was the case for their single-rate counterparts, multi-rate vanilla derivative payoffs (17.1) are rarely traded themselves; rather, they constitute building blocks for traded securities that are (additive) collections of multi-rate cash flows. Examples include the multi-rate exotic swaps (Section 5.13.3), e.g. swaps paying CMS spread or digital CMS spreads, and various flavors of curve caps. Also popular are the range accruals, see Section 5.13.4; the following variations are (among many others) covered by the techniques in this chapter:

1. Accrual is based on a single market rate, but payment is linked to a different market rate (floating range accruals).
2. The payment rate is either fixed or a function of a collection of market rates, and the observation rate is a difference of two market rates (CMS spread range accruals).
3. Dual range accruals.
4. Curve cap range accruals.

As mentioned above, we focus our attention on developing "vanilla" models for multi-rate derivatives, i.e. models that steer clear of defining dynamics for the whole yield curve and instead merely aim to specify, in the most direct way possible, the distribution of the collection of swap rates $S_1(t_1), \ldots, S_d(t_d)$. The efficacy of this approach stems from the fact that the value of payoffs in the class (17.1) have limited, if any, dependence on the actual continuous-time dynamics of the yield curve, since only the joint distribution of terminal rate observations will enter valuation formulas. With this in mind, we may distill the essence of various vanilla-type methods to the following steps.

1. One-dimensional (terminal) distributions of all relevant market rates are extracted from market prices of swaptions.
2. All one-dimensional distributions are brought under the same equivalent martingale measure.
3. A dependence structure is imposed on the vector of market rates, while ensuring that market-implied marginal one-dimensional distributions are preserved.
4. Parameters used in the specification of the dependence structure are estimated historically or, if sufficient market information is available, implied from the prices of certain instruments.
5. A suitable numerical method is applied to integrate the payoff against a specified multi-dimensional distribution.

We start our discussion of multi-rate vanilla derivatives valuation with the first two items of this program.

17.2 Marginal Distributions and Reference Measure

The value of a cash flow with the payoff (17.1) at time $t = 0$ is given by

$$V(0) = \mathrm{E}\left(\beta(T_p)^{-1} f\left(S_1(t_1), \ldots, S_d(t_d)\right)\right), \qquad (17.2)$$

where E is the expectation operator for the risk-neutral measure Q. As discussed in Chapter 16, the distribution of each swap rate $S_i(t_i)$ under its annuity measure Q^{A_i} (a measure for which the annuity $A_i(t)$ linked to the rate $S_i(t)$ is a numeraire) can be deduced from market prices of swaptions across strikes. We emphasize that the measures Q^{A_i} are different for different swap rates, and there will generally exist no measure under which all S_i's are martingales. In principle, one can choose any annuity measure and proceed to derive distributions of all rates under that measure — technical tools for this can be developed relatively easily by extending the results of Section 16.6.9. This approach suffers from a certain arbitrariness, and it is typically both more natural and more convenient to work with the T_p-forward measure. Moreover, we already know how to translate a swap rate distribution under the annuity measure into its distribution under the T_p-forward measure, see Section 16.6.9.

Changing to T_p-forward measure in (17.2), we obtain

$$V(0) = P(0, T_p) \, \mathrm{E}^{T_p}\left(f\left(S_1(t_1), \ldots, S_d(t_d)\right)\right). \qquad (17.3)$$

As we recall from Section 16.6.9, the distribution of $S_i(t_i)$ under Q^{T_p} is linked to the (market-implied) distribution of $S_i(t_i)$ under Q^{A_i} via the density relationship

$$Q^{T_p}\left(S_i(t_i) \in ds\right) = \frac{A_i(0)}{P(0, T_p)} \alpha_i(s) Q^{A_i}\left(S_i(t_i) \in ds\right), \qquad (17.4)$$

where the annuity mapping function $\alpha_i(s)$ is defined as

$$\alpha_i(s) = \mathrm{E}^{A_i}\left(\left.\frac{P(t_i, T_p)}{A_i(t_i)}\right| S_i(t_i) = s\right), \qquad i = 1, \ldots, d. \qquad (17.5)$$

In deriving the annuity mapping functions we can follow Chapter 16 and impose a functional relationship between $P(t_i, T_p)/A_i(t_i)$ and $S_i(t_i)$. As we may do so independently for each $i = 1, \ldots, d$, the application of the relevant method(s) from Chapter 16 is straightforward.

The formulas (17.4)–(17.5) are exact as written, but we would virtually always apply an approximation for $\alpha_i(s)$ in (17.5). As far as such approximations are concerned, we should note that the techniques discussed in Chapter

16 will be associated with a certain degree of inconsistency in the multi-rate context. For instance, in the common case where $t_i = T$ for some $T \leq T_p$ and all $i = 1, \ldots, d$, we see that $P(T, T_p)/A_i(T)$ for a given i is a function of all swap rates $S_1(T), \ldots, S_d(T)$. Therefore, the calculation of $\alpha_i(s)$ should in principle involve the dependence structure of all swap rates in the payoff. Although we shall introduce such a dependence structure between all swap rates (see later in Section 17.3) to calculate the value of the derivative cash flow, we typically do *not* use this dependence structure when estimating $\alpha_i(s)$ by (17.5). Instead, we would normally content ourselves with the simpler methods of Chapter 16, such as the linear TSR model. While acknowledging the inconsistency we, among others, realize considerable practical advantages of separating the specification of measure changes via (17.4)–(17.5) from that of the dependence structure; consequently, we adopt this approach throughout this chapter.

With marginal distributions of each $S_i(t_i)$ in (17.3) specified by (17.4), the value $V(0)$ will be strongly sensitive to the dependence structure imposed by the model on the random variables $S_1(t_1), \ldots, S_d(t_d)$. Specifying and controlling such dependence is at the heart of the problem of valuing multi-rate vanilla cash flows, with the so-called *copula method* being a popular choice for the job. We introduce this method next.

17.3 Dependence Structure via Copulas

17.3.1 Introduction to Gaussian Copula Method

The copula approach, popularized in financial applications by its widespread usage in credit derivatives modeling, is a method of constructing a joint distribution of random variables consistently with pre-specified one-dimensional marginal distributions. While a thorough treatment of the subject is well beyond the scope of this book — the reader is referred to Nelsen [2006] for that — we proceed to present salient points in the next few sections.

To warm up, let us start with the so-called *Gaussian copula method*, the most common, and easily understood, type of the copula methods in use. We have already seen a particular application of this method in Section 16.7. Let us assume that one-dimensional cumulative distribution functions (CDFs) $\Psi_1(\cdot), \ldots, \Psi_d(\cdot)$ have been given, and we are tasked with constructing a multi-dimensional random vector $(X_1, \ldots, X_d)$ with a measure of control over the dependence of the random variables X_i, but constrained so that each variable X_i has CDF $\Psi_i(\cdot)$, $i = 1, \ldots, d$. The Gaussian copula method accomplishes this by specifying

$$X_i = \Psi_i^{-1}\left(\Phi\left(Z_i\right)\right), \quad i = 1, \ldots, d, \tag{17.6}$$

where $\Phi(\cdot)$ is the standard one-dimensional Gaussian CDF and $(Z_1, \ldots, Z_d)$ is a multi-dimensional, normalized[1] Gaussian vector with the correlation matrix R. Clearly the CDF of each X_i thus defined is $\Psi_i(\cdot)$ (see Section 3.1.1.1), and the correlation matrix R provides a way to control the co-dependence structure in the vector.

Let us denote the joint CDF of $(X_1, \ldots, X_d)$ as constructed above by $\Psi_{\text{gauss}}(x_1, \ldots, x_d)$, and the joint d-dimensional (Gaussian) CDF of $(Z_1, \ldots, Z_d)$ by $\Phi_d(z_1, \ldots, z_d; R)$. Then it follows from (17.6) that

$$\Psi_{\text{gauss}}(x_1, \ldots, x_d) = \Phi_d\left(\Phi^{-1}\left(\Psi_1(x_1)\right), \ldots, \Phi^{-1}\left(\Psi_d(x_d)\right); R\right). \quad (17.7)$$

From the joint CDF of $(X_1, \ldots, X_d)$ we easily obtain the joint probability density function (PDF) as

$$\psi_{\text{gauss}}(x_1, \ldots, x_d) \quad (17.8)$$

$$= \frac{\partial^d}{\partial x_1 \ldots \partial x_d} \Psi_{\text{gauss}}(x_1, \ldots, x_d) \quad (17.9)$$

$$= \frac{\partial^d}{\partial z_1 \ldots \partial z_d} \Phi_d(z_1, \ldots, z_d; R)\Big|_{z_i = \Phi^{-1}(\Psi_i(x_i)), \forall i} \times \prod_{i=1}^{d} \frac{\Psi_i'(x_i)}{\Phi'\left(\Phi^{-1}\left(\Psi_i(x_i)\right)\right)}$$

$$= \phi_d\left(\Phi^{-1}\left(\Psi_1(x_1)\right), \ldots, \Phi^{-1}\left(\Psi_d(x_d)\right); R\right) \times \prod_{i=1}^{d} \frac{\psi_i(x_i)}{\phi\left(\Phi^{-1}\left(\Psi_i(x_i)\right)\right)},$$

where $\phi(z)$ and $\phi_d(z_1, \ldots, z_d; R)$ are the one- and d-dimensional Gaussian PDFs, respectively, and $\psi_i(x_i)$ is the one-dimensional PDF of X_i, $i = 1, \ldots, d$.

With the joint PDF $\psi_{\text{gauss}}(x_1, \ldots, x_d)$ of $(X_1, \ldots, X_d)$ available, the undiscounted[2] value of a derivative with a payoff $f(X_1, \ldots, X_d)$ is given by the multi-dimensional integral

$$V = \int \cdots \int f(x_1, \ldots, x_d)\, \psi_{\text{gauss}}(x_1, \ldots, x_d)\, dx_1 \ldots dx_d. \quad (17.10)$$

Another, sometimes more useful, expression may be obtained from (17.6), as the derivative security with the payoff $f(X_1, \ldots, X_d)$ can also be considered to have the payoff $f(\Psi_1^{-1}(\Phi(Z_1)), \ldots, \Psi_d^{-1}(\Phi(Z_d)))$. Therefore its value is given by

$$V = \int \cdots \int f\left(\Psi_1^{-1}\left(\Phi(z_1)\right), \ldots, \Psi_d^{-1}\left(\Phi(z_d)\right)\right) \phi_d(z_1, \ldots, z_d; R)\, dz_1 \ldots dz_d. \quad (17.11)$$

---

[1] I.e. $\text{E}(Z_i) = 0$, $\text{Var}(Z_i) = 1$, $i = 1, \ldots, d$.

[2] In this chapter we will often consider undiscounted values of derivatives, as we mostly work with cash flows that pay at a single payment time T_p. As it should always be clear whether discounting is applied or not, we do not introduce new notation.

The Gaussian copula method can be implemented easily, is well-understood, and widely used. Still, it suffers from its share of problems, chief among them being its limited control over the shape of the joint distribution, a point that we shall address in more detail later in this chapter.

17.3.2 General Copulas

To develop co-dependence structures more general than those implied by the Gaussian copula method, consider first rewriting (17.7) as

$$\Psi_{\text{gauss}}(x_1, \ldots, x_d) = C_{\text{gauss}}(\Psi_1(x_1), \ldots, \Psi_d(x_d); R),\qquad(17.12)$$

where

$$C_{\text{gauss}}(u_1, \ldots, u_d; R) \triangleq \Phi_d\left(\Phi^{-1}(u_1), \ldots, \Phi^{-1}(u_d); R\right).$$

The function $C_{\text{gauss}}(u_1, \ldots, u_d; R)$ is easily seen to define a specific multi-dimensional distribution function (as defined in, for example, Nelsen [2006]) for a vector of d random variables with marginal distributions that are all *uniform* on $[0, 1]$. It turns out that this concept can be generalized nicely.

Definition 17.3.1. *Consider a function $C : [0,1]^d \to [0,1]$. $C(u_1, \ldots, u_d)$ is said to be a d-dimensional copula function if it defines a valid joint distribution function for a d-dimensional vector of random variables, with each variable being uniformly distributed on $[0,1]$.*

The requirement that a copula defines a joint distribution function for a vector of uniformly distributed random variables puts a number of strong constraints on the form of the function C. For instance, it is clear that C must be increasing in its arguments. Also, if the i-th argument of C is 0, C itself must return 0, irrespective of the values of the remaining $d - 1$ arguments. Further, if all but the i-th argument of C are set to 1, C must return the i-th argument itself. The last two simple relations (which we invite the reader to verify) can be summarized as follows:

$$C(u_1, u_2, \ldots, u_{i-1}, 0, u_{i+1}, \ldots, u_d) = 0,\qquad(17.13)$$
$$C(1, 1, \ldots, 1, u_i, 1, 1, \ldots, 1) = u_i.\qquad(17.14)$$

To give a few introductory examples, notice that a particularly simple copula arises if all d uniform variables underlying the copula are independent. The resulting *independence copula* C_{ID} is

$$C_{\text{ID}} = \prod_{i=1}^{d} u_i.\qquad(17.15)$$

To state the copula C_{D} that defines *perfect dependence*, introduce d uniform random variables $U_1, U_2, \ldots, U_d$ and set $U_1 = U_2 = \ldots = U_d$. In this case we get (with P being a probability measure)

$$C_{\mathrm{D}}\left(u_1, u_2, \ldots, u_d\right) = \mathrm{P}\left(U_1 \leq u_1, U_2 \leq u_2, \ldots, U_d \leq u_d\right)$$
$$= \mathrm{P}\left(U_1 \leq u_1, U_1 \leq u_2, \ldots, U_1 \leq u_d\right)$$
$$= \mathrm{P}\left(U_1 \leq \min_{i=1,\ldots,d} u_i\right)$$
$$= \min_{i=1,\ldots,d} u_i. \tag{17.16}$$

For obvious reasons, the *perfect anti-dependence* copula C_{AD} can only be stated for the case $d = 2$. In this case, we write $U_2 = 1 - U_1$ for two uniform random variables U_1, U_2, and get

$$C_{\mathrm{AD}}\left(u_1, u_2\right) = \mathrm{P}\left(U_1 \leq u_1, U_2 \leq u_2\right)$$
$$= \mathrm{P}\left(U_1 \leq u_1, 1 - U_1 \leq u_2\right)$$
$$= \mathrm{P}\left(1 - u_2 \leq U_1 \leq u_1\right)$$
$$= \left(u_1 + u_2 - 1\right)^+. \tag{17.17}$$

The anti-dependence copula in (17.17) cannot be extended to $d > 2$, but we can still define a function

$$G_{\mathrm{AD}}\left(u_1, u_2, \ldots, u_d\right) = \left(\sum_{i=1}^{d} u_i + 1 - d\right)^+, \tag{17.18}$$

such that $G_{\mathrm{AD}} = C_{\mathrm{AD}}$ for the case $d = 2$. Although G_{AD} defined this way is not itself a copula for $d \geq 3$, it turns out that this function can be used to bound any valid copula function. Specifically, one can prove the following result.

Theorem 17.3.2. *Any valid d-dimensional copula function C must satisfy the* Frechet *bounds*

$$G_{\mathrm{AD}}\left(u_1, u_2, \ldots, u_d\right) \leq C\left(u_1, u_2, \ldots, u_d\right) \leq C_{\mathrm{D}}\left(u_1, u_2, \ldots, u_d\right).$$

The real strength of the copula function concept originates with the fact that it allows us to separate co-dependence information from marginal distributions. Specifically, given a copula function $C(u_1, \ldots, u_d)$ and a collection of marginal CDFs $\Psi_1(x_1), \ldots, \Psi_d(x_d)$, we can construct a d-dimensional joint distribution function $\Psi_C(x_1, \ldots, x_d)$ with marginals $\Psi_1(x_1), \ldots, \Psi_d(x_d)$ by a formula similar to (17.12),

$$\Psi_C\left(x_1, \ldots, x_d\right) = C\left(\Psi_1(x_1), \ldots, \Psi_d(x_d)\right). \tag{17.19}$$

The simple proof of the fact that $\Psi_C(x_1, \ldots, x_d)$ defined by (17.19) is a true d-dimensional distribution function is left to the reader. By the so-called *Sklar's theorem*, the opposite is also true: for any d-dimensional distribution function there exists a copula function such that the joint distribution function can be represented in the form (17.19).[3]

The joint PDF ψ_C associated with the CDF Ψ_C in (17.19) is given by (compare to (17.9))

$$\psi_C (x_1, \ldots, x_d) = \frac{\partial^d}{\partial x_1 \ldots \partial x_d} \Psi_C (x_1, \ldots, x_d)$$

$$= c (\Psi_1(x_1), \ldots, \Psi_d(x_d)) \prod_{i=1}^{d} \psi_i(x_i), \qquad (17.20)$$

where

$$c (u_1, \ldots, u_d) = \frac{\partial^d}{\partial u_1 \ldots \partial u_d} C (u_1, \ldots, u_d), \qquad (17.21)$$

is the *copula density* and $\psi_i(\cdot)$'s are the marginal PDFs. For the Gaussian copula, the copula density is given by

$$c_{\text{gauss}} (u_1, \ldots, u_d; R) = \frac{\phi_d \left(\Phi^{-1} (u_1), \ldots, \Phi^{-1} (u_d) ; R \right)}{\prod_{i=1}^{d} \phi (\Phi^{-1} (u_i))}. \qquad (17.22)$$

17.3.3 Archimedean Copulas

With technical background material now out of the way, let us proceed to examine some concrete examples of copula functions, beyond the Gaussian class. One choice that is quite popular in the copula literature is the so-called *Archimedean* class of copulas. This class requires specification of a *generator function* $\omega\colon [0, 1] \to \mathbb{R}$ satisfying

$$\lim_{x \to 0} \omega(x) = +\infty, \quad \omega(1) = 0, \quad \omega'(x) < 0, \quad \omega''(x) > 0.$$

From a generator function, a corresponding Archimedean copula $C_{\text{arch}}(u_1, \ldots, u_d; \omega)$ can be defined by the relation

$$C_{\text{arch}} (u_1, \ldots, u_d; \omega) = \omega^{-1} \left(\sum_{i=1}^{d} \omega (u_i) \right).$$

It is a trivial exercise to show that $C_{\text{arch}}(u_1, \ldots, u_d; \omega)$ is indeed a copula function.

The generator function is often indexed with a parameter, specifying a parametric family of Archimedean copulas. Of particular note are the following two families:

---

[3]If the marginal distribution functions $\Psi_1, \Psi_2, \ldots, \Psi_d$ are all continuous, then the copula function is *unique*.

- *Clayton copula:*

$$\omega_{\text{clayton}}\left(u;\theta\right) = u^{-\theta} - 1, \quad \theta > 0.$$

- *Gumbel copula:*

$$\omega_{\text{gumbel}}\left(u;\theta\right) = \left(-\ln u\right)^{\theta}, \quad \theta > 0.$$

The corresponding copulas are given by

$$C_{\text{clayton}}\left(u_1,\ldots,u_d;\theta\right) = \left(\sum_{i=1}^{d} u_i^{-\theta} - d + 1\right)^{-1/\theta},$$

$$C_{\text{gumbel}}\left(u_1,\ldots,u_d;\theta\right) = \exp\left(-\left(\sum_{i=1}^{d}\left(-\ln u_i\right)^{\theta}\right)^{1/\theta}\right).$$

In the special case of $\theta = 1$, the Gumbel copula becomes

$$C_{\text{gumbel}}\left(u_1,\ldots,u_d;1\right) = \prod_{i=1}^{d} u_i,$$

which is the independence copula C_{ID} introduced earlier.

A quick graphing exercise shows that as θ is raised, both the Clayton and Gumbel copulas assign increasing probability mass around the point $(0,\ldots,0)$; in terms of the joint distribution of the market rates, this corresponds to an increase in the probability of a joint down-move of the rates.

17.3.4 Making Copulas from Other Copulas

With the Archimedean copulas introduced in the previous section, we can use the parameter θ of both the Clayton or Gumbel copula to control the probability of a joint down-move of interest rates, but we have no direct control over other moves of interest such as a joint *up*-move of the rates. This is easily fixed, however, by an application of the following general result.

Lemma 17.3.3. *If $C(u_1,\ldots,u_d)$ is a copula, then the function obtained by reflecting C in dimension i,*

$$\overline{C}(u_1,\ldots,u_i,\ldots,u_d;\{i\}) \triangleq C(u_1,\ldots,1,\ldots,u_d) - C(u_1,\ldots,1-u_i,\ldots,u_d), \tag{17.23}$$

is also a copula for any $i = 1,\ldots,d$. The density of the reflected copula is given by

$$\overline{c}(u_1,\ldots,u_i,\ldots,u_d;\{i\}) = c(u_1,\ldots,1-u_i,\ldots,u_d).$$

Proof. Trivial consequence of the fact that if U follows a uniform $[0, 1]$ distribution, then so does $1 - U$. □

By repeated application of the lemma, it is easy to see that it generalizes to multiple indices. Specifically, if we denote by $\overline{C}(\ldots; \{i_1, \ldots, i_M\})$ a function obtained by repeating the mapping (17.23) for all i_m, $m = 1, \ldots, M$, this is still a copula. Focusing on the two-dimensional case $d = 2$ and choosing the Clayton copula for concreteness, we have

$$\overline{C}_{\text{clayton}}(u_1, u_2; \theta; \{1, 2\}) = C_{\text{clayton}}(1, 1; \theta) - C_{\text{clayton}}(1, 1 - u_2; \theta)$$
$$- C_{\text{clayton}}(1 - u_1, 1; \theta) + C_{\text{clayton}}(1 - u_1, 1 - u_2; \theta),$$

and now the parameter θ controls the probability of a joint *up*-move in the two market rates. In the copula $\overline{C}_{\text{clayton}}(u_1, u_2; \theta; \{1\})$ the parameter θ controls the joint probability of an up-move of the first rate and a down-move in the second rate, and in the copula $\overline{C}_{\text{clayton}}(u_1, u_2; \theta; \{2\})$, the parameter θ controls the joint probability of a down-move of the first rate and an up-move in the second rate.

Another way of creating copulas uses the observation that a convex combination of copulas is also a copula.

Lemma 17.3.4. *Let us denote* $u = (u_1, \ldots, u_d)^\top$, *and let there be given* M *different d-dimensional copulas* $C_1(u), \ldots, C_M(u)$, *as well as a collection of non-negative weights* $w_1, \ldots, w_M$ *such that* $\sum_{m=1}^{M} w_m = 1$. *The linear combination, or* mixture,

$$C_{\text{mix}}(u) = \sum_{m=1}^{M} w_m C_m(u),$$

is also a copula.

Proof. Trivial. □

One can interpret mixture copulas as representations of the idea that different dependence structures of the random variables are realized in different states of the world, with these states having probabilities w_m, $m = 1, \ldots, M$. To give a simple example, consider a Gaussian copula setting where there are two states of the correlation matrix: R_{hi} ("excited state") and R_{normal} ("normal state"). Assuming that the probabilities of these states are w_{normal} and $w_{\text{hi}} = 1 - w_{\text{normal}}$, respectively, we would define

$$C_{\text{multigauss}}(u_1, \ldots, u_d; w_{\text{normal}}, w_{\text{hi}}, R_{\text{normal}}, R_{\text{hi}})$$
$$= w_{\text{normal}} C_{\text{gauss}}(u_1, \ldots, u_d; R_{\text{normal}}) + w_{\text{hi}} C_{\text{gauss}}(u_1, \ldots, u_d; R_{\text{hi}}). \tag{17.24}$$

If, say, R_{hi} is supposed to reflect an unlikely crash state, we could set $R_{\text{hi}} \gg R_{\text{normal}}$ (in, e.g., an element-wise sense) and $w_{\text{hi}} \ll w_{\text{normal}}$.

To motivate the next method for constructing copulas, we note that for any function $q(u)$ we have a trivial identity

$$u = q(u) \times \frac{u}{q(u)}.$$

Hence, if we have two (2-dimensional) copulas $C_1(u,v)$ and $C_2(u,v)$ and functions $q(u)$, $p(v)$ such that $q(1) = p(1) = 1$, then the function

$$C(u,v) = C_1(q(u),p(v)) C_2\left(\frac{u}{q(u)}, \frac{v}{p(v)}\right)$$

would satisfy requirement (17.14), as

$$C(u,1) = C_1(q(u),1) C_2\left(\frac{u}{q(u)},1\right) = q(u)\frac{u}{q(u)} = u,$$

$$C(1,v) = C_1(1,p(v)) C_2\left(1,\frac{v}{p(v)}\right) = p(v)\frac{v}{p(v)} = v.$$

While we have not demonstrated that C is a copula, it turns out that this is the case if certain conditions are imposed on q (and p). The relevant result (suitably extended to dimension d) for these so-called *product copulas* can be found in Liebscher [2008]:

Theorem 17.3.5. *Let $C_1, \ldots, C_M$ be d-dimensional copulas, and let $q_{m,i}$: $[0,1] \to [0,1]$, $m = 1, \ldots, M$, $i = 1, \ldots, d$, be functions that are either strictly increasing or identically equal to 1. Suppose that $\prod_{m=1}^{M} q_{m,i}(u) = u$ for $u \in [0,1]$, $i = 1, \ldots, d$, and $\lim_{u \to 0+} q_{m,i}(u) = q_{m,i}(0)$. Then*

$$C(u_1, \ldots, u_d) = \prod_{m=1}^{M} C_m(q_{m,1}(u_1), \ldots, q_{m,d}(u_d))$$

is a copula.

In this theorem, consider taking $M = 2$, $C_1 = \prod_{i=1}^{d} u_i$ (the independence copula), $C_2 = C_{\text{gauss}}(u_1, \ldots, u_d; R)$ (Gaussian copula with correlation matrix R), $q_{1,i}(u) = u^{1-\theta_i}$, and $q_{2,i}(u) = u^{\theta_i}$ for $\theta_i \in [0,1]$, $i = 1, \ldots, d$. We then obtain a copula that we find particularly useful for multi-rate derivatives:

Corollary 17.3.6. *Let R be a $d \times d$ correlation matrix and $\theta = (\theta_1, \ldots, \theta_d)^\top \in [0,1]^d$ a d-dimensional vector of parameters. Then the* power Gaussian *function*

$$C_{\text{PG}}(u_1, \ldots, u_d; R, \theta) \triangleq \left(\prod_{i=1}^{d} u_i^{1-\theta_i}\right) C_{\text{gauss}}\left(u_1^{\theta_1}, \ldots, u_d^{\theta_d}; R\right) \qquad (17.25)$$

is a copula.

Section 17.4.3 demonstrates that the power Gaussian copula provides a parsimonious, yet flexible way of specifying dependence structure for CMS spread options.

17.4 Copula Methods for CMS Spread Options

We recall from Section 5.13.3 that the payoff of a spread option is given by

$$(S_1(T) - S_2(T) - K)^+$$

at time $T_p \geq T$, where $S_1(T)$, $S_2(T)$ are two swap rates of different tenors fixing at time T. The (undiscounted) value of a spread option is therefore given by

$$V(0; T, K) = \mathrm{E}^{T_p}\left((S_1(T) - S_2(T) - K)^+\right). \qquad (17.26)$$

CMS spread options (Section 5.13.3) are, arguably, the most liquid of the multi-rate vanilla derivatives. Euro- and USD-denominated spread options are traded among brokers, assuring reasonable visibility into market prices.

17.4.1 Normal Model for the Spread

Market quotes of CMS spread options often come in the form of implied Normal (also known as Bachelier, Gaussian or, simply, basis-point) volatilities of the spread. As we have already seen in Section 14.4.3, the implied Normal spread volatility $\sigma_N(T, K)$ (for a given pair of swap rates) is defined by equating the undiscounted market price of a spread option to its Normal model price at the given volatility, i.e.

$$V_{\mathrm{mkt}}(0; T, K) = c_N\left(0, \mathrm{E}^{T_p}(S_1(T) - S_2(T)); T, K; \sigma_N(T, K)\right),$$

where $c_N(t, S; T, K; \lambda)$ is the Normal pricing formula with volatility λ defined by (7.16).

A few features of this formula are worth pointing out. First, the Normal model, rather than the Black model, is used as a convention for quoting spread options because the spread process $S_1(T) - S_2(T)$ can become negative, something which is disallowed in a log-normal setting. Indeed, spread options that correspond to zero strike, $K = 0$, are among the most liquid — and for the Black model, zero-strike implied volatility is simply not defined. Second, we notice that in order to back out the implied volatility, we must use the *convexity-adjusted* forward of the spread $\mathrm{E}^{T_p}(S_1(T) - S_2(T))$. When trading with each other, dealers therefore need to agree on the convexity adjustments of the underlying swap rates before they can agree on the implied volatility. Finally, we note that in practice the implied volatility of a particular CMS spread will always depend on both the strike and the expiry, reflecting the fact that market distributions of spreads are not perfectly Gaussian. Much of the work in modeling spread options is focused on properly capturing the market-implied distribution of the spread.

The Normal model provides a convenient common language for quoting spread options, but it is not well-suited for risk management. Apart from the standard risk management issues that arise when using models with strike-indexed parameters (see the related discussion in Section 16.1.1), we highlight the fact that strike-indexed spread volatilities generate no explicit link to the marginal distribution parameters — including volatilities — of the underlying swap rates. As a consequence, spread options will, unreasonably, show no vega (volatility sensitivity, see Section 8.9) to swaption parameters, making them difficult, or even impossible, to hedge. A related issue is the absence of direct information useful for the pricing of *other* payoffs that depend on the same two swap rates. For example, it is unclear how to translate a strike-dependent spread volatility into a parameter that could be used to value a spread option with non-standard gearing, i.e. a payoff

$$(S_1(T) - gS_2(T) - K)^+, \tag{17.27}$$

where $g \neq 1$.

The two issues above can be addressed with a certain degree of success by copula methods. We shall describe this approach momentarily, but let us note that if our ultimate goal is to build dynamic term structure models that are consistent with the prices of vanilla spread options (and European swaptions, of course), we will ultimately need to abandon the copula approach — see Section 17.8 and beyond.

17.4.2 Gaussian Copula for Spread Options

Arguably, the simplest copula-based spread option valuation method is obtained by specifying a two-dimensional Gaussian copula, a parameterization that depends on a single correlation parameter ρ. A spread option is then valued using the following procedure.

1. For each of the swap rates $S_i(T)$, $i = 1, 2$, a market-implied density under its own annuity measure, $\psi_i^{A_i}(x)$, is derived from swaption prices. We may perform this derivation non-parametrically (see Section 16.6.9) or rely on a vanilla model calibrated to swaption prices.
2. Using Proposition 16.6.4, each $\psi_i^{A_i}(x)$, $i = 1, 2$, is converted into the PDF $\psi_i^{T_p}(x)$ of the i-th swap rate under the T_p-forward measure .
3. For a given correlation ρ, the joint probability density function $\psi^{T_p}(x_1, x_2; \rho)$ of $(S_1(T), S_2(T))$ is defined by (17.20), where the copula density $c_{\text{gauss}}(\cdot)$ for the Gaussian copula is given by (17.22). (In this case, notice that R is a 2×2 matrix with 1 on the diagonal and ρ off-diagonal.)
4. The payoff $(x_1 - x_2 - K)^+$ is integrated against the density $\psi^{T_p}(x_1, x_2; \rho)$ to obtain the undiscounted model price of the spread option:

$$V_{\text{mdl}}(0; T, K, \rho) = \int \int (x_1 - x_2 - K)^+ \psi^{T_p}(x_1, x_2; \rho) \, dx_1 \, dx_2.$$

See Sections 17.6.1, 17.6.2 for details on numerical implementation.

To calibrate the model to market, we would also perform the following step:

5 Find $\rho = \rho(T, K)$, *the implied spread correlation*, such that the model price of the spread option matches its market price,

$$V_{\mathrm{mdl}}(0; T, K, \rho(T, K)) = V_{\mathrm{mkt}}(0; T, K).$$

Clearly, with ρ fixed, changes to marginal distributions of the two swap rates would affect their joint distribution via (17.20), ultimately impacting the spread option value. Also, for a given ρ we can integrate any payoff, including (17.27), against the joint density, allowing us to value all payoffs linked to the two swap rates consistently. Hence, as advertised the copula method can overcome several of the issues we identified earlier for the simple Normal spread model. On the other hand, were we to calibrate (as in Step 5) the Gaussian copula model to market values of spread options with a fixed maturity date but different strikes, we would typically obtain a *different* value of the implied spread correlation for each strike, an effect sometimes known as the *correlation smile*. In other words, a simple Gaussian copula has insufficient flexibility to capture market-observed distributions of CMS spreads. To elaborate a bit further on this, in Figure 17.1 we show typical shapes of Normal volatilities of spreads across strikes as implied by different values of the Gaussian copula correlation ρ. We see that ρ basically can only shift the implied spread option volatility smile in parallel, making it impossible to match the market-implied volatility smile listed in the figure. To properly match the market, we apparently need to go beyond Gaussian copulas and look for more flexible alternatives.

Before starting our search for an alternative copula, let us briefly clarify one point about Gaussian copulas. While the implied Gaussian copula correlation is often used to characterize the dependence of the swap rates, it is important to realize that this parameter is not well-defined unless marginal distributions of the swap rates (under some common measure!) are clearly specified. Above we defined the implied correlation by i) implying the marginal distributions under the annuity measures from swaptions; and ii) transplanting these distribution into the T_p-forward measure using a given annuity-mapping function (e.g. one defined by a the linear TSR model). Were we to change any of these "ingredients", we would need to use a different correlation in the copula to obtain the same spread option value. For example, the following marginals will also lead to reasonable definitions of the implied spread correlation (all under the T^p-forward measure):

1. The marginal distribution for each swap rate $S_i(T)$, $i = 1, 2$, is log-normal with mean $\mathrm{E}^{T_p}(S_i(T))$, and volatility set equal to the implied Black volatility of the at-the-money (strike $= S_i(0)$) swaption.
2. The marginal distribution for each swap rate $S_i(T)$, $i = 1, 2$, is log-normal with mean $\mathrm{E}^{T_p}(S_i(T))$, and volatility set equal to the implied Black volatility of the at-the-money (strike $= \mathrm{E}^{T_p}(S_i(T))$) CMS cap.

Fig. 17.1. Implied Normal Spread Volatility

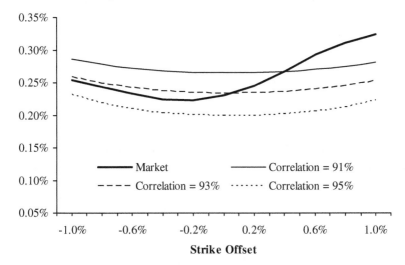

Notes: Implied Normal spread volatility for a 5 year spread option on the difference between 10 year and 2 year swap rates. The x-axis shows the strike as an offset to the CMS-adjusted forward value of the spread. The "Market" data was observed in November 2007. The spread volatility curves implied by a Gaussian copula are shown for three levels of copula correlation (91%, 93%, and 95%), assuming market-implied marginal distributions per Section 17.4.2.

3. The marginal distribution for each swap rate $S_i(T)$, $i = 1, 2$, is Gaussian with mean $\mathrm{E}^{T_p}(S_i(T))$, and volatility equal to the implied Normal volatility of the at-the-money (strike $= S_i(0)$) swaption.

4. The marginal distribution for each swap rate $S_i(T)$, $i = 1, 2$, is Gaussian with mean $\mathrm{E}^{T_p}(S_i(T))$, and volatility equal to the implied Normal volatility of the at-the-money (strike $= \mathrm{E}^{T_p}(S_i(T))$) CMS cap.

This, far from exhaustive, list of alternatives is intended to give the reader a flavor of issues that can arise when two parties are communicating price information in terms of implied spread correlations — clearly they would have to be very precise about all relevant details to avoid misunderstandings[4]. Communication issues aside, the choice of conventions for specifying the meaning of implied Gaussian copula correlations largely comes down to personal preferences and consistency with the rest of one's modeling setup. We leave it to the reader to ponder pros and cons of various approaches; not

---

[4]To demonstrate the dependence of implied correlation on the marginal distribution of swap rates, Appendix 17.A uses a setting with displaced diffusion processes to quantify the effect on ATM spread option prices due to changes in the swap rate volatility skew.

surprisingly, the way we defined implied spread correlations to begin with is our personal choice. It is worth pointing out, however, that in the last case listed above (and *only* in the last case), the spread correlations can be extracted from the implied spread volatilities directly by simple algebraic manipulations. If we denote the Gaussian copula correlation in case 4 by $\rho_N(T, K)$, the relevant at-the-money CMS cap volatilities by $\sigma_{N,i}$, $i = 1, 2$, and the normal spread volatility by $\sigma_N(T, K)$, then it is easily seen that

$$\sigma_N (T, K)^2 = \sigma_{N,1}^2 + \sigma_{N,2}^2 - 2\sigma_{N,1}\sigma_{N,2}\rho_N(T, K).$$

17.4.3 Spread Volatility Smile Modeling with the Power Gaussian Copula

After our small detour into various definitions of implied spread correlations, we now proceed with the task of identifying a copula that would allow us to match the market-implied spread volatility smile as closely as possible. As we saw in Figure 17.1, the Gaussian copula provides us with the ability to change the overall level of the implied spread volatility smile, but lacks controls over its slope or curvature. Mixtures of Gaussian copulas as in (17.24) allow for more flexibility, but ultimately only provide the mechanism to control the curvature of the implied spread volatility smile and not its slope, a fact that is easy to verify experimentally. While we can consider adding standard Archimedean copulas from Section 17.3.3, either stand-alone, "reflected" as in Lemma 17.3.3, or as parts of mixtures, we generally find that this does not provide sufficient flexibility either. On the other hand, the power Gaussian copula (17.25), despite its parsimonious specification, turns out to provide direct control over the relevant features of the spread volatility smile.

In the two-dimensional case, the Gaussian power copula is given by

$$C_{\mathrm{PG}} (u, v; \rho, \theta_1, \theta_2) = u^{1-\theta_1}v^{1-\theta_2}C_{\mathrm{gauss}} \left(u^{\theta_1}, v^{\theta_2}; \rho\right), \qquad (17.28)$$

where ρ is a correlation coefficient and $\theta_1, \theta_2 \in [0, 1]$. We would expect the correlation ρ to move the implied spread volatility smile up and down, just as in the (pure) Gaussian case of Figure 17.1. It turns out that the remaining two parameters θ_1, θ_2 provide good control over the slope and curvature of the smile. Starting from $\theta_1 = \theta_2 = 1$, the base Gaussian case, we find that as θ_1 decreases from 1 towards 0, the spread smile rotates counter-clock-wise, and as the parameter θ_2 decreases, the smile rotates in the opposite direction. The effects are clearly visible in the first graph in Figure 17.2. By decreasing both parameters at the same time curvature is added to the smile, and a good fit to market volatilities can be achieved, as the second graph in Figure 17.2 demonstrates.

Fig. 17.2. Implied Normal Spread Volatility

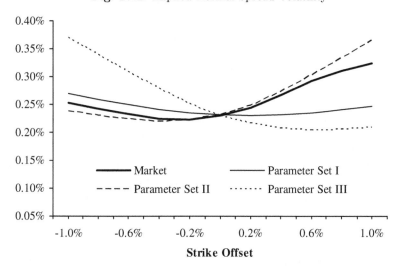

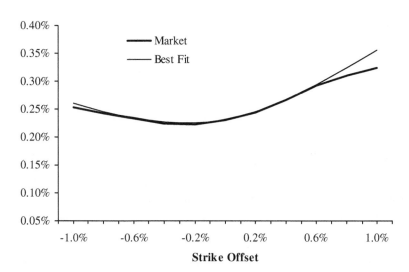

Notes: Implied Normal spread volatility for the spread option in Figure 17.1 in the power Gaussian copula (17.28). Three parameter scenarios as well as best fit parameters from Table 17.1 are shown.

17.4.4 Copula Implied From Spread Options

As we have shown, the power Gaussian copula is capable of reproducing a wide range of market-observed shapes of spread volatility smiles. Yet, clearly, it cannot reproduce at least some of the spread volatility smiles exactly (as,

	Set I	Set II	Set III	Best Fit
ρ	92.6%	95.6%	95.8%	95.5%
θ_1	100.0%	90.0%	100.0%	91.0%
θ_2	100.0%	100.0%	90.0%	99.0%

Table 17.1. Power Gaussian copula (17.28) parameter sets for Figure 17.2.

for example, seen in the wings of the graph in Figure 17.2). Nor, in theory, could the same feat be accomplished by any other copula function with a finite number of parameters. It is, then, natural to wonder whether it is possible to come up with a copula that would reproduce market volatilities (or, equivalently, values) of spread options *exactly* for all values of spread strikes. Of course, spread options are never traded in the whole continuum of strikes, yet the question remains valid in the idealized case of knowing the full distribution of the spread. We call such a copula a *spread-implied copula*.

As we write this book there are no definitive results on spread-implied copulas, so our treatment will necessarily be brief. From general dimensionality analysis it is clear that, in general, there should be a continuum of copulas that would match a set of spread option values of all strikes (and, of course, given marginal distributions of individual swap rates). It is also clear that some collections of spread option values are fundamentally incompatible with given marginal distributions. As a somewhat trivial example, consider marginal distributions that correspond to non-random swap rates (i.e. marginal PDFs are delta functions). Then, clearly, most exogenously specified values of spread options will be incompatible with such marginals irrespective of what dependence structure we would specify.

While it is relatively easy to come up with some examples where spread option values are inconsistent with marginal distributions, the precise results on the existence of spread-implied copulas are unknown to us. Pragmatically, however, there are a number of constructive algorithms for creating "candidate" spread-implied copulas; upon construction these functions can be verified to be true copulas (or not, as the case might be).

One such algorithm, recently proposed by Austing [2010] in the context of a related problem in FX cross-smile modeling, uses the fact that the joint distribution of the swap rates $S_1(T)$, $S_2(T)$ can be obtained from the prices of the so-called *best-of-calls options*, i.e. options with the payoff

$$\max\left((S_1(T) - K_1)^+, (S_2(T) - K_2)^+\right)$$

for all strikes K_1 and K_2. Importantly, prices of such best-of call options can be parameterized consistently with given values of options on each individual swap rate (for all strikes) and all spread options. We refer the reader to the original paper for details, and instead take a somewhat different tack.

Of central importance to our discussion is the result that links a copula function to values of spread options that we develop later in the book, see Corollary 17.6.2. For convenience we pre-announce it here; the formula (17.38) states that the spread option values $V(0; T, K)$ are given by

$$V(0; T, K) = \int_{-\infty}^{\infty} \left(1_{\{x>0\}} - C\left(\Psi_1(x), \Psi_2(x - K) \right) \right) dx - \mathrm{E}^{T_P}(S_2(T)) - K$$
(17.29)

for all values of strike K. Here Ψ_1 and Ψ_2 are the marginal CDFs of the swap rates $S_1(T)$ and $S_2(T)$ and C is the copula function. If spread option values $V(0; T, K)$ are given for all K, then we can treat (17.29) as an (integral) equation for the unknown copula function C.

One can imagine a number of approaches for attacking this equation. We can, for example, discretize $V(0; T, \cdot)$ and $C(\cdot, \cdot)$ on a grid to obtain a linear system for the grid values of the copula function. The constraints on C to be a copula are then given by certain linear inequalities, and the resulting problem can be solved by linear algebra methods. One should be mindful that (17.29) does not introduce enough constraints to give a unique solution, so they should be supplemented with other, exogenous, conditions. We leave it to the reader to explore these ideas.

As (17.29) is underspecified, another line of attack would involve parameterizing C by a one-dimensional family and then solving for the parameter function. For example, one can take an Archimedean copula from Section 17.3.3 with an unspecified parameter function $w(\cdot)$ that one then would solve for (numerically) from (17.29). In the same vain, one can take a product copula from Theorem 17.3.5 with one of the $q_{m,i}(\cdot)$ unspecified and solve for it from (17.29).

A rather simple alternative to these, admittedly quite computationally expensive, methods is based on the idea of mixing two copulas together with the weight that is a function of (essentially) spread option strike. The resulting function is not guaranteed to be a copula (something that should be checked post-construction) but this deficiency is somewhat compensated by a rather simple numerical algorithm, an algorithm that we now proceed to outline.

Let us choose two copulas, $C_{\mathrm{lo}}(u, v)$ and $C_{\mathrm{hi}}(u, v)$, and let us define the "copula" C by

$$C(u, v; \alpha(\cdot)) = \alpha \left(\Psi_1^{-1}(u) - \Psi_2^{-1}(v) \right) C_{\mathrm{lo}}(u, v)$$
$$+ \left(1 - \alpha \left(\Psi_1^{-1}(u) - \Psi_2^{-1}(v) \right) \right) C_{\mathrm{hi}}(u, v), \quad (17.30)$$

where $\alpha(\cdot)$ is an unknown mixing weight. Substituting this expression into (17.29) we obtain a simple equation on the weight function $\alpha(K)$ that we can solve to yield

$$\alpha(K) = \frac{V(0; T, K) - V_{\mathrm{hi}}(0; T, K)}{V_{\mathrm{lo}}(0; T, K) - V_{\mathrm{hi}}(0; T, K)},$$

where V_{lo} and V_{hi} are the spread option values that correspond to the copulas C_{lo} and C_{hi}. The copulas C_{lo} and C_{hi} should be chosen so that the spread option values V are spanned by V_{lo} and V_{hi}; for example we can take the anti-dependence copula C_{AD} (see (17.17)) as C_{lo} and the perfect dependence copula C_{D} (see (17.16)) as C_{hi}, or we can take Gaussian copulas with sufficiently low/high correlations. The right choice of C_{lo} and C_{hi} will guarantee that $C(u, v) \in [0, 1]$; also by construction the marginals conditions (17.13)–(17.14) are always satisfied. Where C may fail to be a real copula is in being a true two-dimensional distribution function; whether this is satisfied or not will have to be checked *post-factum*.

17.5 Rates Observed at Different Times

In Section 17.4 we assumed that all swap rates fix at the same time T. This assumption is, however, not required for the copula method to work, although certain complications do arise for securities with multiple fixing dates. To illustrate, consider a floating range accrual which, according to the definition in Section 5.13.4, pays the amount

$$S_1(T) \times \frac{\sharp \{t \in [T, T + \tau] : S_2(t) \in [l, u]\}}{\sharp \{t \in [T, T + \tau]\}}$$

at time $T_p = T + \tau$, where $\sharp\{\cdot\}$ is the number of business days that satisfy the specified trigger condition. Clearly, a range accrual can be decomposed into a series of floating digital options, i.e. contracts with the payoff

$$S_1(T) \times 1_{\{S_2(t) \in [l, u]\}},$$

for $t \in [T, T + \tau]$, paid at T_p. The (undiscounted) value of such digital is

$$V_{\mathrm{digi}}(0) = \mathrm{E}^{T_p} \left(S_1(T) \times 1_{\{S_2(t) \in [l, u]\}} \right).$$

The distributions of $S_1(T)$, $S_2(t)$ in their annuity measures can be mapped into the distributions under the T_p-forward measure using standard techniques, at which point one could, in principle, apply the copula method. The main complication here is not so much the mechanics of the copula method, but rather the meaning of the parameters for the dependence structure. If we take the Gaussian copula as an example, the correlation parameter of the copula would specify the dependence between $S_1(T)$ and $S_2(t)$ — two swap rates observed at *different* times. Clearly we cannot use the same correlation parameter as we would use to characterize the dependence between $S_1(T)$ and $S_2(T)$ (i.e. when the rates are observed on the same date)[5]. Instead,

---

[5]This is most easily seen by assuming that the two rates are in fact the same. In this case the correlation between $S_1(T)$ and $S_2(T)$ is obviously 1, whereas the correlation between $S_1(T)$ and $S_2(t)$ would be less than 1 as the increment $S_2(t) - S_2(T)$ would typically be only weakly dependent on $S_1(T)$.

the correct correlation parameter should originate from the *terminal* co-dependence between the two rates $(S_1(T), S_2(t))$ and should reflect both the correlation of rates observed at the same time *and* their inter-temporal de-correlation (which is typically quite significant).

One can attempt to deal with the issue above by specifying copula correlations (or, in general, copula parameters) that are functions of time t (in addition to them being functions of time T). Clearly, this is not particularly satisfactory as a large number of parameters would need to be kept and updated. Independent marking of such parameters would impose a heavy operational burden and, importantly, could introduce hard-to-trace arbitrage possibilities into the model. A better alternative, in our view, is to maintain only one copula correlation that corresponds to the dependence of the two rates $(S_1(T), S_2(T))$ observed on the same date, and then devise rules that would link other correlations to this "anchor" correlation. As we have done in the past, we may look at term structure models (that, by definition, are self-consistent in this regard) for inspiration. By resorting to approximations we limit the applicability of the method to only relatively small mismatches between the observation dates (about a year or so, probably) — for anything longer we would strongly recommend a direct application of a suitable multi-factor term structure model.

One reasonable approximation can be derived by assuming that $S_2(t) - S_2(T)$ is independent from $S_1(T)$ — a very respectable approximation — and that we can approximate the dynamics of $S_2(t)$ by a one-factor Gaussian mean-reverting process with constant volatility and mean reversion parameter $\varkappa_{S_2}$, the same approximations that we employed in Section 16.8.6. Proceeding as in that section, we obtain (compare to (16.118))

$$\text{Corr}\,(S_1(T), S_2(t)) = \text{Corr}\,(S_1(T), S_2(T)) \times \sqrt{\frac{1 - e^{-2\varkappa_{S_2} T}}{1 - e^{-2\varkappa_{S_2} t}}}.$$

With this parameterization, correlation of $(S_1(T), S_2(T))$, with the two rates observed at the same time, defines the overall level of correlations for rates observed on *different* dates, while the mean reversion $\varkappa_{S_2}$ defines speed of further de-correlation arising from fixing date mismatches, providing a parsimonious yet flexible description of the whole universe of various correlations.

17.6 Numerical Methods for Copulas

Let us now turn our attention to issues of numerical implementation of valuation methods for copulas. The (undiscounted) value of a derivative with the payoff $f(S_1(t_1), \ldots, S_d(t_d))$ paid at time T_p, where $S_i(t_i)$ is a swap rate observed at time t_i, $i = 1, \ldots, d$, is equal to

$$V(0) = \mathrm{E}^{T_p} \left(f\left(S_1(t_1), \ldots, S_d(t_d) \right) \right). \qquad (17.31)$$

Letting $\psi(x_1, \ldots, x_d)$ be the joint probability density of the swap rates $S_1(t_1), \ldots, S_d(t_d)$ under the T_p-forward measure[6], then the value can be represented as an integral

$$V(0) = \int_{-\infty}^{\infty} \cdots \int_{-\infty}^{\infty} f(x_1, \ldots, x_d) \, \psi(x_1, \ldots, x_d) \, dx_1 \ldots dx_d.$$

If the dependence structure between the swap rates is defined by a copula $C(u_1, \ldots, u_d)$ then, according to (17.20),

$$V(0) = \int_{-\infty}^{\infty} \cdots \int_{-\infty}^{\infty} f(x_1, \ldots, x_d)$$

$$\times \, c\left(\Psi_1(x_1), \ldots, \Psi_d(x_d) \right) \left(\prod_{i=1}^{d} \psi_i(x_i) \right) dx_1 \ldots dx_d, \quad (17.32)$$

where the copula density $c(u_1, \ldots, u_d)$ is defined by (17.21) and where we have denoted marginal PDFs and CDFs of the swap rates under the T_p-forward measure by $\psi_i(x)$ and $\Psi_i(x)$, $i = 1, \ldots, d$, respectively. The formula (17.32) is the basic valuation formula for the copula method; changing to variables $u_i = \Psi_i(x_i)$ we can rewrite it as another useful formula,

$$V(0) = \int_{0}^{1} \cdots \int_{0}^{1} f\left(\Psi_1^{-1}(u_1), \ldots, \Psi_d^{-1}(u_d) \right) c(u_1, \ldots, u_d) \, du_1 \ldots du_d.$$

$$(17.33)$$

17.6.1 Numerical Integration Methods

A quadrature rule approximates an integral with a finite sum of the values of the integrand over a suitably chosen grid covering the domain of integration. While not strictly required, the integration grid is often chosen to be a direct product of one-dimensional integration grids, so that we have something like

$$\int \cdots \int g(y_1, \ldots, y_d) \, dy_1 \ldots dy_d$$

$$\approx \sum_{m_1=1}^{M_1} \cdots \sum_{m_d=1}^{M_d} \mu_{m_1, \ldots, m_d} g(y_{1, m_1}, \ldots, y_{d, m_d}) \quad (17.34)$$

for the grid $\{y_{1, m_1}\} \times \ldots \times \{y_{d, m_d}\}$ and weights $\{\mu_{m_1, \ldots, m_d}\}$. The presence of nested sums makes (17.34) impractical in high dimensions, as the number of points grows exponentially in the number of dimensions ("curse

---

[6]Note we drop the superscript T_p for notational convenience for the duration of this section.

of dimensionality"). However, in the practically important case where d is small (say, $d = 2$ or 3), the integrals in (17.32) or (17.33) can be computed quite efficiently with various schemes of the type (17.34). A good selection of methods is reviewed in Press et al. [1992], and many schemes are implemented in numerical software packages. With such pre-canned routines readily available, numerical integration for the copula method may therefore seem as straightforward as calling a suitable black-box procedure for the integral we are interested in; however, a robust, efficient implementation requires a bit more thought.

The first decision we need to make is which of the two integration formulas (17.32) and (17.33) to use. In (17.33) the limits of integration are finite, which simplifies discretization, and the marginal PDFs and CDFs $\psi_i(x)$, $\Psi_i(x)$ are not required when evaluating the integrand. On the other hand, (17.33) requires an efficient algorithm for calculating the inverses of marginal CDFs $\Psi_i^{-1}(u)$. Ultimately, whether (17.32) or (17.33) is most convenient will often depend on the specifics of the model at hand. For concreteness, we here choose (17.33) for our discussion.

Inverse CDFs that appear in (17.33) are rarely available in closed form and typically must be calculated numerically. For the sake of efficiency, these inverse CDFs should always be pre-computed before the main integration starts. This could be (for a given dimension i) as simple as caching $v_j = \Psi_i(\xi_j)$ for ξ_j on a given grid $\xi_1 < \ldots < \xi_J$ that spans the domain of $\Psi_i(\cdot)$. In the main integration computation one might then approximate

$$\Psi_i^{-1}(u) \approx \frac{v_{n+1} - u}{v_{n+1} - v_n}\xi_n + \frac{u - v_n}{v_{n+1} - v_n}\xi_{n+1},$$

with n such that $v_n \leq u < v_{n+1}$. A more refined approach would augment this by inserting extra points in the intervals $[\xi_n, \xi_{n+1}]$ for which gaps $v_{n+1} - v_n$ are larger than a pre-specified tolerance $\epsilon > 0$.

With the inverse CDFs pre-computed and stored in caches, each evaluation of the integrand in (17.33) would consist of d lookups in inverse CDF caches, together with a (usually quite straightforward) evaluation of the payoff $f(\cdot)$ and a computation of the copula density function $c(\cdot)$. As the copula density function in most cases is known analytically, cache lookups will often dominate the evaluation time, suggesting that the integration grid be organized in such a way that cache lookups are efficient.

As we sometimes prefer not to use adaptive integration schemes (see Section 23.2.1 for explanation), we need to pay attention to the smoothness properties of the integrand, in particular the payoff f (as the copula density function c is usually smooth enough). Clearly, for most payoffs f would be either discontinuous (digital options on CMS spread, say) or, if continuous, then not differentiable (put and call options on CMS spread, say). Many relevant strategies for dealing with non-smooth payoffs are discussed in Chapter 23, and we urge the reader to get acquainted with the material in that chapter before proceeding with an actual implementation. For the purposes of our discussion here, we just observe that in the integral (17.33) or, rather, the generic scheme (17.34), we often must treat the innermost integration (summation) differently from the outer integrals. To explain, let us set $d = 2$ and consider the nested integral

$$\int_0^1 \left(\int_0^1 g\left(u_1, u_2\right) \, du_1 \right) du_2,$$

where we use the short-hand notation

$$g\left(u_1, u_2\right) = f(\Psi_1^{-1}(u_1), \Psi_2^{-1}(u_2))c(u_1, u_2).$$

If g is non-smooth, numerical computation of the inner integral

$$\int_0^1 g\left(u_1, u_2\right) \, du_1$$

will generally require us to identify the points where $g(\cdot, u_2)$ is singular and either include them in the integration grid or increase the density of grid points around these points (see Chapter 23). It is therefore critical that we use an integration scheme that allows us complete freedom in locating the integration nodes[7]. On the other hand, when we calculate

$$\int_0^1 G\left(u_2\right) \, du_2,$$

where $G(u_2)$ is (the numerically calculated value of) the integral $\int_0^1 g(u_1, u_2) \, du_1$, the integrand G is often smooth, as the singularities of g would have been integrated out in the inner integral. Hence, for the outer integration, we can use fast schemes suitable for smooth integrands. For integrating over a finite interval (such as $[0, 1]$), the *Gauss-Lobatto* quadrature[8] (see Kythe and Schäferkotter [2004]) is a good choice.

While integrating in one of the dimensions may cure singularities, this cannot be the case if the non-smooth features of the payoff are defined solely by the "outer" variable; in the example above that would be the case if we had $g(u_1, u_2) = 1_{\{u_2 > K\}}$ or something similar. This situation can be handled by switching the order of integration and integrating in variable u_2 first. For most cases, in fact, integrating in one particular dimension gives a smoother function than in any other, so an advanced integration routine would have logic to determine the dimension on which to perform the innermost quadrature. Of course there are situations when the payoff

---

[7]A trapezoidal scheme is a possible candidate. Most Gaussian quadrature rules, however, are not ideal, since their integration grids are not directly user-specifiable, but emerge as roots of a particular function.

[8]Sometimes also called *Gauss-Legendre* quadrature after the name of the family of polynomials whose roots define the integration grid.

is non-smooth in both directions; careful handling of singularities in both integrals is then required.

Before concluding our remarks on integration methods for copulas, we briefly consider the specialization of the method for Gaussian copulas (see Section 17.3.1). For $C = C_{\text{gauss}}$, we recall the alternative valuation formula (17.11),

$$V(0) = \int_{-\infty}^{\infty} \cdots \int_{-\infty}^{\infty} f\left(\Psi_1^{-1}\left(\Phi\left(z_1\right)\right), \ldots, \Psi_d^{-1}\left(\Phi\left(z_d\right)\right)\right)$$
$$\times \phi\left(z_1, \ldots, z_d; R\right) dz_1 \ldots dz_d.$$

Using this representation can lead to numerical improvements, since each nested integral is now represented as an integral over a Gaussian density. For such integrals, the Gauss-Hermite quadrature method (already mentioned in Section 12.3.4.3; see also Press et al. [1992]) is often very efficient. Of course, we would recommend this quadrature rule only if the integrand is sufficiently smooth (e.g., for outer integrals). Also, our recommendation to cache inverse CDFs still apply.

17.6.2 Dimensionality Reduction for CMS Spread Options

The discussion in the previous section is generic and applies to arbitrary multi-rate payoffs. In the important special case of CMS spread options, we can achieve further efficiencies of numerical implementation by reformulating the problem as one involving only *one*-dimensional integrals. The following proposition (following Dhaene and Goovaerts [1996] and Berrahoui [2005]) contains the relevant result.

Proposition 17.6.1. *Let us denote*

$$\gamma(x, K) = \frac{d}{dx} \Psi\left(x, x - K\right),$$

where $\Psi(x_1, x_2)$ is the joint CDF of $(S_1(T), S_2(T))$ under Q^{T_p}. Then, the undiscounted value of a spread option as defined by (17.26) is given by

$$V(0; T, K) = \int_{-\infty}^{+\infty} x\gamma(x, K)\, dx - \int_{-\infty}^{+\infty} x\psi_2(x)\, dx - K, \qquad (17.35)$$

where $\psi_2(x)$ is the one-dimensional marginal PDF of $S_2(T)$ under Q^{T_p}.

Proof. We have

$$V(0; T, K) = \int_{-\infty}^{+\infty} \left(\int_{-\infty}^{x_1 - K} (x_1 - x_2 - K) \, \psi(x_1, x_2) \, dx_2 \right) dx_1$$

$$= \int_{-\infty}^{+\infty} x_1 \left(\int_{-\infty}^{x_1 - K} \psi(x_1, x_2) \, dx_2 \right) dx_1$$

$$- \int_{-\infty}^{+\infty} (x_2 + K) \left(\int_{x_2 + K}^{+\infty} \psi(x_1, x_2) \, dx_1 \right) dx_2.$$

Recall that $\psi(x_1, x_2) = \frac{\partial^2}{\partial x_1 \partial x_2} \Psi(x_1, x_2)$. Therefore

$$V(0; T, K) = \int_{-\infty}^{+\infty} x_1 \left(\frac{\partial}{\partial x_1} \Psi(x_1, x_1 - K) - \frac{\partial}{\partial x_1} \Psi(x_1, -\infty) \right) dx_1$$

$$- \int_{-\infty}^{+\infty} (x_2 + K) \left(\frac{\partial}{\partial x_2} \Psi(+\infty, x_2) - \frac{\partial}{\partial x_2} \Psi(x_2 + K, x_2) \right) dx_2.$$

Note that

$$\frac{\partial}{\partial x_1} \Psi(x_1, -\infty) = 0, \qquad \frac{\partial}{\partial x_2} \Psi(+\infty, x_2) = \psi_2(x_2).$$

Then

$$V(0; T, K) = \int_{-\infty}^{+\infty} x_1 \frac{\partial}{\partial x_1} \Psi(x_1, x_1 - K) \, dx_1$$

$$+ \int_{-\infty}^{+\infty} (x_2 + K) \frac{\partial}{\partial x_2} \Psi(x_2 + K, x_2) \, dx_2 - \int_{-\infty}^{+\infty} x \psi_2(x) \, dx - K. \quad (17.36)$$

We also have

$$\gamma(x, K) = \frac{\partial}{\partial x_1} \Psi(x, x - K) + \frac{\partial}{\partial x_2} \Psi(x, x - K)$$

and

$$\int_{-\infty}^{+\infty} x \gamma(x, K) \, dx = \int_{-\infty}^{+\infty} x \frac{\partial}{\partial x_1} \Psi(x, x - K) \, dx$$

$$+ \int_{-\infty}^{+\infty} x \frac{\partial}{\partial x_2} \Psi(x, x - K) \, dx.$$

By substituting $x = x_1$ in the first integral and $x = x_2 + K$ in the second, the proposition now follows from (17.36). □

In the case when the joint CDF is generated by a copula $C(u_1, u_2)$, i.e. when $\Psi(x_1, x_2) = C(\Psi_1(x_1), \Psi_2(x_2))$, we have

$$\gamma(x, K) = \frac{\partial}{\partial u_1} C(\Psi_1(x), \Psi_2(x - K)) \, \psi_1(x)$$

$$+ \frac{\partial}{\partial u_2} C(\Psi_1(x), \Psi_2(x - K)) \, \psi_2(x - K).$$

Finally, let us present the result of Proposition 17.6.1 in a somewhat different form, already used in the discussion in Section 17.4.4.

Corollary 17.6.2. *The undiscounted value $V(0; T, K)$ at time $t = 0$ of a spread option with strike K at expiry T as defined by (17.26) is given by*

$$V(0; T, K) = \int_{-\infty}^{\infty} \left(1_{\{x>0\}} - \Psi(x, x - K)\right) dx - \mathrm{E}^{T_P}(S_2(T)) - K. \quad (17.37)$$

Alternatively it can be expressed in terms of the copula function,

$$V(0; T, K) = \int_{-\infty}^{\infty} \left(1_{\{x>0\}} - C\left(\Psi_1(x), \Psi_2(x - K)\right)\right) dx - \mathrm{E}^{T_P}(S_2(T)) - K. \quad (17.38)$$

Proof. Integrating by parts we obtain

$$\int_{-\infty}^{\infty} x\gamma(x, K)\, dx = \int_{-\infty}^{\infty} x \left(\frac{d}{dx}\Psi(x, x - K)\right) dx$$

$$= \int_{-\infty}^{0} x\, d\Psi(x, x - K) - \int_{0}^{\infty} x\, d\left(1 - \Psi(x, x - K)\right)$$

$$= -\int_{-\infty}^{0} \Psi(x, x - K)\, dx + \int_{0}^{\infty} \left(1 - \Psi(x, x - K)\right) dx,$$

and (17.37) follows. The result (17.38) follows from the definition of the copula function.   □

17.6.3 Dimensionality Reduction for Other Multi-Rate Derivatives

The formula (17.35) serves as a base for deriving similar one-dimensional representations for other derivatives. For example, differentiating the definition

$$V_{\text{spread}}(0; T, K) = \mathrm{E}^{T_P}\left((S_1(T) - S_2(T) - K)^+\right)$$

with respect to the strike and exchanging the order of the expected value operator and differentiation, we obtain

$$\mathrm{E}^{T_P}\left(1_{\{S_1(T)-S_2(T)>K\}}\right) = -\frac{\partial}{\partial K}V_{\text{spread}}(0; T, K).$$

On the left-hand side we have the value of a *digital* spread option, and the expression on the right-hand side may be computed as the one-dimensional integral obtained by differentiating (17.35) with respect to K,

$$\mathrm{E}^{T_P}\left(1_{\{S_1(T)-S_2(T)>K\}}\right) = -\frac{\partial}{\partial K}V_{\text{spread}}(0; T, K)$$

$$= -\int_{-\infty}^{+\infty} x\frac{\partial\gamma(x, K)}{\partial K}\, dx + 1.$$

To obtain further results, we first generalize Proposition 17.6.1 to spread options with arbitrary gearings.

Corollary 17.6.3. *Let $a_1, a_2 > 0$. Then*

$$\mathrm{E}^{T_p}\left((a_1 S_1(T) - a_2 S_2(T) - K)^+\right)$$

$$= \int_{-\infty}^{+\infty} x \left(\frac{d}{dx}\Psi\left(x/a_1, (x - K)/a_2\right)\right) dx$$

$$- \int_{-\infty}^{+\infty} x \left(\frac{d}{dx}\Psi\left(+\infty, x/a_2\right)\right) dx - K, \quad (17.39)$$

where $\Psi(x_1, x_2)$ is, as before, the joint CDF of $(S_1(T), S_2(T))$ under Q^{T_p}.

We can now differentiate the formula (17.39) with respect to a_1 and a_2, yielding one-dimensional integral representations for the values of derivatives with the payoffs

$$S_1(T)1_{\{a_1 S_1(T) - a_2 S_2(T) > K\}}, \quad S_2(T)1_{\{a_1 S_1(T) - a_2 S_2(T) > K\}}. \quad (17.40)$$

We leave the detailed derivation of these integrals as an exercise to the reader, and just note that the payoffs in (17.40) are those of floating digital spread options. These options are not only important by themselves but are also components of floating spread range accruals, as explained in Section 17.5. As long as the payment rate of the range accrual is either S_1 or S_2 (i.e. equal to one of the rates that define the spread), the value of this security can be obtained by one-dimensional integration. We notice that the specification (17.40) also includes (non-spread) floating digitals, as we can produce the payoff

$$S_1(T)1_{\{S_2(T) > K\}}$$

by setting $a_1 = 0, a_2 = -1$.

The valuation expression for a floating digital is, in fact, easy — and instructive — to derive directly. The key here is the following important lemma.

Lemma 17.6.4. *If the distribution of $S_1(T), S_2(T)$ under Q^{T_p} is given by the joint CDF $\Psi(x_1, x_2)$ with the copula function $C(u_1, u_2)$, then the conditional CDF of $S_2(T)$ given $S_1(T)$ is directly computable from the copula,*

$$\mathrm{Q}^{T_p}(S_2(T) < x_2 | S_1(T) = x_1) = \frac{\partial C}{\partial u_1}(\Psi_1(x_1), \Psi_2(x_2)),$$

where Ψ_1, Ψ_2 are the marginal CDFs of S_1 and S_2.

Proof. We have

$$Q^{T_P}(S_2(T) < x_2|S_1(T) = x_1)$$
$$= \frac{E^{T_P}(\delta(S_1(T) - x_1)1_{\{S_2(T)<x_2\}})}{E^{T_P}(\delta(S_1(T) - x_1))} = \frac{\partial\Psi(x_1, x_2)/\partial x_1}{\psi_1(x_1)},$$

where $\delta(\cdot)$ is the Dirac delta function and ψ_1 is the PDF of S_1. On the other hand,

$$\frac{\partial\Psi(x_1, x_2)}{\partial x_1} = \frac{\partial C(\Psi_1(x_1), \Psi_2(x_2))}{\partial x_1} = \frac{\partial C(\Psi_1(x_1), \Psi_2(x_2))}{\partial u_1}\psi_1(x_1),$$

and the result follows. $\square$

Conditioning the payoff of the floating digital on $S_1(T)$ and applying the lemma, we obtain

$$E^{T_P}(S_1(T)1_{\{S_2(T)>K\}}) = \int_{-\infty}^{\infty} x Q^{T_P}(S_2(T) > K|S_1(T) = x)\psi_1(x)\, dx$$
$$= \int_{-\infty}^{\infty} x\left(1 - \frac{\partial C}{\partial u_1}(\Psi_1(x), \Psi_2(K))\right)\psi_1(x)\, dx.$$

$$(17.41)$$

Remark 17.6.5. The result of Lemma 17.6.4 provides an alternative, perhaps more elegant, route to the proof of Proposition 17.6.1 (or, more generally, Corollary 17.6.3). To see this, we write for the spread option value $V(0; T, K)$,

$$V(0; T, K) = E^{T_P}\left(S_1(T)1_{\{S_1(T)-S_2(T)-K\geq 0\}}\right)$$
$$- E^{T_P}\left(S_2(T)1_{\{S_1(T)-S_2(T)-K\geq 0\}}\right)$$
$$- KE^{T_P}\left(1_{\{S_1(T)-S_2(T)-K\geq 0\}}\right).$$

Conditioning the first term on $S_1(T)$ and the second on $S_2(T)$ (and the third one on either $S_1(T)$ or $S_2(T)$) and using the result of Lemma 17.6.4, a one-dimensional integral representation for the spread option is obtained.

17.6.4 Dimensionality Reduction by Conditioning

Reducing the dimensionality of integrals is often an effective technique for improving computational performance and/or for extending the domain of applicability of direct integration methods. For some payoffs this can be achieved by application of the method in Sections 17.6.2 and 17.6.3; others require different approaches. In Lemma 17.6.4 we demonstrated a particular application of the principle of *conditioning*, an idea that has quite general applicability. The gist of the method is simple: if we can calculate (or approximate) in closed form the expectation of the payoff conditioned on some b-dimensional subset of the d variables in the payoff definition, then we can reduce the dimension of the valuation integral from d to b.

Gaussian random variables are particularly amenable to conditioning methods, courtesy of Lemma 14.6.5 that we have already applied in the context of Brownian bridge calculations, and for calculating swaption values in a two-dimensional Gaussian model (see Section 12.1.6.1). The lemma helps us to calculate the distribution of one Gaussian random variable conditioned on another; importantly, the conditional distribution remains Gaussian. Let us demonstrate how the lemma could be applied to the problem of dimensionality reduction, by considering a simple floating digital payoff[9]

$$S_1(T)1_{\{S_2(T)>K\}}, \tag{17.42}$$

where S_1 and S_2 are two swap rates, for notational simplicity assumed to pay at T. The log-normal model would specify

$$S_i(T) = \widetilde{S}_i e^{\sigma_i Z_i - \sigma_i^2/2}, \quad i = 1, 2, \tag{17.43}$$

where $\widetilde{S}_i = \mathrm{E}^T(S_i(T))$ are CMS-adjusted forward swap rates, σ_i are un-scaled[10] log-normal volatilities of the swap rates, and the vector $(Z_1, Z_2)^\top$ is Gaussian with zero mean, unit variance and correlation ρ (all under Q^T). The (undiscounted) value of the derivative with payoff (17.42) is given by

$$V(0) = \widetilde{S}_1\mathrm{E}^T\left(e^{\sigma_1 Z_1 - \sigma_1^2/2}1_{\{Z_2>\ln(K/\widetilde{S}_2)/\sigma_2+\sigma_2/2\}}\right). \tag{17.44}$$

A direct evaluation of (17.42) would require two-dimensional integration, but it is easy to see that by applying the result of Lemma 14.6.5 we can reduce the dimension of the integral to one. In particular, if we condition on Z_2, we have,

$$V = \widetilde{S}_1\mathrm{E}^T\left(\mathrm{E}^T\left(e^{\sigma_1 Z_1 - \sigma_1^2/2}1_{\{Z_2>\ln(K/\widetilde{S}_2)/\sigma_2+\sigma_2/2\}}\middle|Z_2\right)\right)$$
$$= \widetilde{S}_1\mathrm{E}^T\left(1_{\{Z_2>\ln(K/\widetilde{S}_2)/\sigma_2+\sigma_2/2\}}\mathrm{E}^T\left(e^{\sigma_1 Z_1 - \sigma_1^2/2}\middle|Z_2\right)\right). \tag{17.45}$$

Since

$$Z_1|Z_2 \sim \mathcal{N}\left(Z_2\rho, 1 - \rho^2\right), \tag{17.46}$$

we can evaluate the conditional expected value analytically,

$$\mathrm{E}^T\left(e^{\sigma_1 Z_1 - \sigma_1^2/2}\middle|Z_2\right) = e^{\sigma_1\rho Z_2 - \sigma_1^2\rho^2/2}, \tag{17.47}$$

which gives us

$$V(0) = \widetilde{S}_1 e^{-\sigma_1^2\rho^2/2}\mathrm{E}^T\left(e^{\rho\sigma_1 Z_2}1_{\{Z_2>\ln(K/\widetilde{S}_2)/\sigma_2+\sigma_2/2\}}\right).$$

---

[9]Note that in Section 17.6.3 we already derived a one-dimensional integral representation for this payoff, but we use it anyway to demonstrate the main idea of the method we develop here.

[10]Not annualized, i.e. not divided by $\sqrt{T}$.

Now we need to integrate the modified payoff against the distribution of Z_2 only, which is a one-dimensional integration. Of course, in the log-normal model, this will give the same result as the formula (17.41).

As should be clear from the above, successful applications of conditioning methods rely to a large degree on the availability of the appropriate analytical tools such as Lemma 14.6.5. For general copulas we can calculate conditional CDFs quite easily, as Lemma 17.6.4 demonstrated. This can take us a long way, but to gain analytical tractability, we may ultimately need to consider a narrower range of copulas. In particular, if one is content to use Gaussian copulas, or combinations thereof, then the techniques discussed above can be applied virtually unchanged. For example, conditional CDFs in Gaussian copulas are available in closed form by direct application of Lemma 14.6.5. We leave the (simple) derivation of this generic result to the reader, while focusing here on the applications of the method to some specific payoffs.

As we recall, the Gaussian copula method essentially replaces (17.43) with a more generic expression

$$S_i(T) = \Lambda_i(Z_i), \quad i = 1, 2, \tag{17.48}$$

where $\Lambda_i(x) \triangleq \Psi_i^{-1}(\Phi(x))$ are "mapping functions" from Gaussian variates to market rates. We have already shown (Lemma 17.6.4) how to calculate the value of the payoff (17.42) by conditioning on S_1. Conditioning on S_2, as done for the Gaussian case in (17.45), is also straightforward. Specifically, we obtain

$$V(0) = \mathrm{E}^T \left(1_{\{\Lambda_2(Z_2) > K\}} \mathrm{E}^T \left(\Lambda_1(Z_1) | Z_2 \right) \right),$$

and we observe that it is necessary to calculate, or approximate, $\mathrm{E}^T(\Lambda_1(Z_1)|Z_2)$. A simple approximation may be obtained by replacing $\Lambda_1(\cdot)$ with a quadratic function:

$$\Lambda_1(x) \approx \Lambda_1(0) + \Lambda_1'(0)x + \frac{1}{2}\Lambda_1''(0)x^2,$$

so that we would have

$$\mathrm{E}^T \left(\Lambda_1(Z_1) | Z_2 \right) \approx \Lambda_1(0) + \Lambda_1'(0)\mathrm{E}^T \left(Z_1 | Z_2 \right) + \frac{1}{2}\Lambda_1''(0)\mathrm{E}^T \left(Z_1^2 | Z_2 \right),$$

where, from (17.46),

$$\mathrm{E}^T \left(Z_1 | Z_2 \right) = \rho Z_2, \quad \mathrm{E}^T \left(Z_1^2 | Z_2 \right) = 1 + \rho^2(Z_2^2 - 1).$$

This could be refined slightly by expanding not around $x = 0$ but around $x^* = \Lambda_1^{-1}(\widetilde{S}_1)$, where $\widetilde{S}_1$ is the forward CMS-adjusted swap rate.

A more general approach of expanding $\Lambda_1(x)$ into a Taylor series of arbitrary order is also possible and not much more work, as all required terms of the type $\mathrm{E}^T(Z_1^r|Z_2)$, $r \geq 1$, are easily available from (17.46). Finally, a related, and very accurate, method is based on approximating $\Lambda_1(x)$ with a truncated cosine series. If we choose a range $[-z_{\max}, z_{\max}]$ such that $Q^T(Z_1 \in [-z_{\max}, z_{\max}]) = 1 - \epsilon$ for small $\epsilon > 0$ (for example we can take $z_{\max}$ equal to 3 or 4), then, since $\Lambda_1(x)$ is typically continuous, we can approximate it for $x \in [-z_{\max}, z_{\max}]$ arbitrarily closely by a sum of the type

$$\Lambda_1(x) \approx \mathrm{Re}\left(\sum_{m=0}^{M} w_m e^{\lambda_m x - \lambda_m^2/2}\right), \tag{17.49}$$

where

$$\lambda_m = \frac{\pi i m}{2 z_{\max}}, \quad m = 0, \ldots, M,$$

and $i = \sqrt{-1}$. The weights w_m could be quickly computed by an inverse discrete cosine transform, see Press et al. [1992]. The required conditional expected value is then given by (see (17.47))

$$\mathrm{E}^T\left(\Lambda_1(Z_1) \middle| Z_2\right) \approx \mathrm{Re}\left(\sum_{m=0}^{M} w_m e^{\lambda_m \rho Z_2 - \lambda_m^2 \rho^2/2}\right).$$

The expansion methods do not always work; for example, a payment amount of a floating digital could be a function of the rate S_1 rather than equal to the rate itself. To consider a typical example, let us analyze the expectation

$$V(0) = \mathrm{E}\left((S_1 - u)^+ 1_{\{S_2 > K\}}\right),$$

essentially a call option on S_1 conditioned on S_2 being above some level. Conditioning on Z_2 we see that we need to calculate the expected value

$$\mathrm{E}\left((\Lambda_1(Z_1) - u)^+ \middle| Z_2\right). \tag{17.50}$$

In the case where S_1 is log-normal, this conditional expected value would present no difficulties and would be given by the Black formula (with some parameters dependent on Z_2). Matters are more complicated in the general case, however. To proceed, we recall that $Z_1 = \rho Z_2 + \bar{\rho}\xi$, $\bar{\rho} = (1 - \rho^2)^{1/2}$, where ξ is a standard Gaussian random variable independent of Z_2. Hence, for a fixed value of Z_2, the task of evaluating (17.50) reduces to that of calculating

$$\mathrm{E}\left((\Lambda_1(a + b\xi) - u)^+\right) \tag{17.51}$$

for some a, b.

To proceed further, it is important to recall that the values of options of the type $e(v) = \mathrm{E}((\Lambda_1(\xi) - v)^+)$ are easily available to us, as these are just the option values in the marginal model we use for the rate S_1. With that in mind, we can calculate the option value in (17.51) by replicating the payoff with $e(v)$, $v \in \mathbb{R}$, following the ideas of Section 16.6.1. However, it is not clear if we can achieve substantial computational savings — the original aim of the conditioning method — along this route as the replication method requires calculation of an integral. To save computational effort, suppose we limit ourselves to a single strike in the replication. A natural choice of that strike is such v that the two payoffs

$$(\Lambda_1(a + bx) - u)^+, \quad (\Lambda_1(x) - v)^+$$

have a discontinuity at exactly the same point x. Such a point is given by

$$v^* = \Lambda_1(x^*), \quad x^* = \left(\Lambda_1^{-1}(u) - a\right)/b.$$

Then, by matching the slope of both payoffs at the critical point x^*, we obtain a single-strike "replication",

$$(\Lambda_1(a + bx) - u)^+ \approx \frac{b\Lambda_1'(a + bx^*)}{\Lambda_1'(x^*)}(\Lambda_1(x) - \Lambda_1(x^*))^+$$

which, when applied to the conditional expected value in (17.50), gives us the following approximation

$$E\left((\Lambda_1(Z_1) - u)^+ \mid Z_2\right) \approx \overline{\rho}\frac{\Lambda_1'(\rho Z_2 + \overline{\rho}x^*(Z_2))}{\Lambda_1'(x^*(Z_2))}e(\Lambda_1(x^*(Z_2))).$$

where the critical point $x^*(Z_2)$ depends on Z_2 and is given by

$$x^*(Z_2) = \left(\Lambda_1^{-1}(u) - \rho Z_2\right)/\overline{\rho}.$$

The accuracy of the method obviously depends on how far the mapping function $\Lambda_1(\cdot)$ is from identity.

17.6.5 Dimensionality Reduction by Measure Change

Methods based on conditioning are not the only choice for dimensionality reduction. Another useful approach is based in the idea of performing calculations under a different measure to simplify the payoff. The main technical tool here is the following specialization of the Girsanov theorem (see Theorem 1.5.1).

Lemma 17.6.6. *If X is a d-dimensional Gaussian vector with mean μ and covariance matrix Σ, v is a d-dimensional vector and $f(\cdot)$ is a function $\mathbb{R}^d \to \mathbb{R}$, then*

$$E\left(e^{v^\top (X-\mu) - v^\top \Sigma v/2}f(X)\right) = \widehat{E}\left(f(X)\right),$$

where in measure $\widehat{Q}$, vector X has mean $\mu + \Sigma v$ and covariance matrix Σ.

Proof. See Section 3.5 of Karatzas and Shreve [1991].   □

To see how this method works, let us continue with the payoff (17.42) and return to the log-normal model (17.43) for the moment. Applying Lemma 17.6.6 to the problem (17.44), we immediately obtain the following one-dimensional representation:

$$V(0) = \tilde{S}_1 \widehat{E} \left(1_{\{Z_2 > \ln(K/\tilde{S}_2)/\sigma_2 + \sigma_2/2\}} \right), \tag{17.52}$$

where Z_2 under $\widehat{Q}$ is Gaussian but has a different drift,

$$Z_2 \sim \mathcal{N} \left(\sigma_1 \rho, 1 \right).$$

As desired, the problem has been reduced to that of a one-dimensional integration (in fact, in this case, the expectation is available in closed form). While this example is rather simple, many other payoffs are amenable to the same type of treatment.

As was the case for the conditioning method, the measure change method extends to the general Gaussian copula setup. In the model (17.48), with the approximation (17.49), we have for the value of the payoff (17.42),

$$V(0) = E^T \left(\Lambda_1(Z_1) 1_{\{\Lambda_2(Z_2) > K\}} \right)$$

$$\approx \mathrm{Re} \left(E^T \left(\left(\sum_{m=0}^{M} w_m e^{\lambda_m Z_1 - \lambda_m^2/2} \right) 1_{\{\Lambda_2(Z_2) > K\}} \right) \right)$$

$$= \mathrm{Re} \left(\sum_{m=0}^{M} w_m E^T \left(e^{\lambda_m Z_1 - \lambda_m^2/2} 1_{\{\Lambda_2(Z_2) > K\}} \right) \right).$$

Applying Lemma 17.6.6 to each term in turn, we obtain

$$V(0) \approx \mathrm{Re} \left(\sum_{m=0}^{M} w_m \widehat{E}^m \left(1_{\{\Lambda_2(Z_2) > K\}} \right) \right), \tag{17.53}$$

where under measure $\widehat{Q}^m$, Z_2 is Gaussian with variance 1 and the mean $\lambda_m \sigma_1 \rho$ for each $m = 1, \ldots, M$. While it may seem strange to have Gaussian random variables with a (purely) imaginary mean, each term $\widehat{P}^m(\Lambda_2(Z_2) > K)$ in (17.53) is actually well-defined as an analytic continuation of the Gaussian CDF into a complex domain. Let us demonstrate on a simple example. Let Z be a standard Gaussian random variable. Then

$$E^T \left(e^{iZ+1/2} 1_{\{Z > K\}} \right) = \frac{1}{\sqrt{2\pi}} \int_K^{\infty} e^{ix+1/2} e^{-x^2/2} \, dx$$

$$= \frac{1}{\sqrt{2\pi}} \int_K^{\infty} e^{-(x-i)^2/2} \, dx$$

$$= \frac{1}{\sqrt{2\pi}} \int_{K-i}^{\infty-i} e^{-z^2/2} \, dz,$$

where the last integral is understood to be over the contour $\{x - i :$ $x \in [K, \infty)\}$ in the complex plane and is well-defined, see Chapter 7 of Abramowitz and Stegun [1965].

The method of (17.53) requires M one-dimensional integrations, which is an improvement — as long as M is not too large — over one *two*-dimensional integration that would be required for a standard valuation of (17.42).

17.6.6 Monte Carlo Methods

Let us return to the general problem of calculating the value of a multi-rate payoff in (17.31). We observe again that if the dimensionality d of the payoff is higher than 3 (after potential dimensionality reductions by conditioning and measure change methods), nested numerical integration of the type (17.34) becomes unattractive compared to direct Monte Carlo simulation of the d-dimensional joint distribution. A particularly simple Monte Carlo scheme is obtained if the copula used is Gaussian (with correlation matrix R). Recalling (17.6), it should be clear that we can calculate the (undiscounted) value of the derivative as

$$V(0) \approx \frac{1}{N} \sum_{n=1}^{N} f\left(\Psi_1^{-1}\left(\Phi\left(Z_{n,1}\right)\right), \ldots, \Psi_d^{-1}\left(\Phi\left(Z_{n,d}\right)\right)\right), \qquad (17.54)$$

where $\mathbf{Z}_1, \ldots, \mathbf{Z}_N$, with $\mathbf{Z}_n = (Z_{n,1}, \ldots, Z_{n,d})$, are N independent samples from a d-dimensional Gaussian distribution.

The case of a non-Gaussian copula is conceptually similar. Given a copula function $C(u_1, \ldots, u_d)$, assume momentarily that we know how to generate a random sample for the vector $\mathbf{U} = (U_1, \ldots, U_d)$, where each U_i has a uniform distribution on $[0, 1]$, and the dependence structure is given by the copula C,

$$Q^{T_p}\left(U_1 < u_1, \ldots, U_d < u_d\right) = C\left(u_1, \ldots, u_d\right). \qquad (17.55)$$

Then, the value of the derivative in (17.33) is given by

$$V(0) \approx \frac{1}{N} \sum_{n=1}^{N} f\left(\Psi_1^{-1}\left(U_{n,1}\right), \ldots, \Psi_d^{-1}\left(U_{n,d}\right)\right), \qquad (17.56)$$

where $\mathbf{U}_n = (U_{n,1}, \ldots, U_{n,d})$ has the same distribution as $\mathbf{U}$ for each $n = 1, \ldots, N$, and all $\mathbf{U}_n$ are independent. Calculations with formulas (17.54) or (17.56) are straightforward; we only remind the reader that inverse CDFs $\Psi_i^{-1}(u)$ should be pre-computed and cached before the main simulation.

The success of the numerical implementation of the scheme (17.56) hinges on our ability to simulate a random sample from a given copula as in (17.55). We have demonstrated how to do this for the Gaussian copula, so let us consider the Archimedean copulas that were also introduced in Section 17.3. The simulation algorithm for a bivariate Archimedean copula with the generator function $\omega(\cdot)$ can be based on the following result from Nelsen [2006].

Lemma 17.6.7. *Let (U_1, U_2) be a random vector with uniform marginals and joint distribution function $C_{\text{arch}}(u_1, u_2; \omega(\cdot))$. Define two new random variables*

$$R = \omega(U_1) / (\omega(U_1) + \omega(U_2)), \quad F = C_{\text{arch}}(U_1, U_2; \omega(\cdot)).$$

Then, the joint distribution function of (R, F) is given by

$$P(R < r, F < f) = r \times A_\omega(f), \quad A_\omega(f) \triangleq 1 - \omega(f)/\omega'(f).$$

Here, R and F are independent and R is uniformly distributed on $[0, 1]$.

With the help of this lemma, a sample (U_1, U_2) from an Archimedean copula can be generated with the following algorithm:

1. Simulate two independent random variables R and W, uniformly distributed on $[0, 1]$.
2. Set $F = A_\omega^{-1}(W)$, where $A_\omega(f) = 1 - \omega(f)/\omega'(f)$.
3. Set $U_1 = \omega^{-1}(R\omega(F))$ and $U_2 = \omega^{-1}((1 - R)\omega(F))$.

A multi-dimensional extension of this algorithm exists; see Wu et al. [2006] for details.

From our basic ability to simulate Gaussian and Archimedean copulas, we can devise simulation schemes for the "aggregate" copulas outlined in Section 17.3.4. First, let us consider the reflection method of Lemma 17.3.3. If $(U_1, \ldots, U_i, \ldots, U_d)$ is a sample from some copula $C(u_1, \ldots, u_d)$, then, clearly, $(U_1, \ldots, 1 - U_i, \ldots, U_d)$ is a sample from the reflected copula $\overline{C}(u_1, \ldots, u_d; \{i\})$; it follows that simulating reflected copulas is straightforward.

Simulation of a convex linear combination of copulas as given by Lemma 17.3.4 is also easy. To state the algorithm, let $\mathbf{U}^m = (U_1^m, \ldots, U_d^m)$, $m = 1, \ldots, M$, be a collection of independent samples from the copulas $C_{m,}$; and let W be a discrete random variable with the distribution $P(W = m) = w_m$, $m = 1, \ldots, M$. If W is independent of all $\mathbf{U}^m$, then the sample

$$(U_1, \ldots, U_d) = \mathbf{U} \triangleq \mathbf{U}^W = \sum_{m=1}^M \mathbf{U}^m 1_{\{W=m\}}$$

is a sample from the mixture copula

$$\sum_{m=1}^M w_m C_m(u_1, u_2, \ldots u_d).$$

Finally, let us turn to the product copulas as defined by Theorem 17.3.5. While they may appear more complicated than mixture copulas, product copulas are, in fact, quite straightforward to simulate. To present the basic idea, we first observe that if random variables X_m, $m = 1, \ldots, M$, are independent, then

$$P\left(\max_{m=1,\ldots,M} X_m < x\right) = P\left(X_1 < x, \ldots, X_M < x\right) = \prod_{m=1}^{M} P\left(X_m < x\right).$$

(17.57)

As above, let $\mathbf{U}^m = (U_1^m, \ldots, U_d^m)$, $m = 1, \ldots, M$, be a collection of independent samples from the copulas C_m. As pointed out by Liebscher [2008], if we define

$$\mathbf{U} = (U_1, \ldots, U_d), \quad U_i \triangleq \max_{m=1,\ldots,M} q_{m,i}^{-1}(U_i^m), \quad i = 1, \ldots, d,$$

it follows from (17.57) that

$$P\left(U_1 < u_1, \ldots, U_d < u_d\right)$$

$$= P\left(\max_{m=1,\ldots,M} q_{m,1}^{-1}(U_1^m) < u_1, \ldots, \max_{m=1,\ldots,M} q_{m,d}^{-1}(U_d^m) < u_d\right)$$

$$= \prod_{m=1}^{M} P\left(U_1^m < q_{m,1}(u_1), \ldots, U_d^m < q_{m,i}(u_d)\right),$$

which is the product copula we wished to produce. We leave it to the reader to write down a step-by-step simulation scheme for the product copula, using the result above.

17.7 Limitations of the Copula Method

The copula method has gained widespread acceptance for multi-rate derivatives, in large part due to the ease with which a multivariate distribution consistent with market-observed marginal distributions can be constructed and parameterized. Before the reader starts assuming that copulas are a panacea, we should warn that copula applications have their share of limitations. First and foremost, we emphasize that copulas generally do not result in a dynamic model for the yield curve, nor are they consistent with the most popular classes of such models. As we have seen, the copula method allows us to easily ascribe a joint terminal distribution to a collection of CMS rates, for the purpose of pricing European multi-rate options. In the general case it is difficult, if not impossible, to find a dynamic term structure model that would be consistent with the joint terminal distributions produced by the copula method. To see why this might be problematic, consider a path-dependent exotic security for which a multi-rate vanilla is part of the payoff specification; in such a case, application of a copula for the "embedded" vanilla option would inherently be inconsistent with the dynamic model itself. Such an inconsistency may, among other ills, cause internal arbitrages in the model and nonsensical hedge ratios.

Also, before one gets carried away by the simplicity and convenience of copula parameterizations, it should be remembered that copulas in practical use are not usually chosen for their links to observed financial relationships, but instead for their technical properties and ease of implementation. While we (as should be clear to the reader by now) always welcome a dose of pragmatism in financial modeling, choosing tools simply because they are easy to use, rather than because they make sense, is obviously a problematic idea. A related issue is the fact that many copulas have parameters which are either devoid of meaning, or are consistently misinterpreted by users. For instance, the parameters in a Gaussian copula, the entries in the correlation matrix R, are obviously *not* the actual correlations of the interest rates being modeled[11] — instead, the (linear) correlations between rates in the copula will depend on the marginal distribution of the rates. In effect, the meaning of the copula parameters (the matrix R) will change when the marginal distributions of the rates change. See Appendix 17.A for a concrete demonstration of this effect.

As a last warning about copulas, let us note that a typical copula uses just a few parameters to capture the often complex dependence structure of various rates over periods of time. In practice, the parameterization of any model almost invariably defines the rules for parameter interpolation: if a is a particular model parameter, then it is natural for the users to use constant or linear interpolation in a, either explicitly in a risk management system or implicitly in their heads, to fill in values between observations. In a copula setting, this can become a problem as naive interpolation of parameters across option expiries may lead to a distorted picture of how the dynamic dependence of two rates evolves[12]. Potential problems of this kind are touched upon in Section 17.5.

Although care must be taken to avoid some of the pitfalls above, copulas clearly have a place in modeling of vanilla derivatives and, in any case, have managed to become a de-facto standard for some important products, such as European CMS spread options. Our message in this section is simply that the limitations of the method should be clearly understood, to ensure that it is applied effectively and appropriately. On this note, let us stop our discussion of copulas and look at alternative ways of introducing dependence between market rates.

---

[11]Unless the marginal distributions of the rates are Gaussian, of course.

[12]At the time of writing, the Gaussian copula correlations implied from at-the-money spread options were largely independent of the expiry, an observation that is inconsistent with predictions of almost all multi-factor term structure models. Some observers attribute this market feature to the proliferation of copulas with time-independent parameters.

17.8 Stochastic Volatility Modeling for Multi-Rate Options

As described in Chapter 16, a stochastic volatility model is often our preferred choice for pricing of single-rate derivatives such as swaptions and CMS-linked products. In the context of multi-rate derivatives, having the distribution of each rate described by a stochastic volatility model gives us an opportunity to define co-dependence between these rates by techniques other than the copula method. Broadly, if each swap rate involved in the payoff of a given multi-rate derivative has its own asset process and its own stochastic variance process (such as in (16.75)–(16.76)), then the co-dependence structure between rates can be controlled by correlating the Brownian motions that drive the asset and stochastic variance processes.

A stochastic volatility model for a given swap rate is often formulated in the annuity measure specific to that rate, in which case a translation into a common measure will be required in the multi-rate setup. Leaning on the general discussion in Section 17.2, we choose as common measure the forward measure associated with the payment date T_p. Conveniently, the problem of translation of dynamics from annuity to forward measures has already been considered in Chapter 16 where two different approaches were suggested. We consider both in turn.

17.8.1 Measure Change by Drift Adjustment

In Section 16.6.11 we derived a change of drifts associated with a shift of measure for SDEs driving a stochastic volatility model. Applying Proposition 16.6.8 to each swap rate, we obtain the following dynamics for a collection of d swap rates $(S_1(\cdot), \ldots, S_d(\cdot))$ in the T_p-forward measure,

$$dS_i(t) = \lambda_i \varphi_i \left(S_i(t)\right) \sqrt{z_i(t)} \left(dW_i^{T_p}(t) + v^{S_i}(t) \, dt\right), \tag{17.58}$$

$$dz_i(t) = \theta \left(1 - z_i(t)\right) dt + \eta_i \psi_i \left(z_i(t)\right) \left(dZ_i^{T_p}(t) + v^{z_i}(t) \, dt\right), \quad z_i(0) = 1, \tag{17.59}$$

where $i = 1, \ldots, d$ and the different parameters are explained in Section 16.6.11. Individual swap rate parameters $(\lambda_i, \varphi_i(\cdot), \eta_i)$ are obtained by the standard European swaption calibration for each swap rate (Section 16.1.4), and the drifts $v^{z_i}(t)$, $v^{S_i}(t)$ follow by the measure-change arguments of Proposition 16.6.8 (see also Corollary 16.6.9 for an important special case). The dependence structure between the swap rates may then be defined by correlations

$$\left\langle dW_i^{T_p}(t), dW_j^{T_p}(t) \right\rangle = \rho_{i,j} \, dt, \quad \left\langle dZ_i^{T_p}(t), dZ_j^{T_p}(t) \right\rangle = R_{i,j} \, dt, \quad i, j = 1, \ldots, d,$$

where we keep asset/volatility correlations for each swap rate at zero, i.e.

$$\left\langle dW_i^{T_p}(t), dZ_i^{T_p}(t) \right\rangle = 0, \quad i = 1, \dots, d,$$

to replicate the setup of Proposition 16.6.8 exactly. Alternatively, we can calibrate these correlations together with other marginal parameters (which would require a small extension of Proposition 16.6.8). Valuation of multi-rate derivatives in the model (17.58) requires Monte Carlo simulation, as the dimensionality of the model is high and the complexity of drifts does not lend itself easily to closed-form approximations.

17.8.2 Measure Change by CMS Caplet Calibration

The need to perform Monte Carlo simulation in the model (17.58) does not automatically render the approach unsuited for practical purposes — the Monte Carlo simulation of a d-asset stochastic volatility model is fairly quick, especially for important special cases of $d = 2, 3$. Nevertheless, it is a drawback. One way to develop more efficient schemes relies on the approach in Section 16.6.10, where we suggested translating stochastic volatility parameters from the annuity measure to the forward measure by first pricing CMS caplets in the model defined in the annuity measure, and then calibrating a new stochastic volatility model in the T_p-measure to these prices. See in particular (16.77) and the surrounding discussion. We remind the reader that this approach to a measure change is largely ad-hoc.

Let us assume that parameters have been suitably adjusted to incorporate the measure translation for each swap rate $S_i(t)$, $i = 1, \dots, d$. We therefore have available a CMS-adjusted forward rate $\widetilde{S}_i = \mathrm{E}^{T_p}(S_i(T))$, and a triple of T_p-measure parameters (λ_i, b_i, η_i) for each rate, where we dropped tildes from the notation of (16.77) for improved readability. A d-rate model can then be formulated by correlating all the driving Brownian motions,

$$dS_i(t) = \lambda_i \left(b_i S_i(t) + (1 - b_i) \widetilde{S}_i \right) \sqrt{z_i(t)} \, dW_i^{T_p}(t), \quad S_i(0) = \widetilde{S}_i, \quad (17.60)$$

$$dz_i(t) = \theta \left(1 - z_i(t) \right) dt + \eta_i \sqrt{z_i(t)} \, dZ_i^{T_p}(t), \quad z_i(0) = 1,$$

$i = 1, \dots, d$, with the correlations defined by a $2d \times 2d$ correlation matrix R in the block form

$$\mathrm{Corr} \left(\begin{pmatrix} dW^{T_p}(t) \\ dZ^{T_p}(t) \end{pmatrix}, \begin{pmatrix} dW^{T_p}(t) \\ dZ^{T_p}(t) \end{pmatrix} \right) = R, \quad R \triangleq \begin{pmatrix} R^{WW} & R^{WZ} \\ (R^{WZ})^\top & R^{ZZ} \end{pmatrix},$$

$$(17.61)$$

where $W^{T_p}(t) = (W_1^{T_p}(t), \dots, W_d^{T_p}(t))^\top$, $Z^{T_p}(t) = (Z_1^{T_p}(t), \dots, Z_d^{T_p}(t))^\top$, and the matrices R^{WW}, R^{WZ}, R^{ZZ} are $d \times d$. We emphasize that the parameters (λ_i, b_i, η_i), $i = 1, \dots, d$, are not obtained by a standard European swaption calibration for each swap rate, but by the more complicated two-step calibration described in Section 16.6.10.

With the joint dynamics of all swap rates under the same measure, the model (17.60) presents a straightforward extension of a one-factor displaced

Heston model to d dimensions. For standard payoffs such as spread options or, more generally, options on the weighted average of d rates, (17.60) is simple enough for us to derive efficient closed-form approximations by Markovian projection methods, an exercise that we postpone to Appendix A. For more complicated derivatives we can instead resort to Monte Carlo simulation. Each individual swap rate can be efficiently simulated using the methods from Section 9.5, for instance the Quadratic-Exponential (QE) discretization scheme for stochastic variance of Section 9.5.3.3, and the simplified Broadie-Kaya algorithm of Section 9.5.5.2 for the swap rate. To correlate different swap rates, as well as swap rate variances (and swap rates to variances of other swap rates) we just draw correlated Gaussian random variables to drive the discretization schemes[13]. For non-linear schemes such as QE, we note that this approach is not exact as the correlation between increments of swap rates and variances will not be exactly equal to the correlations of driving Gaussian variables. Nevertheless, numerical tests on the QE scheme show that this seemingly naive approximation is often of very good quality even for relatively coarse time discretizations. It is worth pointing out, however, that construction of accurate discretization schemes for multi-dimensional Heston-style stochastic volatility SDEs is an area of ongoing research.

17.8.3 Impact of Correlations on the Spread Smile

Apart from $\langle dW_i^{T_p}(t), dZ_i^{T_p}(t) \rangle$, $i = 1, \ldots, d$, correlation parameters (17.61) in the model (17.60) do not affect the marginal distributions of each rate. They are, however, expected to affect the joint distribution of the rates; so, in a way, they define a "copula" function for the swap rates. To gain some intuition for the impact of various correlation parameters on the joint distribution, let us for illustrative purposes consider a model suitable for a CMS spread option with payoff

$$(S_1(T) - S_2(T) - K)^+, \text{ paid at } T_p, \tag{17.62}$$

i.e., a version of (17.60) with $d = 2$. The correlation matrix then has the form,

$$R = \begin{pmatrix} 1 & R_{12}^{WW} & R_{11}^{WZ} & R_{12}^{WZ} \\ R_{12}^{WW} & 1 & R_{21}^{WZ} & R_{22}^{WZ} \\ R_{11}^{WZ} & R_{21}^{WZ} & 1 & R_{12}^{ZZ} \\ R_{12}^{WZ} & R_{22}^{WZ} & R_{12}^{ZZ} & 1 \end{pmatrix},$$

where only the parameters R_{12}^{WW}, R_{12}^{WZ}, R_{21}^{WZ}, R_{12}^{ZZ} are "free" in the sense of not affecting the marginal distributions of the two swap rates.

---

[13]If the QE scheme is used, whenever a uniform random variable U is required, we would write $U = \Phi(Z)$ with Z being a standard Gaussian random variable. This way, the QE scheme can be driven solely by Gaussian random variables.

Of the various entries in R, the "spot-spot" correlation R_{12}^{WW} is the most obvious in its effect on option value. Specifically, as the spread option value depends strongly on the variance of the swap rate difference, increasing (decreasing) the correlation between the Brownian motions driving the rates will decrease (increase) the option value. In the model (17.60), R_{12}^{WW} is typically the primary determinant of the spread option value or, equivalently, the overall level of the spread volatility smile (see Section 17.4.1). The effect of other correlation entries in R is more subtle and hard to grasp without resorting to numerical experiments. To conduct such an experiment, let us look at a CMS spread option with expiry $T=5$ years, with the model parameters in Table 17.2.

	Rate 1	Rate 2
CMS-adjusted swap rate $\widetilde{S}_i$	4.97%	4.60%
Volatility λ_i	11.8%	13.2%
Skew b_i	100%	70%
Mean reversion of variance θ	10%	10%
Volatility of variance η_i	120%	120%

Table 17.2. Model Parameters for Heston model for CMS Spread

The base case for the correlation matrix is given in (17.63).

$$R = \begin{pmatrix} 100\% & 95\% & -25\% & -25\% \\ 95\% & 100\% & -20\% & -25\% \\ -25\% & -20\% & 100\% & 95\% \\ -25\% & -25\% & 95\% & 100\% \end{pmatrix}. \qquad (17.63)$$

As is evident from Figure 17.3, the "vol-vol" correlation R_{12}^{ZZ} will move the overall level of the volatility smile up and down; after correcting for this level effect with the "spot-spot" correlation R_{12}^{WW}, increasing "vol-vol" correlation will allow one to add curvature to the spread volatility smile. Also, in Figure 17.3 we see that "spot1-vol2" correlation R_{12}^{WZ} affects the slope of the spread volatility smile. Interestingly, for this particular set of parameters, the "spot2-vol1" correlation R_{21}^{WZ} has very little impact on the spread volatility smile (as a consequence, the effect is not shown).

17.8.4 Connection to Term Structure Models

As we discussed previously in Section 17.7, making the copula method of Section 17.3 consistent with a typical dynamic term structure model is a difficult objective that may not be possible to achieve effectively. In this respect, multi-rate vanilla models of the type (17.60) have a clear edge, as stochastic volatility is our preferred method of adding volatility smile

Fig. 17.3. Implied Normal Spread Volatility

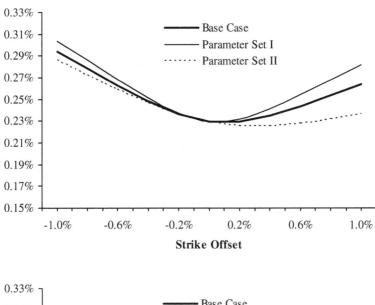

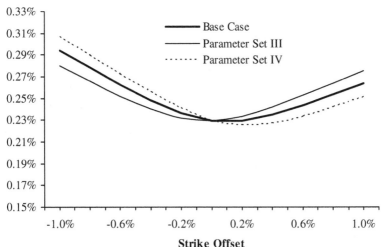

Notes: Implied Normal spread volatility for the spread option (17.62) in the model described in Section 17.8.3, in particular in Table 17.2 and equation (17.63). The correlation scenarios shown in the graphs are listed in Table 17.3. "Strike offset" is the difference between the strike and the expected value of the CMS spread.

capabilities to full term structure models. In particular, the multi-stochastic volatility LM model of Section 15.7, when applied to spread options, takes the

	Base Case	Set I	Set II	Set III	Set IV
R_{12}^{WW}	95%	92%	97%	95%	95%
R_{12}^{WZ}	-25%	-25%	-25%	-30%	-20%
R_{21}^{WZ}	-25%	-25%	-25%	-25%	-25%
R_{12}^{ZZ}	95%	97%	93%	95%	95%

Table 17.3. Correlation Parameter Sets for Figure 17.3

form (17.60) (see Proposition 15.7.1), and hence the methods of Section 17.8 could be applied.

As multi-stochastic volatility models have yet to enter the mainstream, a more pressing task would be the construction of efficient methods for spread option pricing in simpler term structure models, e.g. ordinary (mono) stochastic volatility LM models. A crude approach for this was introduced in Section 14.4.3.2; the next section discusses several refinements.

17.9 CMS Spread Options in Term Structure Models

There are several reasons why we would want to efficiently calculate values of multi-rate derivatives — and in particular CMS spread options — in a term structure model. For example, for pricing an exotic derivative on underlyings that involve multi-rate payoffs we may wish to check how closely the term structure model values the underlying compared to the market; any observed differences can be used to correct the price of the exotic produced by the term structure model (see Chapter 21 for more on this topic). Another need arises in calibration, as we sometimes include CMS spread options in the calibration set as a source of market-implied correlations between swap rates (see Section 14.5.9). The efficiency requirements imposed by both of these applications typically rule out Monte Carlo methods, so here we seek to develop closed-form approximations for a few specific term structure models.

17.9.1 Libor Market Model

We first consider Libor market (LM) models, as these are often used as workhorses for pricing exotics on multi-rate underlyings. For simplicity, let us focus on a CMS spread option with the payoff (17.62), although similar techniques can be applied to other payoffs.

In Section 14.4.3.2 we presented a simple method for pricing CMS spread options in the LM model, based on a Gaussian approximation to the spread process (and little concern for the effects of measure changes). Here we develop a more sophisticated, and more accurate, approach that utilizes copula techniques. In fact, there is not much left to do at this point, as we have already developed almost all the "ingredients" we need for the method. In Section 16.6.6 we derived the annuity mapping function consistent with the LM model; using the machinery of Section 16.6.9, the mapping function could be turned into a CDF of each swap rate in the T_p-forward measure. Thus, according to Section 17.4, all that remains to do is to find a copula that is consistent with the dependence structure of the swap rates in the LM model. Considering, for concreteness, the stochastic volatility LM model specification (14.15)–(14.16) (also see (16.59)), we recall the results of Section 14.4.3.1 and, in particular, (14.34) that tell us that the instantaneous correlation between two swap rates in the stochastic volatility LM model is, in fact, deterministic[14] and is the same as in a displaced log-normal LM model. In a displaced log-normal LM model, swap rates are approximately functions of Gaussian variables (see Appendix 17.A), so we can approximate the dependence structure of two swap rates in the SV-LM model by a Gaussian copula. The correlation to be used in the copula may then be approximated as the term correlation of the swap rates $\rho_{\text{term}}(0, T)$, as given by (14.35). We formalize this discussion as a proposition.

Proposition 17.9.1. *The undiscounted value $V(0)$ of a CMS spread option with the payoff (17.62) in the stochastic volatility LM model (14.15)–(14.16) is approximately given by the two-dimensional Gaussian copula integral*

$$V(0) \approx \int \int \left(\left(\Psi_1^{T_p} \right)^{-1} (\Phi(z_1)) - \left(\Psi_2^{T_p} \right)^{-1} (\Phi(z_2)) - K \right)^+$$
$$\times \, \phi(z_1, z_2; \rho_{\text{term}}(0, T)) \, dz_1 dz_2,$$

where $\phi(z_1, z_2; R)$ is a two-dimensional Gaussian density with correlation R, $\rho_{\text{term}}(0, T)$ is given by (14.35), and $\Phi(z)$ is the standard Gaussian CDF.

Remark 17.9.2. In Proposition 17.9.1, $\Psi_i^{T_p}(s)$ are the T_p-forward measure CDFs of the swap rates $S_i(T)$, $i = 1, 2$. From Proposition 16.6.4 these CDFs can be obtained from the annuity-measure swap rate CDFs $\Psi_i^{A_i}(s)$ and the annuity mapping functions $\alpha_i(s)$, $i = 1, 2$. The annuity mapping function $\alpha_i(s)$ may be computed as in Proposition 16.6.3 and the CDFs $\Psi_i^{A_i}(s)$, $i = 1, 2$, can be obtained by differentiating (in strike) European swaption values in the LM model, per formulas in Proposition 14.4.3 or Section 15.2.

The result in Proposition 17.9.1 trivially generalizes to general multi-rate cash flows. Of course, for spread options specifically, we can make the algorithm more efficient by utilizing one-dimensional integration formulas from Section 17.6.2. If the speed of this approach is too slow, we can trade accuracy for speed by utilizing some of the ideas of Section 17.8.2. Recall that in the SV-LM model, the distribution of each $S_i(T)$ in its corresponding annuity measure is given by the SV model. If we can approximate the marginal

---

[14]This is a consequence of using just a single stochastic variance scaling applied to all Libor rates.

distributions of $S_1(T)$, $S_2(T)$ in Q^{T_p} by the SV model as well, then the distribution of $S_1(T) - S_2(T)$ could be quite effectively approximated by the same distribution, see Appendix A, leading to a closed-form approximation to the spread option value. Section 16.6.10 outlines the numerical approach to the measure change that preserves the SV distribution class; in a pinch, we can just reuse the SV parameters from the annuity measure distributions while adjusting the forward swap rates with the CMS convexity adjustments,

$$S_i(0) \to E^{T_p}(S_i(T)), \quad i = 1, 2. \tag{17.64}$$

While Section 16.6.10 warns against performing the measure shift in the SV model by the sole change of the forward (17.64), the impact of such laissez-faire approach on the quality of CMS spread option valuation in LM models is likely to be muted, given all the other approximations we are making on the way.

Finally, let us note that Antonov and Arneguy [2009] present an alternative approximation idea that is based on working in a measure in which the spread $S_1(t) - S_2(t)$ is a martingale. While such a measure cannot be easily characterized by a numeraire, it can still be defined by the drift change in the Brownian motions driving the model's SDEs. We refer the interested reader to the source paper, as the approach is too involved to be described here in a few lines.

17.9.2 Quadratic Gaussian Model

Having dealt with the LM models, we now turn our attention to the multi-factor quadratic Gaussian (QG) models of Section 12.3. Interestingly, the most productive approach here is quite different from that used for the LM models — reflecting the different approaches for European swaption pricing in the two models, see Section 12.3.4. For the purpose of elaborating on this observation, we continue examining the CMS spread option (17.62).

As a start, we recall that in a QG model, a swap rate $S(T)$ is a deterministic function of the state vector $z(T)$, $S(T) = S(T, z(T))$. In one of the approximations to the swap rate we developed in Section 12.3.4, we replaced the function with a quadratic form of the state vector, see (12.92). Let us denote the quadratic approximations to the two rates involved in the payoff (17.62) by $S_{1,q}(T, z)$ and $S_{2,q}(T, z)$,

$$S_{i,q}(T, z) = z^\top \gamma_{S_i} z + h_{S_i}^\top z - E^A \left(z(T)^\top \gamma_{S_i} z(T) + h_{S_i}^\top z(T) \right) + S_i(0), \ i = 1, 2.$$

Then, the undiscounted value of the spread option is given approximately by

$$V(0) \approx E^{T_p} \left((S_{1,q}(T, z(T)) - S_{2,q}(T, z(T)) - K)^+ \right).$$

Two points should now be obvious. One is that the difference of two quadratic forms $S_{1,q}(T, z) - S_{2,q}(T, z)$ is itself a quadratic form in z,

$$S_{1,q}(T,z) - S_{2,q}(T,z) = z^\top (\gamma_{S_1} - \gamma_{S_2}) z + (h_{S_1} - h_{S_2})^\top z$$
$$- \mathrm{E}^A \left(z(T)^\top (\gamma_{S_1} - \gamma_{S_2}) z(T) + (h_{S_1} - h_{S_2})^\top z(T) \right) + S_1(0) - S_2(0).$$

Another is that the distribution of $z(T)$ in the T_p-forward measure is known — it is Gaussian with a known mean $m^{T_p}(0, T, 0)$ and covariance matrix $\nu^{T_p}(0, T, 0)$, see Proposition 12.3.4. Hence, the problem of pricing a spread option in the QG model is almost identical to the problem of pricing a European swaption, and any of the methods of Section 12.3.4.3 could be applied (with the most efficient method probably being the two-dimensional integration method in Theorem 12.3.7). Further details are available in Piterbarg [2009b].

17.A Appendix: Implied Correlation in Displaced Log-Normal Models

17.A.1 Preliminaries

The purpose of this appendix is to briefly examine how marginal distributions affect spread option prices in a Gaussian copula. We shall consider only marginal distributions originating from displaced log-normal dynamics, so first let us recall that a (one-dimensional) process of the form

$$dX(t) = \lambda \left((1 - b)X_0 + bX(t) \right) dW(t), \quad X(0) = X_0,$$

with $b > 0$ has the solution

$$X(T) = \frac{X_0}{b} \left(\exp \left(-b^2 \lambda^2 T/2 + b\lambda W(T) \right) - 1 + b \right).$$

Given this result, let us consider two swap rates $S_1(t)$ and $S_2(t)$ in the T-forward measure, with

$$\mathrm{E}^T (S_1(T)) = \mathrm{E}^T (S_2(T)) = S_0 > 0,$$

and set, for $b > 0$,

$$S_i(T) = \frac{S_0}{b} \left(\exp \left(-b^2 \lambda_i^2 T/2 + b\lambda_i Z_i \sqrt{T} \right) - 1 + b \right), \quad b > 0, \quad i = 1, 2, \tag{17.65}$$

where Z_1 and Z_2 are two standard Gaussian random variables with constant correlation ρ. As $S_1(T)$ and $S_2(T)$ are monotonic functions of correlated Gaussian variables, it is clear[15] that their dependency is generated by a two-dimensional Gaussian copula with correlation parameter ρ.

---

[15] A copula is easily shown to be invariant with respect to monotonic transformations of the underlying variables.

From the definition (17.65) it follows that

$$\text{Var}\,(S_i(T)) = \frac{S_0^2}{b^2}\left(e^{b^2 \lambda_i^2 T} - 1\right), \quad i = 1, 2,$$

and

$$\text{Cov}\,(S_1(T), S_2(T)) = \frac{S_0^2}{b^2}\left(e^{\rho b^2 \lambda_1 \lambda_2 T} - 1\right). \tag{17.66}$$

Therefore

$$\text{Corr}\,(S_1(T), S_2(T)) = \frac{e^{\rho b^2 \lambda_1 \lambda_2 T} - 1}{\sqrt{e^{b^2 \lambda_1^2 T} - 1}\sqrt{e^{b^2 \lambda_2^2 T} - 1}}. \tag{17.67}$$

We notice that $\text{Corr}\,(S_1(T), S_2(T))$ is, of course, not equal to ρ, but instead is given by a more complicated expression that in most practically relevant cases will decrease in b, for fixed λ_1 and λ_2. Based solely on this observation, we might therefore expect that spread option prices would increase in the skew parameter b.

17.A.2 Implied Log-Normal Correlation

The correlation in (17.67) is not a particularly market-oriented way of characterizing spread option value. As we have seen earlier in this chapter, a better measure may be to use implied spread volatility. To offer a slightly different perspective, in this appendix we instead work with an implied log-normal correlation ρ_{LN}, defined as the value of the copula correlation ρ that will match the ATM spread option[16] in the true model, after b has been set to 1 (but with ATM option prices kept constant). Examining how ρ_{LN} depends on the skew b will give us a convenient scalar measure of how skew affects co-dependence in a spread option setting.

To compute ρ_{LN}, first consider the zero-strike payout

$$V(0) = \text{E}^T((S_1(T) - S_2(T))^+).$$

Writing $V(0) = \text{E}^T(S_1(T)(1 - S_2(T)/S_1(T))^+)$ shows (after a measure change, see Section 17.6.5) that

$$V(0) = \frac{S_0}{b}\,(2\Phi(d) - 1), \quad d = \frac{\sigma b}{2}\sqrt{T}, \tag{17.68}$$

where $\sigma = \sqrt{\lambda_1^2 + \lambda_2^2 - 2\rho\lambda_1\lambda_2}$. We note in passing that when $b = 1$ this formula is known as the *Margrabe formula*, see Margrabe [1978]. We also note that, for $i = 1, 2$,

---

[16] We can extend the definition to handle non-ATM strikes, but this shall not be needed for the purposes of this appendix.

$$\mathrm{E}^T \left((S_i(T) - S_0)^+ \right) = \frac{S_0}{b} \left(2\Phi(y_i) - 1 \right), \quad y_i = \frac{\lambda_i b}{2} \sqrt{T}. \tag{17.69}$$

Suppose now that we observe implied Black ATM volatility for $S_1(T)$ and $S_2(T)$ to be σ_1 and σ_2, respectively. Using the Black option pricing formula, for $i = 1, 2$ we therefore must have

$$\mathrm{E}^T \left((S_i(T) - S_0)^+ \right) = S_0 \Phi \left(\frac{1}{2} \sigma_i \sqrt{T} \right) - S_0 \Phi \left(-\frac{1}{2} \sigma_i \sqrt{T} \right)$$

$$= S_0 \left(2\Phi \left(\frac{\sigma_i}{2} \sqrt{T} \right) - 1 \right). \tag{17.70}$$

Suppose also that we have best-fit b to some value different from 1. To preserve ATM option prices, we equate (17.69) and (17.70),

$$\frac{1}{b} \left(2\Phi \left(\frac{\lambda_i b}{2} \sqrt{T} \right) - 1 \right) = 2\Phi \left(\frac{\sigma_i}{2} \sqrt{T} \right) - 1, \quad i = 1, 2.$$

We can solve these equations in closed form for λ_i to yield

$$\lambda_i = \frac{2}{b\sqrt{T}} \Phi^{-1} \left(\left(\Phi \left(\frac{\sigma_i}{2} \sqrt{T} \right) - \frac{1}{2} \right) b + \frac{1}{2} \right), \quad i = 1, 2. \tag{17.71}$$

In most cases, this equation results in $\lambda_1 \approx \sigma_1$ and $\lambda_2 \approx \sigma_2$. Assuming that the copula correlation in our model has been set to ρ, by our definition of ρ_{LN}, we get from (17.68) that

$$\frac{1}{b} \left(2\Phi(d) - 1 \right) = 2\Phi \left(\frac{1}{2} \sqrt{\sigma_1^2 + \sigma_2^2 - 2\rho_{LN}\sigma_1\sigma_2} \sqrt{T} \right) - 1,$$

or

$$\sqrt{\sigma_1^2 + \sigma_2^2 - 2\rho_{LN}\sigma_1\sigma_2} = \frac{2}{\sqrt{T}} \Phi^{-1} \left(\frac{1}{2b} \left(2\Phi(d) - 1 \right) + \frac{1}{2} \right), \tag{17.72}$$

from which we can extract an analytical expression for $\rho_{LN} = \rho_{LN}(b, \sigma_1, \sigma_2, \rho, T)$. We omit further details in the interest of brevity, but just notice the curious fact that $\rho_{LN}(b, \sigma_1, \sigma_1, 0.5, T) = 0.5$, independently of b, σ_1, T.

17.A.3 A Few Numerical Results

Going forward, we assume that log-normal volatilities are $\sigma_1 = \sigma_2 = 25\%$. First, we set $T = 10$ years and $\rho = 0.75$. Figure 17.4 shows ρ_{LN} as a function of b. Notice the implied correlation ρ_{LN} here decreases in b; in other words, tilting the skew downwards (i.e. lowering b) has the effect of increasing the effective spread option correlation.

Next, we freeze $b = 0.25$ and $\rho = 0.75$ and let T vary. Figure 17.5 shows the resulting effects on ρ_{LN}. The effect of skew on implied correlation ρ_{LN} is

Fig. 17.4. Log-Normal Correlation ρ_{LN}

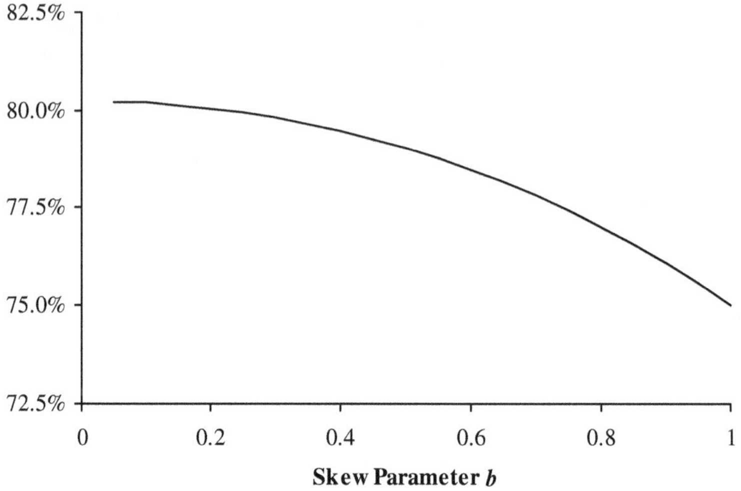

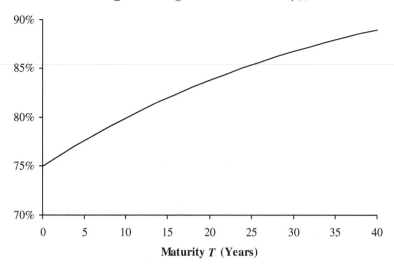

Notes: Model parameters are: $T = 10$, $\rho = 0.75$, and $\sigma_1 = \sigma_2 = 0.25$.

Fig. 17.5. Log-Normal Correlation ρ_{LN}

Notes: Model parameters are: $b = 0.25$, $\rho = 0.75$, and $\sigma_1 = \sigma_2 = 0.25$.

clearly quite sensitive to maturity T, with the difference $|\rho - \rho_{LN}|$ increasing in T.

While we do not show the results, we note that lowering volatility (σ_1 and σ_2) will have qualitatively the same effect as lowering maturity. In low-volatility regimes (and for short-dated options), the practice of assuming that the market-implied ATM spread option correlation is independent of skew may therefore be defensible.

18

Callable Libor Exotics

Having discussed relatively simple vanilla securities in Chapters 16 and 17, we now move to the other extreme of the product spectrum and consider issues of valuation, calibration and risk management of *callable Libor exotics* (CLEs). CLEs were defined in Section 5.14 and constitute the most complicated class of interest rate derivatives traded in the market. The material in this chapter deals with CLEs in general; later chapters will take a more in-depth look at the idiosyncrasies of some of the most popular and/or challenging securities inside the CLE class.

18.1 Model Calibration for Callable Libor Exotics

Due to their inherent complexity, CLEs will have non-trivial dependencies on the dynamics of market rates and will require sophisticated term structure models for valuation and risk management. As discussed in earlier chapters, computation of non-vanilla prices in such models virtually always requires Monte Carlo simulation, which in turn introduces complications in determining the optimal exercise rules for CLEs. We shall spend a good portion of this chapter on a detailed discussion of algorithms for this particular problem. However, before a model can even be considered for CLE valuation, it needs to be calibrated, i.e. the *volatility structure* of the model needs to be parameterized to match available market information relevant for valuation.

At this point we should emphasize that CLEs and other exotic interest rate derivatives are different from many of vanilla securities considered previously in that their values are not observable in the market, a consequence of the fact that most exotic derivatives are sold to clients rather than traded between dealers. This fact leads to fundamental differences in the way models are used for valuation of vanilla vs. exotic derivatives. Whereas for many vanilla derivatives a model is primarily used to generate a hedging strategy and to perform inter- and extrapolation on market-observable prices, for exotics the values are fundamentally derived from a model[1]. Given the absence of market prices of exotics themselves, models for these securities have to be calibrated indirectly, to other market (and perhaps also non-market) information that is deemed relevant for the class of exotics under consideration. Speaking very loosely, the purpose of a model for CLEs can therefore be characterized as performing a sophisticated extrapolation of information from a series of "spanning" vanilla markets to compute a meaningful exotics price, something we already alluded to in Section 14.5.5.

18.1.1 Risk Factors for CLEs

In order to make sure that a CLE model captures as much relevant market information as possible, it is important to understand what we can actually rely on when calibrating a model. There are essentially three sources of potentially relevant information. The first, and arguably the most important, source is the market prices of liquid vanilla interest rate derivatives. The second source is historical information about market quantities such as volatilities and correlations of various market rates that may not be available for observation directly from quoted market prices. The third, and somewhat more amorphous, source is the modeler's beliefs on what constitutes reasonable behavior of the model parameters. The last category, for example, includes views on how time-stationary or how smooth model parameters should be.

To perform model calibration, we typically choose particular targets from all three sources above, and ultimately set model parameters in a way to match those targets in our model. In order to be able to choose relevant targets, it is important to identify risk factors that affect the valuation of the particular CLE in question. This part of the analysis is necessarily product specific, so let us look at a particular example to demonstrate some salient points. We choose a reasonably representative (albeit not very complicated) example, namely a callable inverse floater, or CIF, with exercise dates

$$T_1 < T_2 < \ldots < T_{N-1}$$

and structured coupon

$$C_n = \min\left(\max\left(6\% - L_n(T_n), 0\%\right), 4\%\right),$$

where $L_n(t)$ is a forward Libor rate fixing at T_n. In the language of Section 5.14, we see that the coupon strike is equal to 6%, the cap is at 4% and the floor is at 0%. It should be clear that the coupon can be decomposed into a portfolio of a long floorlet with strike 6% and a short floorlet with strike 2%. Hence, the callable inverse floater can be thought of as a Bermudan-style option on a combination of floors and a Libor leg.

---

[1] A practice sometimes known as *mark-to-model*.

The underlying swap of this CIF consists of a collection of floorlets, which are vanilla derivatives with observable market prices. Market-implied volatilities[2] as always serve as a convenient representation of these market values. For classification purposes that will be clear in a moment, we call such volatilities *spot volatilities*, as we can observe them on the valuation date, or "on the spot", from the market. Clearly, these volatilities affect the value of the exotic swap that underlies the CIF and hence have a direct impact on the value of the CIF. For that reason we should include them as targets for model calibration. Importantly, as caplet volatilities virtually always exhibit a volatility smile, volatilities for both 2% and 6% strikes are potentially relevant.

While the underlying exotic swap has no dependencies on spot volatilities of market rates other than those underlying the floorlets inherent in the CIF, the callability structure of the CIF nevertheless introduces additional dependence on other vanilla derivatives. To see this, suppose that we, say, ignored all exercise CIF dates but one, in which case the CIF would degenerate into a European-style option to enter the underlying swap. Even though the underlying swap is not vanilla (i.e. not a standard fixed-for-floating rate swap), it is clearly related to one and our single-call CIF is therefore related to the European swaption on that swap. More generally, the (multi-call) CIF will depend on the (spot) implied volatility of the swap rate that fixes on T_i, and runs for the period $[T_i, T_N]$, for each $i = 1, \ldots, N - 1$. It is less clear what strike we should use to define this implied volatility; while all strikes are in fact relevant, as shall be clear later, sometimes one may choose to simplify and just use the ATM swaption volatility.

To summarize our discussion so far, we have argued that the value of the CIF depends on market-implied spot Libor rate volatilities for all expiries until T_{N-1} (of two different strikes), and spot swap rate volatilities for those swaptions for which expiry plus tenor is equal to T_N — the so called *core*, *diagonal*, or *coterminal* swaptions. When calibrating a model to price this CIF, we would seek (as a minimum) to calibrate it to these market volatilities.

While there are no more obvious spot volatility targets that should be included in CIF calibration, there are other volatility-related quantities that will affect the valuation of the CIF and other similar CLEs. Let us imagine that we have arrived at exercise date T_n, $n < N - 1$, and assume that the CIF has not been exercised in the past. At this point in time, the option holder needs to decide whether to exercise and receive the underlying value $U_n(T_n)$ (see (5.24)) or whether to continue to hold on to the CIF (with a view of perhaps exercising it later). The time T_n value of the remaining part of the underlying swap depends on caplet volatilities as observed at time T_n. As these volatilities will be known only at time T_n, we call them *forward volatilities* of Libor rates. Likewise, the option to hold a Bermudan-style

---

[2]As expressed by, say, Black volatilities.

option on the underlying $U_{n+1}(T_n)$, will depend on forward volatilities of swap rates, specifically core swaption volatilities observed at time T_n. As the exercise decision at time T_n evidently will depend on the forward volatility structure at time T_n, the time 0 value of the CIF will depend on it as well.

At time 0, forward volatilities are generally unknown random variables, but any model will impose certain dynamics on the volatility structure of interest rates, which will have a direct impact on the model value of the CIF (and other CLEs). Ideally, we should make sure that the model projections for forward volatilities are in line with market-implied information. The fact that such information is typically either not available or difficult to extract adds significantly to the challenge of model calibration for CLEs. Frequently, it will be necessary to lean on historical data and to impose exogenous assumptions that the model builder might feel are financially reasonable.

In Section 13.1.8.1 we identified inter-temporal correlations — i.e. correlations of core swap rates observed on different fixing dates — as value drivers for Bermudan swaptions. Inter-temporal correlations are closely linked to the forward volatilities discussed above, and the two concepts can often be used interchangeably. To demonstrate, note that the (inter-temporal) correlation between two core swap rates, $S_{n,N-n}(T_n)$ and $S_{m,N-m}(T_m)$ $(n < m)$ is given by

$$\text{Corr}\,(S_{n,N-n}(T_n), S_{m,N-m}(T_m)) = \text{Cov}\,(S_{n,N-n}(T_n), S_{m,N-m}(T_m))$$
$$\times \text{Var}\,(S_{n,N-n}(T_n))^{-1/2}\,\text{Var}\,(S_{m,N-m}(T_m))^{-1/2}\,.$$

As $S_{m,N-m}(T_m) - S_{m,N-m}(T_n)$ is often only mildly dependent on $S_{m,N-m}(T_n)$ and $S_{n,N-n}(T_n)$, we can rewrite this as

$$\text{Corr}\,(S_{n,N-n}(T_n), S_{m,N-m}(T_m))$$
$$\approx \text{Cov}\,(S_{n,N-n}(T_n), S_{m,N-m}(T_m))\,\text{Var}\,(S_{n,N-n}(T_n))^{-1/2}$$
$$\times\,(\text{Var}\,(S_{m,N-m}(T_n)) + \text{Var}\,(S_{m,N-m}(T_m) - S_{m,N-m}(T_n)))^{-1/2}$$
$$= \text{Corr}\,(S_{n,N-n}(T_n), S_{m,N-m}(T_n))$$
$$\times\left(1 + \frac{\text{Var}\,(S_{m,N-m}(T_m) - S_{m,N-m}(T_n))}{\text{Var}\,(S_{m,N-m}(T_n))}\right)^{-1/2}$$

and we see that, with (spot) correlation $\text{Corr}(S_{n,N-n}(T_n), S_{m,N-m}(T_n))$ and (spot) variance $\text{Var}(S_{m,N-m}(T_n))$ fixed, the inter-temporal correlation is directly linked to the forward variance (square of forward volatility) $\text{Var}(S_{m,N-m}(T_m) - S_{m,N-m}(T_n))$. This equivalence will play a role when we discuss the local projection method in Section 18.4.

Another aspect of the model behavior worth emphasizing here is the volatility smile dynamics imposed by the model. In Sections 8.8 and 16.1.1 we considered the hedging impact of joint moves in rates and volatility smiles. For CLEs, volatility smile dynamics affect not only the hedging strategy, but also the valuation of CLEs, due to the forward volatility effect we discussed earlier. We notice that for many CLEs the effective coupon strikes at the exercise boundary often end up being deeply in- or out-of-the-money, since interest rate levels at which option exercise is optimal are usually significantly different from the levels of rates at inception of the trade.

18.1.2 Model Choice and Calibration

As argued above, the value of a typical CLE depends strongly on volatility smile dynamics and on market (spot) implied volatilities for a wide selection of vanilla options, often across a range of strikes. As such the appropriate choice of a model for CLE valuation should typically involve the following criteria.

- Ability to calibrate to a large collection of vanilla options across expiries, tenors, and strikes.
- Reasonable and controllable dynamics of the volatility structure.
- Multi-factor interest rate dynamics, especially for CLEs on multi-rate underlyings.

A combination of these requirements often rules out simpler, low-dimensional models, especially for more complicated CLEs. We typically recommend using either Libor market, multi-factor quasi-Gaussian, or multi-factor quadratic Gaussian models. While simpler low-dimensional models might succeed in fitting spot volatility information for selected Libor and swap rates, they often achieve this only by using values of model parameters that would imply unrealistic evolution of the volatility structure and, therefore, unreasonable pricing of CLEs (but see Section 18.4).

Another observation favors models that can calibrate to a large collection of European swaptions. A swap rate can be seen to be, approximately, a weighted average of Libor rates (see Section 14.4.2). Hence, an implied European swaption volatility contains some amount of information on market-consistent correlations between Libor (and other swap) rates. Extracting this information is complicated, but in a sense is what a Libor market model calibrated to, say, the whole swaption grid (including caps) is designed to do. In the same spirit, a model calibrated to all swaptions and caps can be thought of as giving us the best available *implied* forward volatility structure, i.e. market-consistent information of what the most likely behavior of the volatility structure through time would be.

Another dimension to consider here is the volatility smile dynamics. Obviously, for the model to generate unspanned stochasticity in volatility movements, stochastic volatility (SV) version of the model would be required. Given the discussion in Section 8.8 and the importance for CLE pricing of controlling the dynamics of the volatility smile, SV models typically have a clear edge over their local volatility counterparts in CLE pricing applications.

With the choice of the model settled, calibration typically proceeds by identifying proper targets and fitting them with model parameters. For Libor market models we, in fact, have already covered relevant issues quite extensively, and the reader is advised to revisit Section 14.5.5. For other types of multi-factor models, the issues are similar and can be resolved in the same spirit. The mechanics of calibration for relevant models are covered in their respective chapters. For example, smile calibration of LM models is discussed in Section 15.2.

18.2 Valuation Theory

18.2.1 Preliminaries

While various measures could be used for CLE valuation, for concreteness we choose to use the spot Libor measure Q^B with the discrete money market numeraire $B(t)$ defined on a tenor structure

$$0 = T_0 < T_1 < \ldots < T_N, \quad \tau_n = T_{n+1} - T_n.$$

For notational simplicity, we let $E(\cdot)$ be the expected value operator in measure Q^B.

We recall from Section 5.14 that a callable Libor exotic is a Bermudan-style option on an exotic swap that specifies an exchange of structured coupons C_n for Libor rates L_n, fixing on T_n and paying on T_{n+1}, $n = 1, \ldots, N - 1$. We denote by X_n the net payment seen by the structured coupon receiver,

$$X_n = \tau_n \times (C_n - L_n(T_n)). \tag{18.1}$$

Also, we let $U_n(t)$ be the n-th exercise value, i.e. the value of all future payments if the callable Libor exotic is exercised at time T_n. Clearly

$$U_n(t) = B(t) \sum_{i=n}^{N-1} E_t \left(B(T_{i+1})^{-1} X_i \right). \tag{18.2}$$

For completeness we set

$$U_N(t) \equiv 0.$$

If a callable Libor exotic is exercised on T_n, the holder will receive $U_n(T_n)$, the present value of the remainder of the underlying exotic swap.

Finally, let $H_n(t)$ be the value at time t of a callable Libor exotic where exercise opportunities have been restricted to the dates $\{T_{n+1}, \ldots, T_{N-1}\}$. In particular, $H_0(0)$ is then the time 0 value of the CLE. Each H_n is called a *hold value*, since $H_n(T_n)$ is the value of the choice of *not* exercising on date T_n, i.e. continuing to "hold" the derivative. We must necessarily have

$$H_0(t) \geq H_1(t) \geq \ldots \geq H_{N-2}(t). \tag{18.3}$$

18.2.2 Recursion for Callable Libor Exotics

Let $\mathcal{T}_n$ be a set of all stopping time indices that exceed n, $n \in \{0, \ldots, N-1\}$, i.e. a set of random variables taking values in the set $\{n+1, \ldots, N\}$ such that for any k and any $\xi \in \mathcal{T}_n$,

$$\{\xi = k\} \in \mathcal{F}_{T_k}.$$

The sequence of random variables $H_0(T_0), \ldots, H_{N-1}(T_{N-1})$ (as in Section 18.2.1) defines the Snell envelope for the sequence of (discounted) exercise values,

$$H_n(T_n) = B(T_n) \sup_{\xi \in \mathcal{T}_n} \mathrm{E}_{T_n} \left(B(T_\xi)^{-1} U_\xi(T_\xi) \right), \quad n = 0, \ldots, N-1. \quad (18.4)$$

By the general theory of optimal stopping (see Chapter 1.10), the random time index η_n that maximizes the right-hand side of (18.4) is given by

$$\eta_n(\omega) = \min \{k > n : U_k(T_k) \geq H_k(T_k)\} \wedge N, \quad (18.5)$$

and we set

$$\eta \triangleq \eta_0 = \min \{k \geq 1 : U_k(T_k) \geq H_k(T_k)\} \wedge N.$$

With this definition, the value of a callable contract can be re-written as

$$H_0(0) = \mathrm{E} \left(B(T_\eta)^{-1} U_\eta(T_\eta) \right) = \mathrm{E} \left(\sum_{n=\eta}^{N-1} B(T_{n+1})^{-1} X_n \right). \quad (18.6)$$

The Hamilton-Jacobi-Bellman equation that corresponds to the optimal stopping problem (18.4) can be solved by backward induction. In particular, we have, for $n = N-1, \ldots, 1$,

$$H_{n-1}(T_{n-1}) = B(T_{n-1}) \mathrm{E}_{T_{n-1}} \left(B(T_n)^{-1} \max \left(H_n(T_n), U_n(T_n) \right) \right), \quad (18.7)$$

subject to the terminal condition $H_{N-1} \equiv 0$. The recursion starts at the final time $n = N-1$ and progresses backward in time until we obtain the desired time 0 security value $H_0(0)$.

The financial meaning of the recursion above is straightforward. If a callable contract H_0 has not been exercised up to and including time T_n, then it is worth the hold value $H_n(T_n)$. If the callable contract is exercised at time T_n its value is equal to $U_n(T_n)$. Assuming optimal exercise, the value of the callable Libor exotic H_0 at time T_n is the maximum of the two,

$$\max \left(H_n(T_n), U_n(T_n) \right).$$

The value of this payoff at time T_{n-1} is then

$$B(T_{n-1}) \mathrm{E}_{T_{n-1}} \left(B(T_n)^{-1} \max \left(H_n(T_n), U_n(T_n) \right) \right).$$

But clearly, this is the value of the CLE that can only be exercised at times T_n and beyond, i.e. of the CLE H_{n-1}, as specified in (18.7).

18.2.3 Marginal Exercise Value Decomposition

Before discussing techniques to numerically implement the valuation equations of the previous section, let us briefly review an important decomposition result for CLEs that follows from the recursion (18.7). After a slight rewrite, we obtain

$$H_{n-1}(T_{n-1}) - B(T_{n-1})\mathrm{E}_{T_{n-1}}\left(B(T_n)^{-1}H_n(T_n)\right)$$
$$= B(T_{n-1})\mathrm{E}_{T_{n-1}}\left(B(T_n)^{-1}\left(U_n(T_n) - H_n(T_n)\right)^+\right).$$

Note that

$$B(T_{n-1})\mathrm{E}_{T_{n-1}}\left(B(T_n)^{-1}H_n(T_n)\right) = H_n(T_{n-1}),$$

so that

$$H_{n-1}(T_{n-1}) - H_n(T_{n-1}) = B(T_{n-1})\mathrm{E}_{T_{n-1}}\left(B(T_n)^{-1}(U_n(T_n) - H_n(T_n))^+\right).$$

Taking discounted expectations to time 0 we obtain

$$H_{n-1}(0) - H_n(0) = \mathrm{E}\left(B(T_n)^{-1}\left(U_n(T_n) - H_n(T_n)\right)^+\right)$$

and, summing up from $n = 1$ to $N - 1$,

$$H_0(0) - H_{N-1}(0) = \sum_{n=1}^{N-1} \mathrm{E}\left(B(T_n)^{-1}\left(U_n(T_n) - H_n(T_n)\right)^+\right).$$

Since $H_{N-1}(0) = 0$ and $H_0(0)$ is the time 0 value of the CLE, we have established the following proposition, which is essentially inspired by an integral representation of an American option from Jamshidian [1992], as presented in Proposition 1.10.7.

Proposition 18.2.1. *The time 0 value $H_0(0)$ of a callable Libor exotic is equal to the sum of European options on the difference between the exercise and hold values at all exercise dates,*

$$H_0(0) = \sum_{n=1}^{N-1} \mathrm{E}\left(B(T_n)^{-1}\left(U_n(T_n) - H_n(T_n)\right)^+\right). \tag{18.8}$$

In Proposition 18.2.1 each of the terms $\mathrm{E}(B(T_n)^{-1}(U_n(T_n) - H_n(T_n))^+)$ can be interpreted as a "marginal" exercise value, i.e. as the incremental value of having an exercise at time T_n. The total CLE value is then equal to the sum of the marginal exercise values.

18.3 Monte Carlo Valuation

If a model admits a low-dimensional Markovian representation, then PDE methods are available for valuation of CLEs, and the backward recursion (18.7) is easy to implement, see Section 2.7.4. The situation is more complicated with Monte Carlo based models; fortunately, a range of methods for approximate solutions of optimal exercise problems in Monte Carlo have been developed. The mechanics of the scheme have been broadly outlined in Section 3.5.4, and we now proceed to discuss implementation details for the CLE class.

18.3.1 Regression-Based Valuation of CLEs, Basic Scheme

We start with a basic regression-based method for estimating the value of a CLE. As we recall from the discussion of Section 3.5.4, the regression-based LS (for *Least Squares*) scheme builds on the idea that the expected value of a random variable conditioned on information at time T can be calculated in a Monte Carlo simulation by regressing the random variable against simulated state variables of the model at time T.

To make matters precise, we introduce some notation. Let $\zeta(t) = (\zeta_1(t), \dots, \zeta_q(t))^\top$ be a q-dimensional vector process of *regression variables*, to be defined later. For a given Monte Carlo path ω, let us denote the value of a random variable X on that path by $X(\omega)$. Suppose K paths $\omega_1, \dots, \omega_K$ are generated. For a random variable X, we denote by

$$\mathcal{R}_T(X)$$

the results of regression of the K-dimensional vector $(X(\omega_1), \dots, X(\omega_K))$ on the $K \times q$ matrix of regression variable observations at time T, $(\zeta(T, \omega_1), \dots, \zeta(T, \omega_K))^\top$, i.e.

$$\mathcal{R}_T(X) = \zeta(T)^\top \beta,$$

where the q-dimensional column vector β is obtained by, for instance[3], solving the minimization problem

$$\left\| \left(X(\omega_1) - \zeta(T, \omega_1)^\top \beta, \dots, X(\omega_K) - \zeta(T, \omega_K)^\top \beta \right) \right\|^2 \to \min. \quad (18.9)$$

This least-squares problem can be solved in closed form as explained in Section 3.5.4; to link our discussion here to the results of that section we simply need to set $\zeta_i(t) = \psi_i(x(t))$, $i = 1, \dots, q$, where ψ's and x's are defined in Section 3.5.4. For future reference, we denote the solution vector β to (18.9) by

---

[3]We discuss the details of the implementation of the regression algorithm later, in Section 18.3.10.

$$\mathcal{C}\left(\mathcal{R}_T(X)\right),$$

with the notation meant to be read as "coefficients of the regression of X", so that

$$\mathcal{R}_T(X) = \mathcal{C}\left(\mathcal{R}_T(X)\right)^{\top} \zeta(T). \tag{18.10}$$

As it turns out, there are several possible LS schemes for CLE valuation. The most basic one is based on the idea of simply replacing the conditional expected value operator $\mathbf{E}_T$ in (18.7) with the *regression operator* $\mathcal{R}_T$ introduced above, i.e. to write

$$\widetilde{H}_{n-1}\left(T_{n-1}\right) = \mathcal{R}_{T_{n-1}}\left(\frac{B(T_{n-1})}{B(T_n)} \max\left(\widetilde{H}_n(T_n), U_n(T_n)\right)\right), \tag{18.11}$$

for $n = N-1, \ldots, 1$, where $\widetilde{H}_n$ is an approximation to the true hold value H_n. This approach was originally suggested in Carrière [1996] and Tsitsiklis and Roy [2001]. While we shall later describe better LS schemes than (18.11), let us nevertheless take some care in documenting all the steps necessary to apply it. Some of the steps are shared with the standard (non-callable) Monte Carlo valuation algorithm, but we list them anyway for completeness.

1. Choose and calibrate a term structure model (such as the LM model).
2. Decide on what to use for the regression variables process $\zeta(t)$ (we will have more to say about this later).
3. Simulate K paths $\omega_1, \ldots, \omega_K$. For the LM model in particular, each ω_k represents one simulated path of all core Libor rates.
4. For each path ω_k calculate simulated values of the numeraire $B(T_n, \omega_k)$, $n = 1, \ldots, N-1$.
5. For each path ω_k, calculate[4] the value $U_n(T_n, \omega_k)$ of the underlying exotic swap on all exercise dates $n = 1, \ldots, N-1$.
6. For each path ω_k, calculate the values of the q-dimensional regression variables process ζ on the exercise dates, $\zeta(T_n, \omega_k)$, $n = 1, \ldots, N-1$.
7. Set $\widetilde{H}_{N-1} \equiv 0$.
8. For each $n = N-1, \ldots, 1$
   a) Form a K-dimensional vector $V_n = (V_n(\omega_1), \ldots, V_n(\omega_K))^{\top}$,

$$V_n\left(\omega_k\right) = \frac{B(T_{n-1}, \omega_k)}{B(T_n, \omega_k)} \max\left(\widetilde{H}_n\left(T_n, \omega_k\right), U_n\left(T_n, \omega_k\right)\right) \tag{18.12}$$

   for $k = 1, \ldots, K$.

---

[4] In this basic scheme we implicitly assume that underlying value U_n at time T_n can be calculated in closed form from the simulated yield curve at time T_n (or, more generally, from the model state variables observed on and *before* time T_n). This is a strong restriction which we relax later.

b) Calculate

$$\widetilde{H}_{n-1}(T_{n-1}) = \mathcal{R}_{T_{n-1}}(V_n)$$

by regressing $V_n(\omega_k)$ against the matrix of regression variables observed on date T_{n-1}, using (18.9).

9. At this point we have computed $\widetilde{H}_0(T_0)$ which is an estimate of the value of the CLE. Return it.

Note that the last iteration in Step 8 ($n = 1$) involves regression on the values of $\zeta(T_0)$, $\widetilde{H}_0(T_0) = \mathcal{R}_{T_0}(V_1)$. As $T_0 = 0$, $\zeta(T_0)$ is not random, so the regression here degenerates into a simple average[5] of $V_1(\omega_1), \ldots, V_1(\omega_K)$.

There are a number of shortcomings in the above scheme that make it poorly suited for industrial-strength pricing of CLEs. We list them below as they shall guide us in systematically building more refined versions of the algorithm.

1. In Step 5, there is an assumption that exercise values $U_n(T_n)$ can be computed in closed form from information available at time T_n. While this is possible for, say, simple Bermudan swaptions, more complicated exotic swaps will generally violate this assumption.
2. The use of "regressions upon regressions", i.e. the fact that we apply $\mathcal{R}_{T_{n-1}}$ in (18.11) to (a function of) a regressed value $\widetilde{H}_n(T_n)$, could lead to significant biases building up as the scheme marches backward in time.
3. In general we cannot state whether $\widetilde{H}_n$'s are low- or high-biased estimates of H_n's.

18.3.2 Regression for Underlying

To improve our basic LS scheme, we start out by examining the assumption of the basic LS scheme that exercise value at time t can be evaluated in closed form from time t state variables (typically the simulated yield curve and the state of stochastic volatility parameters). As mentioned earlier, vanilla fixed-for floating-rate swaps underlying Bermudan swaptions certainly satisfy this assumption, as their values can be obtained by discounting projected cash flows on a simulated yield curve at time t. In principle, a number of other exotic swaps could fit under the assumption as well. For example, for a callable inverse floater, the underlying swap is a collection of simple Libor rate options, the values of which could be calculated from simulated market data at times T_n, $n = 1, \ldots, N-1$, by applying option pricing formulas developed earlier in this book. While such a scheme is indeed possible, it would come at a significant computational cost of having to invoke option valuation formulas multiple times for each simulated path, i.e. easily thousands of

---

[5] A naive numerical implementation of regression will not have this property, see Section 18.3.10.

times overall. Moreover, closed-form caplet/swaption/CMS option pricing is rarely exact in term structure interest rate models; embedded into a backwards recursion algorithm these errors may potentially build up and skew the pricing results for the CLE.

Fortunately, it turns out that the extension of the basic scheme to arbitrary underlying swaps is simple. Specifically, we can use formula (18.2) which states that the exercise value $U_n(T_n)$ is equal to the conditional expected value of all (net) coupons paid after T_n. As we have already introduced the regression operator as a numerical proxy for conditional expected value, all we need to do now is to approximate $U_n(T_n)$ with the *regressed* value of all (net) coupons paid after T_n. Of course a coupon paid at time T_{i+1} is always[6] measurable with respect to $\mathcal{F}_{T_{i+1}}$ or, equivalently, can be calculated from the knowledge of the simulated model state variables up to and including T_{i+1}. So, all coupon values are known once a given path is simulated, and we can extend the basic scheme to arbitrary underlyings with the following two modifications. First, we replace Step 5 with

5a. For each path ω_k, calculate the values of all net coupons $X_n(T_n, \omega_k)$, $n = 1, \ldots, N-1$.

Second, we must replace the formula (18.12) in Step 8 of the basic scheme with

$$V_n(\omega_k) = \frac{B(T_{n-1}, \omega_k)}{B(T_n, \omega_k)} \max\left(\tilde{H}_n(T_n, \omega_k), \tilde{U}_n(T_n, \omega_k)\right), \quad k = 1, \ldots, K,$$
(18.13)

where

$$\tilde{U}_n(T_n) = \mathcal{R}_{T_n}\left(\sum_{i=n}^{N-1} \frac{B(T_n)}{B(T_{i+1})} X_i\right).$$
(18.14)

We can write (18.14) in a backward-recursive format,

$$Y_n = \frac{B(T_n)}{B(T_{n+1})}(X_n + Y_{n+1}), \quad \tilde{U}_n(T_n) = \mathcal{R}_{T_n}(Y_n), \quad n = N-1, \ldots, 1,$$
(18.15)

where we start from $Y_N \equiv 0$. In this form it fits nicely into the backward recursion of the basic LS scheme.

Interestingly, we can rewrite (18.2) (for $t = T_n$) as

$$U_n(T_n) = E_{T_n}\left(B(T_n)B(T_{n+1})^{-1}(X_n + U_{n+1}(T_{n+1}))\right),$$
(18.16)

which gives raise to an *alternative backward scheme* for $\tilde{U}_n(T_n)$,

$$\tilde{U}_n(T_n) = \mathcal{R}_{T_n}\left(B(T_n)B(T_{n+1})^{-1}\left(X_n + \tilde{U}_{n+1}(T_{n+1})\right)\right), \quad n = N-1, \ldots, 1.$$
(18.17)

---

[6]The value of a coupon must obviously be known at the time it is paid.

While (18.16) is trivially equivalent to (18.2) due to the additivity of the conditional expected values and the iterated conditional expectations property, (18.17) is *not* equivalent to (18.15). In (18.17), on step n we add (discounted) coupon X_n to an already-regressed value of future coupons $\widetilde{U}_{n+1}(T_{n+1})$, whereas in (18.15) we sum up values of un-regressed coupons from $N-1$ to n and then regress them to obtain $\widetilde{U}_n$. The difference between the two schemes originates with the fact that the regression operator does not satisfy an equivalent to the iterated conditional expectations property[7] (see footnote 3 of Chapter 4). An inquisitive reader may ask whether one scheme is better (in the sense of producing smaller bias for the value of the CLE) than the other. The answer is not straightforward. While (18.15) avoids applying regression to the output of other regressions and thus could be expected to produce lower bias, empirical evidence suggests that no scheme is universally better than the other for all CLEs. We consequently recommend a flexible implementation that can use both schemes.

18.3.3 Valuing CLE as a Cancelable Note

Using regressions for the underlying is not the only approach that extends the basic scheme in Section 18.3.2 to arbitrary underlying swaps. An alternative scheme is based on the idea of representing a CLE as a *cancelable note*. To describe this in more detail, let us denote

$$G_n(t) = H_n(t) - U_n(t),$$

and obtain from (18.7) and (18.16) that

$$
\begin{aligned}
G_{n-1}&(T_{n-1}) \\
&= H_{n-1}(T_{n-1}) - U_{n-1}(T_{n-1}) \\
&= \mathrm{E}_{T_{n-1}}\left(B(T_{n-1})B(T_n)^{-1}(\max(H_n(T_n), U_n(T_n)) - (X_{n-1} + U_n(T_n)))\right) \\
&= \mathrm{E}_{T_{n-1}}\left(B(T_{n-1})B(T_n)^{-1}\left((H_n(T_n) - U_n(T_n))^+ - X_{n-1}\right)\right) \\
&= \mathrm{E}_{T_{n-1}}\left(B(T_{n-1})B(T_n)^{-1}\left(-X_{n-1} + G_n(T_n)^+\right)\right),
\end{aligned}
\tag{18.18}
$$

$n = N, \ldots, 1$, where we for uniformity of notation have introduced a "fake" coupon at time zero,

$$X_0 = 0.$$

We see that the value $G_0(0)$ is the value of the swap that pays (net) coupons $-X_n$ on dates T_n, $n = 0, \ldots, N-1$, plus the right to cancel it, at zero cost, on any of the exercise dates $T_1, \ldots, T_{N-1}$. In fact, as explained in Section 5.14, it is this structure (a cancelable structured note) that a bank usually sells to clients. While the representation of the cancelable note as a CLE plus

---

[7] A similar issue arises with the regressed hold values, as discussed in Section 18.3.4.

the underlying non-callable swap is often convenient for risk management, for the purposes of valuation it is, in fact, often useful to consider the original format.

The LS version of (18.18) is, naturally, given by

$$V_n(\omega_k) = \frac{B(T_{n-1}, \omega_k)}{B(T_n, \omega_k)} \left(-X_{n-1}(\omega_k) + \tilde{G}_n \left(T_n, \omega_k\right)^+ \right), \quad k = 1, \ldots, K,$$

followed by

$$\tilde{G}_{n-1}(T_{n-1}) = \mathcal{R}_{T_{n-1}}(V_n), \tag{18.19}$$

for each $n = N, \ldots, 1$. The starting point is given by $\tilde{G}_N(T_N) \equiv 0$. We trust the reader should have no problem amending the basic scheme to use the cancelable note representation.

Interestingly, by linearity (which *does* hold for the regression operator, unlike the tower property), we can rewrite (18.19) as

$$\tilde{G}_{n-1}(T_{n-1}) = -\mathcal{R}_{T_{n-1}} \left(B(T_{n-1})B(T_n)^{-1} X_{n-1} \right)$$
$$+ \mathcal{R}_{T_{n-1}} \left(B(T_{n-1})B(T_n)^{-1}\tilde{G}_n(T_n)^+ \right), \quad n = N, \ldots, 1.$$

In other words, to get the value of the cancelable note at time T_{n-1}, start with its value $\tilde{G}_n(T_n)$ at time T_n, apply the optimal cancelability condition $\tilde{G}_n(T_n) \to \tilde{G}_n(T_n)^+$, discount to T_{n-1}, regress on $\zeta(T_{n-1})$, and then add the time T_{n-1} value of the $(n-1)$-th coupon. If X_{n-1} is actually $\mathcal{F}_{T_{n-1}}$-measurable, as is often the case (exceptions include range accrual coupons and coupons that depend on rates observed in-arrears, i.e. at the end of the observation period rather than the beginning), the scheme simplifies a bit more,

$$\tilde{G}_{n-1}(T_{n-1}) = -P(T_{n-1}, T_n) X_{n-1}$$
$$+ \mathcal{R}_{T_{n-1}} \left(B(T_{n-1})B(T_n)^{-1}\tilde{G}_n(T_n)^+ \right), \quad n = N, \ldots, 1. \tag{18.20}$$

Once the value $\tilde{G}_0(0)$ has been calculated, the estimate of the CLE value $H_0(0)$ can be obtained via $H_0(0) = \tilde{G}_0(0) + U_0(0)$, where U_0 may, if not available in closed form, be calculated via a (standard) Monte Carlo algorithm for (18.2). In estimating U_0, we would normally reuse the paths ω_k, $k = 1, \ldots, K$ that were used for the computation of $\tilde{G}_0(0)$.

18.3.4 Using Regressed Variables for Decision Only

Our second criticism of the basic LS scheme of Section 18.3.1 focused on the issue that the values that are regressed at time T_{n-1} themselves come as the result of a regression at time T_n. Such compounded regression could lead to substantial biases. Interestingly, a small modification of the algorithm reduces this bias significantly.

Let us work with the cancelable note scheme (18.20), although the same idea can easily be applied to the basic CLE scheme. Going back to the expression (18.6) for the value of the CLE expressed as the sum of all coupons paid post optimal exercise, we note that a similar formula holds for cancelable notes,

$$G_0(0) = -\mathrm{E}\left(\sum_{n=0}^{\eta-1} B\left(T_{n+1}\right)^{-1} X_n\right) \qquad (18.21)$$

(recall that $X_0 = 0$). The exercise index η here is the same as in (18.6), since the optimal cancel time for the cancelable note G is the same as the optimal exercise time for the CLE H. We rewrite the formula as

$$G_0(0) = -\mathrm{E}\left(\sum_{n=0}^{N-1} B\left(T_{n+1}\right)^{-1} X_n 1_{\{\eta>n\}}\right),$$

and note that

$$1_{\{\eta>n\}} = \prod_{i=1}^{n} 1_{\{G_i(T_i)>0\}}. \qquad (18.22)$$

Let us define V_n's recursively,

$$V_n = B(T_n)B\left(T_{n+1}\right)^{-1}\left(-X_n + 1_{\{G_{n+1}(T_{n+1})>0\}}V_{n+1}\right), \quad n = N-1,\ldots,0, \qquad (18.23)$$

with $V_N = 0$. Then

$$V_0 = -\sum_{n=0}^{N-1} B(T_{n+1})^{-1}\left(\prod_{i=1}^{n} 1_{\{G_i(T_i)>0\}}\right) X_n$$

and, computing the expected value and using (18.22), we obtain that

$$G_0(0) = \mathrm{E}\left(V_0\right).$$

Moreover,

$$G_n(T_n) = \mathrm{E}_{T_n}\left(V_n\right). \qquad (18.24)$$

We note that the recursion for V_n involves the value of the cancelable note for exercise decisions only, through the indicators $1_{\{G_i(T_i)>0\}}$, whereas the coupon values that are added up, X_i's, are never regressed. Following Longstaff and Schwartz [2001], we can take advantage of this observation by defining a new approximation $\widehat{G}_n(T_n)$, $n = 1,\ldots,N-1$, to the true value of the cancelable note by

$$\widehat{G}_{n-1}(T_{n-1}) = B(T_{n-1})B(T_n)^{-1}\left(-X_{n-1} + 1_{\{\widetilde{G}_n(T_n)>0\}}\widehat{G}_n(T_n)\right),$$
(18.25)

$$\widetilde{G}_{n-1}(T_{n-1}) = \mathcal{R}_{T_{n-1}}\left(\widehat{G}_{n-1}(T_{n-1})\right),$$
(18.26)

defined backwards for $n = N, \ldots, 1$. The first equation here comes from (18.23) and the second from (18.24). We emphasize that regression is only used to establish the exercise indicator functions in (18.25), i.e. whether and where to exercise on each path. In (18.24), $\widehat{G}_0(0) = \widehat{G}_0(0, \omega)$ is the (random) accumulation of discounted coupons up to exercise, and our estimate of the true value of the cancelable note G_0 is given by $\widetilde{G}_0(0)$, i.e. the simple average of the realizations of $\widehat{G}_0(0, \omega_1), \ldots, \widehat{G}_0(0, \omega_K)$. Clearly, this is the numerical equivalent of the (unconditional) expected value in (18.21). We find that the scheme (18.25)–(18.26) typically has significantly less bias than the naive scheme (18.20).

While (18.25)–(18.26) is the scheme that we recommend for most applications, Egloff et al. [2007] introduce a "blend" of (18.25)–(18.26) and (18.20) with a tunable parameter that can be optimized over to select a scheme with the lowest bias. To briefly outline this idea, we note that the scheme (18.25)–(18.26) only uses un-regressed values of coupons while (18.20) always uses regressed values. A "blended" scheme uses the first few coupons that are unregressed, while the rest are regressed. For example, since we can write

$$G_0(0) = -\mathbb{E}\left(\sum_{n=0}^{m-1} B(T_{n+1})^{-1}X_n 1_{\{\eta>n\}} + B(T_m)^{-1}G_m(T_m)1_{\{G_m(T_m)>0\}}\right)$$

for any $m = 1, \ldots, N-1$, we can replace (18.25) with

$$\widehat{G}_{n-1}(T_{n-1}) = -\sum_{i=n-1}^{\rceil n+m-1\lceil -1}\left(\frac{B(T_{n-1})}{B(T_{i+1})}\left(\prod_{j=n}^{i}1_{\{\widetilde{G}_j(T_j)>0\}}\right)X_i\right)$$

$$+\mathcal{R}_{T_{n-1}}\left(\frac{B(T_{n-1})}{B(T_{\rceil n+m-1\lceil})}\left(\prod_{j=n}^{\rceil n+m-1\lceil}1_{\{\widetilde{G}_j(T_j)>0\}}\right)\widetilde{G}_{\rceil n+m-1\lceil}(T_{\rceil n+m-1\lceil})\right)$$

coupled with (18.26), where we have denoted $\rceil l\lceil \triangleq l \wedge N$. We would then run the regression algorithm for different m's (reusing the paths, of course) and choose the value of m that would give us the highest value[8]. We recover (18.20) for $m = 1$ and (18.25)–(18.26) for $m = +\infty$.

18.3.5 Regression Valuation with Boundary Optimization

The regression-based method can sometimes be improved with methods similar to those of the parametric boundary optimization discussed in Sec-

---

[8]Jumping a bit ahead, we note that it is essential to use this optimization in conjunction with an independent post-simulation, see Section 18.3.6.

tion 3.5.2, an idea popularized by Bender et al. [2006]. Starting with, for example, scheme (18.25)–(18.26), we fix a collection of trigger thresholds $\mathbf{h} = (h_1, \ldots, h_{N-1})$ and rewrite (18.25)–(18.26) as (note dependence on the triggers in the indicator functions)

$$\widehat{G}_{n-1}\left(T_{n-1}, \mathbf{h}\right) = B(T_{n-1})B(T_n)^{-1}\left(-X_{n-1} + 1_{\left\{\widetilde{G}_n(T_n, \mathbf{h}) > h_n\right\}}\widehat{G}_n\left(T_n, \mathbf{h}\right)\right),$$

$$\widetilde{G}_{n-1}\left(T_{n-1}, \mathbf{h}\right) = \mathcal{R}_{T_{n-1}}\left(\widehat{G}_{n-1}\left(T_{n-1}, \mathbf{h}\right)\right),$$

$n = N - 1, \ldots, 1$. For a given set $\mathbf{h}$, the scheme produces an estimate $\widetilde{G}_0(0, \mathbf{h})$ of the value of the cancelable note. Then, iterating over $\mathbf{h}$, we find the optimal (highest) value of $\widetilde{G}_0$ and return this as our improved estimate for the cancelable note. Of course, if our original regression was fundamentally sound, we should see the optimal value of $\mathbf{h}$ being very close to $(0, 0, \ldots, 0)$, in which case the trigger iteration adds no value. To the extent, however, that the original LS method produces significantly sub-optimal exercise decisions (e.g. due to a poor choice of regression variables), the trigger iteration may lead to pick-up of substantial additional value.

The search for the optimal value of the triggers can be efficiently organized as a sequence of $N - 1$ one-dimensional optimizations, along the same lines as the algorithm in Section 3.5.3. In particular, we see that $\widehat{G}_{n-1}(T_{n-1}, \mathbf{h})$ depends on $h_n, \ldots, h_{N-1}$ only. Moreover, if for a given n, h_n maximizes the value of $\widetilde{G}_0(0, \mathbf{h})$ then it also maximizes the value of the cancelable note with first $n - 1$ exercise dates removed, i.e. the value $\mathcal{R}_0(\widehat{G}_{n-1}(T_{n-1}, \mathbf{h}))$ (recall that $\mathcal{R}_0(X) = K^{-1}\sum X(\omega_k)$, i.e. just the average of path values of X). Hence, we find the optimal value of the n-th trigger h_n^* via

$$h_n^* = \operatorname*{argmin}_{h_n} \mathcal{R}_0(\widehat{G}_{n-1}(T_{n-1}, (h_n, h_{n+1}^*, \ldots, h_{N-1}^*))), \quad n = N - 1, \ldots, 1,$$

where, slightly abusing notation, we use $(h_n, h_{n+1}^*, \ldots, h_{N-1}^*)$ to denote a vector $\mathbf{h}$ with the last $N - n - 1$ elements fixed to the optimal values found on previous steps (and first $n - 1$ elements irrelevant). The optimization problem above is easy to solve, but we remind the reader of our comments from Section 3.5.3, where we noted that for a finite-path simulation, the objective functions in each of optimization problems will not be smooth, so one should avoid the use of a derivatives-based numerical optimizer.

18.3.6 Lower Bound via Regression Scheme

All the variations of the regression scheme developed so far produce an estimate of the value of the CLE H_0 (or, equivalently, the value of the corresponding cancelable note G_0) but the bias of the estimate is generally unknown. On one hand, the exercise decisions used in the schemes are necessarily suboptimal, as we use estimates, rather than actual values, of hold/underlying variables to define them. This, in isolation, suggests that our estimate should be biased low. But on the other hand, our schemes use the same set of sample paths to estimate the exercise decision as to calculate the values of the security if it is exercised. This could lead to an upward bias in the estimate as some amount of future information can affect our decision to exercise, leading to a "perfect foresight" high-bias, see (3.110).

For risk assessment and for price quotation, it is typically desirable to know the sign of biases in computed security prices. To control the sign of the bias in CLE valuation by regression methods, we can follow the advice of Section 3.5 and use the LS regression scheme to estimate the exercise decision rule only, while using an independent simulation to calculate the value of the CLE *given* that exercise rule. A typical implementation algorithm is outlined below.

1. Run the basic regression-based scheme in Section 18.3.1 or any of the alternatives in Sections 18.3.2–18.3.5.
2. The output of the regression are the regression coefficients for the hold and exercise values at all exercise times, $\mathcal{C}(\widetilde{H}_n(T_n))$, $\mathcal{C}(\widetilde{U}_n(T_n))$, $n = 1, \ldots, N - 1$. See (18.10).
3. Simulate additional K' paths, $\omega'_1, \ldots, \omega'_{K'}$, that are independent from the paths used in the regression scheme.
4. For each path ω'_k, calculate the values of the q-dimensional regression variables process ζ on the exercise dates, $\zeta(T_n, \omega'_k)$, $n = 1, \ldots, N - 1$.
5. For each path ω'_k, calculate an estimate of the exercise index $\widetilde{\eta}$ by

$$\widetilde{\eta}(\omega'_k) = \min\left\{ n \geq 1 : \mathcal{C}\left(\widetilde{U}_n(T_n)\right)^\top \zeta(T_n, \omega'_k) \right.$$
$$\left. \geq \mathcal{C}\left(\widetilde{H}_n(T_n)\right)^\top \zeta(T_n, \omega'_k) \right\} \wedge N. \quad (18.27)$$

In simple terms the exercise is based on the regression estimates of the exercise and hold values obtained in the basic scheme, applied to the new values of the regression variables ζ.

6. Calculate the CLE value as the Monte Carlo value of a knock-in discrete barrier option based on the exercise rule $\widetilde{\eta}$,

$$H_0(0) \approx \frac{1}{K'} \sum_{k=1}^{K'} \left(\sum_{n=\widetilde{\eta}(\omega'_k)}^{N-1} B(T_{n+1}, \omega'_k)^{-1} X_n(\omega'_k) \right). \quad (18.28)$$

The guaranteed low bias of this two-stage scheme comes at an additional cost of simulating and evaluating payoffs on extra K' paths. For the record, we often choose $K' \approx 10,000$ to $100,000$ and $K \approx K'/4$ to $K'/2$. So the cost is not inconsiderable, on average slowing down a valuation by a factor of up to 2 or so. Importantly, this additional cost is not incurred in the evaluation of most risk sensitivities, since we can reuse the exercise boundary in calculations of first-order sensitivities, as we explain in Chapter 24. However, if performance is an issue, it is worth pointing out that we often find the values obtained by the scheme (18.25)–(18.26) (with only a single batch of paths simulated) to be pretty close to those from the more costly two-stage scheme (18.28). In particular, it appears that in many practical situations (18.25)–(18.26) is biased low, even though it is not guaranteed to be so.

It is easy to see the strong connection between the scheme (18.25)–(18.26) and (18.28). If, instead of the independent paths in the second stage of (18.28) we used the same paths as in the regression scheme, i.e. $K' = K$ and $w'_k = w_k$ for $k = 1, \ldots, K$, then the values produced by the two scheme would be exactly the same. We leave the verification of this simple fact to the reader.

18.3.7 Iterative Improvement of Lower Bound

Consider the cancelable note that pays (net) coupons $-X_n$ at T_{n+1}, $n = 0, \ldots, N-1$. Suppose we are given some exercise policy, generally suboptimal. An exercise policy is, in essence, a stopping time index α that specifies when the note is canceled. For technical reasons we want this exercise strategy to be specified not just for the original cancelable note but also for cancelable notes with the first k, $k = 1, \ldots, N - 2$ exercise dates removed. This is most conveniently expressed as a collection of stopping times $\boldsymbol{\alpha} = (\alpha_0, \ldots, \alpha_{N-1})$, with $\alpha_0 = \alpha$, that satisfy the following *exercise policy consistency conditions*,

$$n + 1 \leq \alpha_n \leq N, \quad \alpha_{N-1} \equiv N,$$

and

$$\alpha_n > n + 1 \Rightarrow \alpha_n = \alpha_{n+1}, \quad n = 0, \ldots, N - 2.$$

Let us consider the part of the note that includes coupons $-X_n, \ldots, -X_{N-1}$ only. Then the value of this note, exercised per stopping time α_k, $k \geq n$, is given at time T_n by

$$G_n^{\alpha_k}(T_n) = -B(T_n)\mathrm{E}_{T_n}\left(\sum_{i=n}^{\alpha_k - 1} B\left(T_{i+1}\right)^{-1} X_i\right)$$

$$= -B(T_n)\mathrm{E}_{T_n}\left(\sum_{i=n}^{N-1} B\left(T_{i+1}\right)^{-1} X_i 1_{\{\alpha_k > i\}}\right).$$

We note that $\alpha_k \geq k + 1$, so the coupons $-X_n, \ldots, -X_k$ are always paid. The *optimal* exercise policy $\boldsymbol{\eta} \triangleq (\eta_0, \ldots, \eta_{N-1})$, as defined by (18.5), satisfies the consistency conditions, and we of course have

$$G_n^{\eta_n}(T_n) = G_n(T_n), \quad n = 0, \ldots, N - 1,$$

where on the right-hand side we have the actual values of the remaining parts of the cancelable note. The approximations to the optimal exercise rule we developed in previous sections, such as (18.27), satisfy the consistency conditions as well.

As pointed out previously, for any exercise policy α we have

$$G_n^{\alpha_n}(T_n) \leq G_n(T_n), \quad n = 0, \ldots, N - 1,$$

so an exercise policy gives us a way to obtain a lower bound, namely $G_0^{\alpha_0}(0)$, for the time 0 cancelable note value $G_0(0)$. It turns out that the general theory of optimal stopping tells us how to improve a given exercise policy, i.e. how to find another exercise policy that would be better, in the sense of producing a higher lower bound. The improvements could be iterated, eventually (after $N - 1$ iterations) converging to the optimal exercise policy irrespective of the starting point. This "policy iteration" method was first applied to the pricing of callable derivatives by Bender et al. [2006] and Kolodko and Schoenmakers [2006]. To demonstrate how policy iteration works, for a given policy α, let us define its improvement $\widehat{\alpha} = (\widehat{\alpha}_0, \ldots, \widehat{\alpha}_{N-1})$ by

$$\widehat{\alpha}_n = \min \left\{ k > n : \max_{j \geq k} G_k^{\alpha_j}(T_k) \leq 0 \right\} \wedge N, \quad n = 0, \ldots, N - 1. \quad (18.29)$$

To understand this definition, first note that $G_k^{\alpha_j}(T_k)$ is the time-T_k value of the coupon stream from T_k onwards, exercised per stopping time α_j, i.e. we basically add up the (discounted) coupons $-X_k, \ldots, -X_j$ and thereafter the remaining coupons subject to the original exercise rule. Hence, the improved exercise rules states that we should exercise (cancel the note) at the first date T_k for which holding on to the note under the original exercise policy does not make sense, specifically if the maximum of the remaining value of the note over all original exercise rules that go strictly past T_k is non-positive.

The proof that $\widehat{\alpha}$ is an improvement over α, i.e. that

$$G_0^{\widehat{\alpha}_0}(0) \geq G_0^{\alpha_0}(0)$$

can be found in Kolodko and Schoenmakers [2006]. The improvement could be applied to the policy $\widehat{\alpha}$ as well, and this can be iterated multiple times. After $N - 1$ iterations the exercise policy converges to the optimal one, as also proven in Kolodko and Schoenmakers [2006].

While in theory we can find the optimal exercise strategy by iteration starting from any initial policy — such as a trivial policy $\alpha_n = n + 1$ for any $n = 0, \ldots, N - 1$ — performing more than a few iterations is impractical, as shall be discussed shortly. In practice, a sensible strategy would apply just one round of improvements to an already decent exercise policy, such as (18.27) obtained by the basic scheme of Section 18.3.1 or its variations.

The high numerical cost of the policy iteration stems from the presence of terms like $G_k^{\alpha_j}(T_k)$ in (18.29). These are the time T_k values of the coupon stream exercised per some rule, and are rarely, if ever, available in closed form. One might guess that we could estimate the required conditional expected values via regressions, as we have done for other quantities. Alas, using regressed values in (18.29) in place of true conditional expected values generally leads to *no* policy improvements. Consider, for example, the scheme (18.25)–(18.26), with the corresponding exercise policy defined by

$$\alpha_n = \min\left\{ k > n : \widetilde{G}_k(T_k) \leq 0 \right\} \wedge N, \quad n = 0, \ldots, N-1.$$

Then the iterated exercise policy would base the exercise decision on $\max_{j \geq k} \widetilde{G}_k^{\alpha_j}(T_k)$. However, as α_n's are constructed to be the *optimal* stopping policy for $\widetilde{G}_n(T_n)$'s, we always have

$$\widetilde{G}_k^{\alpha_j}(T_k) \leq \widetilde{G}_k^{\alpha_k}(T_k) = \widetilde{G}_k(T_k),$$

and

$$\max_{j \geq k} \widetilde{G}_k^{\alpha_j}(T_k) = \widetilde{G}_k(T_k)$$

for any k. Thus the "improved" policy $\widehat{\boldsymbol{\alpha}}$ computed from (18.29) would coincide with $\boldsymbol{\alpha}$, the original policy.

As regression methods cannot be used, an alternative is to resort to "brute-force" nested simulation to compute unbiased estimates for the conditional expectations $G_k^{\alpha_j}(T_k)$. The basic idea is simple: if we need to estimate $\mathrm{E}_T(X)$ for some random variable X then, for each Monte Carlo path ω, we simulate a number of additional sub-paths with the simulated model state at time T on path ω as a starting point, and then estimate the conditional expected value on path ω by averaging the values of X realized on all sub-paths[9]. For policy iteration, such sub-paths must be launched for each (outer) simulated path ω_k, $k = 1, \ldots, K$, for each exercise time T_n, $n = 1, \ldots, N-1$, so the computational expense is quite considerable even with a modest number of sub-paths. Bender et al. [2006] recommend using control variates to speed up valuation, and Beveridge and Joshi [2009] list a number of additional suggestions to improve computational performance. Nevertheless, there is no doubt that to keep Monte Carlo simulation error of the conditional expectation low enough for the policy iteration scheme to be effective, a substantial numerical effort is required. As such, we do not recommend routine application of policy iteration in the pricing of CLEs. This, of course, will require that the regression estimates of the exercise policy[10] are sufficiently accurate as is, something that to a large extent depends on how careful we are in choosing the regression variable vector $\zeta(t)$. We turn to this topic in Section 18.3.9, but first we need to address the fundamental problem of how to test whether a given exercise policy is close to optimal in the first place. One useful approach to this problem is to use the estimated

---

[9]For more detailed discussion of nested Monte Carlo simulation, see Section 18.3.8.

[10]Which may include refinements such as that in Section 18.3.5, of course.

exercise policy to construct an *upper bound* for the option price, which, if close to the lower bound, will give us confidence that our exercise policy is close to optimal. Section 18.3.8 below discusses this technique in detail.

18.3.8 Upper Bound

Computing a lower bound for a CLE price is straightforward: pick some exercise policy and price the CLE by Monte Carlo methods. Assuming our computation of the exercise policy did not "cheat" by using information from the Monte Carlo trials subsequently used for valuation, the resulting price estimates will always have a non-positive bias, as the chosen exercise strategy will almost certainly be suboptimal. The lower bound algorithm of Section 18.3.6 was based on precisely such a strategy. To complement a computed lower bound for the CLE, we are now interested in using the lower bound exercise policy to construct an upper bound for the CLE price. Taken together with the lower bound, the upper bound can be used to construct a valid confidence interval for the CLE price; this, in turn, will allow us to assess the quality of the exercise policy.

18.3.8.1 Basic Ideas

Our strategy to construct an upper bound for CLEs will draw directly on the duality results in Section 1.10.2 and the generic upper bound simulation ideas in Section 3.5.5. To formulate these results in the CLE setting, let us start with the description (18.4)–(18.6):

$$H_0(0) = \sup_{\xi \in \mathcal{T}_0} \mathrm{E}\left(B(T_\xi)^{-1} U_\xi\left(T_\xi\right)\right) = \mathrm{E}\left(B(T_\eta)^{-1} U_\eta\left(T_\eta\right)\right). \qquad (18.30)$$

Following Section 1.10.2, let $\mathcal{K}$ denote the space of adapted martingales M for which $\sup_{\xi \in \mathcal{T}_0} \mathrm{E}|M(T_\xi)| < \infty$. For any martingale $M \in \mathcal{K}$, (1.71) demonstrates that

$$H_0(0) \le M(0) + \mathrm{E}\left(\max_{n=1,\dots,N-1}\left(\frac{U_n(T_n)}{B(T_n)} - M(T_n)\right)\right). \qquad (18.31)$$

Also, we know from the duality result (1.72) that this upper bound will become an *equality* provided that M is chosen to be the martingale component of the (supermartingale) deflated value process of the CLE. To emphasize this result, set

$$V_{\mathrm{CLE}}(t) = B(t)\mathrm{E}_t\left(B(T_n)^{-1}\max\left(U_n(T_n), H_n(T_n)\right)\right), \quad t \in (T_{n-1}, T_n], \qquad (18.32)$$

and use the Doob-Meyer decomposition of Section 1.10.2 to write $V_{\mathrm{CLE}}(t)/B(t) = m_{\mathrm{CLE}}(t) - A(t)$, where $m_{\mathrm{CLE}}(t)$ is a martingale and $A(t)$ an increasing predictable process with $A(0) = 0$. Then setting $M(t) = m_{\mathrm{CLE}}(t)$ yields

$$m_{\mathrm{CLE}}(0) + \mathrm{E}\left(\max_{n=1,\dots,N-1}\left(\frac{U_n(T_n)}{B(T_n)} - m_{\mathrm{CLE}}(T_n)\right)\right) = H_0(0). \qquad (18.33)$$

If the underlying model is driven by a vector-valued Q^B-Brownian motion $W(t)$, the martingale representation theorem (Theorem 1.1.4) shows that any martingale M in (18.31) must be of the form

$$M(t) = \int_0^t \sigma(s)^\top \, dW(s), \qquad (18.34)$$

for some adapted vector-process $\sigma(t)$ satisfying the usual conditions required for the stochastic integral to be proper martingale. Clearly, however, if $\sigma(t)$ is chosen arbitrarily, the resulting upper bound computed from (18.31) is likely to be very loose, and probably not particularly useful. While (18.33) is of little immediate practical use (since we do not know the process $V_{\mathrm{CLE}}(t)/B(t)$), it does suggest that for a chosen martingale $M(t)$ in (18.31) to produce a tight upper bound, it needs to be "close" to $m_{\mathrm{CLE}}(t)$.

Several strategies have been proposed for constructing a good martingale $M(t)$. When working in a simple model setup on simple payouts, sometimes one can make inspired guesses for what $M(t)$ should be. For instance, in a one-dimensional Black-Scholes model, Rogers [2001] shows that using the numeraire-deflated European put option price (which is analytically known) as a guess for $M(t)$ generates good bounds for a Bermudan put option price. This approach, however, does not easily generalize to the CLE setting with its more complicated model and exercise payouts.

18.3.8.2 Nested Simulation (NS) Algorithm

Andersen and Broadie [2004] propose a general strategy for generating upper bounds, starting from any approximation $\widetilde{\eta}$ to the optimal exercise strategy η. Typically, this approximation would originate from an LS regression, e.g. as in Section 18.3.6, or from an optimization of a parametric formulation of the exercise rule, as in Section 3.5.2 or Section 19.6.2. In a nutshell, the algorithm in Andersen and Broadie [2004] uses nested simulation — also known as "simulation within a simulation" and already mentioned in Section 18.3.7 — to construct an estimate of the low-bound value process $\widetilde{V}_{\mathrm{CLE}}(t)$ generated from $\widetilde{\eta}$. The martingale component of the numeraire-deflated value of this process is then used as $M(t)$ in (18.31).

To outline the basic nested simulation (NS) algorithm in further detail, let us work on the exercise time line $\{T_1, T_2, \dots, T_{N-1}\}$ and define

$$
\begin{aligned}
M(T_{n+1}) - M(T_n) &= \frac{\widetilde{V}_{\mathrm{CLE}}(T_{n+1})}{B(T_{n+1})} - \mathrm{E}_{T_n}\left(\frac{\widetilde{V}_{\mathrm{CLE}}(T_{n+1})}{B(T_{n+1})}\right) \\
&= \mathrm{E}_{T_{n+1}}\left(\frac{U_{\widetilde{\eta}_n}(T_{\widetilde{\eta}_n})}{B(T_{\widetilde{\eta}_n})}\right) - \mathrm{E}_{T_n}\left(\frac{U_{\widetilde{\eta}_n}(T_{\widetilde{\eta}_n})}{B(T_{\widetilde{\eta}_n})}\right), \qquad (18.35)
\end{aligned}
$$

with $M(0) = \widetilde{V}_{\mathrm{CLE}}(0)$ and $M(T_1) = \widetilde{V}_{\mathrm{CLE}}(T_1)/B(T_1)$. In the second equality of (18.35), we use $\widetilde{\eta}_n$ to denote the restriction of our approximate exercise policy exercise to the index set $\{n+1, \ldots, N\}$, such that $\widetilde{\eta}_0 = \widetilde{\eta}$. For instance, if we use the algorithm in Section 18.3.6, we have (compare to (18.27))

$$\widetilde{\eta}_n = \min \left\{ k \geq n+1 : U_k(T_k) \geq \mathcal{C} \left(\widetilde{H}_k(T_k) \right)^{\top} \zeta(T_k) \right\} \wedge N. \qquad (18.36)$$

For convenience, let us define an $\mathcal{F}_{T_{n+1}}$-measurable exercise indicator

$$\iota(T_{n+1}) = 1_{\{\widetilde{\eta}_n = n+1\}}$$

which will be one at time T_{n+1} if our exercise policy indicates that the CLE should be exercised, and zero otherwise. We can also define hold values

$$\frac{\widetilde{H}_n(T_n)}{B(T_n)} = \mathrm{E}_{T_n} \left(\frac{U(T_{\widetilde{\eta}_n})}{B(T_{\widetilde{\eta}_n})} \right),$$

in which case (18.35) can be rewritten as

$$M(T_{n+1}) - M(T_n) = \iota(T_{n+1}) \frac{U_{n+1}(T_{n+1})}{B(T_{n+1})}$$
$$+ (1 - \iota(T_{n+1})) \frac{\widetilde{H}_{n+1}(T_{n+1})}{B(T_{n+1})} - \frac{\widetilde{H}_n(T_n)}{B(T_n)}. \qquad (18.37)$$

Notice that

$$\frac{\widetilde{V}_{\mathrm{CLE}}(T_{n+1})}{B(T_{n+1})} - \frac{\widetilde{V}_{\mathrm{CLE}}(T_n)}{B(T_n)} = \frac{\widetilde{V}_{\mathrm{CLE}}(T_{n+1})}{B(T_{n+1})} - \mathrm{E}_{T_n} \left(\frac{\widetilde{V}_{\mathrm{CLE}}(T_{n+1})}{B(T_{n+1})} \right)$$
$$- \left(\frac{\widetilde{V}_{\mathrm{CLE}}(T_n)}{B(T_n)} - \mathrm{E}_{T_n} \left(\frac{\widetilde{V}_{\mathrm{CLE}}(T_{n+1})}{B(T_{n+1})} \right) \right)$$
$$= M(T_{n+1}) - M(T_n) - (A(T_{n+1}) - A(T_n)),$$

where we have denoted

$$A(T_{n+1}) - A(T_n) = \frac{\widetilde{V}_{\mathrm{CLE}}(T_n)}{B(T_n)} - \mathrm{E}_{T_n} \left(\frac{\widetilde{V}_{\mathrm{CLE}}(T_{n+1})}{B(T_{n+1})} \right)$$
$$= \iota(T_n) \left\{ \frac{U_n(T_n)}{B(T_n)} - \frac{\widetilde{H}_n(T_n)}{B(T_n)} \right\}$$

with $A(0) = 0$. The second equality follows from the fact that $\widetilde{H}_n(T_n)/B(T_n) = \mathrm{E}_{T_n}(\widetilde{V}_{\mathrm{CLE}}(T_{n+1})/B(T_{n+1}))$. Therefore, we have the following decomposition

$$\frac{\widetilde{V}_{\mathrm{CLE}}(T_n)}{B(T_n)} = M(T_n) - A(T_n).$$

Notice that the process A is *not* an increasing process (and $\widetilde{V}_{\mathrm{CLE}}$ therefore not a supermartingale), since we cannot guarantee that $\widetilde{H}_n(T_n) \geq U(T_n)$: whenever an incorrect exercise decision is made, A decreases.

For the purpose of computing an upper bound, the hold values $\widetilde{H}_n(T_n)$ and $\widetilde{H}_{n+1}(T_{n+1})$ in (18.37) *cannot* be estimated by regression; doing so will introduce unknown biases which will destroy the martingale property of M and, in turn, invalidate the inequality in (18.31). Instead, following Andersen and Broadie [2004] we can launch at times T_n and T_{n+1} Monte Carlo simulations to estimate the two expectations in (18.35). Notice that these "inner" Monte Carlo simulations will be nested inside a main "outer" simulation trial that generate sample paths of U, B, and M, as needed to estimate the expectation on the right-hand side of (18.31).

Now, we insert the martingale defined by (18.37) into the right-hand side of (18.31), which gives rise to a high-biased estimate $H_0^{\mathrm{hi}}(0)$ for the CLE value,

$$H_0^{\mathrm{hi}}(0) = \widetilde{H}_0(0) + \Delta \geq H_0(0), \qquad (18.38)$$

where the *duality gap* Δ is defined as

$$\Delta = \mathrm{E}(D), \quad D \triangleq \max_{n=1,\dots,N-1} \left(\frac{U_n(T_n)}{B(T_n)} - M(T_n) \right). \qquad (18.39)$$

The hold value $\widetilde{H}_0(0) = \widetilde{V}_{\mathrm{CLE}}(0)$ can be estimated bias-free from the given exercise strategy $\widetilde{\eta}$ by standard Monte Carlo methods (see Section 18.3.6), so we focus on providing an estimate of the duality gap Δ. The following *NS algorithm* can be used for this purpose.

1. Simulate K_U paths $\omega_1, \dots, \omega_{K_U}$.
2. For each path ω_k calculate simulated values of the numeraire $B(T_n, \omega_k)$, $n = 1, \dots, N-1$.
3. For each path ω_k, calculate the value $U_n(T_n, \omega_k)$ of the underlying exotic swap on all exercise dates $n = 1, \dots, N-1$.
4. For each path ω_k *and each* T_n, $n = 1, \dots, N-2$, launch K_{nest} independent sub-paths $\{\omega_{k,1}^n, \dots, \omega_{k,K_{\mathrm{nest}}}^n\}$ to time T_N and estimate hold values $\widetilde{H}_n(T_n, \omega_k)$ as

$$\widetilde{H}_n(T_n, \omega_k) \approx \widehat{H}_n(T_n, \omega_k) \triangleq \frac{1}{K_{\mathrm{nest}}} \sum_{j=1}^{K_{\mathrm{nest}}} \frac{U_{\widetilde{\gamma}(j,k,n)}(T_{\widetilde{\gamma}(j,k,n)})}{B(T_{\widetilde{\gamma}(j,k,n)})},$$

where, in slightly labored notation, $\widetilde{\gamma}(j, k, n)$ is the first exercise date for sub-path j, date T_n, and "outer" path k:

$$\widetilde{\gamma}(j, k, n) = \min \left\{ l > n : \iota(T_l, \omega_{k,j}^n) = 1 \right\} \wedge N.$$

5. For each path ω_k, use $\widehat{H}_n(T_n, \omega_k)$, $n = 1, \ldots, N - 1$, to form martingale estimates $\widehat{M}(T_n, \omega_k)$ by substituting $\widehat{H}_n(T_n, \omega_k)$ for $\widetilde{H}_n(T_n, \omega_k)$ in (18.37).

6. For each path ω_k, compute pathwise duality gaps as

$$\widehat{D}(\omega_k) = \max_{n=1,\ldots,N-1} \left(\frac{U_n(T_n, \omega_k)}{B(T_n, \omega_k)} - \widehat{M}(T_n, \omega_k) \right).$$

7. Estimate Δ as $\widehat{\Delta}$, where

$$\widehat{\Delta} = \frac{1}{K_U} \sum_{k=1}^{K_U} \widehat{D}(\omega_k).$$

18.3.8.3 Bias and Computational Cost of NS Algorithm

Having outlined the basic NS algorithm above, our first order of business is to establish formally that the estimator for $H_0^{\text{hi}}(0)$ resulting from the NS algorithm is, in fact, biased high. Our primary concern here is the effect of the usage in Step 4 of nested Monte Carlo estimators $\widehat{H}_n(T_n, \omega_k)$ in place of the true hold values $H_n(T_n, \omega_k)$. The key result is the following.

Proposition 18.3.1. *In the NS algorithm, the estimator for the duality gap is biased high, i.e.*

$$\mathrm{E}\left(\frac{1}{K_U} \sum_{k=1}^{K_U} \widehat{D}(\omega_k) \right) \geq \mathrm{E}(D).$$

Proof. We drop ω_k throughout the proof, such that $\widehat{H}_n(T_n) = \widehat{H}_n(T_n, \omega_k)$ and so forth. By construction, $\widehat{H}_n(T_n)$ is an unbiased estimator for $\widetilde{H}_n(T_n)$, i.e.

$$\widehat{H}_n(T_n) = \widetilde{H}_n(T_n) + e_n,$$

where e_n is a pure-noise error term with mean 0 and standard deviation proportional to $1/\sqrt{K_{\text{nest}}}$. It follows from (18.37) that Step 5 in the NS algorithm will compute

$$\widehat{M}(T_{n+1}) - \widehat{M}(T_n)$$
$$= \iota(T_{n+1}) \frac{U_{n+1}(T_{n+1})}{B(T_{n+1})}$$
$$\quad + (1 - \iota(T_{n+1})) \frac{\widetilde{H}_{n+1}(T_{n+1}) + e_{n+1}}{B(T_{n+1})} - \frac{\widetilde{H}_n(T_n) + e_n}{B(T_n)}$$
$$= M(T_{n+1}) - M(T_n) + (1 - \iota(T_{n+1})) \frac{e_{n+1}}{B(T_{n+1})} - \frac{e_n}{B(T_n)}.$$

By induction, it follows that, for $n = 1, \ldots, N - 1$,

$$\widehat{M}(T_n) = M(T_n) + q_n$$

where $q_n = q_n(\omega_k)$ is a random variable with zero mean (the explicit expression for q_n is irrelevant) for all n. Hence,

$$\widehat{D} = \max_{n=1,\ldots,N-1} \left(\frac{U_n(T_n)}{B(T_n)} - M(T_n) - q_n \right).$$

On the path ω_k, let $\bar{n}(\omega_k)$ be the date index at which $U_n(T_n,\omega_k)/B(T_n,\omega_k) - M(T_n,\omega_k)$ attains it maximum. We can therefore write (again dropping ω_k's)

$$\text{E}\left(\max_{n=1,\ldots,N-1} \left(\frac{U_n(T_n)}{B(T_n)} - M(T_n) - q_n \right) \right)$$
$$\geq \text{E}\left(\frac{U_{\bar{n}}(T_{\bar{n}})}{B(T_{\bar{n}})} - M(T_{\bar{n}}) - q_{\bar{n}} \right)$$
$$= \text{E}\left(\frac{U_{\bar{n}}(T_{\bar{n}})}{B(T_{\bar{n}})} - M(T_{\bar{n}}) \right)$$
$$= \text{E}\left(\max_{n=1,\ldots,N-1} \left(\frac{U_n(T_n)}{B(T_n)} - M(T_n) \right) \right),$$

where the first equality follows from the zero mean of $q_{\bar{n}}$. $\square$

Proposition 18.3.1 demonstrates that our estimate of the duality gap is biased high for finite values of the sub-path sample size K_{nest}. As such, our finite sample estimate for $H_0^{\text{hi}}(0)$ will itself be biased high and have a mean that is *above* the true CLE value, as desired. By increasing K_{nest} we can reduce the bias originating from the finite sample size of sub-paths, and thereby tighten the upper bound. Of course, even in the limit $K_{\text{nest}} \to \infty$ we will still produce an estimator that is biased high; the size of this bias will reflect the quality of our exercise strategy choice, and will only vanish in the (unlikely) event that we manage to use precisely the optimal exercise strategy, i.e. when $\tilde{\eta} = \eta$.

While the basic NS algorithm is quite straightforward, the need for nested simulations makes it numerically expensive: the worst-case workload will be proportional to

$$K_{\text{nest}} \cdot K_U \cdot N^2. \tag{18.40}$$

For comparison, with K' simulation trials the lower bound simulation in Section 18.3.6 has a workload proportional to $K' \cdot N$, plus the work required to estimate the exercise rule in a pre-simulation. In many cases the inner simulations of the NS algorithm can be stopped quickly (due to exercise of the CLE), so in practice the dependence on N in (18.40) is often less than quadratic and sometimes close to linear. Finally, K_{nest} can often be set to a number much smaller than K_U without significantly affecting the quality of the upper bound, and even very small values of K_{nest} (e.g., 50–100 or less) may yield informative results. If one additionally takes advantage of the algorithm refinements discussed in Section 18.3.8.6, it is typically possible to execute an upper bound computation for a long-dated (say, 30 years) CLE in a few minutes. This is still relatively time-consuming, so upper bound computations are often most useful in practice as a way to test the quality of a postulated exercise strategy. We expand on this topic in the next section.

18.3.8.4 Confidence Intervals and Practical Usage

For concreteness, assume that the algorithm in Section 18.3.6 has been used to compute a K'-path lower bound estimate of $V_{\mathrm{CLE}}(0) = H_0(0)$. Let this estimate be denoted $\widehat{V}_{\mathrm{CLE}}$, and let its recorded sample standard deviation be $\widehat{s}_L$. Also assume that an independent simulation of the NS algorithm with K_U outer simulation trials has produced an estimate $\widehat{\Delta}$ for the duality gap, with sample standard deviation $\widehat{s}_U$. Asymptotically, a $100(1 - \gamma)\%$ confidence interval for the true price $V_{\mathrm{CLE}}(0)$ must be *tighter* than

$$\left[\widehat{V}_{\mathrm{CLE}} - u_{\gamma/2} \frac{\widehat{s}_L}{\sqrt{K'}} ; \widehat{V}_{\mathrm{CLE}} + \widehat{\Delta} + u_{\gamma/2} \sqrt{\frac{\widehat{s}_U^2}{K_U} + \frac{\widehat{s}_L^2}{K'}} \right], \qquad (18.41)$$

where $\Phi(u_{\gamma/2}) = 1 - \gamma/2$. As already mentioned in Section 3.5.6, the confidence interval is conservative[11] because of the low bias in the sample estimator $\widehat{V}_{\mathrm{CLE}}$ (i.e., $\mathrm{E}(\widehat{V}_{\mathrm{CLE}}) \leq H_0(0)$) and the high bias in $\widehat{V}_{\mathrm{CLE}} + \widehat{\Delta}$, which originates in part from the nature of the upper bound, and in part from the earlier mentioned additional high bias introduced by the finite sample size of the inner simulations (see Proposition 18.3.1).

As noted in Section 3.5.6, it is not uncommon that the upper and lower bounds for the option price often are roughly symmetric around the true value, so in the event that we have computed both bounds, the obvious point estimate

$$\widehat{V}_{\mathrm{CLE}} + \frac{1}{2}\widehat{\Delta} \qquad (18.42)$$

will often give better price estimates than either the upper or lower bound alone.

Upper bound simulation algorithms can typically be expected to be both more involved and/or more expensive than lower bound simulation methods. In many cases, the best use of the upper bound simulation algorithm will therefore be to test whether postulated lower bound exercise strategies are tight or not. Specifically, starting from some guess for the exercise strategy, we can produce confidence intervals using (18.41) to test whether the lower

---

[11] In addition to random Monte Carlo error, simulation of some models may also involve a systematic error stemming from the time-discretization of the model dynamics. Such discretization errors are not accounted for in (18.41), nor in any previous argument. In most cases, however, the discretization bias will be negligible relative to the random Monte Carlo error.

bound estimate is of good quality, in which case the confidence interval can be made tight by using large values of K_U and K' (as well as the number of inner simulation trials, K_{nest}). In case the lower bound estimator is deemed unsatisfactory, we can iteratively refine it, by altering the choice of basis functions, say, until the confidence interval is tight. Importantly, such tests can often be done at a high level, covering entire classes of payouts and/or models. Once an exercise strategy has been validated for a particular product or model, day-to-day pricing of callable securities can be done by the lower bound method, with only occasional runs of the upper bound method needed (e.g., if market conditions change markedly). If upper bound methods are predominantly used in this fashion, the fact that they may sometimes be computationally intensive[12] becomes less punitive.

18.3.8.5 Non-Analytic Exercise Values

The observant reader might have noticed that the NS algorithm outlined in Section 18.3.8.2 assumes (in Steps 3, 4, and 6) that exercise values are directly computable at each maturity and each state of the world. A similar assumption was made in the basic LS regression algorithm of Section 18.3.1, but relaxed later, in Section 18.3.2. To handle CLEs that involve exotic swap underlyings for which the values are not easily computable, we can modify the upper bound simulation algorithm in at least two ways.

Our first approach is straightforward, and based on the representations

$$\frac{U_n(T_n)}{B(T_n)} = \mathrm{E}_{T_n}\left(\sum_{i=n}^{N-1} \frac{X_i}{B(T_{i+1})}\right),$$

and

$$\frac{\widetilde{H}_n(T_n)}{B(T_n)} = \mathrm{E}_{T_n}\left(\frac{U_{\widetilde{\eta}_n}(T_{\widetilde{\eta}_n})}{B(T_{\widetilde{\eta}_n})}\right) = \mathrm{E}_{T_n}\left(\frac{\mathrm{E}_{T_{\widetilde{\eta}_n}}\left(\sum_{i=\widetilde{\eta}_n}^{N-1} \frac{X_i}{B(T_{i+1})}\right)}{B(T_{\widetilde{\eta}_n})}\right)$$

$$= \mathrm{E}_{T_n}\left(\frac{\sum_{i=\widetilde{\eta}_n}^{N-1} \frac{X_i}{B(T_{i+1})}}{B(T_{\widetilde{\eta}_n})}\right), \tag{18.43}$$

where the last equality follows from the optional sampling theorem. The relevant expectations can be computed bias-free by launching a nested simulation at time T_n, generating sample paths from time T_n to T_{N-1}. For instance, for the path ω_k and date T_n, we would write

---

[12] Note that in testing the viability of a class of exercise rules through an upper bound simulation, it is often acceptable to work with a reduced set of exercise opportunities — e.g. change a quarterly exercise schedule to an annual one, say — in order to save computation time (see (18.40)).

$$U_n(T_n, \omega_k) \approx \widehat{U}_n(T_n, \omega_k) = \frac{1}{K_{\text{nest}}} \sum_{j=1}^{K_{\text{nest}}} \sum_{i=n}^{N-1} \frac{X_i(\omega_{k,j}^n)}{B(T_{i+1}, \omega_{k,j}^n)},$$

where $\{\omega_{k,1}^n, \ldots, \omega_{k,K_{\text{nest}}}^n\}$ is the set of "inner" sub-paths spawned at time T_n for the "outer" path ω_k. An estimator $\widehat{H}_n(T_n, \omega_k)$ for $\widetilde{H}_n(T_n, \omega_k)$ may be computed from (18.43) the same way, except that only net coupons $X_i(\omega_{k,j}^n)$ *after* exercise will be counted in the accumulation of deflated coupons. With these estimators used in Steps 4 and 6, the NS algorithm in Section 18.3.8.2 may proceed as before. As the nested simulation will produce bias-free (but noisy) estimates for the true exercise and hold values, it follows from Proposition 18.3.1 that the resulting upper bound estimator will still be guaranteed to be biased high.

While using nested simulation in the manner described above does not add to the order of the computational complexity of the upper bound algorithm (it remains as in (18.40)), the need to construct exercise values by simulating coupon streams will obviously add additional noise to the basic algorithm, which in turn will increase the finite sample bias and also widen the confidence interval (18.41) for a given computational budget.

In our second approach to dealing with non-analytic exercise values, we follow the alternative route also taken in Section 18.3.3, and focus on pricing the cancelable note $G_0(0)$, where

$$G_0(0) = H_0(0) - U_0(0).$$

As a starting point, consider the expression (18.21), which we may rewrite as

$$G_0(0) = \sup_{\xi \in \mathcal{T}_0} \mathrm{E}\left(-\sum_{n=0}^{\xi-1} B(T_{n+1})^{-1} X_n \right)$$

$$= \sup_{\xi \in \mathcal{T}_0} \mathrm{E}\left(B(T_\xi)^{-1} J(T_\xi) \right), \quad J(T_\xi) \triangleq -\sum_{n=0}^{\xi-1} \frac{B(T_\xi)}{B(T_{n+1})} X_n. \quad (18.44)$$

Financially, the quantity $J(T_\xi)$ can be interpreted as the payout at the exercise date T_ξ from re-investing all pre-exercise coupons into the numeraire B. Effectively, this formulation removes all pre-exercise cash flows from the expectation for $G_0(0)$, making (18.44) structurally identical to (18.30), but with the exercise value $U_\xi(T_\xi)$ replaced by $J(T_\xi)$. Of course, while $U_\xi(T_\xi)$ may be difficult to compute at time T_ξ in a Monte Carlo simulation, $J(T_\xi)$ is not.

To construct an upper bound for $G_0(0)$, we follow the same principles that lead to (18.35) and construct a martingale M as

$$M(T_{n+1}) - M(T_n) = \mathrm{E}_{T_{n+1}}\left(\frac{J(T_{\tilde{\eta}_n})}{B(T_{\tilde{\eta}_n})} \right) - \mathrm{E}_{T_n}\left(\frac{J(T_{\tilde{\eta}_n})}{B(T_{\tilde{\eta}_n})} \right), \quad (18.45)$$

where $\widetilde{\eta}$ is a given exercise strategy. With $\widetilde{G}_0(0)$ being the lower bound value for $G_0(0)$ computed by using $\widetilde{\eta}$ as the exercise strategy, an upper bound for $G_0(0)$ is (compare to (18.38))

$$G_0^{\text{hi}}(0) = \widetilde{G}_0(0) + \Delta_G \geq G_0(0),$$

where the duality gap Δ_G is defined as

$$\Delta_G = \mathrm{E}(D_G), \quad D_G \triangleq \max_{n=1,\dots,N-1} \left(\frac{J(T_n)}{B(T_n)} - M(T_n) \right).$$

When using Monte Carlo simulation to estimate Δ_G, we can use nested simulation to establish a bias-free estimator for the martingale in (18.45), in the same way as was done for the NS algorithm. By the arguments in Proposition 18.3.1, the resulting estimator for Δ_G will be biased high, i.e. our upper bound is valid. Confidence intervals for $G_0(0)$ (and for $H_0(0) = G_0(0) + U_0(0)$) can be constructed using the principles of Section 18.3.8.4.

18.3.8.6 Improvements to NS Algorithm

In Broadie and Cao [2008], the authors outline a number of improvements to both upper and lower bound simulations. As some of the proposed techniques are fairly involved, we cannot give this paper full justice, but contend ourselves with listing one relatively straightforward trick from Broadie and Cao [2008]. The reader interested in additional techniques should consult Broadie and Cao [2008] directly.

Let us return to the setting of Section 18.3.8.2, and assume that an easily computable lower limit $\underline{H}(T_n)$ exists for the hold value at time T_n,

$$H_n(T_n) \geq \underline{H}(T_n), \quad n = 1, \dots, N-1. \tag{18.46}$$

We comment on how to choose $\underline{H}(\cdot)$ later. Let us also assume that the inequality (18.46) is honored by the given exercise policy $\widetilde{\eta}$, i.e. we assume that exercise will never take place when $U_n(T_n) < \underline{H}(T_n)$. When using a regression approach, we can ensure that this assumption is true by simply writing (compare to (18.36))

$$\widetilde{\eta}_n = \min \left\{ k \geq n+1 : U_k(T_k) \geq \max \left(\underline{H}(T_k), \mathcal{C}\left(\widetilde{H}_k(T_k) \right)^{\top} \zeta(T_k) \right) \right\} \wedge N. \tag{18.47}$$

Modifying an exercise policy to accommodate bounds such as (18.46) is sometimes known as *policy fixing*, see Broadie and Glasserman [2004].

Before stating our next result, let us make the additional assumption that $\underline{H}(t)/B(t)$ is a submartingale in measure Q^B for $t \leq T_N$; this is, for instance, the case when $\underline{H}(t)$ is chosen to be either zero or the price of an asset that pays no cash flows before T_N. With this assumption, we have

$$\frac{\widetilde{H}_n(T_n)}{B(T_n)} = \mathrm{E}_{T_n}\left(\frac{U_{\widetilde{\eta}_n}(T_{\widetilde{\eta}_n})}{B(T_{\widetilde{\eta}_n})}\right) \geq \mathrm{E}_{T_n}\left(\frac{\underline{H}(T_{\widetilde{\eta}_n})}{B(T_{\widetilde{\eta}_n})}\right) \geq \frac{\underline{H}(T_n)}{B(T_n)}. \qquad (18.48)$$

We use this result to show the following proposition, adapted from Broadie and Cao [2008].

Proposition 18.3.2. *Let the martingale $M(t)$ be defined as in (18.35) and (18.37), and assume that $\underline{H}(t)/B(t)$ is a submartingale. Assume also that $U_k(T_k) < \underline{H}(T_k)$ for all $k = l, \ldots, n$, with $1 \leq l \leq n$. Then we have,*

$$M(T_n) = M(T_{l-1}) - \widetilde{H}_{l-1}(T_{l-1})/B(T_{l-1}) + \widetilde{H}_n(T_n)/B(T_n)$$

and

$$U_n(T_n)/B(T_n) - M(T_n) < \widetilde{H}_{l-1}(T_{l-1})/B(T_{l-1}) - M(T_{l-1}). \qquad (18.49)$$

In particular, if $l = 1$ then

$$M(T_n) = \widetilde{H}_n(T_n)/B(T_n), \quad U_n(T_n)/B(T_n) < M(T_n). \qquad (18.50)$$

Proof. First, we notice that exercise will never take place on the interval $[T_l, T_n]$ such that, from (18.37),

$$M(T_n) = M(T_{n-1}) + \frac{\widetilde{H}_n(T_n)}{B(T_n)} - \frac{\widetilde{H}_{n-1}(T_{n-1})}{B(T_{n-1})}$$

$$= M(T_{l-1}) + \sum_{j=l}^{n}\left(\frac{\widetilde{H}_j(T_j)}{B(T_j)} - \frac{\widetilde{H}_{j-1}(T_{j-1})}{B(T_{j-1})}\right)$$

$$= M(T_{l-1}) - \frac{\widetilde{H}_{l-1}(T_{l-1})}{B(T_{l-1})} + \frac{\widetilde{H}_n(T_n)}{B(T_n)}.$$

Using (18.48), we also have

$$\frac{\widetilde{H}_n(T_n)}{B(T_n)} \geq \frac{\underline{H}(T_n)}{B(T_n)} > \frac{U_n(T_n)}{B(T_n)}, \qquad (18.51)$$

which proves (18.49). The results for the special case where $l = 1$ follows from the fact that $M(0) = \widetilde{H}_0(T_0)$. $\square$

Recalling (18.39), it is clear from (18.50) that there is no contribution to the upper bound increment D before the underlying process enters the region where $U_n(T_n) \geq \underline{H}(T_n)$. Moreover, whenever the option is inside a region where the exercise value is less than the lower limit $\underline{H}$, M does not depend on the actual path taken inside the region. This allows for the following straightforward modification to the NS algorithm: whenever at time T_n we observe that $U_n(T_n) < \underline{H}(T_n)$, we simply skip the nested simulations and proceed to the next date T_{n+1}; otherwise we launch K_{nest} sub-simulations and update $M(T_n)$ according to Proposition 18.3.2. When computing $\widehat{D}$ in Step 6, we can ignore all dates where $U_n(T_n) < \underline{H}(T_n)$. For options that are deeply out-of-the money, a substantial number of nested simulations can be avoided in this fashion, leading to significant improvements in computational performance.

The refinement in the Broadie and Cao [2008] algorithm hinges upon our ability to establish a meaningful lower limit submartingale $\underline{H}$. While this must generally be done on a case-by-case basis, one obvious possibility in the standard CLE setting is to set the lower limit to zero. This is a relatively coarse bound, and it may be tempting to sharpen it by observing that, for any $t \geq 0$ and $n, m \geq 0$,

$$H_n(t) \geq H_{n+m}(t) \geq U_{n+m+1}(t),$$

which, assuming we can calculate $U_{n+m+1}(t)$'s analytically (admittedly a strong assumption), shows that a lower limit can be computed as

$$\underline{H}(T_n) = \max_{k=n+1,\ldots,N-1} \left(U_k(T_n) \right).$$

However, while this choice of the lower limit[13] $\underline{H}(\cdot)$ can certainly be used in policy fixing to improve the lower bound, it is generally not a submartingale and therefore *cannot* be used in upper bound computations.

18.3.8.7 Other Upper Bound Algorithms

To avoid the need for nested simulation, it is tempting to return to the representation in (18.34) and contemplate whether one can estimate the optimal choice of $\sigma(t)$ by extracting it from the empirical volatility of (the martingale component of) $\widetilde{V}_{\mathrm{CLE}}(t)/B(t)$. One advantage to this approach is that any errors in the estimation of $\sigma(t)$ will not affect the martingale property of $M(t)$ in (18.34). Starting again from a postulated exercise strategy $\widetilde{\eta}$, Belomestny et al. [2007] use this observation to construct a regression on a set of basis functions to produce an estimate for the function $\sigma(t)$. By applying regression techniques this way, the authors are able to construct a true martingale process $M(t)$, which can be turned into a valid upper bound through (18.31). While the resulting algorithm involves no nested simulation, it requires considerable care in its implementation, in part because the optimal integrand $\sigma(t)$ can be expected to be considerably less regular than the optimal martingale $M(t)$ itself. This, in turn, requires additional thought in the selection of appropriate basis functions for the regression. One possibility advocated in Belomestny et al. [2007] is to include, whenever available, exact or approximate expressions for the diffusion term

---

[13]This limit is essentially equivalent to imposing a so-called "carry" restriction on exercise, a notion that we explore in some detail in Section 18.3.10.2 and Proposition 19.7.1.

in dynamics of several still-alive European options underlying the Bermudan option. This strategy is akin to that of Rogers [2001], and its feasibility depends on the pricing problem at hand. In cases where it does apply, the authors of Belomestny et al. [2007] demonstrate that their method gives good results, with the upper bound often being nearly as tight as that of the nested algorithm in Andersen and Broadie [2004]. They also show how to use their technique to develop a variance-reduced version of the algorithm in Andersen and Broadie [2004].

Additional techniques for computing upper bounds can be found in Broadie and Glasserman [1997] and Glasserman and Yu [2002]. The latter is based purely on regression, but requires strong conditions on regression basis functions, that may be hard to check in practice.

18.3.9 Regression Variable Choice

The single most critical determinant of the performance of the regression-based valuation methods is the choice of the regression variables[14] $\zeta(t) = (\zeta_1(t), \ldots, \zeta_q(t))^\top$. Recall that the values of these variables at time T partly[15] serve to approximate the information contained in the sigma-algebra $\mathcal{F}_T$. Some of this information is relevant to the valuation of a given security and some is not. The closer $\zeta(T)$ approximates the information in $\mathcal{F}_T$ that is relevant for the security, the better the regression method performs, i.e. the smaller is the bias of lower bound estimates of the security value.

18.3.9.1 State Variables Approach

For Markovian models, all information observed at (but not before) time T is encoded in the Markovian state variables, say some (d-dimensional) vector $x(T)$. As discussed earlier in Section 3.5.4, for such models it is often natural take the $\zeta(T)$'s to be deterministic functions of the state variables; these functions should approximately span the set of all functions of the state variables. A good choice is to use monomials of $x(T)$, i.e. functions of the type $\prod_{i=1}^{d} x_i(T)^{p_i}$, which corresponds to using a polynomial basis in the LS regression. One caveat applies: the filtration $\sigma(x(T))$ generated by the state variables at $x(T)$ is clearly smaller than $\mathcal{F}_T$, since the former consists of model information observed at time T only, while the latter contains information for *all* times from 0 to T. For some derivatives the reduction from $\mathcal{F}_T$ to $\sigma(x(T))$ is irrelevant — this includes all callable securities whose coupons are not path-dependent. On the other hand, for securities such

---

[14]While for convenience we write ζ as a function of time t, the regression variables only need to be defined at times when we perform regression, i.e. at exercise dates $T_1, \ldots, T_{N-1}$.

[15]Beyond representing information available at time T, $\zeta(T)$ also serves to define the function space that is obtainable by least-squares regression.

as callable snowballs (Section 5.14.4) whose coupons depend on the whole history of interest rate evolution, the regression variables must be augmented with some that carry information from the past. To do this in a product-independent way, we can, for instance, include state variables from previous times in $\zeta(T)$.

Markovian interest rate models that would allow for such essentially product-independent choice of regression variables include quasi-Gaussian models and quadratic Gaussian models, at least if the number of Markov state variables is not too high. While Libor market (LM) models are Markovian in the set of all Libor forward rates (and any stochastic volatility variables), the dimensionality of the state vector is typically so large that the regression will suffer from numerical problems, an issue we shall discuss further later. One general way to reduce the dimension of the state variable vector is to perform a principal components analysis. To demonstrate this idea in a LM model setting, suppose that we wish to synthesize d "state variables" by forming the first d principal components of the vector of Libor rates. At a given time T_n we have a vector of (centered) still-alive forward Libor rates

$$A = (L_n(T_n) - \mathrm{E}(L_n(T_n)), \ldots, L_{N-1}(T_n) - \mathrm{E}(L_{N-1}(T_n)))^\top.$$

We assume $N - n \geq d$. The term covariance matrix of the rates,

$$c = \mathrm{E}\left(AA^\top\right),$$

can be estimated using methods from Chapter 14. Then, by principal components analysis (see Section 3.1.3) we can find an $(N - n) \times d$ matrix D such that $DD^\top$ is the closest (in the Frobenius norm) rank-d approximation to c. Then

$$A \approx Dx$$

for some d-dimensional vector $x = x(T_n)$, which we recover from the least-squares (regression) problem

$$\|A - Dx\|^2 \to \min,$$

with the solution

$$x = \left(D^\top D\right)^{-1} D^\top A.$$

Doing this for each T_n gives us a set of d approximate state variables for regression on each exercise date.

18.3.9.2 Explanatory Variables

While the state variable approach (with or without principal components dimension reduction) is appealing in its independence of security specifics, it has several shortcomings. First, as mentioned earlier, situations may arise where the information carried in the state variables is inadequate for the security in question (e.g. snowballs or other path-dependent CLEs). Second, there may be cases where there is *too much* information in the state variables, leading to an overabundance of regression variables. For instance, a standard Bermudan swaption turns out (see Section 19.6.1) to primarily depend on the overall level of interest rates, so if, say, the number d of PCA state variables used in an LM model is much larger than 1, many of the regression variables will have little or no explanatory power.

Having too many regression variables not only adds more work to the numerical scheme, it ultimately tends to reduce the quality of the CLE price estimate. Indeed, for a finite budget of simulated paths used in regression, having too many regression variables[16] will induce errors in the regression coefficient estimates and, therefore, in the exercise rule and, ultimately, in the value of the callable security. Moreover, if some of the regression variables add no explanatory power to the regression, their inclusion in the regression may lead to spurious noise and further issues with the estimation quality of the exercise rule[17].

In light of the comments above, in practice it is hard to avoid a careful analysis of each security type, to ensure that neither too little or too much information is contained within the set of regression variables. In such a "product-specific" approach to choosing regression variables, we would aim to choose regression variables in a way that maximizes their explanatory power for the particular security in question.

Similar to the state variables approach, we find it convenient to specify regression variables $\zeta(t)$ as simple functions of so-called *explanatory variables*, which can be thought of as product-specific analogs to state variables of the model. Let us use $x(t) = (x_1(t), \ldots, x_d(t))^\top$ to denote these explanatory variables; the fact that we recycle the notation for x and d from Section 18.3.9.1 should not lead to any confusion. When constructing regression variables from explanatory variables, we typically choose monomials of explanatory variables

$$\prod_{i=1}^{d} x_i(t)^{p_i}, \tag{18.52}$$

up to a low order r, say 3 to 4, so that $\sum_{i=1}^{d} p_i \leq r$. This, of course, amounts to fitting exercise and hold values with polynomials of degree r in explanatory variables. It is worth noting that while higher-order polynomials could give a closer fit, they could also lead to unexpected behavior outside of the range of the points used in fitting. As exercise boundaries may lie far away

---

[16] Glasserman and Yu [2004] determine that the number of paths should grow exponentially in the number of regression variables. Under different assumptions, Moni [2005] shows that the number of paths should grow polynomially. Polynomial rate of growth is also derived in Egloff et al. [2007].

[17] Section 18.3.10 lists a number of regularization approaches that may help guard against some of the issues associated with spurious regression variables.

from typical simulation scenarios, this could lead to poor estimation of the exercise rule.

When selecting explanatory variables, we should look for variables that have high explanatory power in the regression of hold and exercise values of a given CLE. Inevitably, this process is trade-type specific and combines elements of both science and art. While trial and error is always needed, we can make a few recommendations. Generally, we always prefer using financially meaningful explanatory variables such as various market rates or values of (simple) market instruments, and find it convenient to distinguish three classes of potential explanatory variables. The first class contains the market variables that directly drive the values of coupons. For example, if a coupon is linked to a CMS spread between two rates, we should include the spread as a potential explanatory variable. The second class contains those variables that describe the past; these variables are only relevant for CLEs with path-dependent coupons. For instance, recall that in a snowball the value of a current coupon depends on the values of the previous coupons, so we would recommend including the current coupon in the set of potential explanatory variables. Finally, the last class contains those market variables that are thought to drive the exercise decision. Inspired by the case of a simple Bermudan swaption, we would often include a variable responsible for the overall level of the yield curve. In particular, for the regression on date T_n we would typically include a swap rate that fixes at T_n and spans all periods to the last exercise (a so-called "core" swap rate). We might also include a variable that reflects the slope of the yield curve on the exercise date; a front Libor rate, i.e. a Libor rate that fixes on the exercise date, is a good choice for this.

After collecting potential explanatory variables from all three classes, we would typically proceed to analyze the list and try to reduce it to a manageable number of variable, such as 2 to 4, that we would then use in our regression method. While nothing replaces careful analysis, the selection process can be automated somewhat and done during the actual regression — we discuss this in more details later on, in Section 18.3.10.1.

Whatever variables we choose, we should be careful to always choose variables for regression on T_n that are $\mathcal{F}_{T_n}$-measurable. In plain speak, variables should not be "future-looking", but should be computable by using only the state of the model as observed up to and at time T_n. This requirement, while seemingly technical, is critical to the success of the regression algorithm, as one must not be allowed to "see into the future" when making decisions about exercise.

In selecting explanatory variables we should be thinking about qualitative impact of various variables on exercise/hold values but, fortunately, we do not need to be quantitatively exact in capturing the effects. As long as the general influence is accounted for correctly, the fitting of parametric functions will generally take care of choosing the best scalings and/or linear combinations of explanatory variables[18]. For example, one may decide (as we often recommend) to include in the set of explanatory variables the level of the yield curve, as measured by a core swap rate, as well as the slope of the yield curve, as measured by the difference between the swap rate and some short-tenor Libor rate. For the latter variable, one does not need to explicitly include the difference of the rates as an explanatory variable, since the short-tenor Libor rate will work just as well — the properly weighted difference of the two rates will implicitly be used in the fitting of the polynomials.

18.3.9.3 Explanatory Variables with Convexity

Sometimes the quality of exercise boundary estimation is improved by such seemingly minor changes as using core swap *values* in the set of explanatory variables, rather than core swap rates. In practical applications, we always recommend trying both and seeing which one is better. The reasons for this effect are not always entirely clear, but originate with subtle convexity differences between the two choices. It turns out that convexity, i.e. non-linear dependence between exercise values and the explanatory variables, can be quite important. In particular, instead of using simple rates such as a core swap rate or a front Libor rate, we sometimes find it useful to use *functions* of them as explanatory variables. While this may seem superfluous — after all we will ultimately be applying polynomial functions to explanatory variables to get our regression variables — it turns out that using functional mappings more finely tuned to the features of the trade is often beneficial. The effect is due to the fact that for many CLEs the underlying swap consists of coupons that are options on rates. By matching, roughly, the resulting convexity in value of coupons relative to underlying rates, we can often improve the regression fit. Sometimes this can be accomplished by simply using $(S(T_n) - K)^+$, instead of some rate $S(T_n)$, as an explanatory variable for a CLE whose coupons are strike-K options on the rate $S(T_n)$ (or some other, relatively similar, rate). For a more refined approach we could try to roughly estimate the option value of the remaining coupons as of the exercise date T_n. Even if coupons do not have any optionality, such as for Bermudan swaptions, we may wish to use European options on the core swap rates as explanatory variables to better fit *hold* values, which are always convex due to callability.

If one uses a model that allows for closed-form valuation of European options on rates at any future time, then, for suitably simple CLEs, it may be possible to use the exact exercise value as an explanatory variable. This, however, only works for a subset of models, and the approach can significantly impair the speed of valuation, since one needs to compute European option prices a large number of times (one per path per exercise time per underlying

---

[18] As discussed in Section 18.3.10, a simple pre-normalization of variables before the regression may still be useful in preventing numerical problems.

option). Fortunately, it is not necessary to be particularly precise in matching the curvature of the exercise values with the explanatory variables, and a rough estimate will typically do just fine. For example one can use the Black formula with an approximate volatility to value the options in the underlying, even in non-Black models. Also, if the underlying is a strip of options, one does not need to value all options in the underlying, but just one, e.g. the one in the middle of the strip.

To give an example of an implementation using option-like explanatory variables, consider a callable capped floater. The structured coupon at time T_n is here given by

$$C_n = \min(L_n(T_n) + s, c),$$

received against a Libor rate payment,

$$X_n = \tau_n \times (C_n - L_n(T_n)).$$

We note that the net coupon X_n can be written

$$X_n = \tau_n \times (\min(L_n(T_n) + s, c) - L_n(T_n))$$
$$= \tau_n \times s - \tau_n \times (L_n(T_n) - (c - s))^+,$$

so the payment is a combination of a fixed rate payment with rate s and a call option on the Libor rate with strike $c - s$.

For the first explanatory variable we may use an approximate value of the exercise value on each exercise date

$$x_1(T_n) = \sum_{i=n}^{N-1} \tau_i \times P(T_n, T_{i+1}) \times (s - c_B(T_n, L_i(T_n); T_i, c - s; \sigma_{n,i})),$$

where the Black formula $c_B(\ldots)$ is defined in Remark 7.2.8. The volatility $\sigma_{n,i}$ is the log-normal volatility of the Libor rate $L_i(t)$ over the interval $[T_n, T_i]$ as given by the model. To reiterate, very crude approximations can be used here. For the displaced log-normal Libor market model (see (14.12)) one may use $\sigma_{n,i} = \overline{\lambda}$, where $\overline{\lambda}$ is some average of relative forward Libor model volatilities over time and Libor rates, and for the stochastic volatility version (see (14.16)) one may use

$$\sigma_{n,i} = \overline{\lambda}\sqrt{z(T_n)}. \tag{18.53}$$

For the second explanatory variable, we can use a front Libor rate fixing on an exercise date, as we recommended before. Alternatively, we can use just the approximate value of the first coupon payment,

$$x_2(T_n) = \tau_n \times P(T_n, T_{n+1}) \times (s - c_B(T_n, L_n(T_n), c - s; T_n; \sigma_{n,n}))$$

or the last one,

$$x_2(T_n) = \tau_n \times P(T_n, T_N) \times (s - c_B(T_n, L_{N-1}(T_n); T_{N-1}, c - s; \sigma_{n,N-1})).$$

In conjunction with the first explanatory variable, either choice of $x_2(T_n)$ will capture the effect of changes in the interest rate curve slope.

In the example above, we proposed using an explanatory variable dependent on the stochastic volatility process (see (18.53)). It turns out that this is quite important to do in general, even for CLEs that do not have optionality in the underlying, such as Bermudan swaptions. The stochastic volatility process $z(t)$ can either be incorporated into an explanatory variable in the way of (18.53), or used as a separate explanatory variable.

18.3.10 Regression Implementation

While the selection of the regression variables is the primary key to the success of a regression-based method for CLE valuation, the details of implementation of the numerical algorithm for performing regressions are important as well. The basic regression algorithm does not involve much more than matrix inversion, see Section 3.5.4, but there are a number of ways in which the algorithm could be made more robust. We discuss them in this section.

For future reference, let us quickly fix notations. Throughout, we shall consider a particular date T_n and assume that paths $\omega_1, \ldots, \omega_K$ have been simulated, allowing us to collect a $K \times q$ matrix Z of simulated values of the regression variables $\zeta(T_n)$, with

$$Z_{k,j} = \zeta_j(T_n, \omega_k), \quad k = 1, \ldots, K, \quad j = 1, \ldots, q.$$

Recall that each $\zeta_j(T_n)$ is typically obtained as a particular monomial (18.52) applied to the vector of explanatory variables (or state variables, per Section 18.3.9.1) observed at time T_n. We also assume that we have available a K-dimensional vector of simulated values $Y = (Y_1, \ldots, Y_K)^\top$ that we would like to regress, for example the vector $Y_k = V_n(\omega_k)$, $k = 1, \ldots, K$, from (18.12) in the basic regression scheme of Section 18.3.1. The goal is to find a q-dimensional vector β such that $Z\beta$ approximates Y in some sense, e.g. in the least-squares sense

$$\|Y - Z\beta\|^2 \to \min. \tag{18.54}$$

A (naive) solution to this problem is, of course, given by

$$\beta = (Z^\top Z)^{-1} Z^\top Y. \tag{18.55}$$

18.3.10.1 Automated Explanatory Variable Selection

As we explained in Section 18.3.9.2, we often find ourselves in a situation where we have many potential candidates for the explanatory variables but want to prune the set to keep only the most relevant variables for regression.

One approach here is to analyze potential candidates for explanatory variables and, based on experience, choose the ones that subjectively appear to be the most relevant. Another approach is to try to extract a subset of variables based on numerical criteria of regression fit. It turns out that the latter approach is quite common in econometrics circles, allowing us to draw on known techniques. In the problem of *automatic econometric model selection*, one considers a given time series of data — says investment returns of a hedge fund — and tries to choose macro variables, such as equity returns, oil prices, etc., that contribute the most to explaining the time series data. Conceptually, the solution to the problem is simple: one tries regressing the time series on subsets of potential regression variables and observes which subset of variables provides the best fit. The details of how these trials are conducted are, however, quite important, as a brute-force test of all variable combinations would often be impractical: even with just 10 variables to choose from, the number of potential subsets of variables to check is $2^{10} - 1$. The literature on the subject of automatic econometric model selection is quite extensive[19] so we do not go into much detail here, but merely scratch the surface by demonstrating a simple algorithm.

In an algorithm for automatic pruning of explanatory variables, the first question that needs to be answered is how to measure the quality of fit of a given set of regression variables. In the context of linear regression (18.54), an often-used measure[20] is R^2 ("R-squared") which measures the variance of the residual $Y - Z\beta$ relative to the variance of Y

$$R^2 = 1 - \frac{\|Y - Z\beta\|^2}{\|Y\|^2}.$$

With the solution (18.55), R^2 is equal to

$$R^2 = \frac{Y^\top Z\beta}{Y^\top Y}.$$

The higher R^2, the better the fit is deemed to be. So, in our algorithm, we might choose different subsets of explanatory variables and examine the resulting values of the regression R^2. The results could, say, be used to choose a subset of variables that would give us the highest R^2, subject to a constraint on the maximum size d of the variable vector.

Many search paths through the collection of subsets of explanatory variables are possible. For demonstration, let us consider a special case of the *General-to-Specific* (GETS) approach[21]. Here we start with all potential

---

[19]The interested reader could start with Campos et al. [2005].

[20]A potentially more accurate measure is "modified R-squared", see Campos et al. [2005]. For our applications where the dimension of the dataset (K) is very large, modified R-squared and ordinary R-squared are almost identical, however.

[21]See e.g. www.pcgive.com/pcgets.

explanatory variables included in the regression and calculate the baseline R^2. Then we remove each explanatory variable in turn (always returning previously-removed variables to the set after each regression) and calculate the R^2. Having tried removing all variables, we then select the one that gave us the smallest reduction in R^2 and throw it away. With the potential set of explanatory variables reduced by one, we repeat the procedure and again exclude a variable that gives us the smallest effect on the R^2. We continue until we have either reached a pre-determined number of explanatory variables, or until we have reached some maximum allowed reduction in R^2. With each regression typically being pretty quick to execute, this approach does not affect the overall valuation time much, yet generally gives good results. As one can imagine, however, substantially more sophisticated approaches for variable pruning are possible; we refer the reader to Campos et al. [2005] as a starting point.

While we motivated our discussion of explanatory variable selection with a problem in econometrics, we note that our regression problem is not exactly the same as that typically faced in econometric analysis. While econometricians often try to explain virtually all changes in their time series through changes in explanatory variables — i.e. they seek values of R^2 that are close to one — the problems we are interested in here would typically be characterized by values of R^2 that are much lower than one. This is a consequence of the fact that even the full information set available at time T cannot explain all changes in hold/exercise values after time T. In fact, if in the regression we obtain high values of R^2, this would indicate that there is something wrong with our choice of explanatory variables and they most likely are "future-looking".

18.3.10.2 Suboptimal Point Exclusion

The main point of performing a regression on exercise/hold values is to determine an exercise rule. If for a given path, at a given point in time, we can prove that the exercise can never be optimal then, arguably, we should exclude this point from the regression that defines the exercise rule.

Interestingly, it turns out that in some general cases we can indeed establish situations where exercise is never optimal. Consider a cancelable note on a coupon stream $-X_1, \ldots, -X_{N-1}$, see Section 18.3.3. From (18.18),

$$G_{n-1}(T_{n-1}) = E_{T_{n-1}}\left(B(T_{n-1})B(T_n)^{-1}\left(-X_{n-1} + (G_n(T_n))^+\right)\right)$$
$$\geq -E_{T_{n-1}}\left(B(T_{n-1})B(T_n)^{-1}X_{n-1}\right) \tag{18.56}$$

so if, for a given path ω, $-(E_{T_{n-1}}(B(T_{n-1})B(T_n)^{-1}X_{n-1}))(\omega)$ is positive, then $G_{n-1}(T_{n-1}, \omega)$ is positive and it cannot be optimal to cancel the note at T_{n-1} on that path[22]. Fortunately, in many cases, X_{n-1} is $\mathcal{F}_{T_{n-1}}$-measurable and

---

[22] A more general result is derived in Section 19.7.2.

$$\mathrm{E}_{T_{n-1}}\left(B(T_{n-1})B(T_n)^{-1}X_{n-1}\right) = P(T_{n-1}, T_n)X_{n-1},$$

a quantity easily computable at time T_{n-1}. Since we know the scenarios where it is never optimal to exercise, we can do two things. First, as a simple application of the policy fixing rule (18.47), we can outright forbid exercise in the suboptimal scenarios, even if our regression-based rule instructs us to exercise. Second, as suggested by Beveridge and Joshi [2008], we can exclude those paths ω_k for which $X_{n-1}(\omega_k)$ is negative from participating in the regression fit at time T_{n-1}, potentially improving the fit in the region that matters for the exercise rule.

The idea of excluding "uninteresting" paths could be taken further, albeit based on more practical than theoretical considerations. For example, we may decide to exclude (or at least de-emphasize) paths that are very deeply in or out of the money, since we want most precision around the exercise frontier itself. We formalize this idea in the next section.

18.3.10.3 Two Step Regression

As mentioned in our discussion on convexity in Section 18.3.9.3, simple polynomial functions may not capture to sufficient precision the functional dependence of regressed hold and exercise variables on regression variables. The functional mapping of Section 18.3.9.3 gives us one way of addressing this shortcoming. Another idea is to fit polynomials separately in different regions of values of explanatory variables. For example, with a single explanatory variable, we can split the range of explanatory variable values into a few intervals, and fit polynomials separately in each interval. With many explanatory variables, simple subdivision of the space into intervals quickly becomes impractical. In this case, however, Beveridge and Joshi [2009] suggest a somewhat similar idea of space subdivision based on the moneyness of the derivative in question. They point out that global regression fit of polynomials often works well for values of the regressed variables "in the wings", while for points near the "at-the-money" of the CLE (whatever that might mean), the richer functional structure of the hold and exercise values is often not well-approximated by globally fit polynomials. Beveridge and Joshi [2009] propose a two-step scheme, that we describe in the context of cancelable notes of Section 18.3.3 and, in particular, for the scheme (18.25)–(18.26). First, for a given T_n, we regress the variables $Y_k = \widehat{G}_n(T_n, \omega_k)$ to obtain values $\zeta(T_n, \omega_k)^{\top}\beta$, an approximation to $\widetilde{G}_n(T_n, \omega_k)$, $k = 1, \ldots, K$. Subsequently we can choose those ω_k for which $\widetilde{G}_n(T_n, \omega_k)$ is close to zero, which we consider the definition of "at-the-money" (ATM) for a cancelable note. Specifically, for some threshold value ϵ we set

$$\mathcal{W}^\epsilon = \left\{\omega_k : \left|\zeta(T_n, \omega_k)^{\top}\beta\right| \le \epsilon\right\}.$$

Excluding paths outside of $\mathcal{W}^\epsilon$, we now perform a separate fit of $\widehat{G}_n(T_n, \omega)$, $\omega \in \mathcal{W}^\epsilon$, with the same polynomials; since the set of regression points is different, a new regression coefficient vector β^ϵ is obtained. Finally, we set

$$\widetilde{G}_n(T_n, \omega_k) = \begin{cases} \zeta(T_n, \omega_k)^\top \beta, & \omega_k \notin \mathcal{W}^\epsilon, \\ \zeta(T_n, \omega_k)^\top \beta^\epsilon, & \omega_k \in \mathcal{W}^\epsilon, \end{cases}$$

i.e. we use the values of the original regression for non-ATM points $\omega_k \notin \mathcal{W}^\epsilon$, and a new regression for ATM points $\omega_k \in \mathcal{W}^\epsilon$ only. The value of ϵ could be set to be some (small) fraction of the notional of the derivative, say 5%; ultimately some experimentation here may be required.

Rather than a binary division of regression points into "near" and "far" groups, one could also imagine using some kind of smooth kernel that weighs points in the regression differently, depending on how close one is perceived to be to the exercise point, e.g. by how far away from zero $\widetilde{G}_n(T_n, \omega_k)$ is. One can also refine it through iterations, where the procedure is repeated multiple times.

18.3.10.4 Robust Implementation of Regression Algorithm

In Section 18.3.1 we assumed that the regression operator $\mathcal{R}_T$ at time $T = 0$ reduces to averaging of the values of the random variable it is applied to, because the values of the regression variables $\zeta(0)$ are the same for all paths. This is indeed the case for the true solution of the problem (18.54). If we assume that all the rows of Z are the same and equal to a vector $\varsigma^\top$, then the objective function in (18.54) is equal to

$$\sum_{k=1}^{K} \left(Y_k - \varsigma^\top \beta\right)^2.$$

Differentiating with respect to β_j and setting the derivative to zero we obtain

$$\sum_{k=1}^{K} \left(Y_k - \varsigma^\top \beta\right) \varsigma_j = 0,$$

giving us a regressed solution

$$\varsigma^\top \beta = \frac{1}{K} \sum_{k=1}^{K} Y_k,$$

i.e. the average of Y_k's as advertised. The solution, however, is *not* given by (18.55) as the inverse of $Z^\top Z = K \varsigma \varsigma^\top$ here will not exist[23].

This simple example highlights the danger of using the textbook regression solution (18.55) to the problem (18.54). Beyond the degenerate case

---

[23] On a related note, Rasmussen [2005] suggests starting simulation in the past to ensure sufficient variability in the state space for small times, and shows that this step improves estimation of the exercise boundary.

described above, similar issues will arise whenever the matrix $Z^\top Z$ is ill-conditioned, i.e. close to singular. This can happen either due to outright user error (e.g. an inexperienced user accidentally entering the same explanatory variable twice) or due to subtle near-linear dependencies between explanatory variables. In such cases, the regression problem becomes ill-posed and the numerical solution of the regression problem will be unstable. To counteract this, the user will normally have to add additional structure to the regression problem in order for a robust solution with desirable properties to exist.

To stabilize an ill-posed regression problem, we should first contemplate what would constitute desirable properties for the vector β provided that the regression data matrix Z imposes insufficient constraints on its behavior. A standard approach is to give preference to solutions for β with smaller norms, a choice we can motivate by the observation that if some of the regression coefficients are not constrained by the data, they should be set to zero to ensure that the corresponding regression variables (monomials) do not contribute to the fit. With this in mind, we choose a scalar regularization weight $w_{\text{reg}} > 0$ and replace (18.54) with

$$\|Y - Z\beta\|^2 + w_{\text{reg}} \|\beta\|^2 \to \min. \tag{18.57}$$

Intuitively, we should choose the weight w_{reg} small enough so that the extra term in the objective function does not interfere with the regression objective, yet sufficiently large that the regularization term performs its function. Numerous methods for data-driven selection of w_{reg} have been published in the literature, including the *L-curve method* in Hansen [1992], the *discrepancy principle* in Morozov [1966], and *generalized cross-validation* in Craven and Wahba [1979]. Andersen [2005] contains a review of many of these methods in the setting of yield curve construction. Whatever method is used, the size of w_{reg} should obviously reflect the relative scale of the numbers used in the regression. To examine the scaling issue a bit further, notice that the quadratic (in β) term in the objective function is given by

$$\beta^\top \left(Z^\top Z + w_{\text{reg}} I \right) \beta,$$

where I is the identity $q \times q$ matrix. The sum of squares of all elements in I and $Z^\top Z$ equals q and

$$\text{tr} \left(\left(Z^\top Z \right)^\top \left(Z^\top Z \right) \right),$$

respectively. Consequently it is natural to write

$$w_{\text{reg}} = \epsilon \left(\frac{1}{q} \text{tr} \left(Z^\top Z Z^\top Z \right) \right)^{1/2}, \tag{18.58}$$

where ϵ is a new scale-free constant to be determined. While ideally we should rely on one of the data-driven approaches above, in a pinch we can always try to set ϵ equal to a small number, such as 10^{-4}.

The formal solution to (18.57) is given by

$$\beta = (Z^\top Z + w_{\text{reg}} I)^{-1} Z^\top Y. \tag{18.59}$$

Note that the matrix $Z^\top Z + w_{\text{reg}} I$ is of full rank even if $Z^\top Z$ is not, so the matrix inversion in (18.59) is always well-defined, even when Z is ill-conditioned. The resulting method (which we have used before, in Section 6.4.3) is often called *Tikhonov regularization* or *ridge regression*.

Tikhonov regularization is attractive because it retains a fair amount of intuition as to what happens to the regression coefficients as a result of regularization. We should, however, mention other regularization alternatives, in particular the *pseudo-inverse* or truncated singular value decomposition (TSVD) method. To briefly outline this approach, let us rewrite (18.55) as a system of linear equations on β,

$$M\beta = Z^\top Y, \tag{18.60}$$

where $M = Z^\top Z$. The SVD method, see Press et al. [1992], allows us to decompose the $q \times q$ matrix M into a product of three matrices,

$$M = U \Sigma V^\top,$$

where U (not to be confused with the exercise value notation from earlier in the chapter) and V are $q \times q$ orthogonal matrices (i.e. $U^\top = U^{-1}, V^\top = V^{-1}$) and Σ is a diagonal $q \times q$ matrix. The diagonal elements of Σ are called *singular values* and are ordered by their absolute value (highest first). The decomposition applies even to singular matrices M. In particular, if M is of rank r, $r \le q$, then only the first r diagonal elements of Σ will be non-zero.

The pseudo-inverse of the matrix M is defined by

$$M^+ = V \Sigma^+ U^\top,$$

where Σ^+ is a diagonal matrix with elements

$$\Sigma^+_{i,i} = \begin{cases} 1/\Sigma_{i,i}, & \Sigma_{i,i} \ne 0, \\ 0, & \Sigma_{i,i} = 0. \end{cases}$$

Then we have

$$MM^+ = M^+ M = \text{diag}((1, \ldots, 1, 0, \ldots, 0)^\top),$$

where there are as many 1's on the diagonal as the rank of he matrix M.

The pseudo-inverse allows us to define a solution to (18.60) that always exists. For numerical stability, it is common to modify the solution slightly by choosing a truncation cut-off value $\epsilon > 0$ and defining a diagonal matrix Σ^ϵ by

$$\Sigma^\epsilon_{i,i} = \begin{cases} 1/\Sigma_{i,i}, & |\Sigma_{i,i}| \ge \epsilon |\Sigma_{1,1}|, \\ 0, & |\Sigma_{i,i}| < \epsilon |\Sigma_{1,1}|. \end{cases}$$

Then the solution to (18.60) and, ultimately, the regression problem (18.54), is given by

$$\beta = V \Sigma^{\epsilon} U^{\top} Z^{\top} Y.$$

A possible choice for ϵ is $\epsilon = 10^{-6}$.

To understand better the intuition behind TSVD, let us highlight an interesting connection between Tikhonov regularization and TSVD. Specifically, it can be shown that the Tikhonov solution (18.59) to the regression problem can be written as

$$\beta = V \Sigma^{\text{Tikhonov}} U^{\top} Z^{\top} Y.$$

where Σ^{Tikhonov} is a diagonal $q \times q$ matrix computed from the singular value matrix Σ as

$$\Sigma_{i,i}^{\text{Tikhonov}} = \frac{\Sigma_{i,i}}{\Sigma_{i,i}^2 + w_{\text{reg}}}.$$

We recognize this as a smoothed version of the cut-off matrix Σ^{ϵ} above, with the Tikhonov factor w_{reg} determining how much small singular values get dampened out.

Singular values of widely different magnitudes in the matrix $Z^{\top} Z$ generally cause numerical problems in inversion of the matrix, something that the Tikhonov and TSVD method can help rectify. However, widely different scales of singular values do not necessarily arise only when explanatory variables are poorly chosen (e.g. highly dependent among themselves), but can also emerge if explanatory variables used are themselves of different scales. For example, if one variable is a swap rate measured in the units of a few percent and another is the value of the current coupon measured in the unit of millions of dollars, such scale discrepancy could lead to numerical problems in the regression. The problem is exacerbated by our choice of polynomials as basis functions, as one million to the power of, say, 4 is obviously quite different from one percentage point to the 4th power.

Fortunately, such scaling issues are easy to rectify, as we only need to rescale all variables to the same base before applying the regression. So, instead of the matrix Z we would use $\tilde{Z}$, whose elements are given by

$$\tilde{Z}_{i,j} = \frac{Z_{i,j} - \overline{Z}_{\cdot,j}}{\left(K^{-1} \sum_{k=1}^{K} (Z_{k,j} - \overline{Z}_{\cdot,j})^2 \right)^{1/2}}, \quad i = 1, \dots, K, \quad j = 1, \dots, q, \tag{18.61}$$

where $\overline{Z}_{\cdot,j} = K^{-1} \sum_{k=1}^{K} Z_{k,j}$. This transformation sets all columns in the matrix Z to have zero mean and unit (empirical) standard deviation. The only caveat with (18.61) is that, if applied to a column of Z with constant numbers — a column that is always present as we typically include a constant function in our regression — a division by 0 would occur. One obvious workaround is to simply avoid scaling the constant columns.

Once the standard deviation, as required by (18.61), of each column of Z is calculated, then we can apply another simple method to increase the robustness: we can watch out for points that are well outside a "reasonable" range — say outside of 10 standard deviations for a given column — and exclude them from our regression.

18.4 Valuation with Low–Dimensional Models

Libor market models are often our preferred choice for valuation and risk management of callable Libor exotics. For some CLEs, however, we can use simpler and faster models without sacrificing the benefits of proper calibration and good model dynamics. The trick here is to calibrate a simpler model in a special way, an approach we call the *local projection method.*

18.4.1 Single-Rate Callable Libor Exotics

The local projection method builds on the calibration discussion of Section 18.1 by calibrating a "local", low-dimensional model to the volatility information that we identified as important to the CLE valuation. Information obtained from the market is used directly, and the rest is extracted from a "global", fully-calibrated model such as the Libor market model. The success of the method depends on our ability to identify the relevant volatility information, and how well the local model can calibrate to this information. As a low-dimensional model has only a limited number of parameters, it can only be successfully calibrated for a CLE that depends on a relatively small subset of all available market information.

Callable Libor exotics most amenable to the local projection method are those that have, for each n, $n = 1, \ldots, N - 1$, coupon C_n that is a function of at most a single market rate. We denote such structures *single-rate* CLEs. Examples include Bermudan swaptions, callable inverse floaters, callable CMS capped floaters and fixed-rate callable range accruals. Excluded are CLEs whose coupons depend on spreads between CMS rates, floating-rate callable range accruals, and similar.

The main attraction of using a low-dimensional model is the ability to apply PDE methods for valuation. We have already briefly discussed a relevant pricing scheme, see Section 2.7.4, and the mechanics of the valuation algorithm typically present no special difficulties unless the underlying CLE has path-dependent features. Section 18.4.5 discusses certain PDE techniques that can be used if the path-dependency is sufficiently weak.

18.4.2 Calibration Targets for the Local Projection Method

To start, let us focus on the heart of the local projection method, namely the choice of calibration targets for the local model. Let $\{S_n^1(t)\}_{n=1}^{N-1}$ be the

strip of swap rates that define coupons of a CLE, so that each C_n depends
on $S_n^1(T_n)$, $n = 1, \ldots, N - 1$. We assume that the swap rate S_n^1 has $\mu(n)$
periods, so that $S_n^1(t) = S_{n,\mu(n)}(t)$, where $S_{n,m}(t)$ is the standard notation
for a swap rate fixing on T_n and covering m periods, see e.g. (4.10). For
example, for a callable inverse floater we have $S_n(t) = L_n(t)$ (a Libor rate)
and $\mu(n) = 1$; and for a callable CMS capped floater $\mu(n) \equiv k$, where k is
the number of periods for the underlying CMS rate. In addition, we define a
second strip of swap rates $\{S_n^2(t)\}_{n=1}^{N-1}$ to be the core, or coterminal, swap
rate strip, i.e. $S_n^2(t) = S_{n,N-n}(t)$, $n = 1, \ldots, N - 1$.

The underlying of the CLE, an exotic swap, can be expressed as a
strip of options where the n-th option is written on the rate $S_n^1(T_n)$. Thus,
for a model to reprice the underlying correctly, it should be calibrated to
the market (spot) volatilities of the first swap rate strip, i.e. the implied
volatilities of the European swaptions defined by $\{S_n^1(T_n)\}_{n=1}^{N-1}$. The ability
to match the underlying exotic swap (i.e. the non-callable CLE) is certainly
a prerequisite for any reasonable model for a callable CLE, but, as we have
already seen in Section 18.1.1, we also need to consider that the volatilities
and inter-temporal correlations of core swap rates $\{S_n^2(\cdot)\}_{n=1}^{N-1}$ will affect the
value of the callability feature of the CLE. In light of this, as a starting point
(to be refined later, see Section 18.4.4) we suggest that any low-dimensional
model be calibrated to the following targets:

- The *underlying volatilities*, or swap rate volatilities for $\{S_n^1(T_n)\}_{n=1}^{N-1}$
  that correspond to strikes relevant for the coupons C_n or, in a pinch, to
  at-the-money strikes.
- The *core volatilities*, or swap rate volatilities for $\{S_n^2(T_n)\}_{n=1}^{N-1}$. The
  choice of swaption strikes used to define core volatilities is often not
  straightforward, but at-the-money strikes is a common choice. In some
  cases more advanced methods for strike selection are available, see e.g.
  Section 19.3.
- The *core correlations*, or inter-temporal correlations for $\{S_n^2(T_n)\}_{n=1}^{N-1}$.

While volatilities of swap rates are directly observable from the market,
the inter-temporal correlations are not. This is where we can draw on the
LM (or similar) global model; once it has been calibrated to the market as a
whole, we can calculate the required correlations from the global model. In a
nutshell, the role of the global model is to serve as our "correlation extractor".
The important point here is that by including dynamic information such as
inter-temporal correlations as calibration targets, the local model not only
captures the static information about interest rate volatilities at valuation
time, but also the transition densities and dynamics of the volatility structure,
as seen by a global, fully calibrated and, presumably, realistic model.

18.4.3 Review of Suitable Local Models

The one-dimensional quasi-Gaussian (qG) model developed in Chapter 13 is a natural candidate to consider for the role of the local model in the local projection method for single-rate CLEs. A simple, yet useful, special case setup is based on the version of the qG model with linear local volatility, see Section 13.1.6. The volatility structure of such a qG model is controlled by several time-dependent functions, including the volatility function, the skew function and the mean reversion function. If convexity in the volatility smile is deemed important, the model could be upgraded to the stochastic volatility version in Section 13.2.

Let us first look at the volatility structure specification; we will consider skew and stochastic volatility parameter selection later on. With the volatility function and the mean reversion function discretized over the tenor structure $\{T_n\}_{n=0}^N$, the qG model has $2(N-1)$ independent parameters for volatility calibration. As discussed in Section 13.1.7, the volatility parameters can be used to calibrate the model to term volatilities for one of the swap rate strips. The mean reversion can be used to either match the term volatilities for the second swap rate strip (Section 13.1.8.2), *or* the inter-temporal correlations (Section 13.1.8.3). As such, the one-factor qG model is not large enough to match all three sets of calibration targets identified in Section 18.4.2 above. In some situations, however, this might be acceptable as some securities may turn out to depend only weakly on one of the three calibration targets. For example, for shorter-dated CLEs, inter-temporal correlations may not affect the CLEs value all that much. Likewise, if the underlying has options on the rates $\{S_n^1(T_n)\}_{n=1}^{N-1}$ that are deep in or out of the money, this set of calibration targets can potentially be dropped. Of course, all such decisions must be supported by extensive testing, which fortunately is easy to do as we always have the global (LM or similar) model to benchmark against.

For derivatives where all three sets of calibration targets are important, a one-factor qG model will not suffice[24], and we ideally need to move to models with more stochastic factors. One particularly simple choice is here the two-factor Gaussian model, see Section 12.1.4, which has enough degrees of freedom to match all volatility targets. In this model, some of the time-dependent parameters can be chosen to be constant to make the dynamics of the volatility structure implied by the model more realistic, see Lemma 12.1.11.

The disadvantage of the two-factor Gaussian model is, of course, its lack of control over the volatility smile, so calibration to the volatility targets will require us to identify a single strike per swaption. Improved smile fits can be accomplished by using two-factor versions of either the quasi-Gaussian model (Chapter 13), the affine model, or the quadratic Gaussian model (Chapter

---

[24] Although we can always try to increase the range of applicability of the qG model with some of the techniques of Chapter 21.

12). While all these models are different in some regards, the underlying philosophy and calibration methods will be quite similar.

For models that are sufficiently rich to incorporate volatility skew/smile effects (such as local volatility or stochastic volatility qG models) , we also need to select the market information to which we wish to calibrate skew and smile parameters. Normally we would extend one of the swaption strips, $\{S_n^1\}$ or $\{S_n^2\}$, to multiple strikes for this purpose (for mechanics of calibration see e.g. Section 13.2.3). The choice of the strip is typically driven by an analysis of relative importance of the two sets of smiles to the value of the CLE. It is difficult to state any firm general guidelines here, but we can observe that it is often fairly easy to match the underlying exotic swap value by a judicious choice of a single strike per maturity in the swaption strip $\{S_n^1\}$. On the other hand, it is often difficult to establish which strikes are the most relevant for the "callability" value. Given this, it is often reasonable to use whatever skew/smile parameters we have at our disposal to improve the broad fit of implied core swaption volatilities (the strip $\{S_n^2\}$) at multiple strikes per maturity. If we only have skew, but not smile, parameters, we can use these to match *two* volatilities at each maturity, or to match the slope of the volatility smile at a given strike. The latter could be important if the underlying structured coupons are not simple European options but, for example, of digital or range-accrual type, in which case it is the slope of the volatility smile, and not the overall level, that drives the underlying value. In this case we might, in fact, want to use skew to calibrate the underlying, rather than the callability, value.

18.4.4 Defining a Suitable Analog for Core Swap Rates

When we in Section 18.4.2 looked for the elements of the volatility structure that are relevant for a callable Libor exotic security, we argued that the callability value is driven by the volatilities of core swap rates $S_n^2(t) = S_{n,N-n}(t)$, since a CLE is related to a standard Bermudan swaption. This argument, clearly, has limitations of its applicability. For instance, in Section 19.4 we study Bermudan swaptions on amortizing swaps and show that the most relevant European swaptions in this case are not the standard core European swaptions, but swaptions with tenors based on the durations of the underlying amortizing swap.

In light of this, let us try to refine the selection of the volatility targets relevant for the callability option of a CLE. As a starting point we can use the idea that the local model should match the values of *European options on exercise values* $U_n(T_n)$, $n = 1, \ldots, N-1$. While market values of options on $U_n(T_n)$ could be hard to come by, we can linearize the underlying $U_n(T_n)$ of the CLE and use the resulting rate as a replacement for the core swap rate that should be used in volatility and correlation calibration.

Using a LM model as a backdrop for our analysis, the exercise values $U_n(T_n)$, for each $n = 1, \ldots, N-1$, are functions of the vector of primary

Libor rates

$$\mathbf{L}(T_n) = (L_n(T_n), \ldots, L_{N-1}(T_n))^\top$$

observed on the date T_n, i.e.,

$$U_n(T_n) = f_n(\mathbf{L}(T_n)), \quad n = 1, \ldots, N-1.$$

Linearizing this expression, we obtain

$$U_n(T_n) \approx f_n(\mathbf{L}(0)) + \nabla f_n(\mathbf{L}(0)) \left(\mathbf{L}(T_n) - \mathbf{L}(0)\right),$$

where $\nabla f_n(x)$ is the (row vector) gradient of $f_n(x)$. Hence, the value of the European option on the underlying,

$$\mathrm{E}\left(B(T_n)^{-1} \left(U_n(T_n)\right)^+\right),$$

can be approximated with

$$\mathrm{E}\left(B(T_n)^{-1} \left(\nabla f_n(\mathbf{L}(0))\mathbf{L}(T_n) - \left(\nabla f_n(\mathbf{L}(0))\mathbf{L}(0) - f_n(\mathbf{L}(0))\right)\right)^+\right).$$
(18.62)

We can therefore argue that the most relevant "interest rate" is

$$R_n(T_n) = \nabla f_n(\mathbf{L}(0))\mathbf{L}(T_n) = \sum_j w_{n,j} L_j(T_n), \quad w_{n,j} \triangleq \frac{\partial f_n}{\partial L_j}(\mathbf{L}(0)).$$

Being a linear combination of Libor rates, $R_n(T_n)$ is not, strictly speaking, a market swap rate. However, the volatility of the rate $R_n(T_n)$ can be approximated in a Libor market model (along the same lines as in Section 14.4.2), as well as in local models we may wish to use. Therefore, we can easily use the term volatilities of $R_n(T_n)$, $n = 1, \ldots, N-1$, as volatility targets in place of core swap rates.

The underlying $U_n(t)$ typically consists of options on market rates. The derivatives $\partial f_n/\partial L_j$ can then be computed quite easily with, say, Black-type approximations to option values. Volatilities that should be used in these calculations are the forward (as observed at time T_n) volatilities. Needless to say, high degree of precision is not necessary in these calculations.

To see how consistent the method defined above with our recommendations in Section 18.4.2, let us apply it to a standard Bermudan swaption. For a payer swap with coupon K we have

$$U_n(T_n) = \sum_{i=n}^{N-1} \tau_i \left(L_i(T_n) - K\right) P\left(T_n, T_{i+1}\right)$$

$$= \sum_{i=n}^{N-1} \tau_i \left(L_i(T_n) - K\right) \left(\prod_{k=n}^{i} \frac{1}{1 + \tau_k L_k(T_n)}\right).$$

Hence,

$$f_n(x) = \sum_{i=n}^{N-1} \tau_i (x_i - K) \left(\prod_{k=n}^{i} \frac{1}{1 + \tau_k x_k} \right),$$

and

$$\frac{\partial f_n}{\partial L_j}(x) = \tau_j \prod_{k=n}^{j} \frac{1}{1 + \tau_k x_k} - \frac{\tau_j}{1 + \tau_j x_j} \sum_{i=j}^{N-1} \tau_i (x_i - K) \left(\prod_{k=n}^{i} \frac{1}{1 + \tau_k x_k} \right).$$

Thus

$$w_{n,j} = \tau_j P(0, T_n, T_{j+1}) (1 - U_j(0)/P(0, T_j)).$$

If we compare these weights with those obtained by decomposition of the swap rate into a sum of Libor rates via the "freezing" techniques in Section 14.4.2, we see that they are roughly the same, up to a constant scaling. Thus

$$R_n(T_n) = \sum_{j=n}^{N-1} w_{n,j} L_j(T_n)$$

is quite close to the (scaled) core swap rate $S_{n,N-n}(T_n)$. Therefore, for standard Bermudan swaptions, using $R_n(T_n)$ to define volatility calibration targets should be approximately consistent with the standard method of using core swap rates. When applied to non-standard (e.g., amortizing) Bermudan swaptions, this method produces results that are similar to what we propose later in Section 19.4.

It should be clear that the choice of calibration targets has carries significant impact on the value of a CLE in a local model. Equally important, it also defines the "basis" for vegas (volatility sensitivities), i.e. the set of swaption volatilities to which the CLE is sensitive; hedging of volatility exposure in local model would therefore, as a practical matter, only be done with the swaptions included in the calibration strips. Using a particular swaption for calibration implies the dependence of the CLE value to the volatility of that swaption; conversely, omitting a swaption from the calibration set makes the CLE value (in the local model) insensitive to its volatility. This, of course, is not wholly realistic as even simple CLEs (including Bermudan swaptions) would, when priced in a global model, typically show sensitivity to volatilities of *all* swaptions whose total maturity is no greater than the CLE maturity[25]. In some sense, vegas from a local model (to a subset of swaptions) can be thought of as an aggregation of vegas from the global model. Some traders may in fact prefer such an aggregated view as it (seemingly) simplifies the job of vega hedging.

---

[25] For more detail on this topic, see Chapter 26.

18.4.5 PDE Methods for Path-Dependent CLEs

As mentioned earlier, one attraction of the local projection method is the fact that the resulting model state can often be represented by a low-dimensional state vector $x(t)$. If the dimension of $x(t)$ is less than 3 or 4, this will often allow us to state the CLE value as the solution to a PDE, a problem that can be attacked by the finite difference methods in Chapter 2. While callability is easy to handle (see Section 2.7.4), most path-dependent CLEs are outside the scope of finite difference methods. Exceptions do exist, however, if the path-dependency is sufficiently mild. We show some examples of this below. Of course, even in those cases where a PDE solution is technically possible, one should contemplate whether a local projection model is fundamentally suitable for the path-dependent derivative in question. In particular, the basic single-rate CLE calibration strategies may need adjustment to better capture the path-dependent feature of the payout.

18.4.5.1 CLEs Accreting at Coupon Rate

One particular class of path-dependent CLEs that is amenable to PDE methods has its path-dependency confined to the CLE *notional* only, see Piterbarg [2002]. A prime example of such a CLE is a callable Libor exotic accreting at a coupon rate, see Section 5.14.5, which is the example we consider here. Recall that a CLE is defined by its structured coupon C_n that is fixed at time T_n and paid at time T_{n+1}, $n = 1, \ldots, N-1$. In the standard CLE, the notional of the coupon is constant, or at least deterministic, and has been factored out from the definition in Section 18.2.1. For a coupon-accreting CLE, the notional to which the coupon rate and the Libor rate are applied at time T_{n+1} is equal to the notional at time T_n times an accretion factor that depends on C_n.

Formally, we replace (18.1) with

$$X_n = D_n \tau_n \left(C_n - L_n(T_n) \right),$$

where $D_1 = 1$ and

$$D_n = D_{n-1} \times \left(1 + \tau_{n-1} C_{n-1} \right), \quad n = 2, \ldots, N-1. \tag{18.63}$$

A coupon-accreting CLE is defined as a Bermudan-style option to enter, on date T_n, the remaining part of the underlying, i.e. an exotic swap with the value (at time $t \leq T_n$),

$$U_n(t) = B(t) \sum_{i=n}^{N-1} E_t \left(B(T_{i+1})^{-1} X_i \right).$$

The backward-induction scheme (18.7), (18.16) is trivially extended,

$$U_n(T_n) = \tau_n D_n \left(C_n - L_n(T_n)\right) P\left(T_n, T_{n+1}\right)$$
$$+ B(T_n)E_{T_n}\left(B(T_{n+1})^{-1}U_{n+1}\left(T_{n+1}\right)\right), \tag{18.64}$$
$$H_n(T_n) = B(T_n)E_{T_n}\left(B(T_{n+1})^{-1}\max\left(U_{n+1}\left(T_{n+1}\right), H_{n+1}\left(T_{n+1}\right)\right)\right). \tag{18.65}$$

As written, the scheme cannot be directly implemented in a PDE solver because of path-dependency under the expected value operator in (18.64). However, by employing the method of *similarity reduction*, the scheme can be rewritten in a way amenable to a PDE representation.

Dividing both sides of (18.64)–(18.65) by D_n, and using the fact that D_n is $\mathcal{F}_{T_n}$-measurable, we get

$$U_n'(T_n) = \tau_n \left(C_n - L_n(T_n)\right) P\left(T_n, T_{n+1}\right)$$
$$+ B(T_n)E_{T_n}\left(B(T_{n+1})^{-1}\frac{D_{n+1}}{D_n}U_{n+1}'\left(T_{n+1}\right)\right),$$
$$H_n'(T_n) = B(T_n)E_{T_n}\left(B(T_{n+1})^{-1}\frac{D_{n+1}}{D_n}\max(U_{n+1}'(T_{n+1}), H_{n+1}'(T_{n+1}))\right),$$

where we have defined

$$U_n'(T_n) = \frac{U_n(T_n)}{D_n}, \quad H_n'(T_n) = \frac{H_n(T_n)}{D_n}, \quad n = 1, \ldots, N - 1.$$

From (18.63),
$$\frac{D_{n+1}}{D_n} = 1 + \tau_n C_n,$$

where $1 + \tau_n C_n$ is $\mathcal{F}_{T_n}$-measurable. Therefore, the factor D_{n+1}/D_n can be pulled out from inside the expected value operator, to give us

$$U_n'(T_n) = \tau_n \left(C_n - L_n(T_n)\right) P\left(T_n, T_{n+1}\right)$$
$$+ (1 + \tau_n C_n) B(T_n)E_{T_n}\left(B(T_{n+1})^{-1}U_{n+1}'\left(T_{n+1}\right)\right), \tag{18.66}$$
$$H_n'(T_n) = (1 + \tau_n C_n) B(T_n)$$
$$\times E_{T_n}\left(B(T_{n+1})^{-1}\max\left(U_{n+1}'\left(T_{n+1}\right), H_{n+1}'\left(T_{n+1}\right)\right)\right). \tag{18.67}$$

These equations are used for $n = N-1, \ldots, 0$, with $U_N'(T_N) = H_N'(T_N) = 0$. We have the following result.

Proposition 18.4.1. *For a coupon-accreting CLE, $U_n'(T_n)$ and $H_n'(T_n)$ can, for each $n = 0, \ldots, N$, be written as deterministic functions of the model state variables model at time T_n.*

Proof. The proof is by induction. The statement is trivially true for $n = N$. To prove the induction step, we assume it is true for $n + 1$. We note that

$$B(T_n)E_{T_n}\left(B(T_{n+1})^{-1}U_{n+1}'\left(T_{n+1}\right)\right)$$

and

$$B(T_n)\mathrm{E}_{T_n}\left(B(T_{n+1})^{-1}\max\left(U'_{n+1}\left(T_{n+1}\right),H'_{n+1}\left(T_{n+1}\right)\right)\right)$$

can be written as functions of the state variables at time T_n by applying the PDE rollback scheme to

$$U'_{n+1}\left(T_{n+1}\right),\quad\max\left(U'_{n+1}\left(T_{n+1}\right),H'_{n+1}\left(T_{n+1}\right)\right)$$

(which are functions of the state variables at time T_{n+1} by the induction hypothesis). The accreting factor $(1+\tau_n C_n)$ and the marginal coupon $\tau_n(C_n-L_n(T_n))P(T_n,T_{n+1})$ are functions of the state variables at time T_n as well. The proposition is proved. $\square$

The new scheme (18.66)–(18.67) in fact looks just like (18.64)–(18.65) for a unit-notional CLE, with one modification: on each backward induction step, the values of U' and H' are rescaled by the "marginal" accreting notional $(1+\tau_n C_n)$. The key fact here is that this factor is known at time T_n.

18.4.5.2 Snowballs

Certain other path-dependent CLEs can be valued by PDE methods by introducing extra state variables, along the lines of Section 2.7.5. Among the more popular CLEs for which this method is applicable are the snowball swaps and callables, see Chapter 5. In a snowball, the structured coupon at time T_n is a function of the structured coupon at time T_{n-1} (and rates at time T_n). As an example, recall the basic structure from Section 5.13.5 with the coupon at time T_n given by

$$C_n=(C_{n-1}+s_n-g_n\times L_n(T_n))^+,\quad n=1,\ldots,N-1,$$

with C_0 being a fixed initial coupon.

We can value snowball swaps and callables by PDE method after the introduction of an extra state variable $I(t)$ defined to be the current coupon,

$$I(t)=\sum_{n=0}^{N-1}1_{\{t\in[T_n,T_{n+1})\}}C_n.$$

The backward recursion for the exercise value at time T_n then reads,

$$U_n(T_n)=\tau_n\left(I(T_n)-L_n(T_n)\right)P\left(T_n,T_{n+1}\right)+\mathrm{E}_{T_n}\left(\frac{B(T_n)}{B(T_{n+1})}U_{n+1}\left(T_{n+1}\right)\right).$$
(18.68)

Hence we obtain the following continuity condition for a given value of $I(T_n)=I$ (where T_n- is a time immediately prior to T_n),

$$U_n\left(T_n-,I\right)=\tau_n\left(I-L_n(T_n)\right)P\left(T_n,T_{n+1}\right)$$
$$+U_n\left(T_n,(I+s_n-g_n\times L_n(T_n))^+\right),$$

combined with the following one-period rollback scheme,

$$U_n(T_n, I) = \mathbb{E}_{T_n}\left(\frac{B(T_n)}{B(T_{n+1})} U_{n+1}(T_{n+1}-, I)\right), \qquad (18.69)$$

$n = N-1, \ldots, 1$. For the hold value, the continuity condition is

$$H_n(T_n-, I) = H_n\left(T_n, (I + s_n - g_n \times L_n(T_n))^+\right), \qquad (18.70)$$

where

$$H_n(T_n, I) = \mathbb{E}_{T_n}\left(\frac{B(T_n)}{B(T_{n+1})} \max\left(U_{n+1}(T_{n+1}-, I), H_{n+1}(T_{n+1}-, I)\right)\right). \qquad (18.71)$$

The PDE scheme may be implemented by discretizing I over an appropriate range, solving PDEs (18.69)–(18.71) on each I-plane, and interfacing the solutions between I-planes at times T_n, $n = 1, \ldots, N-1$ using (18.68)–(18.70). The details of implementation follow the general plan of Section 2.7.5, and we do not repeat them here.

19

Bermudan Swaptions

After our general discussion of callable Libor exotics in the previous chapter, we now turn our attention to an important subset of the generic CLE class, the Bermudan swaptions. Bermudan swaptions are among the most liquid exotic interest rate derivatives, and the demands they place on accuracy, fidelity and performance of term structure models have driven many advances in interest rate modeling. While the ideas and methods from the previous chapter all apply to Bermudan swaptions, the simpler structure of Bermudan swaptions compared to general CLEs allows us to considerably deepen our analysis of valuation and risk management methods.

19.1 Definitions

As defined in Section 5.12, a Bermudan swaption is a callable Libor exotic with the coupon paying a fixed rate, $C_n = k$, $n = 1, \ldots, N-1$. Alternatively, we can consider it a Bermudan-style option to enter a simple fixed-for-floating swap. The fixed rate k is often referred to as the *strike* of the Bermudan swaption. Exercise dates of a Bermudan swaption are typically[1] a subset of a tenor structure $\{T_n\}_{n=0}^N$ that defines the underlying swap. In a standard structure, exercise is restricted to the dates $\{T_n\}_{n=s}^{N-1}$ where $s \geq 1$; as we explained in Section 5.12, the period up to T_s is known as the lockout or no-call period. Recall that a Bermudan swaption on, say, a 10 year swap with a 2 year lockout period (at inception) is known as a "10 no-call 2", or "10nc2", Bermudan swaption. For convenience (and without loss of generality) we assume in most of this chapter that all $\{T_n\}_{n=1}^{N-1}$ are, in fact, exercise dates. If the Bermudan swaption is exercised at time T_n, the exercise value, for a payer swap, is given by

$$U_n(t) = \sum_{i=n}^{N-1} \tau_i P(t, T_{i+1}) (L_i(t) - k), \qquad (19.1)$$

---

[1] But see Sections 19.4.7 and 19.4.8 below for exceptions.

where k is the fixed rate. We note that $U_n(t)$ can be written as

$$U_n(T_n) = A_n(T_n)\left(S_n(T_n) - k\right),$$

where $A_n(t) \triangleq A_{n,N-n}(t)$ is the annuity, and $S_n(t) \triangleq S_{n,N-n}(t)$ is the swap rate for the swap into which one can exercise at time T_n (see notations (4.8), (4.10)). The definition of hold values carries over unchanged from Chapter 18.

19.2 Local Projection Method

As Bermudan swaptions are liquid and their volume is relatively high, the performance advantages of PDE methods over Monte Carlo simulation lead many market participants to value Bermudan swaptions in low-factor Markovian models, using either finite difference grids or trees. A sound framework for the usage of low-dimensional Markovian models is provided by the local projection method for single-rate CLEs that we discussed in Section 18.4. The method takes a particularly simple form for Bermudan swaptions, as the underlying swaps have no optionality and only the volatility parameters of core swap rates $\{S_n(\cdot)\}_{n=1}^{N-1}$ are relevant. As we discussed before (in Section 13.1.8.1), we can view a Bermudan swaption as the option to choose the "best" among a collection of swap values observed on different dates. This implies that a Bermudan swaption value is driven by *core volatilities*, or volatilities of the core swap rates $\{S_n(T_n)\}_{n=1}^{N-1}$, and *core correlations*, or inter-temporal correlations of the core swap rates $\{S_n(T_n)\}_{n=1}^{N-1}$. Alternatively, of course, we can think of forward volatilities in place of inter-temporal correlations as the source of "exotic" risk in Bermudan swaptions, see Section 18.1.1.

The relative simplicity of the dependence of Bermudan swaptions on the volatility structure allows us to use models as simple as the one-factor Gaussian model (Section 10.1.2) for valuation and risk management. The time-dependent volatility is typically calibrated to core swaption volatilities, while the mean reversion is calibrated to inter-temporal correlations of core swap rates (see Section 13.1.8.3); these correlations could, for instance, be extracted from an LM model. In practice, it is not unusual to skip the last step — since Bermudan swaptions have been traded well before LM models (or other practical multi-factor models) were invented, a market practice has developed whereby the mean reversion of a Gaussian (or similar) model is used essentially as a free parameter, rather than implied from a global model. This practice continues today, with mean reversions often set to match the "market" prices of Bermudan swaptions that are sometimes observable, or quasi-market prices such as independently-produced averages of dealer-submitted prices of a few typical structures[2]. Another fairly popular

---

[2] At the time of writing this is done by Markit, see www.markit.com.

choice would set mean reversions to match caplet volatilities, although using caplets for mean reversion calibration is rather arbitrary and can sometimes lead to odd mean reversion curves (see for instance the discussion in Section 10.1.2.3). The practice can, however, perhaps be justified if caplets are used as hedging instruments for inter-temporal correlation or forward volatility; technical details are available in Section 13.1.8.2.

Turning to the issue of volatility smile, we recall that the Gaussian model basically has no control over it, and the model can only be calibrated to one[3] volatility per expiry T_n, $n = 1, \ldots, N - 1$. A one-factor quasi-Gaussian (qG) model with local volatility (Section 13.1) would constitute an improvement, as it will also allow to capture the slopes of volatility smiles of core swaptions, in addition to volatilities at specific strikes. Finally, the stochastic volatility version of the one-dimensional qG model (Section 13.2) would essentially allow for (best-fit) calibration to all core swaption volatility smiles across all strikes. On balance, the local volatility qG model is probably sufficient for effective risk management of Bermudan swaptions, although we would of course choose the SV version if available computing power permits. Finally, a two-factor quadratic Gaussian model of Section 12.3 is also a viable choice for Bermudan swaption pricing.

While we are on the subject of the model choice, let us briefly comment on the discussion around what number of factors is appropriate for a Bermudan swaption model. While the usage of single-factor, or essentially single-factor models such as the qG model, for Bermudan swaption valuation is widespread, some argue that single-factor models significantly underprice Bermudan swaptions. The basic claim is that higher de-correlation in rates has a positive impact on Bermudan swaption prices (as a Bermudan swaption is a "best-of" option on swap rates) and two- and multi-factor models intrinsically are able to de-correlate rates more than a single-factor model (where the instantaneous correlations between moves in all forward rates is always one). There are a number of flaws in this argument, starting with the fact that the correlations relevant for Bermudan swaptions are the *inter-temporal* correlations, which can be easily manipulated in a one-factor model through the choice of mean reversion. In addition, when comparing one- and multi-factor rates models, it is obviously important that calibration to European swaptions is unaffected by changes in the number of factors. Careful analysis in Andersen and Andreasen [2001] of Bermudan swaption pricing in a one- and a two-factor Gaussian models shows that, if the models are calibrated in consistent fashion to core European swaptions, the two-factor model price is in fact slightly *lower* than the one-factor price. Experiments with LM models with different numbers of Brownian motions, but all calibrated to the full swaption grid, confirm this analysis.

---

[3]For more information on *which* volatility to calibrate to, see Section 19.3 below.

The reason for the slight decrease in Bermudan swaption price as a function of the number of factors may seem puzzling at first and is, indeed, a rather subtle effect. While there are numerous factors in play (see Andersen and Andreasen [2001] for a full analysis), one important observation is that *forward* volatility in a low-factor model generally is higher than in a multi-factor model, as long as both models are in calibration with the European swaption market. A technical explanation for this phenomenon can be found in Appendix 19.A of this chapter; loosely speaking the effect stems from the fact that a one-factor model will imply a lower time 0 instantaneous forward volatility term structure than will a multi-factor model, a relationship that must be reversed as time progresses to preserve overall variance.

19.3 Smile Calibration

Let us now discuss the issues of smile calibration in more detail. For concreteness, we consider the linear local volatility version of the quasi-Gaussian model from Section 13.1. As we have seen before, the model has enough flexibility to match the level and the slope of the volatility smile for each of the core swap rates. The market volatility smile is, of course, not close to linear, so we often seek to match the volatility of a *particular strike* exactly, or as closely as possible, while roughly matching the overall slope of the volatility smile.

The simplest approach to choosing the strikes that define core volatilities for calibration involves using at-the-money (ATM) strikes for each core swap rate. This rather crude approach is still in use for (we assume) historical reasons, as Bermudan swaptions started trading well before pronounced market smiles developed in interest rate markets, and probably even before the non-ATM points of the swaption volatility cube became liquid enough to keep track of them. Proponents of this approach sometimes rely on hedging arguments, as volatility exposure of a Bermudan swaption is often vega hedged using the most liquid European swaptions — which happen to be at-the-money. Yet another possible justification for the ATM strike choice notes that using the same volatilities for Bermudan swaptions of different strikes (seemingly) ensures consistency of valuation across Bermudan swaptions of different strikes. In reality, however, the ATM strike choice leads to *in*consistent valuation between European and Bermudan swaptions: if one uses a Bermudan swaption model calibrated to ATM swaptions, and applies it to a Bermudan swaption with a non-ATM strike and just a *single* exercise date, the value is going to be different from the value of the same derivative priced as a European swaption. Clearly, this is a strongly undesirable feature of the ATM strike calibration idea.

To ensure consistency between the Bermudan swaption and its underlying core European swaptions, it suffices to set the calibration strike equal to that of the Bermudan swaption itself. That is, if the fixed rate of the

Bermudan swaption is k, then for each expiry T_n one uses the volatility of the appropriate core European swaption that corresponds to the strike k. This method automatically ensures that a European swaption valued as a single-exercise Bermudan swaption has exactly the same value in the model as in the market. The fact that all swaptions we can exercise into are priced exactly is intuitively appealing, and also ensures that certain rational bounds for the Bermudan swaption price will not be violated. Indeed, if $V_{\mathrm{swaption},n}(0;k)$ is the price of a k-strike, T_n-expiry swaption on a swap that matures at time T_N, then clearly[4] the k-strike Bermudan swaption price $V_{\mathrm{Berm}}(0;k)$ at time 0 must satisfy (compare to (18.3))

$$V_{\mathrm{Berm}}(0;k) \geq \max_{n=1,\ldots,N-1} V_{\mathrm{swaption},n}(0;k), \qquad (19.2)$$

where we as mentioned earlier have assumed that exercise can take place at all T_n, $n \geq 1$. If our model fundamentally matches all swaption prices inside the max-operator on the right-hand side of (19.2), then pricing the Bermudan swaption in, say, a finite difference grid will always return a Bermudan swaption that satisfies (19.2). We notice as an aside that the (non-negative) difference between the left- and right-hand sides of (19.2) is sometimes known in trader jargon as the *Bermudanality* of the Bermudan swaption.

While enforcing consistency with European swaptions is useful, the idea of at-the-strike calibration is not a panacea, as Bermudan and European swaptions can behave quite differently. For instance, Bermudan swaptions have other "interesting", e.g. high-convexity, points in the swap rate dimension than just the underlying strike k, the most important of which is the exercise boundary, i.e. the swap rate level (for each time T_n) at which the decision to exercise the swaption switches to the decision to hold. The importance of this point is clearly seen from the marginal exercise value decomposition (18.8), as it corresponds to "strikes" of European options in the representation of the value of a Bermudan swaption as a sum of European options. Hence, a third calibration option available is to use the swap rate volatilities that correspond to the exercise boundary on each of the exercise dates. As was the case for the Bermudan strike method, this method makes valuation of single-exercise Bermudan swaptions consistent with the valuation of (equivalent) European swaptions, since the exercise boundary for a European swaption coincides with its strike. For the same reason, any weighted average of the strike and the exercise boundary would also produce a consistent scheme. We do point out, however, that using calibration strikes other than that of the Bermudan swaption may lead to violations of (19.2).

---

[4]The optimal exercise strategy for a Bermudan swaption must be as least as good as simply picking at time 0 one of the exercise dates and never changing one's mind.

To provide a bit more detail on the idea of using the exercise boundary to select calibration strikes, let us first observe that for any given model, one can determine the exercise boundary as a function of the state variables. To be able to calibrate to European swaption volatilities with the strike at the exercise boundary, one has to be able to translate this "model" exercise boundary into a value of the corresponding core swap rate. Strictly speaking, this can be done unambiguously only in single-factor models, such as the Gaussian model. For the one-factor qG model with its two state variables the boundary is, in fact, represented by a line in a two-dimensional plane of possible values of the x and y state variables; each of the points on this line corresponds, potentially, to a different value of the core swap rate. However, the dependence of the exercise boundary on y is rather mild, a fact that should come as no surprise if one recalls the "auxiliary" nature of the y state variable, see Chapter 13. Hence, for the qG model, we can use the expected value of $y(T_n)$ when converting the exercise boundary for the state variable $x(T_n)$ into the strike for the swap rate $S_n(T_n)$.

The choice of exercise boundary for calibration is, unfortunately, rather inconvenient from the implementation point of view because the exercise boundary information is not available until *after* the valuation algorithm has been run. One can try a recursive scheme where one uses some (e.g., strike-calibrated) volatilities for an initial calibration, values a Bermudan swaption, calculates the exercise boundary, looks up the core volatilities for calibration at this boundary, calibrates the model again, values the Bermudan swaption, and so on. Such a procedure can in fact diverge; thus, one is forced to limit the number of iterations artificially, potentially resulting in unstable risk sensitivities and other problems. Moreover, this scheme in general consumes more computational resources due to the multiple valuations required. For all these reasons the at-the-boundary calibration is of limited use and the at-the-strike volatility calibration method is probably the most reasonable in practice, combining ease of implementation and consistency with European swaptions.

19.4 Amortizing, Accreting, and Other Non-Standard Bermudan Swaptions

A standard (or *vanilla* or *bullet*) Bermudan swaption is characterized by the fact that the notionals on which coupons are paid are all identical, as in (19.1) (where the notionals of all coupons are 1). A relatively popular extension involves making the notional of a Bermudan swaption time-dependent and deterministic, with the exercise value given by

$$\tilde{U}_n(t) = \sum_{i=n}^{N-1} R_i \tau_i P\left(t, T_{i+1}\right) \left(L_i(t) - k\right) \tag{19.3}$$

(compare to (19.1)). Here R_i is the notional for the i-th coupon, $i = 1, \ldots, N - 1$.

If the notional increases with the coupon index, the Bermudan swaption is said to be *accreting*; if it decreases, it is said to be *amortizing*. Other profiles are possible but are much less common. Amortizing Bermudan swaptions are often used as hedges for pools of mortgages, with the amortization feature mimicking prepayments on the pool. Accreting Bermudan swaptions, on the other hand, often appear as a result of issuing "zero coupon" structured notes, i.e. notes with the repayment notional growing over time but paying no coupons during the life of the note, as explained in Section 19.4.6.

Since the notionals in Bermudan swaptions of type (19.3) are still deterministic (even if time-varying), their valuation in a properly calibrated model does not present any particular technical difficulties[5] — but what constitutes "properly calibrated", however, is not always obvious. Of course, in models with global calibration (e.g. LM models), calibration for non-standard Bermudan swaptions is no different from calibration for standard Bermudan swaptions, as model calibration is product-independent by definition. On the other hand, for models requiring local calibration, such as a one-factor Gaussian or a quasi-Gaussian model, calibration for non-standard Bermudan swaptions will require additional analysis. We consider this problem in the next few sections, but it is worth pointing out that, in the opinion of some, making notionals time-dependent pushes Bermudan swaptions across the boundary that separates those securities for which local models are acceptable to use from those for which globally-calibrated multi-factor models are required.

Before commencing on a more detailed analysis, let us first briefly try to understand the basic complications involved in local model calibration for non-vanilla Bermudan swaptions. As established previously, a locally-calibrated model should, as a minimum, be calibrated to the volatilities of core swap rates. For an amortizing Bermudan swaption, say, a core swap rate would correspond to an amortizing swap. Volatilities of amortizing swaps can be extracted from amortizing European swaptions, but the liquidity of such swaptions is significantly poorer than for vanilla European swaptions — in fact, amortizing European swaptions are about as illiquid as Bermudan swaptions themselves. In practice, if one wishes to calibrate a local model to core (amortizing) swaptions, one may need to use a "pre-processing" step to extract amortizing European swaption prices from a model calibrated to liquid vanilla European swaptions, as we do later in Section 19.4.4. Alternatively, one needs to choose a different set of calibration targets in the first place, see Section 19.4.3.

---

[5] Although this is not true for the family of Markov-functional models, see Appendix 11.A in Chapter 11; in fact the difficulty of handling amortizing/accreting Bermudan swaptions is often cited as one of the problems with such models.

19.4.1 Relationship Between Non-Standard and Standard Swap Rates

Regardless of the calibration method ultimately used, it is useful to understand the relationship between non-standard and standard swaps (and, hence, swap rates). To be consistent with (19.3), consider a swap that starts at T_n, ends at T_N, and has a notional schedule $\{R_i\}$. The time t value of such a swap is given by $\tilde{U}_n(t)$ in (19.3). To make some of the formulas below simpler, let us extend the notional schedule by one period and set $R_N = 0$. We denote the annuity and the swap rate that correspond to this non-standard swap by $\tilde{A}_n(\cdot)$ and $\tilde{S}_n(\cdot)$, so that

$$\tilde{A}_n(t) = \sum_{i=n}^{N-1} R_i \tau_i P\left(t, T_{i+1}\right), \quad \tilde{S}_n(t) = \tilde{A}_n(t)^{-1} \sum_{i=n}^{N-1} R_i \tau_i P\left(t, T_{i+1}\right) L_i(t).$$

We would like to decompose the non-standard swap into a linear combination of standard swaps. Such a decomposition is, however, not unique and could be done in a multitude of ways, potentially using any of the standard swaps with starting date on or after T_n, and final payment date on or before T_N. To narrow down the problem, we note that we here are ultimately interested in establishing the volatility of the non-standard rate $\tilde{S}_n(\cdot)$ over the period $[0, T_n]$. Since the values of standard European swaptions provide us with the information on the volatilities of swap rates *only* over the period from time 0 to their start dates, we should focus only on standard swaps that start on T_n; this choice makes the decomposition unique.

Let us denote the value of a standard swap starting at T_n and covering m periods by $V_{n,m}(t)$, and the corresponding annuity and swap rate by $A_{n,m}(t)$ and $S_{n,m}(t)$ (see (4.8), (4.10)). In light of the discussion above, we want to find weights $\{v_{n,m}\}$, $m = 1, \ldots, N - n$, such that

$$\tilde{U}_n(T_n) = \sum_{m=1}^{N-n} v_{n,m} V_{n,m}(T_n).$$

Note that only swaps starting at time T_n are used in the right-hand side of this expression. Matching terms to (19.3) we obtain

$$v_{n,m} = R_{n+m-1} - R_{n+m}, \quad m = 1, \ldots, N - n, \tag{19.4}$$

so that

$$\tilde{U}_n(T_n) = \sum_{m=1}^{N-n} \left(R_{n+m-1} - R_{n+m}\right) V_{n,m}(T_n).$$

After some algebraic manipulations, we obtain the following relationship for the swap rates,

$$\tilde{S}_n(T_n) = \sum_{m=1}^{N-n} w_{n,m}(T_n) S_{n,m}(T_n), \tag{19.5}$$

where (recall that we set $R_N = 0$)

$$w_{n,m}(T_n) = (R_{n+m-1} - R_{n+m}) \frac{A_{n,m}(T_n)}{\widetilde{A}_n(T_n)}.$$

While the swap weights $v_{n,m}$ are deterministic, the swap *rate* weights $w_{n,m}(T_n)$ are not. For a qualitative discussion, however, we note that the weights $w_{n,m}(T_n)$ can be approximated reasonably well by their values at time 0,

$$\widetilde{S}_n(T_n) \approx \sum_{m=1}^{N-n} w_{n,m}(0)S_{n,m}(T_n), \quad w_{n,m}(0) = (R_{n+m-1} - R_{n+m}) \frac{A_{n,m}(0)}{\widetilde{A}_n(0)}.$$
$$(19.6)$$

From the expression (19.6) it follows that the volatility of the non-standard swap rate is a function of volatilities of all standard swap rates with a given expiry (T_n in our case), and of their correlations. Putting correlations aside for a moment, observe that to price a non-standard Bermudan swaption, in principle one needs to calibrate the model to volatilities of standard rates with all expiries ($T_1, \ldots, T_{N-1}$) and all maturities, something a low-dimensional local model will virtually never be able to do. Below, we discuss two possible strategies for going forward.

19.4.2 Same-Tenor Approach

Perhaps the simplest calibration approach for non-standard Bermudan swaptions is to simply pretend that they are standard Bermudan swaptions and set up the model calibration accordingly. So, for expiry T_n, one would choose a European swaption on the $(N - n)$-period swap as the calibration instrument. While easy to implement, the merits of this approach are obviously somewhat wanting, and we do not recommend it. Nevertheless, it is instructive to investigate the issues that would come up if we adopted this scheme. As an example, consider an amortizing Bermudan swaption and a one-factor model calibrated to standard swaptions of the same tenor as the core amortizing swaptions. As a thought experiment, suppose that we increase mean reversion in the model, while keeping it calibrated to our calibration swaption set. In this case, the core amortizing swaption prices would increase, a simple consequence of our decomposition (19.6) and the fact that shorter-tenor (standard) swap rate volatilities increase as a function of mean reversion when volatility of a longer-tenor swap rate is kept fixed, see the discussion in Section 13.1.8.1. As a consequence, mean reversion would affect not only the inter-temporal correlations that are important for Bermudan swaptions, but would also affect the volatilities of core swap rates. In the context of a local projection method, we would then face a dilemma as to which targets to calibrate the mean reversion to: the inter-temporal correlations or the prices of amortizing European swaptions (the latter, just like the former, would be available from a global model). Of course it would be highly unlikely that both calibration targets would imply the same mean reversion.

Volatility smile calibration presents another challenge for the same-tenor approach. For instance, if one chooses a particular strike of the non-standard swaption to calibrate to (e.g. the fixed rate of the non-standard swap), which strike for the *standard* swaption would that correspond to?

19.4.3 Representative Swaption Approach

The idea of the representative swaption approach is to choose a standard swap that approximates the non-standard swap in some reasonable sense, and then to calibrate the Bermudan swaption model to the market-implied volatilities of swaptions on these standard swaps, one per exercise date.

One can define a "representative" swap in many ways. For example, a fairly simple *PVBP matching* method chooses the standard swap whose PVBP (Present Value of a Basis Point, see Section 5.5) matches the PVBP of the non-standard swap most closely. In this case, for expiry T_n, we would choose the tenor μ_n of the standard swap by

$$\mu_n = \operatorname*{argmin}_m \left\{ \left| R_n A_{n,m}(0) - \tilde{A}_n(0) \right| \right\} \tag{19.7}$$

(note that we use the notional of the first period of the swap $\tilde{U}_n$ to scale the PVBP of the standard swap). While somewhat simplistic, the method actually turns out to be reasonably robust for some non-standard Bermudan swaptions. We proceed to improve it and make more rigorous, which will also help us identify not just the right tenor, but also the right strike for the standard calibration swaptions.

We work in the context of a one-factor Gaussian model to demonstrate the main idea, although the method is not tied to a particular model. Let us fix a start date T_n and note that the value of (any) swap at time T_n is a function of the Gaussian short rate state $x = x(T_n)$ on that date. Let $\tilde{U}_n(x)$ be the value of the non-standard swap $\tilde{U}_n(T_n)$, as a function of the short rate state. Note that $\tilde{U}_n(x)$ depends on the mean reversion, but not the volatility parameter of the model, as follows from bond reconstruction formulas (Proposition 10.1.7). Define $V(x; R, q, m)$ to be the value of a standard swap starting on T_n as a function of x, with constant notional R, fixed rate q, and covering m periods. In departure from our normal conventions, we here allow m to be any real number and not necessarily an integer; we interpret a value of, say, $m = 5.3$ as 5 full periods plus three tenths of the sixth period. The rationale for allowing for fractional periods will become clear shortly.

Now, we have three parameters that define the standard swap, R, q, and m. In the *payoff matching* method, we choose the three parameters by matching the level, slope and curvature of the swap payoffs as functions of the state variable,

$$V(x_0; R, q, m) = \widetilde{U}_n(x_0), \tag{19.8}$$

$$\frac{\partial}{\partial x} V(x_0; R, q, m) = \frac{\partial}{\partial x} \widetilde{U}_n(x_0), \tag{19.9}$$

$$\frac{\partial^2}{\partial x^2} V(x_0; R, q, m) = \frac{\partial^2}{\partial x^2} \widetilde{U}_n(x_0), \tag{19.10}$$

where the expansion point x_0 is the expected value of $x(T_n)$ (or close to it). In the parameterization of Section 10.1.2, it suffices to set $x_0 = 0$. The system of equations (19.8)–(19.10) is easy to solve (numerically) in the one-factor Gaussian model; let us denote the solution by R_n^*, q_n^*, m_n^*.

Even though we fix the parameters of the standard swap by local conditions around x_0, numerical experiments show that the swap that solves (19.8)–(19.10) tends to match that of the non-standard swap across a large range of state values $x(T_n)$, suggesting considerable robustness. In addition, even though the functions V and $\widetilde{U}_n$ depend on mean reversion, numerical experiments show that the best-fit parameters R_n^*, q_n^*, m_n^* are only mildly sensitive to mean reversion.

Incorporating the payoff matching method into a volatility calibration routine is quite easy, since the choice of the best-fitting standard swaps is independent of volatility, which allows us to identify the calibration targets *before* we commence on the volatility calibration. Moreover, the strike of the calibration swaptions are produced automatically as part of the payoff matching routine, facilitating calibration to a particular point of the observed volatility smile.

Before discussing application of the representative swaption idea to accreters and amortizers, let us briefly motivate our usage of fractional swap tenors. If tenors are restricted to an integer number of periods, then a perturbation of the market data, e.g. when shifting a yield curve to calculate an interest rate delta, could potentially alter the solved-for number of periods m by plus or minus one period. Hence, restricting m to be an integer would potentially introduce a discontinuity in the calibrated model parameter — and therefore in the value of the Bermudan swaption — as a function of market data. Such a discontinuity would be purely artificial and, as explained at length in Chapter 23, highly undesirable for stability of risk sensitivities. By allowing fractional tenors, we eliminate these problems. Of course, to complete the volatility calibration, we need to know implied volatilities of swaptions with fractional tenors, but these could be obtained by (smoothly!) interpolating implied volatilities of swaptions with integer-valued tenors.

Now, let us see how the representative swaption method works for an amortizing Bermudan swaption, i.e. a Bermudan swaption with decreasing notionals $R_1 \geq R_2 \geq \ldots \geq R_{N-1}$. We note that, according to (19.4), all weights $v_{n,m}$ in the decomposition of the amortizing swap into standard swaps are positive, $v_{n,m} \geq 0$, $m = 1, \ldots, N - n$. It follows that both the PVBP matching and payoff matching methods produce a standard swap whose tenor is some average of tenors of the standard swaps in the basket; in particular, the resulting standard swap have a final maturity that is *shorter* than the amortizing swap. This is an intuitive result, and leads to an amortizing Bermudan swaption being sensitive to interest rate volatilities of standard swaps V_{n,m_n^*}, $n = 1, \ldots, N - 1$, with $n + m_n^* \leq N$ for any $n = 1, \ldots, N - 1$.

The situation is different for accreting Bermudan swaptions, i.e. when $R_1 < R_2 < \ldots < R_{N-1}$. According to (19.4),

$$v_{n,m} < 0, \quad m = 1, \ldots, N - n - 1,$$
$$v_{n,m} > 0, \quad m = N - n.$$

So, an accreting swap decomposes into a standard swap of the maximum tenor $N - n$ and maximum notional R_{N-1}, *minus* a basket of swaps of smaller tenors. The PVBP $\widetilde{A}_n(0)$ of an accreting swap is larger than the PVBP of a standard swap of the matching tenor (times starting notional R_n) so the PVBP matching method would calculate an optimal tenor m^* that is *longer* than the tenor of the amortizing swap, $m^* > N - n$. The same would be true of the payoff matching method as well. This is, of course, rather problematic, as our calibration method would suggest that an accreting Bermudan swaption is sensitive to volatilities of swaptions with total length (the sum of expiry and tenor) exceeding the final maturity of the Bermudan swaption. This would be in direct contradiction to what, say, a globally-calibrated LM model would suggest, as in the latter the price of any derivative is fully determined by the volatility structure of Libor rates that fix before the final maturity of the derivative, and this volatility structure, in turn, is fully determined by the volatilities of swaptions with total length less than the final maturity.

The reader may ask why we are getting reasonable results for amortizing swaptions and unreasonable ones for accreting swaptions. A bit of reflection reveals that the discrepancy originates with the single-factor assumption that we made implicitly in the PVBP matching method, and explicitly in the payoff matching method. For the amortizing swap, the decomposition resulted in a basket of standard swaps with positive weights, a basket that can be reasonably well-hedged with a single swap of average tenor. In the accreting case, our decomposition resulted in a *spread* position in standard swaps: long a long-tenor swap and short a basket of short-tenor swaps. Our one-factor methods suggest hedging this spread position with a single (very) long-dated swap — perfectly reasonable in a one-factor world, but not in actual reality.

While sensibly hedging the spread position in an accreting swap with a single standard swap is not possible, things improve markedly if we allow usage of *two* standard swaps in the hedge. In particular, we may then take as one of our hedges a long position in the T_N-maturity swap with maximum (R_{N-1}) notional, and construct the second swap hedge by PVBP (or payoff) matching the remaining short basket of swaps $\sum_{m=1}^{N-n-1} v_{n,m} V_{n,m}$. As all weights in the short basket are of the same (negative) sign, a single standard swap would often provide a good hedge. Thus, to get a reasonable calibration scheme for an accreting Bermudan swaption, we would need to calibrate to two standard European swaptions per expiry, both of which would have their final payment date on or before the final payment date of the Bermudan swaption. Of course, we would find it difficult to accurately calibrate a one-factor Markovian model to two swaptions per expiry, and would likely need to move on to more elaborate models with additional factors.

In conclusion, we note that the two-swaps approach works universally for Bermudan swaptions with arbitrary notional schedules. To apply it, for each n we would combine swaps with positive weights $v_{n,m}$ into one basket, and swaps with negative weights $v_{n,m}$ into another basket. Then we would represent each basket by one standard swap by the procedures discussed above, yielding the calibration swaption targets for that expiry.

19.4.4 Basket Approach

The discussion above suggests that the pricing of at least some Bermudan swaptions with non-standard notional schedules is best done in multi-factor models. Still, one-factor Markovian models are highly popular due to their performance advantages, and the desire to use them even in situations where they might be overstretched is often considerable. Consequently, rather creative ways of using one-factor models for non-standard Bermudan swaptions have been developed, resulting in a family of approaches that we here all categorize as *basket methods*.

The basket methods generally split the valuation of a non-standard Bermudan swaption into two stages. During the first stage, some model is used to calculate values of core non-standard European swaptions. During the second stage, a one-factor model is calibrated to these values of non-standard core European swaptions, and subsequently used to compute the value of the non-standard Bermudan swaption. Various method differ in how the values of non-standard European swaptions are calculated. One perfectly sound method uses a globally calibrated model such as the LM model for the task, resulting in a *local projection method for non-standard Bermudan swaptions*. We have discussed the local projection method in various flavors often enough, so we trust the reader with filling in remaining details. Instead, we review some alternatives for how to execute the first stage of the basket method.

For concreteness, let us focus on the first non-standard European swaption underlying the Bermudan swaption, i.e. the option expiring at T_1 on a swap that covers $N - 1$ periods. As follows from (19.4), this non-vanilla European swaption can be interpreted as an option on a basket of standard swaps, all starting on T_1 but with different maturities. Hence, to compute the price of the non-standard swaption, one can use a model calibrated to the volatilities of options on such swaps, as well as relevant swap rate correlations. Notice that the standard swaptions involved here all form a "row" of the swaption grid (see Section 5.10), as they all share the same expiry but have different tenors. A one-factor mean-reverting Gaussian (or quasi-Gaussian or quadratic Gaussian) model can be calibrated to this set of European swaptions, although the calibration will be different from our standard procedure. In particular, the prices of all swaption targets depend on the model volatility function over the *same* interval $[0, T_1]$ and, thus, the short rate volatility function (e.g. $\sigma_r(\cdot)$ in the notation of Proposition 10.1.7) cannot be used if we want to match each swaption volatility exactly. Upon reflection, it should be clear that we instead can use the time-dependent mean reversion function ($\varkappa(\cdot)$ in the notation of Proposition 10.1.7) as our main calibration "knob", since the pricing of a T_1-expiry swaption on a swap that covers m periods will depend on the mean reversion function over the period $[T_1, T_{1+m}]$. Hence, a sequential mean reversion calibration is possible: after calibrating to the first m standard swaptions, the $(m + 1)$-th is matched by changing the mean reversion function[6] over the time interval $t \in [T_{m+1}, T_{m+2}]$, for $m = 0, \ldots, N - 2$. The (constant) level of volatility over the first period could be set arbitrarily, as its scaling effect would be compensated by the mean reversion calibration. However, it is advisable to keep it at a "typical" value of, say, 1% so the calibrated mean reversions would also remain in a "typical" region, as the numerical implementation of the model might not cope well with extreme values of mean reversion.

To summarize, the basket method for a mean-reverting one-factor short rate model works like this. First, for each row of the swaption grid that corresponds to an exercise date T_n, $n = 1, \ldots, N - 1$, of the non-vanilla Bermudan swaption, we fit separate instances of the one-factor model by sequentially calibrating the mean reversion function to all relevant swaptions in the row. For each T_n, the relevant instance of the model is then used to compute the price of the T_n-expiry non-vanilla European swaption that the Bermudan swaption can be exercised into. Finally, we calibrate the model once more by setting its short rate volatility function to match the prices of the non-standard European swaptions established in the previous step. In the final calibration, we would typically keep the mean reversion fixed, either at a user-specified level or (ideally) at a level that makes inter-temporal correlations of core swap rates match those coming from a global model; see Section 19.2 for additional discussion.

So far we have side-stepped the issue of what strikes to choose for various calibrations in the basket scheme above. It is fair to say that this choice is a non-trivial problem. Various ad-hoc schemes could be imagined, such as using

---

[6]It is probably advisable to impose smoothness constraints on the time-dependent mean reversion function while performing such calibration.

the standard swaption of the same relative moneyness as the non-standard one, but they are rarely entirely satisfactory. Using a quasi-Gaussian or quadratic Gaussian model (or, even better, a stochastic volatility extension of these models) is obviously preferable to, say, using a Gaussian model, as the former models will allow us to calibrate to the volatility smile at more than a single strike, thereby alleviating somewhat the strike selection problem.

Another issue that we should touch on concerns correlations. By using a one-factor model for establishing non-standard European swaption values, we are implicitly assuming high[7] correlations between standard swaps in the portfolio that replicates the exercise value of the swaption. This is not necessarily as constraining as it might appear to be, as in reality these swaps are indeed highly correlated. Still, we may want to contemplate methods to somehow incorporate into our procedure observations about correlation extracted, say, from historical analysis. One possible route for this would be to apply approaches inspired by basket valuation methods from equities modeling. For example, we could (rather crudely) value a non-standard European swaption by the Black formula on the (non-standard) swap rate whose volatility is obtained by moment matching[8]. This method would need to approximate the weights in the decomposition (19.5) as being deterministic (although they are not). We can also use more advanced approaches, such as the copula methods, or even SV methods, that we developed for multi-rate derivatives in Chapter 17, allowing us to incorporate volatility smile information into the basket valuation.

A few final comments on the method developed in this section are in order. First, it is worth pointing out that most of the basket methods are consistent with the way standard Bermudan swaptions are valued. Specifically, if we apply these methods to standard Bermudan swaptions, we obtain the same price as if we had used the "standard" valuation method of Section 19.2. Second, notice that the method produces volatility sensitivities for non-standard Bermudan (and European, for that matter) swaptions that tend to be intuitive and reasonable. In particular, each underlying European non-standard swaption will show sensitivities only to the correct row of the swaption grid and to ordinary swaptions on swaps with maturities that do not exceed the maturity of the swap in the corresponding non-standard swaption. For an accreting European swaption in particular with, say, an expiry T_n and swap maturity T_N, the sensitivities will be negative for all standard swaptions with expiry T_n and swap maturities T_i, $i = n+1, \ldots, N-1$, and will be positive for a swaption with expiry T_n and swap maturity T_N, thus faithfully representing the accreting swap rate as a "spread".

---

[7]Term correlations between swap rates in one-factor models are not exactly 100% due to time-dependence in parameters and presence of mean reversion.

[8]Appendix 19.B gives a quick tour of the classical moment matching ideas.

19.4.5 Super-Replication for Non-Standard Bermudan Swaptions

The replication method of Proposition 8.4.13 links the value of a European option with a non-standard payoff to that of a portfolio of standard European options. Not only does that give us a way to value a non-standard derivative, it also allows us to fully hedge it in a model-independent way. Such static replication results are extremely convenient, and much research in derivatives pricing theory have been directed towards the search for static hedges for exotic derivatives, see e.g. Andersen et al. [2002]. Unfortunately, the availability of truly model-free static replication methods for non-European options is an exceedingly rare phenomenon and no such results are known to exist for Bermudan swaptions. Interestingly, however, Bermudan swaptions with non-standard notional schedules can be *super*-replicated in a model-independent way, in the sense that for any given non-standard Bermudan swaption we can find a portfolio of standard Bermudan swaptions that would dominate the value of the non-standard Bermudan swaption in all states of the world. Moreover, in some cases the difference in value between the non-standard Bermudan swaption and its super-replicating portfolio can be quite small. While not as convenient as a replicating portfolio, the super-replicating portfolio has several practical uses. First, the value of the portfolio provides a hard no-arbitrage bound for the value of the non-standard Bermudan swaption, and any modeling procedure (including calibration rules, choice of strikes, etc.) can be checked against this bound. Second, if the upper bound provided by the portfolio is known to be relatively tight (as is the case for, e.g., amortizing Bermudan swaptions, as we shall see shortly), then the super-replicating portfolio can be used directly for valuation purposes, perhaps amended with a small ad-hoc adjustment. More importantly, the super-replicating portfolio can be used as a robust hedge that requires little rebalancing over time.

The easiest way to demonstrate the construction of a super-replicating portfolio is by example. Consider first an amortizing Bermudan swaption. For concreteness, assume it is a 10 year (10y) Bermudan swaption with exercises every year, starting in year 1. Suppose the initial notional is 10 and it decreases by 1 every year. The super-replicating portfolio then consists of nine standard Bermudan swaptions, all with unit notional: a 10 no-call 1, a 9 no-call 1, ..., a 3 no-call 1, and a 2 no-call 1. To see that this is indeed a super-replicating portfolio, suppose the amortizing Bermudan swaption is exercised at year 5. Then the option holder receives an amortizing 5 year swap, with a starting notional of 5 that decreases by 1 every year; the value of this swap is equal to the sum of standard swaps of tenors 5y, 4y, ..., 1y, each with notional 1. But clearly the value of each of these standard swaps is dominated by the value of a corresponding standard Bermudan swaption in the super-replicating basket.

Another way to test that the super-replicating strategy dominates the amortizing Bermudan swaption is to impose a particular exercise strategy on the portfolio of standard Bermudan swaptions. Specifically, we simply require that all still-alive (i.e. non-expired) standard Bermudan swaptions shall be exercised at the same time as when the amortizing Bermudan swaption is exercised. A little thought shows that the exercise value obtained from the super-replicating basket is then exactly the same as from the amortizing Bermudan swaption. Hence, the value of the portfolio of standard Bermudan swaptions with this specific exercise rule enforced must precisely equal the value of the amortizing Bermudan swaption. However, as the chosen exercise strategy will generally be sub-optimal for each of the standard Bermudan swaptions in the portfolio, the true value (i.e. the value obtained with optimal exercise) of the basket of standard Bermudan swaptions will be higher than the amortizing Bermudan swaption, and will dominate its value in all states of the world.

To give another example, consider a 10 year accreting Bermudan swaption with an initial notional of 1 that increases by 1 every year. Assuming that exercise can take place annually starting in year 1, the super-replicating portfolio will now consist of a collection of nine 10 year standard unit notional Bermudan swaptions exercisable annually, with lockout periods of 1, 2, ..., 9 years, respectively. To check that this hedge works, let us again assume the accreter is exercised in year 5. Then the holder would receive a swap that can be decomposed into a spot-starting 5 year standard swap, a 4 year standard swap starting in 1y, a 3 year standard swap starting in 2 years and so on, all with notional 1. Again, the value of each of these swaps is dominated by a corresponding standard Bermudan swaption in the super-replicating portfolio.

Super-replicating portfolios for Bermudan swaptions with arbitrary notional schedule always exist, but are rather tedious to write down explicitly (for example, see a related algorithm in Evers and Jamshidian [2005]). The basic idea, however, is quite simple: for any exercise opportunity one needs to ensure that each of the standard swaps into which the exercise value can be decomposed (see (19.4)) is matched by a standard Bermudan swaption in the super-replicating portfolio.

Let us show some numerical results as a way to examine the tightness of the upper value bound produced by the super-replicating portfolio. For the numerical experiments, we throughout use a one-factor quasi-Gaussian model with some reasonable, representative parameters and a yield curve flat at 6%. In Table 19.1 we show values for Bermudan (European) amortizing swaptions of different maturities (with 1y lockout) against the values of corresponding super-replicating portfolios of standard Bermudan (European) swaptions; all contracts are receivers (options on receive-fixed swaps) with 6% strike. The notionals of amortizing swaps decrease linearly from the initial notional indicated in the table by 1 every year. We notice that the upper bounds produced by the super-replicating portfolio are here quite tight, something that appears to hold generally for amortizing Bermudan swaptions.

Maturity	10y	10y	30y	30y
Initial notional	10	10	30	30
Bermudan/European	E	B	E	B
Amortizer value	0.606	0.630	2.180	2.700
Portfolio value	0.614	0.650	2.230	2.830

Table 19.1. The value of an amortizing Bermudan or European swaption vs. the value of a super-replicating portfolio of standard Bermudan/European swaptions for different maturities and contract types.

As a second test we look at a particular amortizing Bermudan swaption across a range of strikes, see Table 19.2. We consider a 30 year amortizing Bermudan swaption with the initial notional of 30, and compare it to the super-replicating portfolio. For reference, the vega (change in Bermudan swaption value to 1% change of Black volatilities of European swaptions) for the 6% Bermudan swaption is about 0.1, with European swaption implied volatilities around 12%. Again, the results from the super-replicating portfolio are quite close to the real option values.

Strike	3%	4%	5%	6%	7%	8%
Amortizer value	0.267	0.632	1.372	2.700	4.640	7.017
Portfolio value	0.355	0.766	1.534	2.830	4.710	7.052

Table 19.2. The value of a 30 year amortizing Bermudan swaption vs. the value of a super-replicating portfolio of standard Bermudan swaptions for different strikes.

Next, we look at the results for accreting swaptions. In Table 19.3 we have the results for accreting receivers with 6% strike with notional accreting at 6% relative rate, across different contract types and maturities. Clearly, the super-replicating portfolio here is substantially more expensive than the accreting Bermudan swaption.

Maturity	10y	10y	30y	30y
Initial notional	1	1	1	1
Bermudan/European	E	B	E	B
Accreter value	0.096	0.113	0.121	0.255
Portfolio value	0.123	0.143	0.280	0.380

Table 19.3. The value of an accreting Bermudan or European swaption vs. the value of a super-replicating portfolio of standard Bermudan/European swaptions for different maturities and contract types.

Finally, we look at a particular 30 year accreting Bermudan swaption with initial notional of 1 across different strikes, see Table 19.4. For each contract, the notional compounds at the rate given by the fixed rate.

Strike	3%	4%	5%	6%	7%	8%
Accreter value	0.021	0.053	0.121	0.255	0.495	0.880
Portfolio value	0.021	0.063	0.164	0.380	0.773	1.400

Table 19.4. The value of a 30 year accreting Bermudan swaption vs. the value of a super-replicating portfolio of standard Bermudan swaptions for different strikes.

Judging by Tables 19.1–19.4, it appears that the super-replicating portfolio tends to produce a tighter bound for amortizers than for accreters. This observation, as it turns out, is not tied to the particular structures and market data used in the tables, but is true in general. To understand why this is the case, recall that the super-replicating portfolio for a non-standard Bermudan swaption will have the exact same value as the non-standard Bermudan swaption if each standard Bermudan swaption in the portfolio is exercised at the same time as the non-standard Bermudan swaption. In other words, the tightness of the upper bound for amortizers therefore suggests that it is optimal to exercise all standard Bermudan swaptions in the super-replicating portfolio at about the same time (for an amortizing Bermudan swaption, we recall that this portfolio consists of Bermudan swaptions with identical lockout periods, but different maturities). However, according to Proposition 19.7.1 proven later in this chapter, arbitrage arguments can be used to show that, at any exercise date, if one does not exercise the standard Bermudan swaption with the shortest tenor among remaining in the basket (i.e. with a remaining 1 year swap in the example above), then one should never exercise *any* of the remaining standard Bermudan swaptions (with tenors 2,3,... years). In light of this result, the tightness of the super-replication bound is therefore not surprising. The same argument does *not* hold for the accreters, because for these structures the super-replicating portfolio consists of standard Bermudan swaptions with different lockout periods, rather than different underlying swap tenors. As such, it is obviously not reasonable to assume that the standard Bermudans will be optimally exercised at the same time.

Finally, let us briefly comment on the *lower* bound for the value of a non-standard Bermudan swaption. While we know of no general results, the carry argument of Section 19.7.2 allows us to show that an amortizing Bermudan swaption with a final notional (i.e. the notional for the final exercise date) of 1 is bounded from below by the standard Bermudan swaption of notional 1, as long as we keep the exercise schedule, strike, etc. unchanged. This holds in a model-independent way. Unfortunately, this lower bound is typically quite loose, except for Bermudan swaptions that amortize slowly.

19.4.6 Zero-Coupon Bermudan Swaptions

Let us momentarily turn to the question of where accreting Bermudan swaptions come from in the first place. Consider an accreting (receiver) Bermudan swaption with the notional defined by

$$R_i = \prod_{j=0}^{i-1}(1 + \tau_j k), \quad i = 1, \ldots, N-1. \tag{19.11}$$

According to (19.3), its n-th exercise value at time T_n is given by

$$\begin{aligned}
\tilde{U}_n(T_n) &= \sum_{i=n}^{N-1} R_i \tau_i P(T_n, T_{i+1})(k - L_i(T_n)) \\
&= \sum_{i=n}^{N-1} \left(\prod_{j=0}^{i-1}(1 + \tau_j k) \right) (P(T_n, T_{i+1})(1 + \tau_i k) - P(T_n, T_i)) \\
&= \sum_{i=n}^{N-1} (R_{i+1} P(T_n, T_{i+1}) - R_i P(T_n, T_i)) \\
&= R_N P(T_n, T_N) - R_n, \tag{19.12}
\end{aligned}$$

where we have used the defining relation $L_i(T_n) = (P(T_n, T_i) - P(T_n, T_{i+1})/(\tau_i P(T_n, T_{i+1}))$.

Now consider a contract in which an investor gives[9] the dealer 1 at time T_0, and the dealer promises to pay the investor the amount of $R_N = \prod_{j=0}^{N-1}(1 + \tau_j k)$ at T_N. The payment of R_N at T_N can be seen as the value of the original investment compounded at the fixed rate k over the time period $[T_0, T_N]$. The contract is essentially a zero-coupon bond with a discretely compounding rate of k. Suppose now that the dealer is granted a Bermudan-style option to cancel the zero-coupon note at any time T_n, $n = 1, \ldots, N-1$, in return for paying the investor the accumulated amount to that date, i.e. R_n. For reasons that should be obvious, the embedded Bermudan option is called a *zero-coupon Bermudan swaption*. The payoff to the dealer upon exercise at time T_n of this option is evidently equal to i) an immediate outflow of R_n; and ii) release from the obligation to pay R_N at time T_N, the value of which is $P(T_n, T_N)R_N$. In other words, the exercise value of a zero-coupon Bermudan swaption, $U_{\text{ZC},n}(T_n)$, is equal to

$$U_{\text{ZC},n}(T_n) = -R_n + R_N P(T_n, T_N),$$

---

[9] Or, equivalently, pays 1 at time T_N in addition to running Libor coupons on a unit notional.

which allows us to deduce from (19.12) that

$$\widetilde{U}_n(T_n) = U_{\mathrm{ZC},n}(T_n), \quad n = 1, \ldots, N-1.$$

Therefore, the value of a zero-coupon Bermudan swaption with rate k is equal to the value of an accreting Bermudan swaption with the notional accreting at rate k, as in (19.11). Our earlier discussion of accreting Bermudan swaptions therefore holds unchanged for zero-coupon Bermudan swaptions.

19.4.7 American Swaptions

Having a time-varying notional schedule is not the only non-standard feature that can be attached to Bermudan swaptions. One relatively common devia-tion from the standard contract permits the option holder to exercise *on any business date*, after a lockout period. Not surprisingly, such swaptions are called *American swaptions*. These are fairly popular in the US as hedges for mortgage bonds, presumably because the American exercise feature might be considered a better hedge for the prepayment behavior of mortgage borrowers.

The coupon-paying nature of the underlying swap makes the definition of an American swaption somewhat complicated. If exercise takes place during a period $[T_n, T_{n+1}]$, then the option holder receives the swap starting at T_{n+1}, i.e. $U_{n+1}(\cdot)$, as well as an "exercise fee" equal to the difference between the Libor rate effective for the period $[T_n, T_{n+1}]$ (i.e. $L_n(T_n)$) and the fixed rate, times the notional and times the remaining time to T_{n+1} (in the appropriate day count convention). In mathematical notations, the exercise value per unit notional at t, $t \in (T_n, T_{n+1}]$, is given by

$$U_n^A(t) = (L_n(T_n) - k)(T_{n+1} - t) + U_{n+1}(t).$$

We emphasize that, as a rule, the time t fee is set to the "accrued current coupon" $(L_n(T_n) - k)(T_{n+1} - t)$, *not* its discounted value from t to the payment date T_{n+1}. This choice, as is true of many others related to contract specification, is made by those who write term sheets (documents outlining details of derivatives contracts) rather than by those responsible for valuation algorithms, and implies that the exercise value will be discontinuous in time,

$$U_n^A(T_n+) \neq U_n^A(T_n) = U_n(T_n). \tag{19.13}$$

Odd as it is, this discontinuity is not the main issue with American swaptions. From a valuation standpoint the biggest problem with American swaptions is the fact that the exercise value at t (if not equal to one of the coupon dates) depends on the value of the Libor rate at $T_n < t$ and is thus *path-dependent*.

For Monte Carlo based valuation methods, the path dependence is not a problem, as the value of the Libor rate is known on each path when estimating the exercise boundary or using it in valuation. In a PDE setting, matters are more complicated. Before describing possible methods, let us comment on the somewhat prevalent view that one can approximate an American swaption with a Bermudan swaption with high frequency of exercise dates.

19.4.7.1 American Swaptions vs. High-Frequency Bermudan Swaptions

Let us choose a particular period $[T_n, T_{n+1}]$ and consider it subdivided into M periods

$$T_n = t_0 < t_1 < \ldots < t_M = T_{n+1}.$$

Then, consider a Bermudan swaption with exercises at $\{t_i\}_{i=1}^M$, versus an American swaption that can be exercised on the same dates. Also, for simplicity assume that the "exercise fee" for an American is in fact, properly discounted, such that the discontinuity in (19.13) is removed. Concentrating only on the exercise value contributions $u_n^A(t)$, $u_n^B(t)$ paid in the period $[T_n, T_{n+1}]$, so that

$$U_n^A(t_m) = u_n^A(t_m) + U_{n+1}(t_m), \quad U_n^B(t_m) = u_n^B(t_m) + U_{n+1}(t_m),$$

a standard Bermudan swaption exercised at t_m gives the holder an exercise value

$$u_n^B(t_m) = \sum_{i=m}^{M-1} (t_{i+1} - t_i) P(t_m, t_{i+1}) \left(\frac{P(t_m, t_i) - P(t_m, t_{i+1})}{(t_{i+1} - t_i) P(t_m, t_{i+1})} - k \right).$$

Here

$$\frac{P(t_m, t_i) - P(t_m, t_{i+1})}{(t_{i+1} - t_i) P(t_m, t_{i+1})}$$

is just the time-t_m forward Libor rate for the period $[t_i, t_{i+1}]$. Using the standard rearrangement of terms, we obtain

$$u_n^B(t_m) = L(t_m, t_m, T_{n+1}) (T_{n+1} - t_m) P(t_m, T_{n+1})$$

$$- k \sum_{i=m}^{M-1} (t_{i+1} - t_i) P(t_m, t_{i+1}), \quad (19.14)$$

with

$$L(t_m, t_m, T_{n+1}) = \frac{1 - P(t_m, T_{n+1})}{(T_{n+1} - t_m) P(t_m, T_{n+1})},$$

where we have used the full notation (4.2) for forward Libor rates. The first term represents payment of a Libor rate covering the period $[t_m, T_{n+1}]$, and the second is a collection of fixed-rate payments on a schedule $\{t_m, \ldots, t_M = T_{n+1}\}$.

Let us contrast this with the exercise value of a true American swaption exercised at t_m,

$$u_n^A(t_m) = L\left(T_n, T_n, T_{n+1}\right)\left(T_{n+1} - t_m\right)P\left(t_m, T_{n+1}\right)$$
$$- k\left(T_{n+1} - t_m\right)P\left(t_m, T_{n+1}\right), \quad (19.15)$$

with

$$L\left(T_n, T_n, T_{n+1}\right) = \frac{1 - P\left(T_n, T_{n+1}\right)}{\left(T_{n+1} - T_n\right)P\left(T_n, T_{n+1}\right)}.$$

Clearly, there are two differences between u_n^B and u_n^A. The first one is the difference in the fixed leg, the payment of the annuity $k\sum_{i=m}^{M-1}(t_{i+1} - t_i)P(t_m, t_{i+1})$ versus a single bullet payment $k(T_{n+1} - t_m)P(t_m, T_{n+1})$. The difference is in the timing of discounting and is normally small, as $T_{n+1} - T_n$ would often be equal to 3 months in the US. Even if one deems this to be an issue, any perceptible difference could be eliminated almost fully by imposing appropriately chosen deterministic exercise fees. The second difference, on the other hand, is much greater, and concerns the difference of *which* Libor rate is applied to the period $[t_m, T_{n+1}]$. For the Bermudan swaption, it is $L(t_m, t_m, T_{n+1})$ and for the American, it is $L(T_n, T_n, T_{n+1})$. Not only will the two Libor rates have different forward values, their different fixing dates $(t_m$ vs. $T_n)$ will affect the amount of volatility each Libor rate experiences over its lifetime. These effects can yield quite significant valuation differences, especially for steeper yield curves and shorter maturities.

19.4.7.2 The Proxy Libor Rate Method

Besides highlighting the fallacy of using a high-frequency Bermudan swaption as a proxy for an American, the analysis above also hints at proper remediation. Indeed, it should be clear that the approximation of an American swaption with a Bermudan swaption suffers the most not from the mismatched exercise frequency, but from the difference in the exercise values in the two contracts, as the Bermudan approximation uses a Libor rate that has the wrong forward value and the wrong volatility. The idea behind the *proxy Libor rate method* involves correcting the forward value/volatility as appropriate, while removing the path-dependence of exercise that hinders the application of backward-induction methods.

Continuing with the notations of the previous section, we define the proxy Libor rate $\widetilde{L}(t_m, t_m, T_{n+1})$ by

$$\widetilde{L}\left(t_m, t_m, T_{n+1}\right) = L\left(0, T_n, T_{n+1}\right)$$
$$+ \frac{\text{Stdev}\left(L\left(T_n, T_n, T_{n+1}\right)\right)}{\text{Stdev}\left(L\left(t_m, t_m, T_{n+1}\right)\right)}\left(L\left(t_m, t_m, T_{n+1}\right) - L\left(0, t_m, T_{n+1}\right)\right). \quad (19.16)$$

Here $\text{Stdev}(X)$ is defined as the Normal (or basis-point, see Remark 7.2.9) term volatility of the rate X, which we may compute from any particular model used for American swaption valuation. The proxy Libor rate enjoys the following properties.

- Its expected value under the T_{n+1}-forward measure is equal to the forward $L(0, T_n, T_{n+1})$, i.e. the forward of the rate used in the "real" American swaption.
- Its (term) volatility is equal to that of $L(T_n, T_n, T_{n+1})$.
- It is a function of the yield curve at time t_m and, unlike for $L(T_n, T_n, T_{n+1})$, its value is available in a backward-induction scheme at t_m.

Having defined the proxy Libor rate, we define a Bermudan swaption whose exercise value approximates the exercise value of an American swaption (compare to (19.14)),

$$u_n^B(t_m) \approx \widetilde{L}(t_m, t_m, T_{n+1})(T_{n+1} - t_m) P(t_m, T_{n+1})$$
$$- k \sum_{i=m}^{M-1} (t_{i+1} - t_i) P(t_m, t_{i+1}).$$

Assuming that the underlying model is Markovian and low-dimensional, a Bermudan swaption with these exercise values can easily be evaluated in a finite difference lattice, and is a close proxy for the "real" American swaption.

19.4.7.3 The Libor-as-Extra-State Method

While having an accurate approximation is a good step forward, it may still be of use to have an exact lattice-based valuation method for American swaptions, especially when assessing the accuracy of various approximations. One approach for this is the "extra state variable" method, where path-dependence is dealt with by back-propagating values of the security in all possible states of the path-dependent state variable, and then applying update conditions between different "slices", see Sections 2.7.5 and 18.4.5. For our purposes here, we define the extra state variable by

$$I(t) = \sum_{n=0}^{N-1} 1_{\{t \in [T_n, T_{n+1})\}} L(T_n, T_n, T_{n+1}).$$

Clearly, the payoff of an American swaption at time t could be expressed as a function of the yield curve observed at t, and the state variable $I(t)$. To elaborate, let $H_n(t, I)$ be the value of an American swaption with exercise dates after T_n at time t given $I(t) = I$. Focusing on the period $[T_n, T_{n+1})$ and assuming the exercise dates fall on the grid $\{t_m\}$ as defined in Section 19.4.7.1, we have the following recursion,

$$H_n(t_m, I)$$
$$= \mathrm{E}_{t_m}\left(\frac{B(t_m)}{B(t_{m+1})} \max\left(u_n^A(t_{m+1}, I) + U_{n+1}(t_{m+1}), H_n(t_{m+1}, I)\right)\right)$$

for $m = M - 1, \ldots, 0$, where we have defined (see (19.15))

$$u_n^A(t_m, I) = (I - k)(T_{n+1} - t_m) P(t_m, T_{n+1}).$$

As $I(T_n) = L(T_n, T_n, T_{n+1})$ is a function of the state variables of the model at T_n, we have that

$$H_n(T_n) = H_n(t_0, L(T_n, T_n, T_{n+1}))$$

is independent of I and only a function of the state variables of the model at T_n. We then use $H_n(T_n)$ to start a recursion for the $(n-1)$ period. The full American value is given by $H_0(0)$.

19.4.8 Mid-Coupon Exercise

Half-way between standard Bermudan swaptions and American swaptions lie Bermudan swaptions that allow exercises on the standard tenor dates $\{T_n\}$ plus a select few — often just one — extra dates per coupon period. At this point, we should note that while we so far for convenience have assumed that fixed and floating payments take place on the same schedule, in reality swaps in many currencies (including USD and EUR) pay floating coupons more frequently than fixed coupons. Taking the US as an example, standard conventions specify that floating rate payments occur every three months (and are linked to three-month Libor rates), while fixed rate payments are made every six months. Exercise dates are often chosen to coincide with floating rate fixing dates, i.e. are spaced three months apart. Having an exercise take place in the middle of a fixed-rate coupon period is not a problem, however, as the value of the remaining part of the fixed-rate coupon is trivial to estimate on the exercise date[10]. Less common are exercise dates in the middle of a floating-rate coupon period. For such contracts we face the same issue as with American swaptions: the exercise value on the exercise date is path-dependent, as it is linked to a fixing of the Libor rate that occurs prior to the exercise date.

PDE pricing of structures with exercise taking place inside a floating rate period involves the same issues as those discussed in Section 19.4.7 for American swaptions, and remediation follows the same path. As an approximation we can use an expression like (19.16) to replace the Libor rate with a proxy Libor rate setting on the exercise date, with the proxy rate constructed to have the same forward value and volatility as the real rate. For exact valuation, the state-variable approach of Section 19.4.7 can be used.

---

[10]We note that exercise in the middle of a fixed-rate period is most often accompanied by an *exercise fee*, a deterministic amount of money payable upon exercise that is agreed upon in advance. The fee is typically calculated to reflect the value of the part of the fixed-rate coupon accrued from the beginning of the period to the exercise date.

19.5 Flexi-Swaps

A Bermudan swaption can be interpreted as a fixed-floating swap with zero
notional and a (single) option to increase the notional to a given level on
any of the exercise dates. Likewise, a cancelable swap can be seen as a swap
of full notional with an option to decrease the notional to zero on any of
the exercise dates; once the option is exercised, the right goes away. More
flexibility in choosing swap notionals is afforded in a so-called *flexi-swap*
(also known as a *chooser swap* or a *band swap*), a swap with multiple options
to change the notional on a given set of exercise dates, subject to certain
constraints. Flexi-swaps are related to the flexi-caps discussed in Section
2.7.6 and are most often used as hedges for so-called *balance-guarantee
swaps*, i.e. swaps with a notional linked to a pool of mortgages. The ability
to gradually decrease the notional in a flexi-swap on each exercise date
allows the option holder to mimic (random) prepayments in the mortgage
pool. We briefly consider flexi-swap valuation in this section.

With a tenor structure $\{T_n\}_{n=0}^{N}$ and a collection of net coupons X_n with
unit notional, fixing at T_n and paying at T_{n+1}, $n = 0, \ldots, N - 1$, we define
a flexi-swap to be a contract that pays a net coupon $X_n R_n$ at time T_{n+1},
where the starting notional R_0 is fixed up-front, and time-T_n notional R_n is
chosen by the holder of the option at time T_n (so that R_n is $\mathcal{F}_{T_n}$-measurable)
for each $n = 1, \ldots, N - 1$, subject to some constraints. The *constraint set*
for the decision at time T_n may include

- Global deterministic bounds, e.g. $R_n \in [g_n^{\text{lo}}, g_n^{\text{hi}}]$.
- Local bounds that are functions of the current notional, e.g. $R_n \in [l_n^{\text{lo}}(R_{n-1}), l_n^{\text{hi}}(R_{n-1})]$.
- Bounds that are function of market data x_n (such as Libor and swap
  rates) at time T_n, e.g. $R_n \in [m_n^{\text{lo}}(x_n), m_n^{\text{hi}}(x_n)]$.

In a general flexi-swap the constraint set for time T_n is the intersection of the global, local and market constraint sets; let us denote this set
$\mathcal{C}_n(R_{n-1}, x_n)$, so that $R_n \in \mathcal{C}_n(R_{n-1}, x_n) \subset \mathbb{R}$. It is common to require that
the notional may only decrease, so that $\mathcal{C}_n(R_{n-1}, x_n) \subset [0, R_{n-1}]$. Of course,
the larger the constraint set is for each date (i.e. the fewer the constraints
enforced), the more expensive the flexi-swap will be.

The valuation of a flexi-swap may proceed by backward induction, while
keeping track of "current notional". To demonstrate, let $V_n(t, R)$ be the time
t of the part of the flexi-swap paying strictly after T_n, given that $R_n = R$.
At time T_{n-1}, the flexi-swap value must be equal to the discounted expected
value of the maximum value at time T_n, with the maximum taken over all
possible choices of the notional. This observation allows us to write down
the backward recursion equation,

$$V_{n-1}(T_{n-1}, R) = P(T_{n-1}, T_n)X_{n-1}R$$

$$+ B(T_{n-1})\mathrm{E}_{T_{n-1}}\left(B(T_n)^{-1} \max_{R' \in \mathcal{C}_n(R, x_n)} \{V_n(T_n, R')\}\right) \quad (19.17)$$

for $n = N, \ldots, 1$, with the terminal condition $V_N(T_N, R) \equiv 0$. The time 0 actual value of the flexi-swap is given by $V_0(T_0, R_0)$. The recursion (19.17) can be implemented in a PDE model by introducing an extra state variable to keep track of the current notional R, along the lines of Sections 2.7.5 and 19.4.7.3.

19.5.1 Purely Global Bounds

Using ideas similar to those from Section 19.4.5, Evers and Jamshidian [2005] demonstrate that a flexi-swap with purely global deterministic bounds can be decomposed exactly into a portfolio of Bermudan swaptions. While theoretically interesting, the replication is sometimes awkward in practice as a typical flexi-swap will decompose into hundreds of Bermudan swaptions; valuing them all one by one is rarely more efficient than just applying the recursion (19.17). On the other hand, when using a local model such as a one-factor qG model, valuing the Bermudan swaptions one by one would allow us to tailor calibration to each individual Bermudan swaption, leading to increased precision in the value of the flexi-swap. The choice of the valuation method will be dictated by the trade-off between calibration accuracy and performance.

19.5.2 Purely Local Bounds

Flexi-swaps that involve non-global constraints generally cannot be replicated with portfolios of Bermudan swaptions. However, in the practically relevant special case of *purely local bounds of scaling type* we may obtain a more efficient valuation formula involving no state variables beyond those driving the yield curve. Specifically, let us assume that only local constraints are enforced and that these are given by the current notional multiplied by lower and upper multipliers, i.e.

$$l_n^{\mathrm{lo}}(R_{n-1}) = \lambda_n^{\mathrm{lo}} R_{n-1}, \quad l_n^{\mathrm{hi}}(R_{n-1}) = \lambda_n^{\mathrm{hi}} R_{n-1}, \quad 0 \le \lambda_n^{\mathrm{lo}} \le \lambda_n^{\mathrm{hi}},$$

for $n = 1, \ldots, N - 1$. To simplify the valuation method (19.17), we make the critical observation that the value of the flexi-swap scales linearly in notional, i.e.

$$V_n(T_n, R) = RV_n(T_n, 1) \quad (19.18)$$

for any T_n, R. This follows from the fact that all coupons scale linearly with notional R, as do all constraints. In particular, as there are no global constraints, our exercise decision at any time T_n is independent of the absolute size of the notional.

An important corollary to (19.18) is that on any step n, the optimal notional choice is of the "all or nothing" type (known in control theory as "bang-bang"). This follows from the fact that

$$\max_{R' \in [\lambda_n^{\mathrm{lo}} R, \lambda_n^{\mathrm{hi}} R]} \{V_n(T_n, R')\} = \max_{x \in [\lambda_n^{\mathrm{lo}}, \lambda_n^{\mathrm{hi}}]} \{V_n(T_n, xR)\}$$

$$= \max_{x \in [\lambda_n^{\mathrm{lo}}, \lambda_n^{\mathrm{hi}}]} \{V_n(T_n, R)x\}$$

whereby the function being maximized, $V_n(T_n, R)x$, is linear in the maximization variable x. Hence the maximum is attained at the boundary of the interval,

$$\max_{R' \in [\lambda_n^{\mathrm{lo}} R, \lambda_n^{\mathrm{hi}} R]} \{V_n(T_n, R')\} = \max \left(V_n(T_n, \lambda_n^{\mathrm{lo}} R), V_n(T_n, \lambda_n^{\mathrm{hi}} R) \right). \quad (19.19)$$

We can use the two observations above to simplify the valuation algorithm. Rewriting (19.17) with the help of (19.19) we get

$$V_{n-1}(T_{n-1}, R) = P(T_{n-1}, T_n)X_{n-1}R$$
$$+ B(T_{n-1})\mathrm{E}_{T_{n-1}} \left(B(T_n)^{-1} \max \left(V_n(T_n, \lambda_n^{\mathrm{lo}} R), V_n(T_n, \lambda_n^{\mathrm{hi}} R) \right) \right).$$

Dividing through by R, using (19.18), and introducing the abbreviated notation $V_n(T_n, 1) = V_n(T_n)$, we obtain the valuation equation

$$V_{n-1}(T_{n-1}) = P(T_{n-1}, T_n)X_{n-1}$$
$$+ B(T_{n-1})\mathrm{E}_{T_{n-1}} \left(B(T_n)^{-1} \max \left(V_n(T_n)\lambda_n^{\mathrm{lo}}, V_n(T_n)\lambda_n^{\mathrm{hi}} \right) \right).$$

Clearly, if $V_n(T_n)$ is positive, then

$$\max(V_n(T_n)\lambda_n^{\mathrm{lo}}, V_n(T_n)\lambda_n^{\mathrm{hi}}) = V_n(T_n) \max(\lambda_n^{\mathrm{lo}}, \lambda_n^{\mathrm{hi}}) = V_n(T_n)\lambda_n^{\mathrm{hi}},$$

and if $V_n(T_n)$ is negative, then

$$\max(V_n(T_n)\lambda_n^{\mathrm{lo}}, V_n(T_n)\lambda_n^{\mathrm{hi}}) = V_n(T_n) \min(\lambda_n^{\mathrm{lo}}, \lambda_n^{\mathrm{hi}}) = V_n(T_n)\lambda_n^{\mathrm{lo}}.$$

In total, we therefore have the ultimate valuation recursion

$$V_{n-1}(T_{n-1}) = P(T_{n-1}, T_n)X_{n-1}$$
$$+ B(T_{n-1})\mathrm{E}_{T_{n-1}} \left(B(T_n)^{-1} V_n(T_n) \left(\lambda_n^{\mathrm{lo}} 1_{\{V_n(T_n)<0\}} + \lambda_n^{\mathrm{hi}} 1_{\{V_n(T_n)>0\}} \right) \right).$$
$$(19.20)$$

The value at time 0 is given by $R_0 V_0(T_0)$; only one PDE plane is required to calculate it, unlike for (19.17) which requires multiple R-planes. We note that the standard cancelable swap valuation recursion is recovered with $\lambda_n^{\mathrm{lo}} = 0$, $\lambda_n^{\mathrm{hi}} = 1$.

19.5.3 Marginal Exercise Value Decomposition

In it instructive to see what the marginal exercise value decomposition of Section 18.2.3 looks like for a flexi-swap with local bounds. We rewrite (19.20) in a slightly different way,

$$V_{n-1}(T_{n-1}) = P(T_{n-1}, T_n)X_{n-1}$$
$$+ B(T_{n-1})\mathrm{E}_{T_{n-1}}\left(B(T_n)^{-1}\left[\lambda_n^{\mathrm{lo}}V_n(T_n)^- + \lambda_n^{\mathrm{hi}}V_n(T_n)^+\right]\right).$$

Then, since

$$V_n(T_n)^+ = V_n(T_n) - V_n(T_n)^-$$

we have

$$V_{n-1}(T_{n-1}) - \lambda_n^{\mathrm{hi}}V_n(T_{n-1})$$
$$= P(T_{n-1}, T_n)X_{n-1} + B(T_{n-1})\mathrm{E}_{T_{n-1}}\left(B(T_n)^{-1}(\lambda_n^{\mathrm{hi}} - \lambda_n^{\mathrm{lo}})(-V_n(T_n)^-)\right)$$
$$= P(T_{n-1}, T_n)X_{n-1} + B(T_{n-1})\mathrm{E}_{T_{n-1}}\left(B(T_n)^{-1}(\lambda_n^{\mathrm{hi}} - \lambda_n^{\mathrm{lo}})(-V_n(T_n))^+\right)$$

and, taking discounted expected values to time 0,

$$V_{n-1}(0) - \lambda_n^{\mathrm{hi}}V_n(0) = \mathrm{E}\left(B(T_n)^{-1}X_{n-1}\right)$$
$$+ \mathrm{E}\left(B(T_n)^{-1}\left(\lambda_n^{\mathrm{hi}} - \lambda_n^{\mathrm{lo}}\right)(-V_n(T_n))^+\right).$$

This holds for $n = 1, \ldots, N$. Weighting the n-th equality with

$$\alpha_n^{\mathrm{hi}} \triangleq \prod_{i=1}^{n-1} \lambda_i^{\mathrm{hi}}$$

(with $\alpha_1^{\mathrm{hi}} = 1$), summing all terms, and observing that

$$\sum_{n=1}^{N} \alpha_n^{\mathrm{hi}}\left(V_{n-1}(0) - \lambda_n^{\mathrm{hi}}V_n(0)\right) = V_0(0) - \alpha_N^{\mathrm{hi}}\lambda_N^{\mathrm{hi}}V_N(0) = V_0(0),$$

we obtain

$$V_0(0) = \sum_{n=1}^{N} \alpha_n^{\mathrm{hi}}\mathrm{E}\left(B(T_n)^{-1}X_{n-1}\right)$$
$$+ \sum_{n=1}^{N} \alpha_n^{\mathrm{hi}}\left(\lambda_n^{\mathrm{hi}} - \lambda_n^{\mathrm{lo}}\right)\mathrm{E}\left(B(T_n)^{-1}(-V_n(T_n))^+\right). \quad (19.21)$$

More generally

$$V_n(T_n) = \frac{1}{\alpha_{n+1}^{\mathrm{hi}}} \sum_{i=n+1}^{N} \alpha_i^{\mathrm{hi}} \mathrm{E}\left(B(T_i)^{-1} X_{i-1}\right)$$

$$+ \frac{1}{\alpha_{n+1}^{\mathrm{hi}}} \sum_{i=n+1}^{N} \alpha_i^{\mathrm{hi}} \left(\lambda_i^{\mathrm{hi}} - \lambda_i^{\mathrm{lo}}\right) \mathrm{E}\left(B(T_i)^{-1}(-V_i(T_i))^+\right). \quad (19.22)$$

19.5.4 Narrow Band Limit

The decomposition (19.21) turns out to be useful to study the flexi-swap when the notional range $|\lambda_n^{\mathrm{hi}} - \lambda_n^{\mathrm{lo}}|$ is small, which is often the case as clients look for cheaper means to hedge their balance-guarantee swaps (recall that narrow range implies less optionality and lower cost). Let $\epsilon = (\lambda_n^{\mathrm{hi}} - \lambda_n^{\mathrm{lo}})/\lambda_n^{\mathrm{hi}}$ be small, and denote by U_n the value of all coupons fixing on or after T_n weighted by α_i^{hi} so that

$$U_n(t) = B(t) \sum_{i=n}^{N-1} \alpha_{i+1}^{\mathrm{hi}} \mathrm{E}_t \left(B(T_{i+1})^{-1} X_i\right).$$

This is the value of the portion of the (amortizing) swap after T_n assuming that on each exercise date the option holder always chooses the multiplier λ_i^{hi}. Then, it follows from (19.22) that to first order in ϵ,

$$\alpha_{n+1}^{\mathrm{hi}} V_n(T_n) - U_n(T_n) = O(\epsilon) \quad (19.23)$$

and we obtain from (19.21) that

$$V_0(0) = U_0(0) + \epsilon \sum_{n=1}^{N} \alpha_{n+1}^{\mathrm{hi}} \mathrm{E}\left(B(T_n)^{-1}(-V_n(T_n))^+\right)$$

$$= U_0(0) + \epsilon \sum_{n=1}^{N} \mathrm{E}\left(B(T_n)^{-1}(-U_n(T_n))^+\right)$$

$$+ \epsilon \sum_{n=1}^{N} \mathrm{E}\left(B(T_n)^{-1}\left[\alpha_{n+1}^{\mathrm{hi}}(-V_n(T_n))^+ - (-U_n(T_n))^+\right]\right).$$

The last line is of the second order in ϵ per (19.23), so that

$$V_0(0) = U_0(0) + \epsilon \sum_{n=1}^{N} \mathrm{E}\left(B(T_n)^{-1}(-U_n(T_n))^+\right) + O(\epsilon^2). \quad (19.24)$$

We recognize the terms in the sum above as European swaptions on the amortizing swap $-U_n(T_n)$, so the value of a narrow-band flexi-swap with local constraints is approximately equal to the underlying amortizing swap plus a strip of European swaptions on the remaining parts of the reverse of the underlying amortizing swap.

19.6 Monte Carlo Valuation

With our discussion so far, we have demonstrated that low-dimensional models, if appropriately calibrated, can be used effectively for Bermudan swaption valuation with PDE or tree-based methods. Still, it is sometimes useful to be able to price Bermudan swaptions with Monte Carlo methods, e.g. to compare prices computed in a low-dimensional model to those of a larger globally-calibrated model (such as the LM model). The mechanics of Monte Carlo valuation follow our discussion in Section 18.3 closely, so here we merely point out the simplifications made possible by the simpler structure of Bermudan swaptions, as compared to the general class of callable Libor exotics (CLEs).

19.6.1 Regression Methods

Bermudan swaptions can be valued by Monte Carlo simulation in straightforward fashion, using the general regression-based methods of Section 18.3. There are, however, a number of shortcuts that are worth pointing out. First, observe that the exercise values of a Bermudan swaption can be calculated directly off the yield curve at the time of exercise, whereby there is no need to use regression methods to estimate exercise values. It follows that we can use the simple algorithm of Section 18.3.1 directly.

Another advantage that Bermudan swaptions enjoy over more complex CLEs is the relative ease with which good explanatory variables can be selected. It is clear that for regressing the hold value of a Bermudan swaption on a given exercise date, the value of the underlying swap is important, suggesting that the overall level of interest rates on each exercise date — as represented by either the swap rate or the value of the swap starting on the exercise date and maturing on the final date of the Bermudan swaption — should always be included in the set of explanatory variables. The slope of the yield curve on each of the exercise dates turns out to be relevant as well, as it is actually the forward-starting swap — that is, the swap that underlies the European swaption expiring on the next exercise date — that impacts the hold value, and the difference in value between a spot-starting and a forward-starting swaps clearly originates with the slope of the yield curve. We can either include the forward starting swap as an additional explanatory variable on each exercise date or, better yet, include the spot Libor rate for the next period on each exercise date. The latter suggestion achieves nearly the same result as the spot-starting swap could be decomposed into an FRA for the next period and a forward-starting swap. Note that empirical evidence shows that it is not advisable to use the forward-starting swap alone as the sole explanatory variable per exercise date as it appears both the level and the slope of the yield curve should be represented in the set, especially in the setting of a multi-factor model. We investigate and compare various concrete exercise strategies in Section 19.6.2 below.

Another observation that can be fruitfully explored in the LS regression algorithm is the fact that prices of European options on the underlying swaps, i.e. European swaptions, could be calculated (or approximated in a computationally efficient manner) in most models of interest. As mentioned in Section 18.3.9.3, usage of (proxies of) European swaptions allows us to better incorporate convexity in the hold values in the regressions, improving the final value estimate. Additionally, we can draw on all tricks and enhancements from Section 18.3.10, including, in particular, policy improvement based on the carry argument of Section 18.3.10.2 (we extend the carry results for Bermudan swaptions in Section 19.7.2).

19.6.2 Parametric Boundary Methods

One hallmark of the LS regression approach is its "semi-automatic" nature: once we have identified some potentially meaningful variables and assumed a particular form of the regression basis functions, we let the regression algorithm work its magic to sort out a reasonable exercise strategy. In contrast, to be effective, the boundary optimization technique introduced in Section 3.5.2 requires more careful thought about the functional form of the exercise boundary. As there is considerable intuition to be gained from the results of the boundary optimization technique, let us spend some time on the application of this method to Bermudan swaption pricing. The material in this section draws on results in Andersen [2000a], where the reader can look up many additional details and numerical results that we do not list in our brief treatment here.

19.6.2.1 Sample Exercise Strategies for Bermudan Swaptions

Perhaps the simplest exercise strategy for a Bermudan swaption (and for many other options with early exercise rights) is to "exercise when the option is sufficiently deep in the money". Mathematically speaking, if $\iota(\cdot)$ is the exercise indicator function[11] then our first proposed strategy is

$$\text{Strategy I:} \quad \iota(T_n) = 1_{\{V_{n,N-n}(T_n) \geq h_I(T_n)\}}, \qquad (19.25)$$

where $V_{n,N-n}(T_n)$ is the underlying swap value (see Section 19.4.1) and $h_I(\cdot)$ is some unknown deterministic function. Assuming that the Bermudan swaption is of the payer type, the swap value $V_{n,N-n}(T_n)$ in (19.25) can be computed directly from the yield curve as

$$V_{n,N-n}(T_n) = A_{n,N-n}(T_n) \left(S_{n,N-n}(T_n) - k \right).$$

It is clear that the function $h_I(\cdot)$ must be strictly non-negative, a constraint that should be checked and enforced in the search for h_I. As discussed in

---

[11]Recall from Section 18.3.8.2 that $\iota(T_n) = 1_{\{U_n(T_n) > H_n(T_n)\}}$ defines the rule for exercising at T_n, assuming we have not exercised previously.

Section 3.5.2, the search for the $N-1$ values $h_I(T_{N-1})$, $h_I(T_{N-2})$, ..., $h_I(T_1)$ can be conducted in backwards fashion from a set of Monte Carlo pre-trials, starting from the known condition $h_I(T_{N-1}) = 0$ and using the fact that a Bermudan swaption with first exercise date T_1 will have the same optimal exercise indicator function at time T_n as will a Bermudan swaption with first exercise date T_n (as long as both are written on swaps with identical coupons and terminal maturity, of course). For each value of n, establishing $h_I(T_n)$ involves a one-dimensional optimization only, to be done either by outright sorting or by a derivatives-free one-dimensional optimizer. We emphasize that all pre-trials should be cached for numerical efficiency; when using Strategy I, for each path it suffices to store at every date T_n, $n = 1, \ldots, N-1$, the intrinsic swap value as well as the numeraire (e.g. the spot numeraire $B(T_n)$ if working in the spot measure), for a total of $2(N-1)$ double-precision numbers per path.

As the function $h_I(T_n)$ tends to be decreasing roughly linearly as a function of T_n, it often will suffice to assume that $h_I(\cdot)$ is piecewise linear on the interval $[T_1, T_{N-1}]$, with a low-dimensional number b of break-points $t_1 < t_2 < \ldots < t_b$, satisfying $t_1 = T_1$ and $t_b = T_{N-1}$. The $b-1$ values $h_I(t_1), \ldots, h_I(t_{b-1})$ can be found by a series of one-dimensional optimizations as described earlier, with the values of $h_I(\cdot)$ at coupon dates $T_1, T_2, \ldots, T_{N-1}$ easily computed by linear interpolation. The piecewise linear representation of the exercise rule not only improves numerical efficiency by reducing the number of optimizations to be performed from $N-1$ to $b-1$, but also makes the overall algorithm more robust by assigning more explanatory value to each quantity that is optimized over. Indeed, the fewer parameters that have to be estimated by optimization, the less Monte Carlo pre-trials are necessary to get a smooth, noise free estimation of the exercise boundary. Andersen [2000a] demonstrates that even very low values of b (e.g. 2 or 3) will often suffice, a consequence of the well-known fact that prices of options with early exercise rights tend to be quite insensitive to the precise location of the exercise barrier.

We shall show some numerical results for Strategy I in (19.25) shortly, but let us first introduce some more advanced strategies. Recalling that Bermudan swaptions cannot be worth less than the most expensive core European swaption (see the bound (19.2)), it is reasonable to contemplate the application of a policy improvement step in (19.25) to enforce this constraint. Let us therefore define

$$V_{\text{swaption},M(n)}^{\max}(T_n) = \max_{i=n+1,\ldots,M(n)} V_{\text{swaption},i}(T_n; k)$$

where $M(n)$ is some n-dependent upper bound for the number of European swaptions to include in the max-operation (and k is the strike, see (19.2)). Then, a second exercise strategy is

Strategy II: $\iota(T_n) = \begin{cases} 1, & V_{n,N-n}(T_n) > \max(h_{II}(T_n), V^{\max}_{\text{swaption},M(n)}(T_n)), \\ 0, & \text{otherwise.} \end{cases}$

$$(19.26)$$

Setting $M(n) = N - 1$ for all n would ensure that our strategy never breaks the hard value bound (19.2), but could also make computation of the strategy computationally expensive, particularly in models where European swaption pricing requires non-trivial work. As typically only the first few European swaptions are candidates for the maximum in (19.26), to cut down on numerical work[12] it may make sense to write

$$M(n) = \min\left(N - 1, n + 1 + m\right), \qquad (19.27)$$

where m is some relatively small integer, e.g. 1 or 2.

A strategy related to (19.26) is

Strategy III: $\iota(T_n) = \begin{cases} 1, & V_{n,N-n}(T_n) > h_{III}(T_n) + V^{\max}_{\text{swaption},M(n)}(T_n), \\ 0, & \text{otherwise.} \end{cases}$

$$(19.28)$$

Strategy III replaces the absolute trigger condition of Strategy I with a relative one, where exercise takes place when the intrinsic value is sufficiently high relative to the most expensive core European swaption. To some extent a Bermudan swaption can be viewed as a multi-factor best-of option (that is, an option to choose the most expensive of several assets, see Section 19.2), and Strategy III allows one to impose the well-known condition that exercise never takes place when the underlying assets are too close to each other, irrespective of their magnitudes[13]. Notice that if we enforce that h_{III} be strictly non-negative, Strategy III would automatically enforce the policy improvement condition in (19.26). By considering multiple component swaptions, Strategies II and III effectively embed more information about the detailed state of the yield curve into the exercise decision than Strategy I. Strategies II and III can thus be expected to be most useful in a multi-factor model[14]. Note that both Strategies II and III can be modified the same way as Strategy I to allow for a piecewise linear representation of the trigger functions h_{II} and h_{III} on some low-dimensional grid $\{t_i\}_{i=1}^b$. Also note that the storage requirements for pre-simulations of Strategies II and III will involve $3(N - 1)$ number per pre-simulation, as we must store on each exercise date T_n i) the numeraire value; ii) the intrinsic swap value; and iii) the maximum core European swaption value.

---

[12] An alternative, and even less expensive, technique to apply policy improvement would rely on the carry argument developed in Section 19.7.2. The resulting bound requires no option price computations.

[13] For a discussion of exercise strategies for best-of options (also known as MAX-options), see Broadie and Detemple [1997].

[14] Indeed, we notice that Strategy I is, in fact, optimal for a 1-factor Markov short rate model. For a 1-factor LM model, however, Strategy I is not optimal (although, as we shall see later, it appears to perform very well).

Strategies I–III all involve only sequences of one-dimensional optimizations to uncover the scalar functions h_I, h_{II}, h_{III}; all optimizations start with the boundary condition $h_I(T_{N-1}) = h_{II}(T_{N-1}) = h_{III}(T_{N-1}) = 0$. Higher-dimensional strategies are possible, too, although they rarely seem worth the extra effort. In Andersen [2000a], the strategies (19.25) and (19.28) are combined into

$$\text{Strategy IV:} \quad \iota(T_n) = \begin{cases} 1, \ V_{n,N-n}(T_n) > \max(h_{IV}^1(T_n), h_{IV}^2(T_n) \\ \qquad\qquad + V_{\text{swaption},M(n)}^{\max}(T_n)), \\ 0, \text{ otherwise,} \end{cases}$$

where now two functions, h_{IV}^1 and h_{IV}^2, have to be determined by optimization. In Andersen [2000a] it is found that this strategy results in no statistically significant pick-up in Bermudan value compared to the simpler strategies above.

19.6.2.2 Some Numerical Tests

To test the exercise strategies outlined above, we shall use simple one- and two-factor log-normal LM models. Specifically, we consider a setting where 3 month Libor rates satisfy

$$dL_k(t)/L_k(t) = O(dt) + \lambda_k(t)^\top dW(t),$$

where $W(t)$ is a vector Brownian motion and where the drift term depends on the probability measure, see Chapter 14. We consider two settings of $\lambda_k(t)$:

$$\text{Scenario A:} \quad \lambda_k(t) = 20\%,$$

$$\text{Scenario B:} \quad \lambda_k(t) = \left(15\%, 15\% - \sqrt{0.009\,(T_k - t)}\right)^\top.$$

In Tables 19.5 and 19.6 are numerical results for various Bermudan swaptions, using several of the strategies outlined earlier. We used an initial forward curve that was flat at 10% (quarterly compounded). In computing the lower bounds in the tables, we used 5,000 pre-trials to establish the trigger functions h_I, h_{II}, and h_{III} on a time line with $b = 4$ break-points; in Strategies II and III, we used $M(n) = N - 1$ for all n. 50,000 independent pricing paths were subsequently drawn to compute the lower bounds for each strategy. The tables also include upper bound duality results, computed from Strategy I using the nested simulation algorithm in Section 18.3.8.2, with $K_U = 750$ outer paths and $K_{\text{nest}} = 300$ inner paths. The 95% confidence interval (CI) listed in the tables were computed as outlined in Section 18.3.8.4.

To comment on the tables, we first notice from Table 19.5 that the duality gap computed from Strategy I is never more than 1–2 basis points, leading

Type	Strike	Strategy I	Strategy II	Strategy III	$\widehat{\Delta}$	95% CI
15M/3M	8%	184.6 (0.1)	184.6 (0.1)	184.6 (0.1)	0.02	184.5 - 184.8
15M/3M	10%	49.1 (0.1)	49.1 (0.1)	48.9 (0.1)	0.02	48.7 - 49.2
15M/3M	12%	8.9 (0.1)	8.9 (0.1)	8.7 (0.1)	0.004	8.5 - 8.9
3Y/1Y	8%	355.6 (0.4)	355.6 (0.4)	355.1 (0.4)	0.07	354.3 - 355.9
3Y/1Y	10%	157.8 (0.5)	157.8 (0.5)	156.8 (0.5)	0.2	156.0 - 158.0
3Y/1Y	12%	61.8 (0.4)	61.8 (0.4)	61.0 (0.3)	0.04	60.2 - 61.7
6Y/1Y	8%	807.2 (0.9)	807.2 (0.9)	808.0 (0.9)	0.23	805.9 - 809.8
6Y/1Y	10%	417.8 (0.9)	417.8 (0.9)	416.9 (0.9)	0.63	13.7 - 418.0
6Y/1Y	12%	212.7 (0.9)	212.7 (0.9)	212.6 (0.9)	0.33	11.4 - 215.2
11Y/1Y	8%	1381.6 (1.6)	1381.6 (1.6)	1380.2 (1.6)	1.33	1378.5 - 1386.3
11Y/1Y	10%	812.9 (1.4)	812.9 (1.4)	813.2 (1.4)	1.26	810.1 - 817.1
11Y/1Y	12%	495.8 (1.5)	495.8 (1.5)	496.7 (1.4)	0.71	495.3 - 502.1
6Y/3Y	8%	493.2 (0.8)	493.7 (0.8)	493.3 (0.8)	0.08	492.3 - 495.7
6Y/3Y	10%	293.6 (0.9)	294.6 (0.9)	293.0 (0.9)	0.65	292.4 - 296.7
6Y/3Y	12%	170.3 (0.8)	170.3 (0.8)	169.9 (0.8)	0.53	168.9 - 172.8

Table 19.5. Upper and lower bound results for the one-factor model in Scenario A. The initial forward curve is flat at 10%, quarterly compounded. All values are computed using Euler-style discretization and are reported in upfront basis points; numbers in parentheses are sample Monte Carlo errors. "Type" refers to the maturity/lockout period of the Bermudan swaption. "$\widehat{\Delta}$" is the upper-lower duality gap estimate and "95% CI" is the 95% confidence interval for the Bermudan swaption price. The computational setup is described in more detail in the text.

Type	Strike	Strategy I	Strategy II	Strategy III	$\widehat{\Delta}$	95% CI
15M/3M	8%	184.0 (0.0)	184.0 (0.0)	184.0 (0)	0.05	183.9 - 184.1
15M/3M	10%	43.3 (0.1)	43.4 (0.1)	43.2 (0.1)	0.06	43.1 - 43.6
15M/3M	12%	5.6 (0.1)	5.6 (0.1)	5.6 (0.1)	0.01	5.5 - 5.7
3Y/1Y	8%	339.7 (0.2)	339.8 (0.2)	339.4 (0.2)	0.4	339.2 - 340.6
3Y/1Y	10%	125.8 (0.3)	125.9 (0.3)	125.7 (0.3)	0.7	125.1 - 127.2
3Y/1Y	12%	36.9 (0.2)	36.8 (0.2)	36.6 (0.2)	0.2	36.4 - 37.6
6Y/1Y	8%	750.2 (0.6)	749.6 (0.6)	751.6 (0.6)	3.7	749.0 - 755.2
6Y/1Y	10%	317.0 (0.7)	315.9 (0.7)	319.4 (0.7)	5.0	315.6 - 323.5
6Y/1Y	12%	127.7 (0.6)	128.0 (0.6)	129.2 (0.6)	2.6	126.5 - 131.6
11Y/1Y	8%	1247.3 (1.2)	1250.9 (1.2)	1253.7 (1.3)	18.1	1245.1 - 1269.0
11Y/1Y	10%	620.8 (1.1)	627.1 (1.1)	633.2 (1.3)	20.8	618.4 - 645.0
11Y/1Y	12%	327.1 (1.2)	331.8 (1.1)	337.0 (1.2)	14.8	324.7 - 345.0
6Y/3Y	8%	444.7 (0.6)	444.4 (0.6)	445.2 (0.6)	0.8	443.6 - 446.6
6Y/3Y	10%	226.9 (0.7)	227.2 (0.7)	227.5 (0.7)	1.2	225.5 - 229.5
6Y/3Y	12%	107.1 (0.6)	107.1 (0.6)	107.6 (0.6)	0.8	105.9 - 109.0

Table 19.6. Upper and lower bound results for two-factor model in Scenario B. All values are in upfront basis points; numbers in parentheses are sample Monte Carlo errors. Labels are identical to those of Table 19.5.

us to conclude that Strategy 1 very accurately captures the correct exercise decision for the model setup in Table 19.5. Supporting this conclusion is the fact that Strategies II and III lead to no statistically significant increase in the Bermudan swaption value. In the two-factor scenario in Table 19.6, the duality gaps are, not surprisingly, wider than for the one-factor case, although still relatively small for most of the contracts examined. Reasonably significant spreads, in the order of 15 to 20 basis points, can be observed for the 11 year contract with 1 year lockout. Intuitively, for the correlation effects introduced by the two-factor model to matter, the exercise period must be quite long; otherwise, even a two-factor model would imply near-perfect correlation of the different swaps the option holder can exercise into. The suboptimality of exercise based on Strategy I for the 11 no-call 1 Bermudan swaption is also reflected in the fact that the more complicated Strategies II and, especially, III here pick up significant additional value relative to Strategy I. In fact, Strategy III produces prices that lie close to the average of the upper and lower bound, suggesting that this strategy is likely quite close to optimal. Using Strategy III (rather than Strategy I) to form an upper bound confirms this: the duality gap for the 11 year contract with 1 year lockout is reduced to 7.3, 6.3, and 3.5 basis points for coupons of 8%, 10%, and 12%, respectively.

While one should not read too much into the limited set of test data presented above, our results do suggest that for models without stochastic volatility, Strategy I is sufficient for short-dated Bermudan swaptions and for models with high forward rate correlation. For longer-dated structures and for multi-factor models, Strategy III is a safer bet. In a LS regression setting, this reinforces our observations of Section 19.6.1 on the importance of including variables that represent both the level and the slope of the yield curve on exercise dates.

19.6.2.3 Additional Comments

A number of papers in the literature elaborate on the analysis in Andersen [2000a]. For instance, in an LM model setting Jensen and Svenstrup [2003] conclude that for Strategy III just setting $m = 1$ in (19.27) typically yields Bermudan swaption values that are indistinguishable from those computed using $M(n) = N - 1$. Jensen and Svenstrup [2003] also compare the parametric boundary optimization technique against an LS regression where the basis functions include the first two powers of the intrinsic swap value and the spot numeraire, as well as their cross product. For an LM model without stochastic volatility, the parametric boundary technique with Strategy III is found to slightly outperform this particular setup of the LS regression. A similar conclusion is reached in Pedersen [1999], where more details on the LS regression for Bermudan swaptions can also be found.

The analysis in Andersen [2000a] (and our discussion in the previous section) concerns itself only with models that contain no stochastic volatility component. Jensen and Svenstrup [2003] examine an LM model with stochastic volatility and conclude that in this case a rather small, but economically significant, duality gap opens up for Strategy III, especially when the volatility of variance is large and/or the mean reversion speed of volatility is low. Not surprisingly, an LS regression where the variance level itself is included in the set of regressors manages to lower this duality gap. In general, for models with stochastic volatility, explicitly specifying the functional form of the exercise boundary seems difficult, and the best approach is typically to use a regression approach to uncover it, as we described in Section 18.3.9.3.

19.7 Other Topics

19.7.1 Robust Bermudan Swaption Hedging with European Swaptions

As we explain in more detail in Chapter 22, risk management of exotic derivatives such as Bermudan swaptions generally involves both delta hedging (offsetting sensitivity to the yield curve by dynamically trading in swaps) and vega hedging (offsetting sensitivity of Bermudan swaptions to changes in volatility). Vega hedging of Bermudan swaptions and other CLEs is typically done by trading in European swaptions, but as transaction costs for options can be relatively high, dealers would prefer hedges that do not require frequent rebalancing. A good example of such a hedge would be the static hedging of CMS-linked derivatives with European swaptions at multiple strikes, as specified by the replication method in Section 16.6.1. The resulting hedge not only needs no rebalancing, but is model-independent (up to the annuity mapping function selection).

For Bermudan swaptions we are not aware of any known model-independent static hedge position in European swaptions, but some insights can be gained from the marginal exercise value decomposition of Section 18.2.3. Even though each of the European options in the decomposition, $E(B(T_n)^{-1}(U_n(T_n) - H_n(T_n))^+)$, is not a standard European swaption, the replication method of Section 16.6.1 tells us that it can easily be represented as a static position of European swaptions over a continuum of strikes. This position, however, is *not* model-independent, as each payoff $(U_n(T_n) - H_n(T_n))^+$ is sensitive (through $H_n(T_n)$) to the model volatilities and, in particular, the *forward volatilities* produced by the model. Also, as volatilities inevitably change over time, the effective payoffs are liable to change, as is therefore the composition of the European swaption portfolio that the Bermudan swaption is decomposed into. This implies that the hedge portfolio will need rebalancing over time, which of course rules it out as a truly static hedging portfolio.

Another approach that could be pursued is the semi-static decomposition of barrier options into European swaptions developed by Andersen et al.

[2002]. While developed specifically for barrier options, the technique of this paper also applies to Bermudan swaptions, as these can be interpreted as barrier options with a knock-in barrier set to the optimal exercise boundary (in fact, we already used this representation in developing valuation algorithms). Unfortunately, this line of attack also fails to produce a static hedge, for the same reasons as for the marginal exercise value decomposition: the hedge portfolio depends strongly on the model-specific volatility structure, and is also likely to need rebalancing over time as volatility moves around randomly.

While no theoretically airtight static hedge for Bermudan swaptions is known (as far as we are aware), various pragmatic strategies — often collectively known as the *portfolio replication approach* — have been more successful. Let us briefly summarize the main idea here, while acknowledging the fact that it can be implemented in many different ways. As an aside, we note that the approach can be applied to many exotic derivatives, although its performance would ultimately depend on the specific risk characteristics of the specific derivative under consideration.

We start by identifying a universe of potential hedging instruments. For a Bermudan swaption of given maturity T_N, we typically would select all European swaptions with expiry + tenor less than or equal T_N, with strikes chosen to span a reasonably wide range. Having identified the hedging instruments, we formulate market data scenarios that we would want the hedge portfolio to cover. These would typically be scenarios of joint moves of the yield curve and the volatility surface. For Bermudan swaptions, it is probably sufficient to choose parallel moves in the yield curve, moves by a pre-specified amount within a given range, although one can add more complicated ones, e.g. the yield curve twists and "bends" suggested by principal components analysis (see Section 14.3.1). Similar types of volatility scenarios could be used, such as parallel and non-parallel shifts across all swaption expiries and maturities.

Suppose we have defined M scenarios and chosen K hedging instruments. Let ΔV_{Berm} denote the M-dimensional vector of value changes of the Bermudan swaption in all M scenarios, and let the vector of value changes of the k-th hedging instrument be ΔV_k, $k = 1, \ldots, K$. Then, on the last step of the portfolio replication method, we look for a vector of weights $\chi = (\chi_1, \ldots, \chi_K)^\top$ such that the portfolio of hedges defined by these weights immunizes the changes in values of the Bermudan swaption in all scenarios. This is usually formalized as a least-squares optimization problem,

$$\left\| \Delta V_{\mathrm{Berm}} - \sum_{k=1}^{K} \chi_k \Delta V_k \right\|^2 \to \min .$$

Variations are possible, including weighting different scenarios differently or adding additional terms to the objective function to express user preferences, such as minimizing the total notional of all swaptions, penalizing excessive use of deep out-of-the-money swaptions, and so forth.

We generally find the portfolio replication method to be an effective risk management tool for Bermudan swaptions. Anecdotal evidence provided by traders suggests that the method often outperforms the standard delta/vega hedging approach relying on "local" sensitivities. It also provides a relatively straightforward way of dealing with the well-known gamma-theta mismatch that plagued many dealers' Bermudan swaption portfolios in the 1980s and 1990s. As it turns out, if one uses the volatility sensitivity information from certain simple models to hedge the vega (volatility sensitivity) of a Bermudan swaption by selling European swaptions, the resulting position will sometimes be short gamma (second order yield curve sensitivity) *and* short time decay (theta). This, however, runs counter to the "standard" Black-Scholes theory in which the gamma and theta balance each other: a long gamma position is always short time decay and makes money in volatile markets and loses money in calm markets; and a short-gamma, long-theta position does the opposite. The unenviable position of being short gamma and short theta will tend to loose money in *all* markets, volatile and calm, and historically resulted in a number of Bermudan swaption book disasters over the years. The portfolio replication method can help resolve the gamma-theta problem by effectively hedging the exposure to forward volatility or to inter-temporal correlation with European swaptions across multiple expiries and tenors[15] — something a globally calibrated LM model would do, for example. On the other hand, vega hedging positions computed in short rate models calibrated according to the views held at the time (where mean reversion was often considered superfluous and either excluded from consideration or linked, directly or indirectly, to volatility as in, e.g., the BDT model from Section 11.1.1) would generally suggest incorrect European swaption hedges for the unobservable volatility positions. For example if one does not explicitly link mean reversion to market values of (off-diagonal) swaptions, the forward volatility exposure could either be not hedged at all, or might (wrongly) be linked to the diagonal European swaptions.

19.7.2 Carry and Exercise

In Section 18.3.10.2 we showed that if on a given exercise date the next (net) coupon of a cancelable note is positive, then it is not optimal to cancel the note on that date. The net coupon of a derivative security is often referred to as its *carry*, so we can state that carry-positive cancelable notes should never be exercised.

---

[15]Of course, when applying the portfolio replication method, we should explicitly link mean reversion to market inputs through the local projection method of Sections 19.2 and 18.4.

Cancelable notes and callable Libor exotics are, of course, intricately linked (see for example Section 18.3.3), so it should come as no surprise that carry-based restrictions on exercise decisions exist for Bermudan swaptions. In fact a result more general than (18.56) holds for cancelable notes and CLEs, but we present it in this chapter, since only for Bermudan swaptions is this more general result actually easy to apply.

We start by recalling that the n-th exercise value of a Bermudan payer swaption can be written as

$$U_n(t) = A_n(t)\left(S_n(t) - k\right).$$

A simple relation follows,

$$U_n(t) = \left[P\left(t, T_n\right) - P\left(t, T_{n+1}\right) - k\tau_n P\left(t, T_{n+1}\right)\right] + U_{n+1}(t)$$

$$= \tau_n P\left(t, T_{n+1}\right)\left[\frac{P\left(t, T_n\right) - P\left(t, T_{n+1}\right)}{\tau_n P\left(t, T_{n+1}\right)} - k\right] + U_{n+1}(t)$$

$$= A_{n,1}(t)\left[S_{n,1}(t) - k\right] + U_{n+1}(t),$$

and, more generally,

$$U_n(t) = A_{n,m}(t)\left[S_{n,m}(t) - k\right] + U_{n+m}(t), \quad m \geq 1, \tag{19.29}$$

where we used the notation $A_{n,m}(t)$ and $S_{n,m}(t)$ for the annuity and the swap rate, respectively, for a swap starting at T_n and covering m periods. Clearly, for the hold values we have $(m \geq 1)$,

$$H_n(t) \geq H_{n+m-1}(t) \geq U_{n+m}(t),$$

hence it follows from (19.29) that

$$U_n(t) \leq A_{n,m}(t)\left[S_{n,m}(t) - k\right] + H_n(t), \quad m \geq 1.$$

Taking a minimum over all m we obtain the following result.

Proposition 19.7.1. *For a given Bermudan payer swaption and a given exercise date T_n, we have*

$$U_n(T_n) - H_n(T_n) \leq \min_{m \geq 1}\left\{A_{n,m}(T_n)\left[S_{n,m}(T_n) - k\right]\right\}, \tag{19.30}$$

and so if any of the swaps that start at T_n and have maturities up to the final maturity of the Bermudan swaption have negative value at T_n, it is never optimal to exercise at time T_n.

Proof. If there exists m such that $S_{n,m}(T_n) - k < 0$ then by (19.30), the exercise value is strictly less than the hold value, and the exercise is not optimal. □

As annuity factors are always positive, it follows from Proposition 19.7.1 above that a Bermudan payer swaption should never be exercised if any swap rate of any still-alive swap is less than the fixed coupon k. A similar result holds for Bermudan receiver swaptions — we trust the reader can derive it himself.

19.7.3 Fast Pricing via Exercise Premia Representation

There are situations when the speed of valuation of Bermudan swaptions is key yet the accuracy could be sacrificed. One example is robust hedging of Bermudan swaptions in Section 19.7.1 where high-precision pricing is not particularly important. Another is the calculation of a *credit value adjustment* (CVA), an adjustment to the value of a derivative that takes into account the possibility that the counterparty could default on its payments. While CVA calculations are outside the scope of this book (see Gregory [2009] for a good overview), in essence the evaluation of CVA requires prices of a portfolio of Bermudan swaptions at many future dates under many simulated market conditions. Again, speed of valuation here is very important. One can speed up valuation by using a simple model, such as the one-factor Gaussian model of Section 10, but even higher performance is often desired. In this section we consider a useful approximation based on the representation of a Bermudan swaption as a stream of coupons paid in the exercise region, an adaptation of the representation we developed in Section 1.10.3 for American options.

Recall the definition of $\iota(\cdot)$, the exercise indicator, from Section 19.6.2.1 and Section 18.3.8.2. Let us denote by $V(t)$ the value of the Bermudan swaption at time t as in (18.32).

Proposition 19.7.2. *The following holds for any $n = 1, \ldots, N - 1$,*

$$
\mathrm{E}_{T_n} \left(\frac{V(T_n)}{B(T_n)} - \frac{V(T_{n+1})}{B(T_{n+1})} \right) = \mathrm{E}_{T_n} \left(\iota(T_n) B(T_{n+1})^{-1} X_n \right)
$$
$$
+ \mathrm{E}_{T_n} \left(\iota(T_n) \left(1 - \iota(T_{n+1}) \right) B(T_{n+1})^{-1} \left(U_{n+1}(T_{n+1}) - H_{n+1}(T_{n+1}) \right) \right),
$$
$$
(19.31)
$$

where we have used the convention that $\iota(T_N) \equiv 1$ and $V(T_N) \equiv 0$. In particular,

$$
V(0) = \mathrm{E} \left(\sum_{n=1}^{N-1} \iota(T_n) B(T_{n+1})^{-1} X_n \right)
$$
$$
+ \mathrm{E} \left(\sum_{n=1}^{N-1} \iota(T_n) \left(1 - \iota(T_{n+1}) \right) B(T_{n+1})^{-1} \left(U_{n+1}(T_{n+1}) - H_{n+1}(T_{n+1}) \right) \right).
$$
$$
(19.32)
$$

Proof. We have that

$$
V(T_n) = \iota(T_n) U_n(T_n) + \left(1 - \iota(T_n) \right) H_n(T_n)
$$

and

$$\frac{V(T_n)}{B(T_n)} - \frac{V(T_{n+1})}{B(T_{n+1})}$$

$$= \iota(T_n)\left(\frac{U_n(T_n)}{B(T_n)} - \frac{V(T_{n+1})}{B(T_{n+1})}\right) + (1 - \iota(T_n))\left(\frac{H_n(T_n)}{B(T_n)} - \frac{V(T_{n+1})}{B(T_{n+1})}\right).$$

Taking expected value conditioned on $\mathcal{F}_{T_n}$ and using the fact that $\iota(T_n)$ is $\mathcal{F}_{T_n}$-measurable and

$$H_n(T_n) = B(T_n)\mathrm{E}_{T_n}\left(\frac{V(T_{n+1})}{B(T_{n+1})}\right)$$

by definition, we get

$$\mathrm{E}_{T_n}\left(\frac{V(T_n)}{B(T_n)} - \frac{V(T_{n+1})}{B(T_{n+1})}\right) = \iota(T_n)\mathrm{E}_{T_n}\left(\frac{U_n(T_n)}{B(T_n)} - \frac{V(T_{n+1})}{B(T_{n+1})}\right).$$

Moreover,

$$\iota(T_n)\left(\frac{U_n(T_n)}{B(T_n)} - \frac{V(T_{n+1})}{B(T_{n+1})}\right)$$

$$= \iota(T_n)\left(\frac{U_n(T_n)}{B(T_n)}\right.$$

$$\left. - \frac{1}{B(T_{n+1})}\left(\iota(T_{n+1})U_{n+1}(T_{n+1}) + (1 - \iota(T_{n+1}))H_{n+1}(T_{n+1})\right)\right)$$

$$= \iota(T_n)\left(\frac{U_n(T_n)}{B(T_n)} - \frac{U_{n+1}(T_{n+1})}{B(T_{n+1})}\right)$$

$$+ \iota(T_n)\left(1 - \iota(T_{n+1})\right)\frac{1}{B(T_{n+1})}\left(U_{n+1}(T_{n+1}) - H_{n+1}(T_{n+1})\right).$$

Then, since

$$\mathrm{E}_{T_n}\left(\frac{U_n(T_n)}{B(T_n)} - \frac{U_{n+1}(T_{n+1})}{B(T_{n+1})}\right) = \mathrm{E}_{T_n}\left(\frac{X_n}{B(T_{n+1})}\right)$$

(see (18.2)), the result (19.31) follows. Finally, the result (19.32) follows by summing up equalities (19.31) for $n = 1, \ldots, N - 1$ and taking the (unconditional) expected value. $\square$

Let us consider the second term on the right-hand side of (19.31). It represents the contribution of those paths that are in the exercise region at time T_n ($\iota(T_n) = 1$) and are in the hold region at time T_{n+1} ($\iota(T_{n+1}) = 0$). One can argue that there are "not too many" of such paths, especially if T_n and T_{n+1} are relatively close. Moreover, quantities that are actually evaluated for those paths, the differences between exercise and hold values $U_{n+1}(T_{n+1}) - H_{n+1}(T_{n+1})$, will be small because the exercise value is close to the hold value on the border between exercise and hold regions (by definition of the exercise boundary). Indeed, in the continuous-exercise limit these terms simply disappear, as should be clear from comparing (19.32) to (1.77). These considerations lead us to suggest an approximation to the value of a Bermudan swaption in which we simply disregard the second sum on the right-hand side of (19.32):

Corollary 19.7.3. *The value of a Bermudan swaption (or, indeed, any callable Libor exotic) is approximately equal to the sum of (net) coupons that are paid only in the exercise region, i.e.*

$$V(0) \approx \mathrm{E}\left(\sum_{n=1}^{N-1} \iota(T_n)B(T_{n+1})^{-1}X_n\right)$$

$$= \mathrm{E}\left(\sum_{n=1}^{N-1} 1_{\{U_n(T_n)\geq H_n(T_n)\}}B(T_{n+1})^{-1}X_n\right). \qquad (19.33)$$

The error of approximation is given by the second term in (19.32); the error will decrease as the frequency of exercise of the Bermudan swaption is lowered.

At this point the reader may recall that we have already derived a similar-looking representation for CLEs, namely the marginal exercise value decomposition of Proposition 18.2.1, where we can rewrite (18.8) as

$$V(0) = \mathrm{E}\left(\sum_{n=1}^{N-1} 1_{\{U_n(T_n)\geq H_n(T_n)\}}B(T_n)^{-1}\left(U_n(T_n) - H_n(T_n)\right)\right). \qquad (19.34)$$

Not surprisingly, (19.33) could also be derived from (19.34) if we observe that, when the Bermudan is "deep in the money" at time T_n then

$$H_n(T_n) = B(T_n)\mathrm{E}_{T_n}\left(B(T_{n+1})^{-1}\max(U_{n+1}(T_{n+1}), H_{n+1}(T_{n+1}))\right)$$
$$\approx B(T_n)\mathrm{E}_{T_n}\left(B(T_{n+1})^{-1}U_{n+1}(T_{n+1})\right)$$
$$= U_{n+1}(T_n) = U_n(T_n) - P(T_n, T_{n+1})X_n,$$

and, therefore, we see that the marginal exercise value decomposition implies that

$$V(0) \approx \mathrm{E}\left(\sum_{n=1}^{N-1} \iota(T_n)B(T_n)^{-1}P(T_n, T_{n+1})X_n\right)$$

$$= \mathrm{E}\left(\sum_{n=1}^{N-1} \iota(T_n)B(T_{n+1})^{-1}X_n\right)$$

which is (19.33) in Corollary 19.7.3.

Everything we have discussed so far is valid for general CLEs. Let us now specialize our setup to Bermudan swaptions, with the net coupon X_n

given by $\tau_n(L_n(T_n) - k)$. The coupon is a function of the Libor rate $L_n(T_n)$; critically, in pretty much all one-factor models (and certainly in the one-factor Gaussian model) the exercise boundary at time T_n could also be parameterized by the same Libor rate and expressed in the form

$$\iota(T_n) = 1_{\{L_n(T_n) \geq h(T_n)\}}$$

for some deterministic function $h(\cdot)$. Then we can rewrite (19.33) as

$$V(0) \approx \sum_{n=1}^{N-1} P(0, T_{n+1}) \tau_n \mathrm{E}^{T_{n+1}} \left((L_n(T_n) - k) 1_{\{L_n(T_n) \geq h(T_n)\}} \right). \quad (19.35)$$

Each term on the right-hand side can be expressed as a combination of caplets and digital caplets on the Libor rate L_n. Notice that the expected value is taken under the T_{n+1}-forward measure, a measure under which the Libor rate is a martingale. In the one-factor Gaussian model the distribution of $L_n(T_n)$ under the T_{n+1}-forward measure is well approximated by the Gaussian distribution, and each term could be evaluated rapidly with just a few applications of the Bachelier formula (7.16). Similar approximations could be derived in many other models. For the Libor market model, in particular, the distributions of Libor rates is often known exactly.

The exercise boundary function $h(\cdot)$ in (19.35) is not known *a-priori* and needs to be found as part of valuation. This can be done efficiently in a backward induction algorithm that utilizes the representation (19.35) for a Bermudan swaption at future times. In particular, we can find the value $h(T_k)$ by solving

$$H_k(T_k)|_{L_k(T_k)=h(T_k)} = U_k(T_k)|_{L_k(T_k)=h(T_k)},$$

where $H_k(T_k)$ is calculated by an analog to (19.35) with $h(T_{k+1}), \ldots, h(T_{N-1})$ already determined from previous steps. The recursion can be accelerated further by search for the exercise boundary only on the subset of exercise dates, with the missing points filled by interpolation (see Ju [1998]). The final algorithm turns out to be quick and robust, and is well-suited for situations where valuation speed is the primary consideration.

19.A Appendix: Forward Volatility and Correlation

When European swaption prices are kept fixed, increasing correlation between forward rates (e.g. by moving from a two- to a one-factor model) will tend to increase forward volatility. There are a number of ways to explain this effect. One approach relies on swaption "triangles" (see Section 20.3), another on the BGM/HJM model formalism. In this appendix we explore the latter.

First, a bit of notation. Let Libor rates $L_k(t) = L_k(t, T_k, T_{k+1})$ satisfy

$$dL_k(t) = O(dt) + \sigma_k(t)\, dW_k(t), \quad k = 1, \dots, N-1,$$

where σ_k's are scalar and the W_k's are correlated Brownian motions. We assume that our calibration is global (see Section 14.5) and therefore the model calibrates properly to caplets (or, equivalently, swaptions on short-tenor swaps). If follows that the quantities

$$\int_0^{T_k} \sigma_k(u)^2\, du \tag{19.36}$$

must be invariants, independent of the correlation between the W_k's.

Consider now the market for short-expiry swaptions — which must also be calibrated in our setup (which is global) — and let us study two different settings of the average correlation between the W_k's, "high" and "low", indicated by appropriate subscripts in what follows. To match the short-expiry, short-maturity swaptions (i.e., caplets), we fundamentally need

$$\sigma_1^{\mathrm{hi}}(0) = \sigma_1^{\mathrm{lo}}(0).$$

On the other hand, to match long-tenor (but still short-expiry) swaptions, we need

$$\sigma_k^{\mathrm{hi}}(0) < \sigma_k^{\mathrm{lo}}(0), \quad k = 2, \dots, N-1, \tag{19.37}$$

a relationship that follows from the fact that swaption volatilities are effectively volatilities of sums of Libor rates. Specifically, as the volatility of a sum increases in correlation, we need to lower the volatility of the "components of the sum" (that is, the Libor rates) to preserve swaption volatilities.

Let us pick some $k > 1$. If we look at (19.36) and (19.37), it is obvious that to satisfy both conditions simultaneously $\sigma_k^{\mathrm{hi}}(t)$ must[16] ultimately "overtake" $\sigma_k^{\mathrm{lo}}(t)$ as t is increased from 0 to T_k. As this holds for all k, it is clear that forward volatilities of both caps and swaptions will, as promised above, be higher in the high-correlation model than in the low-correlation model.

19.B Appendix: A Primer on Moment Matching

19.B.1 Basics

Let there be given d log-normal random variables $X_1, \dots, X_d$, with known distribution parameters m_i and s_i:

$$\ln(X_i) \sim \mathcal{N}\left(m_i, s_i^2\right), \quad i = 1, \dots, d.$$

---

[16]See Figures 1–3 in Andersen and Andreasen [2001] for visual confirmation of this as well as of (19.37).

We assume that the $d \times d$ correlation matrix ρ of logarithms is known,

$$\rho_{i,j} \triangleq \text{Corr}\left(\ln(X_i), \ln(X_j)\right), \quad i,j = 1, \ldots, d.$$

From standard results for log-normal variables, the first two moments of the X_i can be computed as

$$\text{E}\left(X_i\right) = \exp\left(m_i + \frac{s_i^2}{2}\right), \tag{19.38}$$

$$\text{Var}\left(X_i\right) = \text{E}\left(X_i\right)^2 \left(\exp\left(s_i^2\right) - 1\right). \tag{19.39}$$

Also,

$$\text{E}\left(X_i X_j\right) = \text{E}\left(\exp\left(\ln(X_i) + \ln(X_j)\right)\right)$$

$$= \text{E}\left(\exp\left(m_i + s_i Z + m_j + s_j \left(\rho_{i,j} Z + \sqrt{1 - \rho_{i,j}^2} Y\right)\right)\right)$$

where Z and Y are independent standard Gaussian variables, and where we have used the Cholesky decomposition. Therefore, using the result (19.38),

$$\text{E}\left(X_i X_j\right) = \text{E}\left(X_i\right) \text{E}\left(X_j\right) \exp\left(\rho_{i,j} s_i s_j\right). \tag{19.40}$$

We note in passing that therefore (see e.g. (17.66))

$$\text{Cov}\left(X_i, X_j\right) = \text{E}\left(X_i X_j\right) - \text{E}\left(X_i\right) \text{E}\left(X_j\right)$$

$$= \text{E}\left(X_i\right) \text{E}\left(X_j\right) \left(\exp\left(\rho_{i,j} s_i s_j\right) - 1\right). \tag{19.41}$$

Suppose now that we are interested in approximating the moments of the weighted sum

$$\widehat{X} = \sum_{i=1}^{d} w_i X_i, \tag{19.42}$$

where the w_i's are given positive constants. Clearly

$$\text{E}(\widehat{X}) = \sum_{i=1}^{d} w_i \text{E}\left(X_i\right) \tag{19.43}$$

with $\text{E}(X_i)$ given in (19.38). Also,

$$\text{E}(\widehat{X}^2) = \sum_{i=1}^{d} \sum_{j=1}^{d} w_i w_j \text{E}\left(X_i X_j\right)$$

$$= \sum_{i=1}^{d} \sum_{j=1}^{d} w_i w_j \text{E}\left(X_i\right) \text{E}\left(X_j\right) \exp\left(\rho_{i,j} s_i s_j\right)$$

$$= \sum_{i=1}^{d} w_i^2 \text{E}\left(X_i\right)^2 e^{s_i^2} + 2 \sum_{i=1}^{d} \sum_{j=i+1}^{d} w_i w_j \text{E}\left(X_i\right) \text{E}\left(X_j\right) e^{\rho_{i,j} s_i s_j}.$$

$$\tag{19.44}$$

In many applications, we are interested in representing $\widehat{X}$ as being approximately log-normal, i.e. we would like to write

$$\ln(\widehat{X}) \sim \mathcal{N}\left(m_{\widehat{X}}, s_{\widehat{X}}^2\right).$$

Using a moment-matching principle, we would determine $m_{\widehat{X}}$ and $s_{\widehat{X}}$ from the equations

$$\exp\left(m_{\widehat{X}} + \frac{s_{\widehat{X}}^2}{2}\right) = \mathrm{E}(\widehat{X}),$$

$$\exp\left(m_{\widehat{X}} + \frac{s_{\widehat{X}}^2}{2}\right)^2 \exp\left(s_{\widehat{X}}^2\right) = \mathrm{E}(\widehat{X}^2),$$

which can be solved to yield

$$s_{\widehat{X}} = \sqrt{\ln\left(\mathrm{E}(\widehat{X}^2)\right) - \ln\left(\mathrm{E}(\widehat{X})^2\right)}, \quad m_{\widehat{X}} = \ln\left(\mathrm{E}(\widehat{X})\right) - \frac{s_{\widehat{X}}^2}{2}. \quad (19.45)$$

In these formulas $\mathrm{E}(\widehat{X})$ and $\mathrm{E}(\widehat{X}^2)$ should be computed from formulas (19.43) and (19.44), respectively.

Note that if we are willing to relax the requirement that $\widehat{X}$ be approximately log-normal, we can obtain more accurate approximations. A popular choice here is to assume that $\widehat{X}$ is approximately *displaced* log-normal. This introduces one more degree of freedom in the matching distribution (the displacement parameter) which, together with the mean and variance, could be used to match three, rather than two, moments. We leave the details of this for the reader to work out, and in the examples below we stick with simple log-normal moment matching.

19.B.2 Example 1: Asian Option in BSM Model

Let $I(t)$ be some asset following the simple process

$$dI(t)/I(t) = -b(t)\,dt + \sigma(t)\,dW(t), \quad (19.46)$$

where $W(t)$ is a scalar Brownian motion in the risk-neutral measure. For certain weights w_i, we form the weighted average

$$M(T_d) = \sum_{i=1}^{d} w_i I(T_i),$$

on some schedule $0 < T_1 < T_2 < \ldots < T_d$. An Asian option pays

$$V_{\mathrm{Asian}}(T_{pay}) = (M(T_d) - K)^+, \quad T_{pay} \geq T_d,$$

where typically the weights are $w_i = 1/d$ for all i. Standing at time 0, we wish to use moment-matching to model the T_d-observed average $M(T_d)$ as a log-normal variable. From (19.46), it is clear that $I(T_i)$ is a log-normal random variable, since

$$I(T_i) = I(0)l(T_i) \exp \left(-\frac{1}{2} v(T_i)^2 T_i + \int_0^{T_i} \sigma(u) \, dW(u) \right),$$

where we have defined

$$l(T_i) \triangleq \exp \left(-\int_0^{T_i} b(u) \, du \right), \quad v(T_i)^2 \triangleq T_i^{-1} \int_0^{T_i} \sigma(u)^2 \, du.$$

If we define $X_i = I(T_i)$, it follows that, in the notation of Section 19.B.1,

$$m_i = \ln(l(T_i)) - \frac{1}{2} v(T_i)^2 T_i,$$

$$s_i = v(T_i) \sqrt{T_i},$$

and that $M(T_d) = \widehat{X}$, where $\widehat{X}$ is defined in (19.42). To use the results (19.45) it only remains to find the correlation matrix ρ. But clearly

$$\rho_{i,j} = \text{Corr} \left(\int_0^{T_i} \sigma(u) \, dW(u), \int_0^{T_j} \sigma(u) \, dW(u) \right)$$

$$= \frac{\int_0^{\min(T_i,T_j)} \sigma(u)^2 \, du}{\sqrt{\int_0^{T_i} \sigma(u)^2 \, du} \sqrt{\int_0^{T_j} \sigma(u)^2 \, du}}$$

$$= \frac{\min \left(v(T_i)^2 T_i, v(T_j)^2 T_j \right)}{v(T_i) v(T_j) \sqrt{T_i T_j}}$$

$$= \frac{\min \left(v(T_i) \sqrt{T_i}, v(T_j) \sqrt{T_j} \right)}{\max \left(v(T_i) \sqrt{T_i}, v(T_j) \sqrt{T_j} \right)}.$$

Applying (19.43), (19.44), and finally (19.45) then allows us to write, approximately,

$$\ln(M(T_d)) \sim \mathcal{N} \left(m_{\widehat{X}}, s_{\widehat{X}}^2 \right)$$

for computed constants $m_{\widehat{X}}$ and $s_{\widehat{X}}$. Assuming deterministic interest rates, standard Black-Scholes arguments (see Section 1.9) allow us to finally approximate the time 0 option price as

$$V_{\text{Asian}}(0) \approx P(0, T_{pay}) \left(e^{m_{\widehat{X}} + \frac{1}{2} s_{\widehat{X}}^2} \Phi(d_+) - K\Phi(d_-) \right), \tag{19.47}$$

$$d_{\pm} = \frac{\ln \left(e^{m_{\widehat{X}} + \frac{1}{2} s_{\widehat{X}}^2} / K \right) \pm \frac{1}{2} s_{\widehat{X}}^2}{s_{\widehat{X}}} = \frac{m_{\widehat{X}} + \frac{1}{2} s_{\widehat{X}}^2 - \ln(K) \pm \frac{1}{2} s_{\widehat{X}}^2}{s_{\widehat{X}}},$$

where Φ is the Gaussian CDF and $P(0, T_{pay})$ is a risk-free discount factor to time T_{pay}. We note that we may, of course, rewrite this expression in the perhaps slightly more convenient form

$$V_{\text{Asian}}(0) \approx P(0, T_{pay}) \left(\text{E} \left(M(T_d) \right) \Phi(d_+) - K\Phi(d_-) \right), \qquad (19.48)$$

$$d_\pm = \frac{\ln \left(\text{E} \left(M(T_d) \right) / K \right) \pm \frac{1}{2} s_{\widehat{X}}^2}{s_{\widehat{X}}},$$

where

$$\text{E} \left(M(T_d) \right) = \sum_{i=1}^{d} w_i \text{E} \left(I(T_i) \right) = \sum_{i=1}^{d} w_i I(0) l(T_i).$$

Note that this form does not require us to compute $m_{\widehat{X}}$, as only $s_{\widehat{X}}$ is needed.

19.B.3 Example 2: Basket Option in BSM Model

Consider d risk-neutral processes

$$dI_i(t)/I_i(t) = -b_i(t) \, dt + \sigma_i(t) \, dW_i(t), \quad i = 1, \dots, d,$$

where we assume that $\langle dW_i(t), dW_j(t) \rangle = \rho_{i,j} \, dt$. Also consider the payout of a basket option

$$V_{\text{basket}}(T_{pay}) = \left(\widehat{I}(T) - K \right)^+, \quad T_{pay} \geq T,$$

where

$$\widehat{I}(T) = \sum_{i=1}^{d} w_i I_i(T),$$

with the understanding that all basket weights w_i are positive. In the framework of Section 19.B.1, we now set $X_i = I_i(T)$, such that, at time 0, X_i is log-normal with parameters

$$m_i = \ln(l_i(T)) - \frac{1}{2} v_i(T)^2 T, \quad s_i = v_i T \sqrt{T},$$

where we have defined, for $T > 0$,

$$l_i(T) \triangleq \exp \left(-\int_0^T b_i(u) \, du \right), \quad v_i(T)^2 \triangleq T^{-1} \int_0^T \sigma_i(u)^2 \, du.$$

In the notation of Section 19.B.1, clearly $\widehat{I}(T) = \widehat{X}$ and we may proceed as in Example 1 above to find $m_{\widehat{X}}$ and $s_{\widehat{X}}$, at which point the formula (19.47) (or (19.48)) will price the basket option at time 0.

20

TARNs, Volatility Swaps, and Other Derivatives

Having completed our discussion of callable Libor exotics, in this chapter we turn our attention to a few remaining types of exotic interest rate derivatives that are popular in the market. Our analysis gives us the opportunity to provide additional examples of the local projection method introduced in Chapter 18 which, along with the out-of-model adjustment methods in Chapter 21, are the cornerstone techniques for the situations where computational efficiency constraints prohibit the usage of large, globally calibrated models.

20.1 TARNs

20.1.1 Definitions and Examples

As explained in Section 5.15.2, a TARN (Targeted Redemption Note) pays structured coupons in exchange for Libor coupons until the cumulative amount of structured coupon payments exceeds a pre-agreed target, at which point the derivative terminates. While many coupon types could be used in a TARN, we focus our discussion on inverse floating coupons indexed to the Libor rate. Recall (Section 5.13.1) that an inverse floating coupon with strike s, gearing g, a zero floor and no cap is defined as

$$C_n = (s - g \times L_n(T_n))^+, \tag{20.1}$$

with the underlying rate observed (fixed) at time T_n and the coupon paid at T_{n+1}. We shall use the specific structured coupon (20.1) as an example throughout this section; in defining it, we have used the usual notation for spanning Libor rates

$$L_n(t) = L(t, T_n, T_{n+1}) = \frac{P(t, T_n) - P(t, T_{n+1})}{\tau_n P(t, T_{n+1})}, \quad n = 0, \ldots, N-1,$$

and have also introduced a tenor structure

$$0 = T_0 < T_1 < \ldots < T_N, \quad \tau_n = T_{n+1} - T_n.$$

In the TARN, the structured coupon fixed at time T_n is only paid if the sum of coupons fixing before (but not including) time T_n is below a given total return R. Thus, from the investor viewpoint, the value of the TARN at time 0 under is given by

$$V_{\text{tarn}}(0) = \mathrm{E} \left(\sum_{n=1}^{N-1} B(T_{n+1})^{-1} \tau_n \left(C_n - L_n(T_n) \right) 1_{\{Q_n < R\}} \right), \qquad (20.2)$$

$$Q_n = \sum_{i=1}^{n-1} \tau_i C_i, \quad Q_1 = 0,$$

where we, arbitrarily, have used the spot measure numeraire $B(t)$ (and E therefore denotes expectation in measure Q^B). We recall that a TARN typically pays fixed coupons to an investor before the knock-out feature starts; these coupons can be valued separately and are not included in the TARN definition above.

To make the discussion a bit more concrete, let us warm up by considering a typical example. Let the total maturity T_N be 10 years, let the target return R be 3%, and let the strike s and gearing g in (20.1) be 11.5% and 2, respectively. Also suppose the TARN pays annual coupons ($\tau_n = 1$ year). Using a yield curve with continuously compounded yields that grow from 3.5% in 1 day to 6.50% in 10 years and a displaced log-normal LM model with skew parameter 0.6 and calibrated to flat 35% swaption ATM Black volatilities, the value of the TARN with these parameters implies an attractive fixed coupon of 11% in the first year. If the TARN knocks out after the second year (at T_2), the investor would have received 14% return over two years (11% fixed coupon up front plus 3% targeted return), and is repaid the principal upon termination. This scenario comes true provided C_1 is above 3%, which according to (20.1) is equivalent to $L_1(T_1)$ fixing below 4.25%. More generally, the TARN will terminate early if interest rates are low. On the flip side, if, say, the rates go above 5.75% and stay there for the entire 10 year life of the TARN, all coupons C_n pay zero, and the investor receives nothing for 10 years. Yet, he has to pay Libor (by, essentially, forfeiting interest on the principal) for 10 years, so the high-rate scenario is obviously not advantageous to the investor.

For reference, Figure 20.1 plots the probability (in spot measure) of the TARN being alive at future points in time, using the same market data and the model as above. According to the figure, the TARN stays alive for 10 years (bad for the investor) with about 25% probability, and knocks out after the first two years (good for the investor) with about 65% probability. Loosely speaking, the TARN investor therefore makes good money with (risk-neutral) probability of 65%, and loses a significant amount with probability of 25%. This demonstrates how a high leverage inherent in TARNs allows them to pay attractive (i.e., high) coupons in scenarios that favor the investor. The leverage in any particular TARN depends on many factors, but is primarily a function of the target return R, with TARNs having smaller target return R providing higher leverage, *ceteris paribus*.

Fig. 20.1. Probability of TARN Being Alive at Future Years

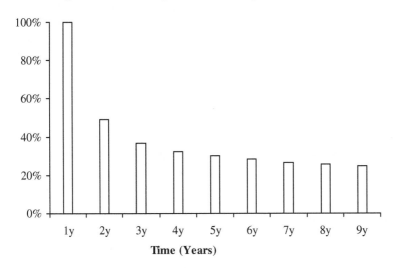

Notes: Model-implied spot measure probability of a TARN being alive after a given number of years. The TARN contract details and the model are described in the text.

20.1.2 Valuation and Risk with Globally Calibrated Models

Using a flexible model (e.g. a Libor market model or a multi-factor quasi-Gaussian model) calibrated to the full swaption volatility grid is always a relatively safe choice for TARN pricing, since a globally calibrated multi-factor model can typically be counted on to capture the majority of all possible market risk factors. As shall be explained later, faithful reproduction of volatility smiles of various Libor rates is important for TARN valuation, so among all possible LM or qG models, we recommend the versions with stochastic volatility (Sections 14.2.5 and 13.3.2), as these models have enough flexibility to provide good fits to volatility smiles for a collection of forward Libor rates.

Pricing TARNs in LM and qG models is conceptually straightforward: as TARNs are purely path-dependent derivatives with no optimal exercise features, standard Monte Carlo simulation techniques apply. However, since the knock-out feature of the TARN will introduce "digital" discontinuities into the payoff, Monte Carlo errors of the contract value and, especially, its risk sensitivities can be quite large, see Section 3.3.1. The number of paths required to get reasonably accurate estimates of risk sensitivities of a derivative with a discontinuous payoff could be high, which may sometimes render the application of a Monte Carlo based market model impractical. We review methods that help us obtain risk sensitivities in Monte Carlo for TARNs later in the book, see Sections 23.2.4, 23.4.4, and 25.2. Ultimately, however, the full power of LM and qG models may not be required for TARNs; despite appearances, TARNs turn out to be relatively simple instruments amenable to treatment by less complex — and more performant — models. We pursue this topic next.

20.1.3 Local Projection Method

In Chapters 18 and 19 we introduced the local projection method and applied it to Bermudan swaptions and other callable Libor exotics. We recall that the method is based on finding a relatively simple, local model that is calibrated to a global model (such as an LM model) in such a way as to approximate the value of the global model for a particular derivative. In particular, the local model should be calibrated to the parts of the global model volatility structure that are relevant to the derivative being valued. Let us apply this approach to TARNs. To start, it is informative to rewrite the TARN value as follows,

$$V_{\mathrm{tarn}}(0) = \mathrm{E}\left(\sum_{n=1}^{N-1} B(T_{n+1})^{-1}\tau_n \left((s - g \times L_n(T_n))^+ - L_n(T_n) \right) \right.$$

$$\left. \times 1_{\left\{ \sum_{i=1}^{n-1} \tau_i (s - g \times L_i(T_i))^+ < R \right\}} \right) \quad (20.3)$$

(with the usual convention that $\sum_{i=1}^{0} = 0$). Scrutinizing the payoff, we notice that it depends on the values

$$\widetilde{L} = (L_1(T_1), L_2(T_2), \ldots, L_{N-1}(T_{N-1}))$$

of Libor rates on their fixing dates *only* (for the discrete money market numeraire $B(t)$ this follows from (4.24)). With the values of Libor rates at intermediate times irrelevant, only the distribution properties of the $(N-1)$-dimensional vector $\widetilde{L}$ must be captured in whatever model we decide to use. Clearly this is a major simplification from a typical valuation problem. Notice, for instance, that a Bermudan swaption would depend on values of Libor rates at various dates on *and before* their fixing dates. A similar principle also holds approximately true for more complicated TARNs linked to swap rates (rather than Libor rates): only the distribution properties of the $(N-1)$-dimensional vector of swap rates observed on their fixing dates needs to be captured. In stating this principle, we have relied on the fact that the dependence on Libor rates through discounting with the spot numeraire is rather mild and has only limited impact on the value of TARNs.

Focusing on the covariance characteristics only (we will deal with volatility smiles later), and assuming log-normal distributions for market rates for the time being, we see that if two models assign the same values to the term variances of Libor rates $\mathrm{Var}(\ln L_n(T_n))$, $n = 1, \ldots, N-1$, and inter-temporal correlations of Libor rates $\mathrm{Corr}(\ln L_n(T_n), \ln L_m(T_m))$, $n, m = 1, \ldots, N-1$, then the values of a TARN in the two models would be the same. With this in mind, we can apply the local projection method as follows. First, we calibrate, say, a Libor market model to the full swaption volatility grid (and, of course, one's views on the proper dynamics of the volatility structure). Second, we use the calibrated LM model to calculate the relevant term volatilities and inter-temporal correlations needed for the TARN. Third, we pick a simpler model and calibrate it to the volatilities and correlations extracted from the LM model. Finally, we use the calibrated local model for valuing the TARN. Of course, when computing risk sensitivities, we would update the volatilities and correlations produced by the global LM model for each shock of market data.

In the procedure above, the local model needed for the third and final steps needs enough flexibility in its volatility structure specification to calibrate to the set of TARN volatility information we identified earlier. Fortunately, the set is not very extensive and, as we have seen before, can be effectively captured even by models as simple as a one-factor Gaussian model, see Sections 13.1.7 and 13.1.8.3. While adequate for capturing the volatility structure, the smile capabilities of the Gaussian model are, however, quite limited, and we shall consider more advanced alternatives below.

20.1.4 Volatility Smile Effects

To investigate the effects of the volatility smile on TARNs, let us consider the TARN value on date T_1 as a function of the Libor rate $L_1(T_1)$:

$$V_{\mathrm{tarn}}(T_1, x)$$
$$= \mathrm{E}\left(B(T_1) \sum_{n=1}^{N-1} B(T_{n+1})^{-1} \tau_n \left(C_n - L_n(T_n) \right) 1_{\{Q_n < R\}} \middle| L_1(T_1) = x \right).$$

We plot $V_{\mathrm{tarn}}(T_1, x)$ as a function of x in Figure 20.2 for the same TARN example and market/model data used in Section 20.1.1. Since $V_{\mathrm{tarn}}(0)$ is given by the integral of $V_{\mathrm{tarn}}(T_1, x)$ over the distribution of $L_1(T_1)$, the features of the payoff $V_{\mathrm{tarn}}(T_1, x)$ highlight the characteristics of the distribution of $L_1(T_1)$ that are important for valuation. Clearly, $V_{\mathrm{tarn}}(T_1, x)$ has an outright discontinuity at a barrier $L_1(T_1) = b_1$ implicitly given by

$$\tau_1 \left(s - gb_1\right)^+ = R, \tag{20.4}$$

and a call-option type singularity (a kink) at s/g. Moreover, values of future coupons are non-linear functions of $L_1(T_1)$, so the payoff $V_{\text{tarn}}(T_1, x)$ is non-linear in x. From the replication argument of Proposition 8.4.13, we recall that a model generally needs to faithfully incorporate the whole distribution of $L_1(T_1)$ as implied from caplet prices across a range of strikes, and not just some summary information such as an implied volatility at a certain strike.

Fig. 20.2. Value of TARN on First Knockout Date

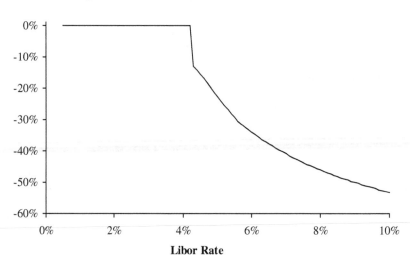

Notes: Value of a TARN on the first knockout date as a function of the spot Libor rate on that date. TARN and model details are given in the text.

Some might argue for focusing all attention on the knockout barrier b_1 and disregarding the rest of the volatility smile, believing that to value a TARN properly, it suffices to choose a model that values a digital caplet with strike b_1 consistently with the market. While this argument has some merit for the first date T_1, it is simply not valid for any subsequent knock-out dates. For instance, it is obvious that the value of $L_2(T_2)$ at which the derivative knocks out would depend on the realized fixing of $L_1(T_1)$, a value that is unknown at time $t = 0$. Since the location of the knock-out barrier at time T_2 is unknown at time $t = 0$, we cannot find a single strike that would faithfully represent relevant features of the volatility smile at T_2; thus a model that only matches the level, or slope, of the implied volatility of $L_2(T_2)$ at a single strike will be inadequate.

From this discussion, and from what we have learned about the local projection method in previous chapters, it is clear that a successful candidate for the local model should have the ability to calibrate to volatility smiles of all Libor rates, in addition to having a low number of state variables and enough flexibility to calibrate to inter-temporal correlations of Libor rates. One reasonable candidate is the one-factor quasi-Gaussian (qG) model with stochastic volatility (13.64). To calibrate this model, we would first fix its mean reversion function to match the inter-temporal correlations of Libor rates $\text{Corr}(\ln L_n(T_n), \ln L_m(T_m))$, $n, m = 1, \ldots, N-1$, as explained in Section 13.1.8.3. Subsequently, we would use the methods of Section 13.2 to calibrate to the volatility smiles of all Libor rates that appear in the payoff formula. With the formulas developed there, the time-dependent local volatility function $\sigma_r(t, x, y)$, and the time-dependent volatility of variance function $\eta(t)$ (see (13.64) for notation) could be chosen to match the implied SV parameters of relevant caplets.

The qG-SV model is a good choice of a local model for TARNs as it has just enough — but not more — flexibility to calibrate to all relevant covariance and smile information. Other suitable model candidates include the one-factor Markov-functional model from Appendix 11.A of Chapter 11, as it could be calibrated in a similar way. Finally, we could also use a two-factor version of the quadratic Gaussian model of Section 12.3. As all these three local models have sufficient flexibility to reproduce the TARN-specific correlation and smile properties of a globally calibrated multi-factor model, we would expect them to produce similar values for TARNs. Still, the models generate volatility smiles using different mechanisms, which may change correlations in subtle ways, as we saw an example of in Appendix 17.A of Chapter 17. While this effect on TARNs is quite minor, it can be significant for other classes of derivatives, see Section 20.2.4 below.

20.1.5 PDE for TARNs

Suppose that we succeeded in applying the local projection method to calibrate a low-dimensional Markov model targeted to TARN valuation. Actual pricing of the TARN structure could then obviously be accomplished by standard Monte Carlo techniques. As low-dimensional models typically allow for particularly efficient path discretization, the resulting scheme would be substantially faster than, say, simulation of a full Libor market model. Even more attractive, if the Markov model has a dimension less than 3 or 4, the local projection method allows for the usage of PDE-based TARN valuation schemes. In a finite-difference setting, the path-dependent nature of TARNs can be dealt with using the now-familiar method of augmenting the state variable space, see for instance Sections 2.7.5 and 18.4.5. In doing so, we shall implicitly assume that all C_n's are non-negative, as is almost universally the case for structured notes.

Let $V(t, I)$ be the value of the TARN at time t assuming that the total accumulated coupon at time t is $I = I(t)$, where we have defined

$$I(t) = \sum_{i=1}^{N-1} \tau_i C_i(T_i) 1_{\{T_i \leq t\}}.$$

We start by initializing the value of the TARN at T_N to 0,

$$V(T_N-, I) = 0. \tag{20.5}$$

Then, for each $n = N - 1, \ldots, 0$, we perform the following steps.

1. Roll back the value of the TARN

$$V(T_n, I) = \mathrm{E}_{T_n} \left(\frac{B(T_n)}{B(T_{n+1})} V(T_{n+1}-, I) \right)$$

   by solving an appropriate model PDE for each value of I.
2. Apply the continuity condition

$$V(T_n-, I) = V(T_n, I + \tau_n C_n)$$

   across I-planes, corresponding to the update of the total return Q_n at time T_n.
3. Add the time T_n coupon times the survival indicator,

$$V(T_n-, I) = V(T_n-, I) + P(T_n, T_{n+1}) \tau_n (C_n - L_n(T_n)) 1_{\{I < R\}}. \tag{20.6}$$

4. Starting from the new terminal condition (20.6), repeat Steps 1 through 3 with $n \to (n-1)$.

The final value is given by $V_{\text{tarn}}(0) = V(0, 0)$.

The discretization scheme based on this algorithm would require specifying bounds, and potentially the discretization grid, for the extra state variable I. The lower bound for I is clearly 0. From (20.6) it follows that the coupon update equation for $I > R$ is trivial so one would think that the upper bound for I should be R. Yet if we look at the continuity condition in Step 2 we see that, on the right-hand side, we may actually need the values of $V(T_n, I)$ for $R < I < R + \tau_n C_n$. Hence the upper bound should be somewhat higher than R and is best determined as

$$R + \max_{n,\omega} \{ \tau_n C_n(\omega) \}, \tag{20.7}$$

where by $C_n(\omega)$ we denoted the value of n-th coupon in the TARN over a realization of an interest rate path ω. For an inverse floater TARN, as well as for many other TARN types, the coupons are globally bounded and the expression in (20.7) makes sense. For TARNs with unbounded coupons this strategy will obviously not work and the global maximum will need to be replaced with a maximum over a set of sufficiently high probability. Needless to say rather crude calculations are sufficient here. For example, if the coupon C_n is a deterministic function of a Libor rate $L_n(T_n)$, then the, say, 99% confidence interval for $L_n(T_n)$ could be established from its forward value and market-observed volatility; this confidence interval on the rate could then be translated into a confidence interval on the coupon.

While trade-specific analysis for the bounds of I is not conceptually difficult, it is always preferable to have a generic scheme that works for a large class of instruments. For example, one could imagine a simple scheme where, prior to solving a PDE, a Monte Carlo simulation with a low number of paths is run and an empirical distribution of I is estimated. Using this distribution, not only the bounds of given probabilistic coverage on I could be established, but we could also use it to set up a discretization grid. For example, we can discretize more finely in the region where realized values of I are dense, and use coarse discretization elsewhere to save calculation time.

Section 5.15.2 lists a few potential tweaks to the standard TARN specification; they could be included in the PDE scheme without much difficulty. For example, to include the "capped at trigger" feature we would replace (20.6) with

$$
\begin{aligned}
V\left(T_{n}-, I\right) =\; & V\left(T_{n}-, I\right) \\
& + P\left(T_{n}, T_{n+1}\right) \tau_{n}\left(\min(C_{n},(R-I)) - L_{n}(T_{n})\right) 1_{\{I<R\}}.
\end{aligned}
$$

And to account for the "make whole" provision we would replace the initialization (20.5) with

$$
V\left(T_{N}-, I\right) = (R-I)^{+}.
$$

20.2 Volatility Swaps

We now shift our attention to volatility swaps introduced in Section 5.16. Valuation of many flavors of volatility swaps is straightforward in Monte Carlo, so a globally-calibrated LM model is a reasonable choice[1]. As always, however, performance considerations suggest that we seek methods that are faster.

---

[1]There is some evidence that many participants in the volatility swap market tend to use fairly naive, low-dimensional models for valuation. As a result, if correlations for the LM model are extracted from spread options, say, the LM model may produce forward volatilities that are lower than the market consensus (see Appendix 19.A for the rationale). Any arbitrages induced by such "segmentation" between the markets for spread options and volatility swaps are hard to exploit in practice, so the market differences can be quite persistent.

20.2.1 Local Projection Method

We recall that the structured coupon paid at T_{n+1} of a typical volatility swap has the form

$$C_n = |S_{n+1}(T_{n+1}) - S_n(T_n)|, \qquad (20.8)$$

where $S_n(t)$, $n = 1, \ldots, N$, are the reference rates of the swap. The payoff of a volatility swap shares certain characteristics with a TARN payoff, and this makes it amenable to the same treatment as what we applied to TARNs. In particular, it is clear from the valuation equation (5.26) that the value of a volatility swap depends on the values of the rates S_n on their fixing dates only. As such, the specific local projection method developed in Section 20.1.3 may be applied to volatility swaps as well. We do not repeat the analysis here, but just emphasize that we can use one-factor models to value volatility swaps as long as we calibrate them to the marginal rate distributions and the correlation structure of $(S_1(T_1), \ldots, S_N(T_N))$. The former typically would come from the market and the latter from a globally calibrated model, e.g., the LM model.

As a properly calibrated one-factor model is appropriate for valuation, one may wonder whether we can use PDE, rather than Monte Carlo, methods. Indeed, this is the case, as each "swaplet" (20.8) in the structured leg of the volatility swap can be valued in a finite difference grid by introducing of an extra state variable to track the "strike" $S_n(T_n)$ in the swaplet payoff. In this particular case, the extra state variable method amounts to calculating, via a PDE, the value of the coupon at time T_n,

$$
\begin{aligned}
V_{\text{swaplet}}(x, T_n) &\triangleq \mathrm{E}_{T_n}^{T_{n+1}} \left(C_n \middle| S_n(T_n) = x \right) \\
&= \mathrm{E}_{T_n}^{T_{n+1}} \left(|S_{n+1}(T_{n+1}) - x| \middle| S_n(T_n) = x \right), \quad (20.9)
\end{aligned}
$$

for a selection of values x (we use T_{n+1}-forward measure in this example). To obtain the time 0 value of the coupon, we can then calculate, again in a PDE, the expected value

$$\mathrm{E}\left(\beta(T_n)^{-1} P(T_n, T_{n+1}) V_{\text{swaplet}}(S_n(T_n), T_n) \right), \qquad (20.10)$$

where $\beta(t)$ is the money market numeraire.

For some versions of the volatility swap payoff, we can go further and derive approximate closed-form expressions. We will discuss this in more detail later but, briefly, the basic approach here is to calculate the value of each swaplet payout with a two-dimensional integration or a suitable approximation for a spread option. In doing so, we rely on a model to pre-calculate the term volatilities and the (inter-temporal) correlation of $S_n(T_n)$ and $S_{n+1}(T_{n+1})$ in (20.8).

20.2.2 Shout Options

As pointed out in Section 5.16.2, volatility swaps often give the receiver of the structured coupons an option to *shout*, i.e. to choose the observation time of the rate $S_{n+1}(\cdot)$ in (20.8). The coupon in (20.8) is then replaced with

$$C_n = |S_{n+1}(\eta_n) - S_n(T_n)|,$$

where the stopping time $\eta_n \in [T_n, T_{n+1}]$ is chosen by the party receiving the coupon[2]. Importantly, the payoff is still paid at time T_{n+1}, even if η_n is strictly less than T_{n+1}. As viewed from time T_n, the option then looks like an American option on an at-the-money straddle, with exercise value $P(\eta_n, T_{n+1})|S_{n+1}(\eta_n) - S_n(T_n)|$; the presence of the discount factor $P(\eta, T_{n+1})$ in the exercise value reflects the fact that payment of the coupon will always take place at time T_{n+1}, irrespective of the time of exercise. Notice that $S_{n+1}(t)$ has almost no drift[3] so Jensen's inequality implies that

$$\begin{aligned} P(\eta_n, T_{n+1})\mathrm{E}_{\eta_n}^{T_{n+1}} \left(|S_{n+1}(T_{n+1}) - S_n(T_n)| \right) \\ \geq P(\eta_n, T_{n+1}) \left| \mathrm{E}_{\eta_n}^{T_{n+1}} \left(S_{n+1}(T_{n+1}) \right) - S_n(T_n) \right| \\ \approx P(\eta_n, T_{n+1}) \left| S_{n+1}(\eta_n) - S_n(T_n) \right|, \end{aligned}$$

i.e. the exercise value is (approximately) dominated by the hold value and the value of the early exercise is negligible. As a consequence, the shout option can safely be ignored for valuation purposes, and the coupons could be valued as if (20.8) were the actual payoff.

The situation is somewhat less clear cut in a reasonably popular case of a *capped coupon* with a shout,

$$C'_n = \min \left(|S_{n+1}(\eta_n) - S_n(T_n)|, c \right) \qquad (20.11)$$

for some $c > 0$. Clearly, if for some $t \in [T_n, T_{n+1})$ the rate $S_{n+1}(t)$ is outside the interval $[S_n(T_n) - c, S_n(T_n) + c]$, then the holder should exercise at that point as he will never get a higher value for the coupon (but if he waits he may end up with a lower value at expiration). So early exercise is optimal in some cases. This may seem like a major complication as any application of Monte Carlo would apparently require the estimation of the optimal exercise rule, potentially requiring the full suite of regression-based methods of Chapter 18. Fortunately the situation is much simpler; we formalize this result as a proposition.

---

[2] Sometimes the coupon is linked to a swap rate that starts at shout time η_n rather than at the end of the period T_{n+1}. This modification to our discussion is easy to incorporate, and we do not consider this case separately.

[3] In the T_{n+1}-forward measure, $S_{n+1}(t)$ will typically have a small convexity-induced drift, as $S_{n+1}(t)$ here represents some CMS rate. However, the period $[T_n, T_{n+1}]$ rarely exceeds one year, and over one year the drift of a CMS rate will normally be quite close to zero.

Proposition 20.2.1. *The value of the American option on a capped straddle with the payoff (20.11) is equal to the value of a straddle with a barrier, so that* $\mathrm{E}_{T_n}^{T_{n+1}}(C_n') = \mathrm{E}_{T_n}^{T_{n+1}}(C_n'')$, *where*

$$C_n'' = c \times 1_{\left\{\max_{t \in [T_n, T_{n+1}]}(|S_{n+1}(t) - S_n(T_n)|) \geq c\right\}}$$
$$+ |S_{n+1}(T_{n+1}) - S_n(T_n)| \times 1_{\left\{\max_{t \in [T_n, T_{n+1}]}(|S_{n+1}(t) - S_n(T_n)|) < c\right\}}.$$

Remark 20.2.2. The proposition tells us that the optimal exercise strategy is known analytically: for the period $[T_n, T_{n+1}]$ one should simply exercise the shout option on the first time t when $S_{n+1}(t)$ hits either of the barriers $S_n(T_n) \pm c$.

Proof. We content ourselves with a sketch of the proof; a more formal argument is developed in Broadie and Detemple [1995]. Let us denote by η_n^c the first hitting time of the double barrier $S_n(T_n) \pm c$,

$$\eta_n^c = \inf\{t \in [T_n, T_{n+1}] : |S_{n+1}(t) - S_n(T_n)| = c\} \wedge T_{n+1}.$$

Then, clearly,

$$\mathrm{E}_{T_n}^{T_{n+1}}(C_n'') = \mathrm{E}_{T_n}^{T_{n+1}}\left(\min\left(|S_{n+1}(\eta_n^c) - S_n(T_n)|, c\right)\right)$$

(we use T_{n+1}-forward measure for valuation here). On one hand,

$$\mathrm{E}_{T_n}^{T_{n+1}}\left(\min\left(|S_{n+1}(\eta_n^c) - S_n(T_n)|, c\right)\right)$$
$$\leq \mathrm{E}_{T_n}^{T_{n+1}}\left(\min\left(|S_{n+1}(\eta_n) - S_n(T_n)|, c\right)\right) \quad (20.12)$$

because, by definition, η_n is the *optimal* stopping time for the American capped straddle. On the other hand, for each $t \in [T_n, T_{n+1}]$,

$$\mathrm{E}_t^{T_{n+1}}\left(\min\left(|S_{n+1}(\eta_n^c) - S_n(T_n)|, c\right)\right) \geq \min\left(|S_{n+1}(t) - S_n(T_n)|, c\right)$$

as Figure 20.3 demonstrates. Therefore,

$$\mathrm{E}_{T_n}^{T_{n+1}}\left(\min\left(|S_{n+1}(\eta_n) - S_n(T_n)|, c\right)\right)$$
$$= \mathrm{E}_{T_n}^{T_{n+1}}\left(\int_{T_n}^{T_{n+1}} \min\left(|S_{n+1}(t) - S_n(T_n)|, c\right) \delta(\eta_n - t)\, dt\right)$$
$$\leq \mathrm{E}_{T_n}^{T_{n+1}}\left(\int_{T_n}^{T_{n+1}} \mathrm{E}_t^{T_{n+1}}\left(\min\left(|S_{n+1}(\eta_n^c) - S_n(T_n)|, c\right)\right) \delta(\eta_n - t)\, dt\right)$$
$$= \mathrm{E}_{T_n}^{T_{n+1}}\left(\min\left(|S_{n+1}(\eta_n^c) - S_n(T_n)|, c\right) \int_{T_n}^{T_{n+1}} \delta(\eta_n - t)\, dt\right)$$
$$= \mathrm{E}_{T_n}^{T_{n+1}}\left(\min\left(|S_{n+1}(\eta_n^c) - S_n(T_n)|, c\right)\right), \quad (20.13)$$

where the second-to-last equality follows by the law of iterated conditional expectations and $\mathcal{F}_t$-measurability of $\delta(\eta_n - t)$. Comparing (20.12) and (20.13) we see that

$$\mathrm{E}_{T_n}^{T_{n+1}}(C_n') = \mathrm{E}_{T_n}^{T_{n+1}}(C_n'')$$

and the optimal exercise strategy is actually given by η_n^c, as stated earlier.

$\square$

Fig. 20.3. Value of a Barrier Option on a Capped Straddle

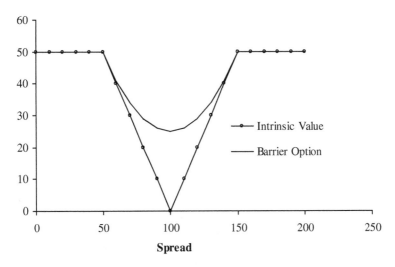

Spread

Notes: The payoff and the present value of a barrier option on a capped straddle, vs. the underlying. The barrier option value dominates the payoff in all states of the world.

Having replaced an American option with a barrier option, we can now value capped volatility swaps in standard Monte Carlo, as no optimal exercise features need to be incorporated. Still, we do have some complications — namely two continuous barriers — that now need to be handled in Monte Carlo. Here the techniques developed in Section 3.2.9 come in handy. Deserving a special mention is the method of Broadie et al. [1997] which replaces a continuously-observed barrier with a discretely-observed one that is shifted by a certain amount, see Theorem 3.2.2. Other methods from Section 3.2.9 could also be used, although all of then assume that the underlying follows a simple process such as a Brownian motion; this is typically not a problem as the dynamics of $S_{n+1}(t)$ for $t \in [T_n, T_{n+1}]$ could be closely approximated as such, irrespective of the underlying model used, since $T_{n+1} - T_n$ tends to be relatively small (one year or less).

PDE methods can also be used for barrier options, with the same trick of using the strike $S_n(T_n)$ as an extra state variable as we discussed in the European case (20.9)–(20.10), only now valuing a (double) barrier option for each value of the strike $S_n(T_n)$. This is probably the best we can do as far as valuation speed is concerned, as it should be obvious that closed-form approximations to values of capped straddle coupons with a shout option would be rather hard to develop.

20.2.3 Min-Max Volatility Swaps

Defined in Section 5.16.3, min-max volatility swaps replace the straddle coupon (20.8) with a coupon that measures the maximum move of a given rate over a given period,

$$C_n = M_n - m_n, \tag{20.14}$$

where

$$M_n = \max_{s \in [T_n, T_{n+1}]} S_n(s), \quad m_n = \min_{s \in [T_n, T_{n+1}]} S_n(s).$$

At first glance the min-max coupon appears significantly more "exotic" than the plain straddle (20.8), to the point that one may wonder if the min-max coupon has significantly different risk characteristics. For example, a superficial analysis could suggest that the min-max coupon has significantly higher "forward skew" exposure, i.e. the exposure to the slope of the volatility smile as observed at the time when the minimum (or maximum) of the process is reached. It is all the more interesting that, in fact, we can show that the two coupons, (20.14) and (20.8), are quite alike.

Let us start by assuming that for $t \in [T_n, T_{n+1}]$, $S_n(t)$ follows a driftless Brownian motion with some constant volatility σ, i.e.

$$dS_n(t) = \sigma \, dW(t),$$

where $W(t)$ is a Brownian motion in the T_{n+1}-forward measure $Q^{T_{n+1}}$. As we have already pointed out, this is not a bad approximation as the period $[T_n, T_{n+1}]$ is often rather short, on the order of one year. By the reflection principle from Section 2.6 of Karatzas and Shreve [1991], we have for any $b \geq S_n(T_n)$,

$$Q_{T_n}^{T_{n+1}} (M_n \geq b) = 2Q_{T_n}^{T_{n+1}} (S_n(T_{n+1}) \geq b). \tag{20.15}$$

As

$$\int_{b_{\min}}^{b_{\max}} 1_{\{x \geq b\}} \, db = \min \left(\max \left(x - b_{\min}, 0 \right), b_{\max} - b_{\min} \right)$$

we obtain, integrating (20.15) over $b \in [S_n(T_n), \infty)$ that

$$E_{T_n}^{T_{n+1}} (M_n - S_n(T_n)) = 2E_{T_n}^{T_{n+1}} \left((S_n(T_{n+1}) - S_n(T_n))^+ \right).$$

Similarly,

$$\mathrm{E}_{T_n}^{T_{n+1}}\left(S_n(T_n) - m_n\right) = 2\mathrm{E}_{T_n}^{T_{n+1}}\left(\left(S_n(T_n) - S_{n+1}(T_n)\right)^+\right)$$

and, adding the last two equations together, we obtain

$$\mathrm{E}_{T_n}^{T_{n+1}}\left(M_n - m_n\right) = 2\mathrm{E}_{T_n}^{T_{n+1}}\left(\left|S_n(T_{n+1}) - S_n(T_n)\right|\right). \qquad (20.16)$$

Therefore, the value of the min-max coupon is (approximately) equal to twice the value of the straddle coupon, and the min-max volatility bond could be valued, and risk managed, in the same way as the standard volatility swap in Section 20.2.

The starting point of our proof, equation (20.15), has an interesting financial interpretation. On the left-hand side we have the value of a "one-touch" option, an option that pays 1 if the underlying process ever touches the barrier b (before T_{n+1}). The equality suggests that this continuously monitored barrier option can somehow be hedged with two European-style digital call options (the right-hand side). To show that this is indeed the case, consider buying two digital calls struck at the barrier. If the underlying process never hits the barrier, both the one-touch and the two digitals expire worthless. On the other hand, if the process touches the barrier, then we sell one of the digital calls and buy one digital put, i.e. an option that pays 1 at T_{n+1} if and only if $S_n(T_{n+1}) \leq b$. The value of the digital call and the digital put are the same due to the (assumed) symmetry of the Brownian motion process for S_n; hence we can trade at zero cost. After the trade, our replicating portfolio consists of one digital call and one digital put, which will produce a payoff of 1 irrespective of the final value of the process $S_n(T_{n+1})$. Note that this is exactly equal to the payoff of the one-touch in this case. Therefore the one-touch and the replicating portfolio have the same payoffs in all states of the world, as claimed. The replicating portfolio (or the inverse of it, a hedging portfolio) is called *semi-static* to reflect the fact that the replicating strategy may involve some (costless) trading activity during the life of the trade.

In deriving (20.16) we represented the min-max payoff as a (continuous) integral of one-touch payoffs. It should then come as no surprise that we can set up a semi-static replicating portfolio for a min-max coupon that starts with two European straddles (see footnote 22 in Section 5.16.3). We leave it as an exercise to the reader to write down the explicit trading strategy for the replication; as a hint we mention that it involves holding at each time $t \in [T_n, T_{n+1}]$ a portfolio with the payoff

$$\left|S_n(T_{n+1}) - M_n(t)\right| + \left|m_n(t) - S_n(T_{n+1})\right|,$$

where $M_n(t)$, $m_n(t)$ are the *running* maximum and minimum,

$$M_n(t) = \max_{s \in [T_n, t]} S_n(s), \quad m_n(t) = \min_{s \in [T_n, t]} S_n(s).$$

The replication of a min-max coupon with two standard straddles is not model-independent, as it relies on approximating the process for the rate $S_n(t)$ with a (driftless) Gaussian process. As a result of this assumption, ATM puts and calls will have identical prices, a relationship often known as *arithmetic put-call symmetry*. The hedging arguments above can be extended to all processes for which arithmetic put-call symmetry holds at a barrier hitting time, i.e. processes for which the distribution of $S_n(T_{n+1})$ observed at any stopping time in $[T_n, T_{n+1}]$ is symmetric. In some settings, it is most useful to assume *geometric put-call symmetry*, which essentially means that the Black volatility smile is symmetric in log-moneyness; a simple example of a process satisfying this assumption is the geometric Brownian motion process without drift, or the (drift-free) Heston model with zero asset/volatility correlation. It is not difficult to prove that a semi-static hedge also exists for this case, although the hedge is somewhat more complicated than two straddles (European options at a full continuum of strikes are required). For further discussions on the topic, see Carr and Lee [2009a] which surveys (and generalizes) the considerable amount of work in the literature on applications of put-call symmetry.

While one can experiment with various assumptions to find more accurate valuation formulas, ultimately the main utility of the result such as (20.16) lies in demonstrating that the risk characteristics of a min-max volatility swap are largely the same as those of a standard volatility swap. Such qualitative understanding is useful when making model selections, even when direct replication arguments no longer work. As a typical example we can mention the *capped* min-max volatility swap, a swap with coupons that have a payoff

$$C_n = \min\left(M_n - m_n, c\right),$$

for some $c > 0$.

20.2.4 Impact of Volatility Dynamics on Volatility Swaps

With the analysis above as background, let us now ponder the question of what is, ultimately, the appropriate model for volatility swaps. While the discussion of Section 20.2.1 has made it clear that (properly calibrated) single-factor models can be safely used, we still need to decide on other features of potential models, such as a faithful reproduction of volatility smile and, perhaps, its dynamics. As always, we look for a model that captures the main risk factors for a given type of derivatives, yet avoids introducing complicated features that may not be relevant.

To show that the model choice is not entirely trivial we point to Figure 20.4. Here, we have plotted the value of volatility swaplets for *fixed-tenor* and *fixed-expiry* volatility swaps (see Section 5.16.1) in three different models. Both swaps are of 10 year maturity and have annual coupons. For the fixed-tenor swap the underlying rate is the 10 year CMS rate; for the fixed-expiry swap it is a 10 year swap rate fixing in 10 years time. For model calibration we use Euro market data from the summer of 2008. The three models are: i) the SV version of the quasi-Gaussian (qG) model as in Section 13.2; ii) the two-factor quadratic Gaussian (QG) model from Section 12.3; and iii) a local volatility version of the qG model, with local volatility a quadratic function of the short rate (not something we normally recommend, see Section 13.1.5) and time. All three models have been calibrated to volatility smiles of the 20 year coterminal swaption strip. To give a sense of market data used in calibration, the vanilla SV model (see Section 16.1.3) used to mark swaptions was set up to have the volatility between 14% and 16%, the skew between -10% and 10% and the volatility of variance between 100% and 200%, with mean reversion of volatility at 20%. All three models have the same mean reversion parameter of 2% in a (loose) attempt to make the inter-temporal correlations of relevant swap rates invariant across models.

We see that the differences in the values of individual coupons for the three models are quite significant. As we have calibrated the models to the same spot market data (volatility smiles) , we must conclude that it is the different dynamics of the volatility structure (and, perhaps, volatility smiles) in the three models that are responsible for the valuation differences. In fact, we will argue that the difference in multi-dimensional distributions of the swap rates in the three models lead to differences in the meaning of the mean reversion parameter which imply different forward volatilities in different models. To understand this better, let us consider the issue of pricing an individual coupon in more details.

Let $S(t)$ be a forward swap rate that corresponds to a swap that starts at time T (with some, unspecified, maturity). Also, define $A(t)$ to be the corresponding annuity. We consider a contract that pays

$$|S(t) - S(u)| \qquad (20.17)$$

at time t, where $0 < u < t \leq T$, a contract commonly called a *forward CMS straddle*. This contract corresponds to a coupon of a fixed-expiry volatility swap, the first graph in Figure 20.4. Let us find an approximate expression for the value of (20.17) in order to study its dependence on the various market quantities. The value can be expressed in the annuity measure induced by S and equals

$$V(0) = A(0)E^A \left(|S(t) - S(u)| / A(t) \right).$$

Using the linear TSR model of Section 16.6.4 we obtain

$$V(0) = A(0)E^A \left(|S(t) - S(u)| \times \left(\frac{1}{A(0)} + \alpha_1 (S(t) - S(0)) \right) \right)$$

for some $\alpha_1 > 0$, so

$$V(0) = V_1 + V_2 + V_3, \qquad (20.18)$$

where we have defined

Fig. 20.4. Values of Volatility Swaplets For Fixed-Tenor and Fixed-Expiry Volatility Swaps

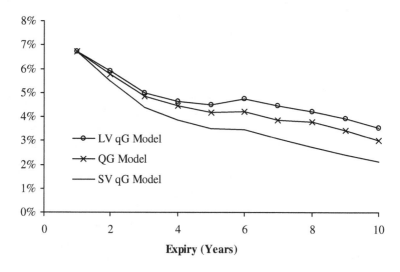

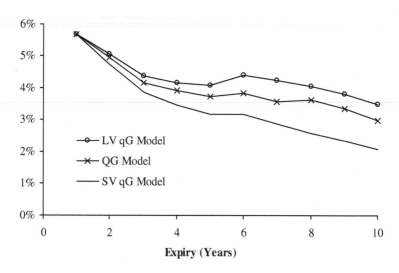

Notes: Values of volatility swaplets for fixed-tenor (first panel) and fixed-expiry (second panel) 10 year maturity annual volatility swaps in three different models, as described in the text.

$$V_1 = (1 - \alpha_1 S(0)A(0)) \, \mathrm{E}^A \left(|S(t) - S(u)| \right),$$
$$V_2 = \alpha_1 A(0) \mathrm{E}^A \left(|S(t) - S(u)| \, (S(t) - S(u)) \right),$$
$$V_3 = \alpha_1 A(0) \mathrm{E}^A \left(|S(t) - S(u)| \, S(u) \right).$$

We shall see that V_1 is linked to the future volatility of the swap rate S, V_2 is linked to the future skew, and V_3 is largely determined by the convexity adjustment as defined by Section 16.6.

Let us take a closer look at the first term V_1. Conditioning on the sigma-algebra at time u we obtain

$$E^A\left(|S(t) - S(u)|\right) = A(0)^{-1}E\left(\beta(u)^{-1}A(u)E_u^A\left(|S(t) - S(u)|\right)\right) \quad (20.19)$$
$$= A(0)^{-1}E\left(\beta(u)^{-1}V_{\text{straddle}}(u, S(u), t)\right),$$

where

$$V_{\text{straddle}}(u, K, t) = A(u)E_u^A\left(|S(t) - K|\right)$$

can be recognized as the time u value of a swaption straddle, i.e. a sum of K-strike payer and receiver European swaptions for delivery at time t of a swap that starts at $T \geq t$, as entered at time u. The quantity $V_{\text{straddle}}(u, S(u), t)$ is then the value at time u of the at-the-money (ATM) straddle. We note that the contract, entered at time 0, that pays

$$A(t)\,|S(t) - S(u)|$$

at time t is known as a *forward swaption straddle*. The forward swaption straddle differs from the forward CMS straddle in (20.17) in that it pays the difference of swap values rather than swap *rates*.

Let us denote by $\sigma_N(u, S(u); t, K)$ the value of the implied basis-point (or Normal) volatility of the swap rate S, as observed at time u for swaptions with expiry t and strike K (and again, on a swap that starts at some $T \geq t$). From the Bachelier pricing formula (7.16), we see that

$$V_{\text{straddle}}(u, S(u), t) = A(u)\sqrt{\frac{2\tau}{\pi}}\sigma_N(u, S(u); t, S(u)), \quad (20.20)$$

so the ATM straddle value is equal to the (scaled) value of the implied basis-point volatility. From (20.19) it follows that

$$V_1 = a_1 E^A\left(\sigma_N(u, S(u); t, S(u))\right)$$

for some constant a_1. As mentioned earlier, the value V_1 is given by the expected value of a future (at-the-money) basis-point volatility of a swap rate S over a period $[u, t]$.

As we have discussed before, the information about such future volatilities is, by and large, not contained in the market data available at time 0, but is mostly driven by the dynamics of a model used for valuation. While calibrated to the same marginal distributions at time 0, the three models used in Figure 20.4 have different dynamics and therefore different forward volatilities. Probably the easiest way to understand it is to recall (see Dupire [1997]) that the price of the contract that pays instantaneous forward *variance* is model-independent (as long as all European options are matched), which implies, by Jensen's inequality, that the price of a contract that pays forward *volatility* will depend on the distribution (mainly variance) of the volatility itself.

We shall discuss forward swaption straddles in more detail in Section 20.3, but first we turn our attention back to the remaining two terms in the decomposition (20.18). We can rewrite the term V_2 as

$$V_2 = a_2 \mathrm{E}^A \left(f \left(S(t) - S(u) \right) \right),$$

where a_2 is some constant and $f(x) = x|x|$. The function $f(x)$ is concave for $x < 0$ and convex for $x > 0$, suggesting that the value V_2 is largely driven by *forward skew*, i.e. the slope of the volatility smile $\sigma_N(u, S(u); t, K)$ at time u. This is most easily seen from the replication result of Proposition 8.4.13, as we have

$$\mathrm{E}^A(f(S(t) - S(u))) =$$
$$\int_0^\infty \left(\mathrm{E}^A \left((S(t) - (S(u) + y))^+ \right) - \mathrm{E}^A \left(((S(u) - y) - S(t))^+ \right) \right) dy,$$

i.e. V_2 is a sum of forward call spreads for different strike offsets y from the time u at-the-money rate $S(u)$. The value of each call spread is largely (to the first order in volatility) determined by the difference in the appropriate implied volatilities,

$$\mathrm{E}^A \left((S(t) - (S(u) + y))^+ \right) - \mathrm{E}^A \left(((S(u) - y) - S(t))^+ \right)$$
$$\approx c \cdot \left(\sigma_N(u, S(u); t, S(u) + y) - \sigma_N(u, S(u); t, S(u) - y) \right),$$

which is clearly related to the slope of the volatility smile as observed at time u. As is the case for forward volatility, forward skew is strongly model-dependent, which helps to explain the differences in forward CMS straddle values in Figure 20.4.

The third term, V_3 in (20.18) provides a relatively small contribution to the value of the forward CMS straddle. To analyze it in more detail, it is convenient to condition on $S(u)$ and use (20.20) to obtain

$$V_3 = a_3 \mathrm{E}^A \left(S(u) \sigma_N(u, S(u); t, S(u)) \right)$$

for some constant a_3. For many models the ATM volatility $\sigma_N(u, S; t, S)$ is a linear function of S,

$$\sigma_N(u, S; t, S) \approx a_4 + a_5 S,$$

so then

$$V_3 = a_3 a_4 S(0) + a_3 a_5 \mathrm{E}^A \left(S(u)^2 \right)$$

and we see that V_3 is mostly influenced by the convexity adjustment of $S(u)$ (see Section 16.6.4). The value of the convexity adjustment is linked to the spot volatility information (volatility smile for options on $S(u)$ as observed at time 0), and models that are calibrated to the (spot) volatility smile on $S(u)$ should give identical values to this term. This is, for example, the case in Figure 20.4 for the fixed-tenor volatility swaps; in the fixed-expiry case the models would typically be calibrated to (spot) volatility smiles on $S(T)$ (T here is the start date for the swap) and may imply different smiles for time u. Anyway, as we already mentioned, numerical experiments show that the V_3 term is not particularly significant.

As the value of the forward CMS straddle is largely defined by the forward volatility term V_1, it may seem puzzling that the values on display in Figure 20.4 are strongly sensitive to the model choice. After all, we used the same mean reversion for the three models (2%), which seemingly implies the same inter-temporal correlations between the swap rates involved, which, as we argued before, should essentially lock in the forward volatilities to the same levels in the three models. However, material differences here arise from the fact that different smile mechanisms in the three models in Figure 20.4 change the meaning of the mean reversion parameter. We have seen a similar effect before in Section 17.A that showed that the effective correlation in a displaced log-normal model depends on the skew. The same happens here: even though the mean reversion parameter is the same, the actual effective inter-temporal correlations (or, equivalently, forward volatilities) are different because of different multi-dimensional distributions of the swap rates involved (even though marginal distributions are nearly identical). To estimate these smile effects on correlation we would face the difficult task of going beyond the Gaussian-type approximations used in, e.g., Section 13.1.8.3.

Let us summarize. The values of forward CMS straddles — i.e. coupons of volatility swaps — depend on the level and shape of volatility smiles at future times. As these are largely defined by the volatility dynamics of the model used, different models can produce significantly different values, even if calibrated to identical (spot) volatility information. In particular the impact of volatility smiles on inter-temporal correlations/forward volatilities is of significant importance, and it is advisable to use models with different mechanisms of smile generation to monitor and control it.

20.3 Forward Swaption Straddles

Besides being closely linked to volatility swaps (see Section 20.2.4), forward swaption straddles are themselves traded as stand-alone products, with most of the demand coming from hedge funds interested in expressing views on future implied volatility (see (20.20)). In this stand-alone traded format forward swaption straddles tend to be relatively short-dated, with typical expiries around 3–5 years. As such, these securities are often treated as vanilla, rather than exotic, derivatives, and it is common to use simple vanilla-type models for their valuation. We proceed to describe a typical approach.

First, to be able to use our standard swap rate notations (4.8), (4.10), we assume that a tenor structure

$$0 < T_0 < T_1 < \ldots < T_N, \quad \tau_n = T_{n+1} - T_n,$$

is given. To introduce matching notation for forward straddles, let

$$V_{n,m}(v; T_k)$$

be the time v value of the forward straddle payoff

$$A_{n,m}(T_n) |S_{n,m}(T_n) - S_{n,m}(T_k)| \tag{20.21}$$

paid at T_n, $k < n$, $n + m < N$. We fix a particular forward straddle; to tie our discussion to Section 20.2.4, we assume that

$$u = T_s, \quad t = T = T_e$$

for some indices[4] $s, e, 0 \le s < e < N$, and that the final payment date of the swap is T_N. So this contract delivers at time T_s the value of an at-the-money straddle with expiry T_e on a swap that starts on T_e and matures on T_N. In other words, the contract has the payoff

$$V_{\text{straddle}}(T_s, S_{e,N-e}(T_s), T_e)$$

paid at T_s or, equivalently, the payoff (20.21) with $k = s$, $n = e$, $m = N - e$. By (20.20), the value at time 0 is equal to

$$V_{e,N-e}(0; T_s) = A_{e,N-e}(0) \sqrt{\frac{2(T_e - T_s)}{\pi}}$$
$$\times \mathrm{E}^A \left(\sigma_{\mathrm{N}}(T_s, S_{e,N-e}(T_s); T_e, S_{e,N-e}(T_s)) \right) \tag{20.22}$$

(the measure Q^A here is actually $\mathrm{Q}^{A_{e,N-e}}$ but we simplify the notation for brevity). As we already mentioned, in most models that we use the at-the-money basis-point volatility $\sigma_{\mathrm{N}}(T_s, S; T_e, S)$ can be approximated to excellent precision as a linear function. This allows us to write

$$\sigma_{\mathrm{N}}(T_s, S_{e,N-e}(T_s); T_e, S_{e,N-e}(T_s)) \approx \sigma_{\mathrm{N}}(T_s, S_{e,N-e}(0); T_e, S_{e,N-e}(0))$$
$$+ \left. \frac{\partial}{\partial S} \sigma_{\mathrm{N}}(T_s, S; T_e, S) \right|_{S = S_{e,N-e}(0)} (S_{e,N-e}(T_s) - S_{e,N-e}(0)).$$

As $S_{e,N-e}$ is a Q^A-martingale we then simply have

---

[4]Here s also stands for "strike setting" and e for "expiry".

$$V_{e,N-e}(0;T_s) \approx A_{e,N-e}(0)\sqrt{\frac{2(T_e - T_s)}{\pi}}\lambda_{e,N-e}(T_s,T_e), \qquad (20.23)$$

where we have abbreviated

$$\lambda_{e,N-e}(T_s,T_e) \triangleq \sigma_N(T_s, S_{e,N-e}(0); T_e, S_{e,N-e}(0))$$

and added subscripts to highlight the rate this volatility corresponds to. As discussed earlier, $\lambda_{e,N-e}(T_s,T_e)$ is the forward volatility of the swap rate $S_{e,N-e}$ over the period $[T_s, T_e]$. While this quantity cannot be observed directly in the market, it can be linked to quantities that can. Indeed, if we approximate the swap rate $S_{e,N-e}$ as following a Gaussian process in measure Q^A, then splitting the total variance of $S_{e,N-e}(T_e)$ over the periods $[0, T_s]$ and $[T_s, T_e]$ we get

$$\lambda_{e,N-e}(T_s,T_e)^2(T_e - T_s) = \lambda_{e,N-e}(0,T_e)^2 T_e - \lambda_{e,N-e}(0,T_s)^2 T_s. \quad (20.24)$$

Clearly, $\lambda_{e,N-e}(0,T_e)$ is observable, since the value of a standard spot (time 0) starting at-the-money straddle with expiry T_e is given by

$$V_{e,N-e}(0;0) \approx A_{e,N-e}(0)\sqrt{\frac{2T_e}{\pi}}\lambda_{e,N-e}(0,T_e).$$

The other term in (20.24), $\lambda_{e,N-e}(0,T_s)$, is the volatility of the swap starting at T_e over the period $[0, T_s]$. Equivalently, it is the volatility implied from the value of an *option on a forward starting swap*; at option expiry T_s the holder has the right to enter into a swap starting at some future time $T_e > T_s$. Such options are not traded (or, rather, not liquid), but note that a forward-starting swap can be represented as a combination of spot-starting swaps. We do the calculation for swap rates:

$$\begin{aligned}
S_{e,N-e}(T_s) &= \frac{P(T_s,T_e) - P(T_s,T_N)}{A_{e,N-e}(T_s)} \\
&= \frac{P(T_s,T_s) - P(T_s,T_N)}{A_{e,N-e}(T_s)} - \frac{P(T_s,T_s) - P(T_s,T_e)}{A_{e,N-e}(T_s)} \\
&= w_1(T_s)S_{s,N-s}(T_s) - w_2(T_s)S_{s,e-s}(T_s), \qquad (20.25)
\end{aligned}$$

where

$$w_1(T_s) = \frac{A_{s,N-s}(T_s)}{A_{e,N-e}(T_s)}, \quad w_2(T_s) = \frac{A_{s,e-s}(T_s)}{A_{e,N-e}(T_s)}.$$

To proceed, we approximate the ratios of PVBPs by their values at time 0,

$$w_1(T_s) \approx w_1 = \frac{A_{s,N-s}(0)}{A_{e,N-e}(0)}, \quad w_2(T_s) \approx w_2 = \frac{A_{s,e-s}(0)}{A_{e,N-e}(0)}, \qquad (20.26)$$

and assume that $S_{s,N-s}(T_s)$, $S_{s,e-s}(T_s)$ are approximately Gaussian with correlation ρ. Then from (20.25) we obtain

$$\lambda_{e,N-e}(0,T_s)^2 \approx w_1^2 \lambda_{s,N-s}(0,T_s)^2$$
$$- 2w_1 w_2 \lambda_{s,N-s}(0,T_s)\lambda_{s,e-s}(0,T_s)\rho + w_2^2 \lambda_{s,e-s}(0,T_s)^2, \quad (20.27)$$

where the volatilities $\lambda_{s,N-s}(0,T_s)$, $\lambda_{s,e-s}(0,T_s)$ are now observable as they correspond to ATM (spot) starting swaptions with expiry T_s on $(N-s)$-period and $(e-s)$-period swaps, respectively. Invoking (20.24) and putting it all together, we have

$$V_{e,N-e}(0;T_s) \approx A_{e,N-e}(0)\sqrt{\frac{2}{\pi}}\Big(\lambda_{e,N-e}(0,T_e)^2 T_e - \big(w_1^2 \lambda_{s,N-s}(0,T_s)^2$$
$$-2w_1 w_2 \lambda_{s,N-s}(0,T_s)\lambda_{s,e-s}(0,T_s)\rho + w_2^2 \lambda_{s,e-s}(0,T_s)^2\big)T_s\Big)^{1/2}, \quad (20.28)$$

where the only unobserved parameter is the correlation ρ. This parameter can, for instance, be estimated from a properly-calibrated LM model or left as an "exotic" parameter for traders to tweak.

As the two relevant swap rates $S_{s,N-s}(T_s)$, $S_{s,e-s}(T_s)$ fix on the same date, the correlation ρ is usually quite high; in fact, it is not uncommon to simply assume that $\rho = 1$. Additionally, sometimes one approximates $A_{e,N-e} \approx T_N - T_e$, $A_{e,s-e} \approx T_s - T_e$, and so forth. This results in the following well-known approximation for forward volatility:

$$\lambda_{e,N-e}(T_s,T_e) = \Big(\lambda_{e,N-e}(0,T_e)^2 \frac{T_e}{T_e - T_s}$$
$$- \Big(\frac{T_N - T_s}{T_N - T_e}\lambda_{s,N-s}(0,T_s) - \frac{T_e - T_s}{T_N - T_e}\lambda_{s,e-s}(0,T_s)\Big)^2 \frac{T_s}{T_e - T_s}\Big)^{1/2}.$$
$$(20.29)$$

While (20.29) may occasionally be useful for back-of-the-envelope computations, (20.27) is still preferable.

The simple expression (20.28) for the value of a forward swaption straddle makes its vega exposure quite transparent. A position in the forward straddle is equivalent to a long position in the spot-starting straddle on the same rate $S_{e,N-e}(T_e)$, minus a spread option on two swap rates $S_{s,N-s}(T_s)$ and $S_{s,e-s}(T_s)$. It is important to realize, however, that the vega hedge suggested by this decomposition is not static: as rates move, the vega of a forward swaption straddle does not change, while the vegas of standard swaptions implicitly used in the decomposition (20.28) do change, and may disappear altogether if the swaptions become sufficiently far in or out of the money. The vega hedge consequently must be rebalanced quite frequently over the life of the forward swaption straddle, often at fairly significant expense. As for other risk sensitivities, the forward swaption straddle has no delta[5] and

---

[5]In the sense that there is no sensitivity to the yield curve, provided that all basis-point (Gaussian) volatilities are kept fixed. If we assume that the basis-point

(almost) no gamma until the time of the strike fix (T_s); so it is, indeed, an instrument with pure volatility exposure.

The formula (20.28) was obtained using Gaussian approximation. In Section 20.2.4, on the other hand, we highlighted the importance of accounting for volatility smile in pricing forward swaption straddles. This, however, does not invalidate (20.28), as we recall that the main issue with the smile is its impact on the meaning of the mean reversion parameter — and, ultimately, the effective correlation in the model. In (20.28), we control de-correlation directly, through an exogenous correlation parameter ρ, which "bundles" smile effects and correlation effects in one parameter.

The task of choosing a reasonable value for ρ is fairly straightforward since, as we recall, forward swaption straddles are usually rather short dated. Still, if we want to study the smile effects separately from correlation, we can extend the model to explicitly include the smile. Many routes could be taken here; let us outline a possible approach. First, we note that the forward swaption straddle value is given by the expected value of the payoff

$$|S_{e,N-e}(T_e) - S_{e,N-e}(T_s)| \tag{20.30}$$

in the annuity measure in which $S_{e,N-e}$ is a martingale. The distribution of $S_{e,N-e}(T_e)$ is known directly in this measure from the swaption values across strikes (recall Chapter 16). The distribution of $S_{e,N-e}(T_s)$ is, however, not known. However, by (20.25) and (20.26) it can be represented as a weighted difference of two swap rates $S_{s,N-s}(T_s)$, $S_{s,e-s}(T_s)$ whose full distributions are, again, observable. So we can rewrite the payoff (20.30) as

$$|S_{e,N-e}(T_e) - w_1 S_{s,N-s}(T_s) + w_2 S_{s,e-s}(T_s)|$$

and then use any of the copula methods from Chapter 16, methods that allow direct inclusion of full marginal distributions of the rates involved. We let the reader fill in the remaining details.

Before concluding, let us mention a few variations of the basic forward swaption straddle product. We already mentioned options on forward starting swaps, which are closely related to forward swaption straddles and can be priced along the same lines as above. Another related contract pays the value of the implied basis-point volatility (for an at-the-money straddle with expiry T_e) at T_s, i.e. a contract with the payoff

$$\sigma_N(T_s, S_{e,N-e}(T_s); T_e, S_{e,N-e}(T_s))$$

at T_s. Its value can be linked to that of a forward swaption straddle as we have

---

volatility surface moves with the yield curve (see discussion in Section 16.1.1 on backbones), then a "shadow delta" could, of course, come into play.

$$\mathrm{E}\left(\beta(T_s)^{-1}\sigma_\mathrm{N}(T_s, S_{e,N-e}(T_s); T_e, S_{e,N-e}(T_s))\right)$$
$$= \mathrm{E}^A\left(\frac{\sigma_\mathrm{N}(T_s, S_{e,N-e}(T_s); T_e, S_{e,N-e}(T_s))}{A_{e,N-e}(T_s)}\right).$$

Apart from some deterministic scaling, the difference from (20.22) is in the convexity term $1/A_{e,N-e}(T_s)$. We can link it to the value of the swap rate $S_{e,N-e}(T_s)$ using methods from Chapter 16. While we are not able to derive a simple formula such as (20.28), copula methods obviously still apply.

21

Out-of-Model Adjustments

When valuing exotic derivatives, like CLEs and TARNs, that are written on an underlying structured swap, it is natural to desire that the structured swap is priced in line with the market[1]. Sometimes such consistency is easy to achieve, as when the term structure model used for exotic derivatives pricing happens to coincide (either exactly or to good approximation) with the vanilla model(s) used to define the "market". For instance, the stochastic volatility versions of the quasi-Gaussian model (see Chapter 13) and the Libor market model (see Chapters 14 and 15) are consistent with the vanilla SV model of Chapter 8 when it comes to European swaption pricing. Such consistency is, however, not always feasible. For instance, when using volatility smile parameterizations such as SABR or SVI for vanilla swaption marking (see Sections 8.6 and 16.1.5), we will typically not be perfectly consistent with any of the standard term structure models above. Similarly, it is hard to imagine a term structure model that would be exactly consistent with some of the copulas we introduced for multi-rate vanilla derivatives pricing in Chapter 17. Even if a sophisticated calibration routine is employed, such lack of consistency generally implies that the vanilla and term structure models will disagree in an economically meaningful way on the prices of structured swaps underlying exotic derivatives, a situation that is typically seen as undesirable.

In this chapter we review various methods to *force* the value of a structured swap in a term structure model to match the market (or vanilla model) value, through outright manipulation of some quantity that affects the derivative price. In selecting the quantity (or quantities) to alter, any of the key "ingredients" in a derivative model price are potential candidates: the model, the market data, and the trade. We consider all three possibilities in what follows. There is a strong ad-hoc flavor to all the methods we present,

---

[1]For exotic swaps that require a model for their valuation, the "market" is often understood in a broad sense to represent the values of underlying exotic swaps in a vanilla model of choice.

and theoretical justification is typically rather weak. Nevertheless, if applied judiciously, risk management and pricing accuracy can sometimes benefit from the methods of this chapter. To be clear, the methods we present are not designed to cover for gross mis-calibrations or mis-valuations of underlying swaps, and should not be used as such. We only (cautiously) endorse them as ways of correcting for "small" mismatches in valuation. While it is difficult to make general statements on how small is small, one should use a combination of common sense, experience and rigorous testing in making the judgment.

21.1 Adjusting the Model

We start out by considering adjustments, where model-derived information is used to adjust the value of an exotic derivative to account for mispricing of the underlying. Here, and throughout the chapter, we denote the coupons of the structured swap used as an underlying for a given derivative by C_n, $n = 1, \ldots, N - 1$, and the exotic derivative itself — a callable Libor exotic or a TARN, for example — by H_0. Note that we therefore depart slightly from our standard notation, whereas we would normally use C_n and H_0 to determine *values* of coupons and exotic derivatives; we do so to distinguish different values calculated by different methods. In particular, the market value of the n-th coupon is denoted by $V_{\mathrm{mkt}}(C_n)$, $n = 1, \ldots, N-1$. The value of the same coupon in the term structure model is denoted by $V_{\mathrm{mdl}}(C_n)$, $n = 1, \ldots, N - 1$.

21.1.1 Calibration to Coupons

As we have seen on numerous occasions in this book, when pricing exotic derivatives, term structure models are typically calibrated to a multitude of European swaptions. Of course, this in itself does not necessarily guarantee that the prices of coupons of the underlying swap in the term structure model would match their market, or vanilla model, prices. A good example here is a fixed-rate callable range accrual (see Section 5.13.4), where the n-th coupon is given by

$$C_n = k \times \frac{1}{T_{n+1} - T_n} \sum_{t \in [T_n, T_{n+1}]} 1_{\{L(t) \in [l, u]\}},$$

with $L(t)$ being some Libor rate. A fixed-rate range accrual coupon is decomposable into a collection of digital options on the Libor rate, and as such can be valued in a vanilla model with slight timing-delay convexity adjustments, see Chapter 16 and in particular Section 16.5. It is likely, however, that $V_{\mathrm{mdl}}(C_n) \neq V_{\mathrm{mkt}}(C_n)$, due to, for example, differences in the treatment of convexity effects or in the volatility smiles implied by the term structure and the vanilla models.

To guarantee that $V_{\mathrm{mdl}}(C_n) = V_{\mathrm{mkt}}(C_n)$ for all $n = 1, \ldots, N - 1$, the model can explicitly be calibrated to the market prices of the underlying coupons. This *extended calibration* method is fairly benign as far as model adjustments go, and could well be considered an extension of the local projection method (see Section 18.4). The ability to calibrate to the prices of underlying coupons relies, of course, on the availability of efficient methods for calculating their values $V_{\mathrm{mdl}}(C_n)$, $n = 1, \ldots, N-1$, in a term structure model used. For most "interesting" coupons, however, closed-form expression are generally unavailable in sophisticated term structure models (e.g. LM-type models), and one has to resort to numerical calculations for calibration, often requiring Monte Carlo simulations. Calibration by Monte Carlo simulation is not something we would typically recommend, but if this approach is chosen nevertheless, some fairly obvious precautions should be taken: all coupons should be computed in the same simulation loop, the simulation seed must be taken to be the same in all calibration iterations as well as for the main valuation, and so on. A body of literature on these so-called *stochastic optimization methods* exists (see e.g. Broadie et al. [2009], Andradóttir [1995], Andradóttir [1996]) and should be consulted before one attempts to use Monte Carlo simulation inside a calibration loop.

For term structure models amenable to PDE methods, calculating coupon values by numerical (PDE) methods for calibration is certainly a plausible strategy. For numerical efficiency, we recommend usage of the forward induction method (see Section 11.3.2.3), rather than the standard backward induction method, when calculating coupon values in a calibration algorithm.

Apart from numerical issues, the extended calibration method has certain other caveats. Calibration to non-standard targets requires special care, as one has to be mindful of using the right parameters in the calibration. For example, the value of the range accrual coupon, being a sum of digital options, is not a monotonic function of volatility, and trying to calibrate volatility of the model to the market prices of range accrual coupons may yield unrealistic volatility levels or fail outright. In this particular case, it is clear that the *skew* of the volatility smile is a primary driver of the range coupon value and, hence, it is the model skews (in a skew-enabled model such as a quasi-Gaussian local volatility model), and not the volatilities, that should be calibrated to the market prices of range accrual coupons.

Even if calibration to coupons does not fail, it may result in a set of model parameters that are inappropriate for valuing other optionality embedded in a given exotic derivative, such as callability. In the callable range accrual example above, had we mistakenly tried to calibrate the volatility of a term structure model to range accrual coupons, the model volatilities could end up being very high or very low, significantly over- or under-estimating the value of callability feature. Similarly, if we were to try to value a CMS spread TARN (see Section 5.15) in a one-factor model by, say, calibrating mean reversion to the underlying CMS spread option values[2], the resulting mean reversion would very likely be inappropriate for valuing the trigger feature of the TARN.

21.1.2 Adjusters

While the extended calibration method can be attractive, brute-force calibration to coupon values may not always be feasible for numerical or other reasons. Fortunately, the problem can be simplified if we recall our main tenet that prudent application of out-of-model adjustments should be limited to correcting for *small* mismatches, in which case we should be able to linearize the problem and solve it with less effort. We call this idea the *adjusters method* after Hagan [2002] who popularized it.

Let ξ be some model parameter — a volatility function, a correlation parameter, a vector of mean reversions or even a yield curve — and ξ_0 be its calibrated value. In general ξ can be represented as a column vector, and to simplify our exposition we assume it is $(N-1)$-dimensional, with the n-th coordinate affecting the value of the n-th coupon. We make the dependence of model prices of various securities on ξ explicit and write $V_{\mathrm{mdl}}(C_n; \xi)$.

Let ξ^* be the solution of

$$V_{\mathrm{mdl}}\left(C_n; \xi^*\right) = V_{\mathrm{mkt}}\left(C_n\right) \text{ for all } n = 1, \ldots, N-1.$$

For all $n = 1, \ldots, N-1$, ξ^* satisfies, to first order,

$$V_{\mathrm{mdl}}\left(C_n; \xi_0\right) + \frac{\partial V_{\mathrm{mdl}}}{\partial \xi}\left(C_n; \xi_0\right)\left(\xi^* - \xi_0\right) \approx V_{\mathrm{mkt}}\left(C_n\right), \tag{21.1}$$

where $\partial V_{\mathrm{mdl}}/\partial \xi$ is a row vector of $\partial V_{\mathrm{mdl}}/\partial \xi_n$, $n = 1, \ldots, N-1$. Hence,

$$\xi^* \approx \xi_0 + \left[\frac{\partial \overline{V}_{\mathrm{mdl}}}{\partial \xi}\left(C; \xi_0\right)\right]^{-1}\left(\overline{V}_{\mathrm{mkt}}(C) - \overline{V}_{\mathrm{mdl}}\left(C; \xi_0\right)\right), \tag{21.2}$$

where we use bars to denote column vectors,

$$\overline{V}_{\mathrm{mdl}}\left(C; \xi\right) \triangleq \left(V_{\mathrm{mdl}}\left(C_1; \xi\right), \ldots, V_{\mathrm{mdl}}\left(C_{N-1}; \xi\right)\right)^{\top},$$

and so forth. In particular,

$$\frac{\partial \overline{V}_{\mathrm{mdl}}}{\partial \xi}\left(C; \xi\right) \tag{21.3}$$

is an $(N-1) \times (N-1)$ matrix whose n-th row is $\partial V_{\mathrm{mdl}}(C_n; \xi)/\partial \xi$.

Before proceeding, let us note that while we for exposition purposes assumed the dimension of the model parameter ξ to be the same as the

---

[2]Needless to say, this is not something that we generally recommend.

number of coupons, this need not be so in actual applications. In particular, if the dimensions do not match, we can think of the equation (21.1) as a linear regression problem and find ξ^* by the appropriate least-squares methods. This procedure has been used many times already in this book, see, for example, Section 6.4.3.

Having identified ξ^*, the adjusted model price of H_0 is given by

$$V_{\text{adj}}(H_0) = V_{\text{mdl}}(H_0; \xi^*),$$

and, expanding to first order and substituting (21.2),

$$V_{\text{adj}}(H_0) \approx V_{\text{mdl}}(H_0; \xi_0) + \frac{\partial V_{\text{mdl}}}{\partial \xi}(H_0)(\xi^* - \xi_0)$$

$$= V_{\text{mdl}}(H_0; \xi_0)$$

$$+ \frac{\partial V_{\text{mdl}}}{\partial \xi}(H_0)\left[\frac{\partial \overline{V}_{\text{mdl}}}{\partial \xi}(C; \xi_0)\right]^{-1}\left(\overline{V}_{\text{mkt}}(C) - \overline{V}_{\text{mdl}}(C; \xi_0)\right).$$

$$(21.4)$$

With these formulas, the adjusters method follows these steps.

1. Given the calibrated model parameter value ξ_0, the unadjusted values of the exotic H_0, and of all the underlying coupons C_n, $n = 1, \ldots, N-1$, are computed.
2. The sensitivities of the value of H_0 and values of the various C_n to the model parameter ξ (at ξ_0) are computed.
3. The matrix of parameter sensitivities in (21.3) is inverted.
4. The adjusted exotic value $V_{\text{adj}}(H_0)$ is calculated via the linear approximation (21.4).

Note that the calibration loop of Section 21.1.1 is now replaced by the calculation of the sensitivity matrix $\partial \overline{V}_{\text{mdl}}(C)/\partial \xi$. Given the actual structure of the problem, this matrix may be known to be of specific form, e.g. diagonal or lower-triangular, further simplifying its evaluation. Moreover, the matrix can often be cached and reused when calculating risk sensitivities, further improving the overall efficiency of the scheme. Other needed quantities such as $\partial V_{\text{mdl}}(H_0)/\partial \xi$, are typically calculated anyway for risk management purposes and should not, as a rule, add to the overall computational burden.

The adjusters method is not restricted to using the underlying coupons as adjusters, but can be applied more broadly. For example, the value of a Bermudan swaption in a model with no volatility smile capabilities could be "adjusted" for the smile by using European swaptions as adjusters. In a sense, we can see the adjustment as a type of a control variate method (see Section 3.4.3 or Chapter 25 below) with the values of the adjusters (coupons or other vanilla instruments) used as controls.

The adjusters method potentially applies to a variety of model parameters, and a key question concerns which parameter should be used as ξ — model volatilities, skews, etc. The answer follows the same logic as in Section 21.1.1: we should apply the method to the parameter(s) that have the most impact on the values of coupons while affecting other features of the model as little as possible. For example, the level of a yield curve often affects coupon values directly and so we can use the yield curve as the adjuster. This case has a special name, the *delta-adjustment method*, and it is similar to some of the approaches we discuss below, such as the spread adjustment method (Section 21.2) and the strike adjustment (Section 21.3.3) methods. Volatilities, too, are often a good choice for adjustment as values of most "interesting" coupons depend on volatilities of relevant rates. Of course the situation could be more complicated as the example of a fixed range accrual coupon in Section 21.1.1 demonstrated; here the skew of the volatility smile, and not its overall level, was the most relevant adjuster. Overall, nothing replaces careful analysis of each type of exotic derivative before the adjusters method is applied.

Before we wrap up our discussion of adjusters we note that the term

$$\frac{\partial V_{\text{mdl}}}{\partial \xi}(H_0) \left[\frac{\partial \overline{V}_{\text{mdl}}}{\partial \xi}(C; \xi_0) \right]^{-1}$$

in (21.4) could be interpreted as the sensitivity of the value of an exotic to the values of the underlying coupons, an interesting measure of sensitivity in its own right.

21.1.3 Path Re-Weighting

In the case of Monte Carlo based models, an approach from Avellaneda et al. [2001] makes it possible to exactly match calibration targets to their desired values, while also correcting for numerical inaccuracies of the valuation method. Let us discuss the idea in some detail. As a start, we denote simulated Monte Carlo paths by ω_i, $i = 1, \ldots, K$. As always, the value estimate of any payoff — be it a zero-coupon bond, a vanilla option, or the coupon C_n — is given by the average of the payoff values associated with each path ω_i, $i = 1, \ldots, K$. Focusing exclusively on the problem of matching the model values of coupons C_n, $n = 1, \ldots, N - 1$, to the market, we denote by C_n^i the value of the n-th coupon, $n = 1, \ldots, N - 1$, along path ω_i, $i = 1, \ldots, K$. Then, the basic Monte Carlo value estimate of the n-th coupon in the model is given by

$$V_{\text{mdl}}(C_n) = \frac{1}{K} \sum_{i=1}^{K} C_n^i. \tag{21.5}$$

The idea of the *path re-weighting method* is to assign non-equal probabilities to the different paths in order to match target values. Let the probability assigned to the path ω_i be p_i, satisfying the standard requirements

$$0 \le p_i \le 1 \quad \forall i = 1, \ldots, K, \tag{21.6}$$

$$\sum_{i=1}^{K} p_i = 1. \tag{21.7}$$

Then the value of the n-th coupon is given by

$$\sum_{i=1}^{K} C_n^i p_i, \tag{21.8}$$

and is a linear function of the vector $p = (p_1, \ldots, p_K)^\top$. Hence, one would expect that it should be fairly straightforward to find a vector p that matches model prices of all coupons to the market,

$$\sum_{i=1}^{K} C_n^i p_i = V_{\mathrm{mkt}}(C_n), \quad n = 1, \ldots, N - 1. \tag{21.9}$$

The resulting "probabilities" can subsequently be reused in the pricing of the exotic derivative.

The problem (21.6), (21.7), (21.9) is under-specified since the number of paths used — which also equals the dimension of the vector p — is typically (much) larger than the number of coupons. Hence, a suitable regularization target is needed if we want to have a unique solution. It is not unreasonable, for example, to try to keep the vector p as close as possible to the equi-weighted probabilities of (21.5). Working with probability distributions, a convenient measure of closeness is the so-called *Kullback-Leibler relative entropy* between the probability vector p and the equi-weighted prior. With this choice of norm, we can formalize the search for p as the following minimization problem:

$$I(p) \triangleq \sum_{i=1}^{K} p_i \ln(p_i) \to \min, \tag{21.10}$$

subject to the linear inequality constraints (21.6), as well as the linear equality constraints (21.7) and (21.9). Proponents of the principle of relative entropy optimization often justify the choice of norm in (21.10) from a perspective of information theory (e.g., as a way to ensure that we do not add information that we do not possess to the problem), but a more standard least-squares norm would likely do just as well[3].

The range of model errors that the path re-weighting method can correct for is limited, since the $\sum_{i=1}^{K} C_n^i p_i$ will obviously always be between the minimum and the maximum path value of C_n, among the K paths. If the target $V_{\mathrm{mkt}}(C_n)$ is outside this range, such a gross mismatch cannot

---

[3]We briefly consider least-squares norms later in the section.

be corrected. Should this situation ever arise, the difference between the model and market values would likely be of such magnitude that the path re-weighting scheme would fundamentally be inappropriate anyway, as we discussed in the beginning of this chapter.

Let us develop the solution to the entropy minimization problem above in a bit more detail. For this purpose, let $\lambda = (\lambda_1, \ldots, \lambda_{N-1})$ be a vector of Lagrange multipliers for the constraints (21.9), and μ a Lagrange multiplier for the total probability constraint (21.7). Then the solution to the following unconstrained problem (the "dual" formulation of the constrained problem, see Cover and Thomas [2006]),

$$\min_{\lambda} \max_{p,\mu} J(p, \lambda, \mu), \qquad (21.11)$$

where (note the negative sign in front of $I(p)$)

$$
\begin{aligned}
J(p, \lambda, \mu) &\triangleq -I(p) \\
&+ \sum_{n=1}^{N-1} \lambda_n \left(\sum_{i=1}^{K} C_n^i p_i - V_{\text{mkt}}(C_n) \right) + \mu \left(\sum_{i=1}^{K} p_i - 1 \right), \quad (21.12)
\end{aligned}
$$

if it happens to satisfy (21.6), would also solve (21.10) subject to (21.6), (21.7) and (21.9).

Proposition 21.1.1. *For a given vector λ, the solution of the inner maximization problem in (21.11) is given by*

$$\mu^* = 1 - \ln(Z(\lambda))$$

and

$$p_i^* = \frac{1}{Z(\lambda)} \exp\left(\sum_{n=1}^{N-1} \lambda_n C_n^i \right), \quad i = 1, \ldots, K, \qquad (21.13)$$

where the partition function $Z(\lambda)$ is given by

$$Z(\lambda) = \sum_{i=1}^{K} \exp\left(\sum_{n=1}^{N-1} \lambda_n C_n^i \right).$$

Proof. The necessary conditions for the inner maximum in (21.11) are given by

$$\frac{\partial J(p^*, \lambda, \mu^*)}{\partial \mu} = 0, \quad \frac{\partial J(p^*, \lambda, \mu^*)}{\partial p_i} = 0, \quad i = 1, \ldots, K,$$

so that we have

$$-\ln(p_i^*) - 1 + \sum_{n=1}^{N-1} \lambda_n C_n^i + \mu^* = 0, \quad i = 1, \ldots, K,$$

and $\sum_{i=1}^{K} p_i^* = 1$. The proposition follows. □

We note that p_i^*'s defined by (21.13) always satisfy (21.6). The distribution of the form (21.13) is known as the *Boltzman-Gibbs distribution* for the
partition function $Z(\lambda)$.

Now, substituting (21.13) into the definition of the objective function
(21.12) we obtain

$$G(\lambda) \triangleq J(p^*, \lambda, \mu^*) = \ln\left(Z(\lambda)\right) - \sum_{n=1}^{N-1} \lambda_n V_{\mathrm{mkt}}\left(C_n\right). \tag{21.14}$$

Now all we need to do is to minimize (21.14), i.e. solve the $(N-1)$-dimensional
optimization problem

$$\lambda^* = \underset{\lambda}{\operatorname{argmin}}\left(G(\lambda)\right). \tag{21.15}$$

Compared to the original formulation (21.10), the dimensionality of the
problem has now been significantly reduced, as normally $N \ll K$. Moreover,
(21.15) is unconstrained, and thus easier to solve by standard optimization
techniques. In addition, it is a "nice" optimization problem as the function
$G(\lambda)$ is globally convex with a single minimum, as stated in the following
proposition.

Proposition 21.1.2. *The function $G(\lambda)$ is globally convex. In particular,
the following holds for all $n, m = 1, \ldots, N - 1$,*

$$\frac{\partial G(\lambda)}{\partial \lambda_n} = \mathrm{E}^\lambda\left(C_n\right) - V_{\mathrm{mkt}}\left(C_n\right), \qquad \frac{\partial^2 G(\lambda)}{\partial \lambda_n \partial \lambda_m} = \mathrm{Cov}^\lambda\left(C_n, C_m\right), \tag{21.16}$$

*where the measure Q^λ is defined on Monte Carlo paths by Boltzman-Gibbs
weights p^* that correspond to λ as per (21.13), i.e. for any random variable
X,*

$$\mathrm{E}^\lambda(X) = \sum_{i=1}^{K} p_i^* X^i,$$

*where X^i is the realization of the random variable X on the i-th path,
$i = 1, \ldots, K$.*

Proof. Let us fix n. By straightforward differentiation,

$$\frac{\partial G(\lambda)}{\partial \lambda_n} = \frac{1}{Z(\lambda)} \frac{\partial Z(\lambda)}{\partial \lambda_n} - V_{\mathrm{mkt}}\left(C_n\right)$$

$$= \frac{1}{Z(\lambda)} \sum_{i=1}^{K} \exp\left(\sum_{j=1}^{N-1} \lambda_j C_j^i\right) C_n^i - V_{\mathrm{mkt}}\left(C_n\right)$$

$$= \sum_{i=1}^{K} p_i^* C_n^i - V_{\mathrm{mkt}}\left(C_n\right).$$

Furthermore,

$$\frac{\partial^2 G(\lambda)}{\partial \lambda_n \partial \lambda_m} = \frac{1}{Z(\lambda)} \frac{\partial^2 Z(\lambda)}{\partial \lambda_n \partial \lambda_m} - \frac{1}{Z(\lambda)^2} \frac{\partial Z(\lambda)}{\partial \lambda_n} \frac{\partial Z(\lambda)}{\partial \lambda_m},$$

and we obtain that

$$\frac{\partial^2 G(\lambda)}{\partial \lambda_n \partial \lambda_m} = \mathrm{Cov}^\lambda (C_n, C_m)$$

by straightforward calculations. The fact that $G(\lambda)$ is globally convex now follows from the representation of the second derivative of G as a covariance matrix in (21.16), and the fact that a covariance matrix is always nonnegative-definite. $\square$

We point out an interesting consequence of (21.16) is that the solution to the optimization problem (21.15) is given by such λ^* that

$$V_{\mathrm{mkt}} (C_n) = \mathrm{E}^{\lambda^*} (C_n), \quad n = 1, \dots, N-1.$$

As the objective function $G(\lambda)$ is globally convex and its first- and second-order derivatives are straightforward to calculate, most non-linear optimization algorithms as discussed in, for example, Section 14.5.7 would work well. For extra performance, specialized methods tuned for convex objective functions, such as the Nesterov-Nemirovskii algorithm (see Nesterov et al. [1994]), could be applied.

The constrained minimization formulation (21.10) as presented in Avellaneda et al. [2001] is not the only possible way to formalize the problem of path re-weighting. For example, we can replace the exact repricing criteria (21.9) by a suitably-defined least-squares target. In particular, denoting by v_n the penalty for violating (21.9) for a given n, $n = 1, \dots, N-1$, the problem can be re-formulated as

$$\sum_{i=1}^{K} p_i \ln (p_i) + \sum_{n=1}^{N-1} v_n \left(\sum_{i=1}^{K} C_n^i p_i - V_{\mathrm{mkt}} (C_n) \right)^2 \to \min, \qquad (21.17)$$

subject to (21.6), (21.7). Not surprisingly, the problem can also be solved by the partition function method along the lines of Proposition 21.1.1, a statement we leave to the reader to verify. Finally, an even simpler quadratic problem could be obtained by replacing relative entropy as an objective function by its second-order Taylor expansion around the equi-weighted probabilities, see e.g. Glasserman [2004], Section 4.5:

$$\sum_{i=1}^{K} (p_i - 1/K)^2 + \sum_{n=1}^{N-1} v_n \left(\sum_{i=1}^{K} C_n^i p_i - V_{\mathrm{mkt}} (C_n) \right)^2 \to \min, \qquad (21.18)$$

subject to (21.6), (21.7). Again, we leave it up to the reader to fill in relevant details.

It is worthwhile pointing out an interesting connection between entropy minimization methods and the problem of calculating risk sensitivities. It turns out that under the (rather unrealistic, admittedly) assumption that market data shocks do not affect generated Monte Carlo paths but only change the right-hand-side values in (21.9), the sensitivities of the exotic derivative to the prices of coupons/market shocks can be deduced via duality arguments from the solutions of the relevant optimization problems (21.10), (21.17) or (21.18). Details can be found in Avellaneda et al. [2001].

Most adjustment methods have undesirable side effects, and path re-weighting is no exception. With non-uniform weights assigned to paths, the prices of zero-coupon bonds in the model may no longer match their market values, with the model then allowing arbitrage. This in principle could be patched up by adding all relevant zero-coupon bonds to the set of constraints (21.9) to match, but, of course, at higher computational cost. Calibration to vanilla options could also unravel — remediation will, once again, involve enlargement of the set of constraints. Again, we remind the reader that over-using methods such as path re-weighting could be dangerous, as it is difficult to control all the consequences if large deviations from the equi-weighted paths are required. Should such situations arise, the model is most likely seriously mis-specified and any valuation results should be treated as suspect.

While introduced here as an adjustment technique, let us finally note that path re-weighting could be interpreted as a variance reduction technique, provided that the option prices we are matching our finite-sample estimates to are known to coincide with the true (infinite-sample) model values. Clearly, the resulting method would have strong similarities to the more familiar technique of control variates (see Chapter 25). Glasserman and Yu [2005] investigate this link further, and prove that the two techniques are essentially identical, for large enough sample sizes. For strict variance reduction purposes, the more straightforward method of control variates is therefore typically preferable.

21.1.4 Proxy Model Method

Suppose we have identified that a given exotic security is sensitive to an "exotic risk" factor. This factor may not be important for the valuation of vanilla securities, and implementing it into a term structure model may result in the model being so complex that analytical approximations used for calibration to the vanilla market fail to be accurate enough. On the other hand, suppose we also have a simpler term structure model that calibrates well to the vanilla market but does not have the required exotic risk factor. The following procedure, which we call the *proxy model method*, is sometimes used to combine the two models to measure the sensitivity to the exotic risk factor. First, we calculate the difference in value of the derivative in the complex model for different values of the exotic risk factor, $\xi = \xi_1$ and

$$\xi = \xi_0,$$

$$\Delta V_{\text{complex}} = V_{\text{complex}}\left(H_0; \xi_1\right) - V_{\text{complex}}\left(H_0; \xi_0\right).$$

Here typically ξ_0 corresponds to the base case, i.e. the value of the exotic risk factor that is more or less consistent with the simplified worldview of the simpler term structure model, and ξ_1 is our view of the actual market-observed value of the risk factor. Next, we calculate the base value of the derivative in the simple model calibrated to the market,

$$V_{\text{simple}}\left(H_0\right).$$

We would like to add $\Delta V_{\text{complex}}$ to $V_{\text{simple}}(H_0)$ to account for the risk factor impact; however, $\Delta V_{\text{complex}}$ is biased due to the problems of calibrating the complex model. To correct for the bias, we calibrate our simple model *to the vanilla prices generated by the complex model* in the two scenarios. Henceforth, we define $V_{\text{simple},0}(H_0)$ and $V_{\text{simple},1}(H_0)$ to be the prices of the derivative in question in the simple model calibrated to the vanilla prices as generated by the complex model with $\xi = \xi_0$ and $\xi = \xi_1$.

As the simple model is insensitive to the exotic risk factor, we would expect

$$\Delta V_{\text{simple}} = V_{\text{simple},1}\left(H_0\right) - V_{\text{simple},0}\left(H_0\right)$$

to solely represent the impact of mis-calibration of the complex model to vanillas on the value of the exotic derivative. Thus, it is not unreasonable to define the adjusted price by

$$V_{\text{adj}}\left(H_0\right) = V_{\text{simple}}\left(H_0\right) + \Delta V_{\text{complex}} - \Delta V_{\text{simple}}.$$

To make the discussion above a bit more concrete, consider the problem of assessing the impact of stochastic volatility de-correlation, which would be the exotic risk factor under consideration, on a callable CMS spread derivative. Suppose we have a suitable model, say an LM model from Section 15.7, which has multiple sources of volatility randomness. We would value the derivative with the correlation of volatilities set to 100% (ξ_0 case), and then set it to some other value that is less than 100% (ξ_1 case) which we obtain by, say, historical estimation. To correct for calibration errors induced by imperfect vanilla approximations, we would calibrate a simpler LM model with a single stochastic volatility factor to the vanilla prices produced by the complex model; the single volatility factor model serves as a model that is (one hopes) sufficiently "similar" to the complex model yet allows for accurate vanilla option approximations. In this case, in addition to the usual European swaptions, we should probably also include CMS spread options in the vanilla market, to make sure we control the extra de-correlation of rates that comes from de-correlating their volatilities.

The method outlined above is rarely accurate enough for trading and risk management purposes, but is useful for qualitative understanding of the impact of certain risk factors, as well as, say, reserve calculations.

21.1.5 Asset-Based Adjustments

Consider as an example an LM model applied to a CMS-style exotic derivative. The CMS convexity adjustment (see Section 16.6) as implied by the LM
model may not be equal to the "market" CMS adjustment (as calculated by,
say, a replication method from Section 16.6.1). This situation might arise in
part because the volatility smiles generated by the LM model differ slightly
from the ones implied by market prices. One way of compensating for the
difference involves changing the trade definition, a method we discuss later
in Section 21.3. Here we, instead, consider a different method in which we
adjust the simulated dynamics of the relevant swap rate(s) in the LM model
(see Van Steenkiste [2009]). The advantage of this *asset-based adjustment
method* is that we are able to not only adjust the overall levels of the relevant
swap rates, but also their volatilities and skews. This, in turn, could further
aid us in closing the valuation gap for the underlying swap of a CMS-based
exotic derivative.

For concreteness, consider a version of the LM model (14.4) with deterministic separable volatility (14.2.4). Suppose a given swap rate[4] $S(t)$ is of
special interest to us, because, say, some coupon of the underlying swap is a
function of it,

$$C = C(S(T)). \tag{21.19}$$

The standard Monte Carlo scheme for the model involves simulating all
Libor rates $\{L_n(T)\}$ per Section 14.6, calculating discount factors from
simulated Libor rates and combining them to calculate the simulated value
of the swap rate $S(T)$. The simulated value of the swap rate is then used to
calculate the simulated value of a coupon in (21.19).

Instead of calculating the swap rate from simulated Libor rates, we can
of course also just simulate $S(t)$ alongside the Libor rates directly. The exact
dynamics of the swap rate under the measure used for simulating Libor rates
(e.g., the spot measure) can be derived by Ito's lemma, or from Proposition
14.4.2 by the appropriate measure change. In particular we have

$$dS(t) = \mu_S(t, \mathbf{L}(t))\, dt + \varphi(S(t)) \sum_{n=1}^{N-1} w_n(t)\lambda_n(t)^\top\, dW^B(t), \tag{21.20}$$

where $\mathbf{L}(t)$ is the vector of all Libor rates at time t, $\mu_S(t, \mathbf{L}(t))$ is the
appropriate drift, $w_n(t)$'s are given by (14.31), and $W^B(t)$ is a Brownian
motion in the spot measure. We could discretize the SDE (21.20) in the
same way we would discretize SDEs for the Libor rates, and simulate $S(t)$
together with the Libor rates; then we can use this simulated value in the
payoff (21.19). Up to the discretization bias and simulation error, the value

---

[4]While we only consider one coupon and one swap rate, it is trivial to extend
our discussion to the standard case of multiple coupons depending on different
rates.

of the derivative computed in this scheme would be the same as in the standard scheme where $S(t)$ is calculated directly from Libor rates.

While the utility of the simulation scheme above by itself is questionable, it gives us a starting point for adjusting the dynamics of $S(t)$ as we see fit. Decoupling the dynamics of $S(t)$ in (21.20) from the dynamics of Libor rates allows us to treat the swap rate as a stand-alone market variable, or "asset" (hence the name of the method), whose model dynamics we can control independently. For example, suppose we would like to shift the mean of the simulated variable $S(T)$ to compensate for the differential of CMS convexity adjustments between the LM model and the market. Then we just use (21.20) with an initial condition shifted by some $c \neq 0$,

$$dS_{\mathrm{adj}}(t) = O(dt) + \varphi(S_{\mathrm{adj}}(t)) \sum_{n=1}^{N-1} w_n(t)\lambda_n(t)^\top \, dW^B(t), \quad S_{\mathrm{adj}}(0) = S(0) + c.$$

When valuing the S-dependent coupon (and the payoff of the entire exotic derivative) we then would use $S_{\mathrm{adj}}(t)$ instead of $S(t)$ in payoff calculations.

The volatility of the swap rate could be adjusted in a similar way; for example we can specify that

$$dS_{\mathrm{adj}}(t) = O(dt) + c\varphi(S_{\mathrm{adj}}(t)) \sum_{n=1}^{N-1} w_n(t)\lambda_n(t)^\top \, dW^B(t), \quad S_{\mathrm{adj}}(0) = S(0),$$

for some volatility adjustment $c > 0$. Or, indeed, we can change the model skew of $S(t)$ by replacing (21.20) with

$$dS_{\mathrm{adj}}(t) = O(dt) + \varphi(aS_{\mathrm{adj}}(t) + b) \sum_{n=1}^{N-1} w_n(t)\lambda_n(t)^\top \, dW^B(t), \quad S_{\mathrm{adj}}(0) = S(0),$$

for some a, b. All three types of dynamics adjustments could, of course, be combined to provide a finer level of control over the distribution of $S(t)$; indeed, possibilities are limitless. The method is not restricted to adjustments of dynamics of swap rates only; we can apply the same trick to the *spread* of two swap rates to ensure that, say, the volatility of a CMS spread in this "adjusted" LM model matches that in the vanilla (multi-rate) model used.

While the asset-based adjustment method is rather flexible, it should be obvious that it comes at a serious cost of introducing arbitrage and making the model internally inconsistent. The swap rate simulated from one of the adjusted SDEs above will no longer equal the "true" swap rate as synthesized from (simulated) Libor rates. As a consequence, a European swaption would have different values in such a model depending on how the payoff is written, see equations (5.10) or (5.11) in Chapter 5. Taking this example to the extreme, one can imagine a trader equipped with such a model selling and buying identical swaptions booked in different formats and generating riskless "profits" on each trade.

21.1.6 Mapping Function Adjustments

Adjusting the dynamics of the swap rate $S(t)$ is not the only way to achieve desired changes to its distribution. As a possible alternative, we can modify its *terminal* distribution directly. In particular, instead of using the swap rate $S(T)$ when calculating the coupon value in (21.19), we would use $S_{\mathrm{adj}}(T)$ defined by

$$S_{\mathrm{adj}}(T) = \Lambda(S(T)).$$

Here $S(T)$ is the model-simulated value of the swap rate, and the *mapping function* $\Lambda(s)$ is chosen in such a way that $S_{\mathrm{adj}}(T)$ has a desired distribution (e.g. one consistent with the swaption market, or perhaps with a particular vanilla model). This approach is sometimes called the *mapping function adjustment*.

When using the mapping function adjustment method, we would not adjust just one swap rate, but all rates used in calculating underlying coupons. If the product requires observations of the swap rate on different dates, we would, of course, use different mapping functions for different observation dates.

Determining the mapping function $\Lambda(s)$ for each required observation of each swap rate is conceptually simple. If $\Psi_{\mathrm{mkt}}(s)$ is the market-implied cumulative distribution function of the swap rate $S(T)$ in the appropriate annuity measure (see Section 16.6.9), and $\Psi_{\mathrm{mdl}}(s)$ is the same for the term structure model that we are adjusting, then we simply set

$$\Lambda(s) = \Psi_{\mathrm{mkt}}^{-1}\left(\Psi_{\mathrm{mdl}}(s)\right).$$

For most models we consider, efficient swaption pricing formulas exist, and the model CDF $\Psi_{\mathrm{mdl}}(s)$ needed here is readily available. Needless to say, the mapping function(s) should be pre-computed and cached before valuing a given trade.

A careful reader will no doubt notice that this type of adjustment has a Markov-functional model flavor (see Appendix 11.A to Chapter 11). It is, however, *not* a full-blown Markov-functional model as, of course, we have made no provisions to retain the arbitrage-free characteristics of the adjusted model. Clearly, the usual caveats to usage of such non-arbitrage-free models apply, and the warnings at the end of Section 21.1.5 should be carefully considered before the mapping approach is used.

21.2 Adjusting the Market

Having finished with adjustments based on models, let us turn to using market data for that purpose. While several sources of market data could be used for adjustment, the most common target is the yield curve, where we can capitalize on the fact that the yield curve is frequently the only parameter that is shared between the term structure model and the vanilla models. The resulting adjustment method is in many ways similar to that in Section 21.1.2, but there are a few twists that warrant a separate discussion.

Continuing with the notations of Section 21.1.2, we first specialize ξ to be the yield curve as used during valuation, and ξ_0 to be the yield curve as fit to the market prices of swaps, etc. (see Chapter 6). We formulate the adjustment problem as finding ξ^* such that

$$V_{\text{mdl}}(C_n; \xi^*) = V_{\text{mkt}}(C_n) \text{ for all } n = 1, \dots, N-1. \tag{21.21}$$

With the view that the impact of the yield curve on the value of a coupon is roughly the same in the two models (vanilla and exotic), we define δ to be the (time-dependent) spread, to be applied to the yield curve, such that

$$V_{\text{vanilla}}(C_n; \xi_0 + \delta) = V_{\text{mdl}}(C_n, \xi_0) \text{ for all } n = 1, \dots, N-1, \tag{21.22}$$

where $V_{\text{vanilla}}(C_n; \xi)$ is the value of coupon C_n in the vanilla model when using yield curve ξ. Then, by approximate linearity and (21.22),

$$V_{\text{mdl}}(C_n, \xi_0 - \delta) \approx V_{\text{vanilla}}(C_n; \xi_0 + \delta - \delta) = V_{\text{mkt}}(C_n),$$

and the approximate solution to (21.21) is given by

$$\xi^* \approx \xi_0 - \delta.$$

Note that solving (21.22) is normally much quicker than solving (21.21) directly.

The valuation of the exotic derivative proceeds with the adjusted yield curve $\xi^* = \xi_0 - \delta$,

$$V_{\text{adj}}(H_0) = V_{\text{mdl}}(H_0, \xi_0 - \delta).$$

We call this the *spread adjustment method*. Notably, the adjusted yield curve should only be applied to the structured leg — the Libor leg, if present, should use the original, unadjusted yield curve. Simultaneous modeling of two yield curves — the adjusted and the original — could follow the deterministic spread approach from Section 15.5.

While the idea behind the method is simple, the need to use multiple yield curves in valuation makes the method somewhat unwieldy, and it is not particularly popular. For derivatives that involve spread options, for obvious reasons we should adjust the slope, rather than the overall level, of the yield curve.

21.3 Adjusting the Trade

Adjusting the trade is probably the most common type of out-of-model adjustments. In this approach, some features of the coupons are changed to line up the values of the adjusted coupons in the term structure model with the values of the original coupons in the vanilla model, i.e. their market values. The adjusted value of the exotic derivative is then calculated by applying the term structure model to a redefined contract with adjusted coupons.

Before discussing a few common approaches, it is worth pointing out the obvious, but sometimes overlooked, point that trade adjustments (and indeed any other type of adjustments) should be performed for each valuation of the trade — and, in particular, for each re-valuation during risk calculations. With trade adjustments in particular, it is tempting to calculate the adjustments once at trade initiation, and then book an adjusted trade in the booking system. However, even if booked trades are re-adjusted periodically, calculated risk measures would be consistently wrong, as they would *not* include the impact of market parameter shocks on the coupon adjustments.

21.3.1 Fee Adjustments

The additive fee adjustment owns its ease of applicability to the additive property of the pricing operator. Suppose we have in mind a payoff A_n to use in adjusting the coupon C_n, with A_n being paid at the payment date of the coupon, often T_{n+1}. Then we can always find a scaling α_n such that

$$\alpha_n V_{\text{mdl}}(A_n) = V_{\text{mkt}}(C_n) - V_{\text{mdl}}(C_n).$$

We define the adjusted coupon C_n^* as C_n plus the adjustment $\alpha_n A_n$. Then, clearly,

$$V_{\text{mdl}}(C_n^*) = V_{\text{mdl}}(C_n + \alpha_n A_n) = V_{\text{mdl}}(C_n) + \alpha_n V_{\text{mdl}}(A_n)$$
$$= V_{\text{mdl}}(C_n) + V_{\text{mkt}}(C_n) - V_{\text{mdl}}(C_n) = V_{\text{mkt}}(C_n).$$

The adjusted value of H_0 is then given by the value, in the model, of H_0^*, where H_0^* is constructed from the adjusted coupons,

$$V_{\text{adj}}(H_0) = V_{\text{mdl}}(H_0^*).$$

This procedure is called the *fee adjustment method* because $\alpha_n A_n$ could be thought of as an extra "fee" that applies to the coupon C_n.

Fee adjustments require calculating $V_{\text{mdl}}(C_n)$ and $V_{\text{mdl}}(A_n)$, but only once as no iterative search is required. As a consequence, the method is computationally quite efficient even for Monte Carlo based models. Of course, for PDE-based models, the forward PDE valuation of coupons should still be favored over the backward PDE.

The simplest form of fee adjustment is the constant fee adjustment, i.e. using $A_n = 1$, paid at the payment date of the coupon C_n. Then the equation simplifies to be an equation on a scalar f_n such that

$$f_n V_{\text{mdl}}(1) = V_{\text{mkt}}(C_n) - V_{\text{mdl}}(C_n). \qquad (21.23)$$

The adjusted coupon is given by $C_n^* = C_n + f_n$. This specialization is slightly faster than the general case as only $V_{\text{mdl}}(C_n)$ needs to be computed.

Another reasonable choice — at least from a computational complexity standpoint — of the adjustment payoff is the coupon itself, $A_n = C_n$. With this choice one would look for α_n such that

$$\alpha_n V_{\text{mdl}}(C_n) = V_{\text{mkt}}(C_n) - V_{\text{mdl}}(C_n), \qquad (21.24)$$

which requires no more effort than finding a constant f_n in (21.23). The adjusted coupon is then given by

$$C_n^* = (1 + \alpha_n) C_n,$$

i.e. the adjustment has the same shape as the original coupon. Sometimes this is called a *multiplicative adjustment*.

The additive and multiplicative fee adjustment methods could be blended. Choosing w_n, $0 \le w_n \le 1$, one can define the adjusted coupon by

$$C_n^* = C_n + (w_n f_n + (1 - w_n) \alpha_n C_n), \qquad (21.25)$$

where f_n, α_n are given by (21.23), (21.24).

21.3.2 Fee Adjustment Impact on Exotic Derivatives

For different fee adjustment methods, the value of the structured swap underlying a given exotic derivative is invariant, by definition. This is not the case for the exotic derivative itself, as different adjustment methods would assign it different values. Generally, such differences originate with an asymmetric impact of a fee on the price of the exotic derivative. Considering a case of a callable derivative, only the changes to the underlying coupons *in the exercise region* will contribute to the price. Conversely, changes to the underlying coupons in the hold region are irrelevant. However, coupon adjustments are calculated to match the integral of the payoff *over the whole of the state space* to the market value. While the integrals of the adjusted exercise value over the whole state space are therefore independent of the type of payoff adjustment type, the same cannot be said for the integrals over the exercise region. To demonstrate, consider a callable inverse floater, i.e. a Bermudan style option to enter a swap to receive an inverse floating coupon $\max(s - gL_n(T_n), 0)$ and pay Libor $L_n(T_n)$. The underlying swap is a sum of net coupons

$$\max(s - (g+1) L_n(T_n), -L_n(T_n)).$$

The exercise value is represented by the solid line in Figures 21.1 and 21.2. The dotted line represents the adjusted exercise value: an additive adjustment in Figure 21.1 and a multiplicative one in Figure 21.2. While the integral of the difference of the dotted and solid lines is the same in both figures, their integrals over the exercise region, as represented by the grey area, are different. In this case, a multiplicative adjustment will assign a higher value to the callable inverse floater.

Fig. 21.1. Additive Adjustment for CLE

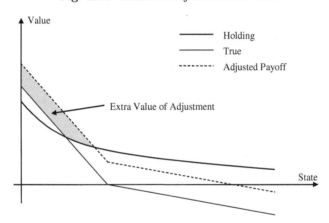

Notes: Effect of the additive fee adjustment on the exercise value of a callable inverse floater. "Holding" denotes the hold value of the callable inverse floater, as a function of the state of the model; "True" denotes the actual payoff of the coupon; and "Adjusted Payoff" represents the payoff adjusted according to the method described.

21.3.3 Strike Adjustment

Many coupon types have a natural "strike" parameter, as a quick recall of the definitions of capped/floored floaters, inverse floaters, etc. in Section 5.13 should confirm. Moreover, the value of a coupon is often a monotonic function of the strike. Denoting the strike by k, the n-th coupon as a function of strike by $C_n(k)$, and the actual value of the strike for the n-th coupon by k_n, we can therefore usually find k_n^* such that

$$V_{\text{mdl}}\left(C_n\left(k_n^*\right)\right) = V_{\text{mkt}}\left(C_n\left(k_n\right)\right) \tag{21.26}$$

for any $n = 1, \ldots, N - 1$. As indicated by the notations, both k_n and k_n^* are coupon-specific, and depend on $n = 1, \ldots, N - 1$. Then, denoting by H_0^* the exotic with the coupon strikes set to k_n^*, $n = 1, \ldots, N - 1$, the adjusted value of the exotic is given by the model price of H_0^*,

Fig. 21.2. Multiplicative Adjustment for CLE

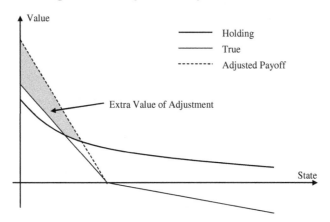

Notes: Effect of the multiplicative fee adjustment on the exercise value of a callable inverse floater. See caption to Figure 21.1 for notations.

$$V_{\text{adj}}\left(H_0\right) = V_{\text{mdl}}\left(H_0^*\right).$$

This procedure is called the *strike adjustment method.*

Other parameters can play the role of the strike in the method. For example, for range accrual coupons, an upper or lower range can be used, as the value of the coupon is monotone in those parameters.

Strike adjustments are more numerically intensive than fee adjustments, since solving (21.26) typically requires multiple calculations of $V_{\text{mdl}}(C_n(s))$ for different values of s. Despite higher computational costs and no discernible theoretical advantage over the fee adjustment method, the strike adjustment method remains popular, perhaps because traders are used to adjusting strikes/barriers for other purposes, such as improvement of risk management of barrier options and adjusting for sampling frequency effects (see e.g. Section 2.5.3, Theorem 3.2.2 and Broadie et al. [1997]).

As with the fee adjustment method, the effect of the strike adjustment on the value of an exotic derivative could be understood by looking at the impact of the adjustments in the relevant part of the state space. Continuing the example from the previous section, Figure 21.3 shows the impact of the strike adjustment on the price of a callable inverse floater.

As a final comment to this chapter, we note that there are undoubtedly many additional ingenious ways of adjusting models, market data and trades that could have been included in this chapter. At the end of the day, however, nothing replaces a good calibration of a well-specified term structure model to the vanilla market. Out-of-model adjustments are useful when applied sparingly, but can easily be abused. For example, it has been rumored that a

Fig. 21.3. Strike Adjustment for CLE

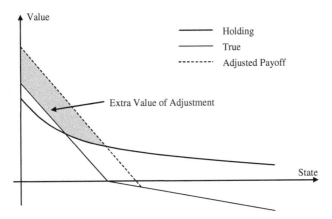

Notes: Effect of the strike adjustment on the exercise value of a callable inverse floater. See caption to Figure 21.1 for notations.

French bank used to risk manage its portfolio of callable CMS spread trades in a one-factor Gaussian model with trade adjustments. Needless to say that this is not something we would recommend. Even in more reasonable applications, the choice of the "right" adjustment could be a delicate exercise and continues to be more art than science.

Part V

Risk Management

22
Introduction to Risk Management

We have reached the point in the book where we are ready to discuss the problem of managing the market risk[1] exposure of interest rate derivatives portfolios. For our purposes here, the topic of primary interest is the quantification and computation of the risk exposure, a task that turns out to be quite challenging and shall require several chapters to cover. First, however, we devote a brief introductory chapter to a high-level overview of the risk management exercise, as practiced by a typical fixed income derivatives trading desk. As part of our analysis, we identify the most common "greeks" (risk sensitivities), and also provide some background on the role these play in hedging and risk management. As we shall see, actual hedging practices tend to deviate considerably from the theoretical ideals of pure delta hedging of Brownian increments (see Section 1.7). We discuss these issues here, and also provide some material on how the risk management and middle office teams in a bank may use market risk exposure information to compute summary statistics for overall risk exposure (the so-called *value-at-risk*), and to perform day-to-day analysis and breakout of portfolio profits and losses. The chapter serves to provide justification for the emphasis on greeks computation in the remainder of the book, and also elaborates on a number of discussions that have cropped up earlier, including Bermudan swaption risk management (Section 19.7.1) and computation of par-point yield curve risk reports (Section 6.4).

---

[1]In addition, derivatives portfolios are exposed to *credit risk*, i.e. the risk that the counterparty to a derivatives transaction will default on its obligations. Management of credit risk is outside the scope of this book.

22.1 Risk Management and Sensitivity Computations

22.1.1 Basic Information Flow

To understand how a trading desk uses a model in practice, it is useful to introduce a bit of notation. Let $\Theta_{mkt}(t)$ be an N_{mkt}-dimensional vector representing the observable market data at time t. For a fixed income desk, the components of $\Theta_{mkt}(t)$ are typically swap and futures rates (for yield curve construction) and cap and swaption prices or implied volatilities at multiple strikes, tenors, and maturities (for volatility calibration). Second, let $\Theta_{prm}(t)$ denote the set of N_{prm} additional parameters that are not directly observed, but are estimated from historical data or are treated as "exotic" constants to be specified directly by the trader. Examples of such parameters include short rate mean reversion parameters, correlation parameterizations, stochastic volatility mean reversion speeds, local volatility parameters (e.g., the CEV power), and so forth. As we have seen in many chapters of this book, the question of how to split $\Theta_{mkt}(t)$ and $\Theta_{prm}(t)$ is often not clear-cut, as one can always attempt to add additional market variables to $\Theta_{mkt}(t)$ to allow us to deduce some of the elements of $\Theta_{prm}(t)$ by direct calibration, in which case these parameters can obviously be removed from $\Theta_{prm}(t)$. As we discussed in Section 14.5.9, one might for instance attempt to eliminate correlation information from $\Theta_{prm}(t)$ by introducing spread option price information into $\Theta_{mkt}(t)$; or one might try to calibrate short rate mean reversion from multiple swaption strips, rather than specify this parameter directly (see for instance Sections 13.1.8.2 and 13.1.8.3). As certain parameters are inherently difficult to extract in a stable and robust manner from market data, in practice it is rarely the case that $\Theta_{prm}(t)$ is completely empty.

Given $\Theta_{mkt}(t)$ and $\Theta_{prm}(t)$, the first step in pricing a derivative security typically involves a calibration procedure, where the vectors $\Theta_{mkt}(t)$ and $\Theta_{prm}(t)$ are turned into a vector[2] of model-appropriate parameters $\Theta_{mdl}(t)$ that contain the discount curve as well as the parameters that control its volatility structure and future dynamics. The calibration may itself require specification of certain control parameters, such as the smoothing weights used in a typical LM model calibration (Section 14.5); for simplicity, we consider these parameters part of $\Theta_{prm}(t)$. The model calibration itself will typically involve at least two steps: the construction of the discount bond curve, followed by calibration of a model for the dynamics of this curve. As the first step can typically be separated completely from the latter, it is informative to break the calibration in two parts, as in Figure 22.1 below. Notice that we here have introduced a pre-processing step where

---

[2]We use the term vector loosely, since some elements of Θ_{prm} (e.g. the discount curve) may be continuous functions.

those elements of $\Theta_{\mathrm{mkt}}(t)$ and $\Theta_{\mathrm{prm}}(t)$ that are relevant[3] for yield curve construction are used to produce a discount bond curve $P(t,T)$, $T \geq t$. Together with the (remaining) elements of $\Theta_{\mathrm{mkt}}(t)$ and $\Theta_{\mathrm{prm}}(t)$, this yield curve is fed to the main model calibration function, which produces $\Theta_{\mathrm{mdl}}(t)$. In any case, we may write

$$\Theta_{\mathrm{mdl}}(t) = C(\Theta_{\mathrm{mkt}}(t); \Theta_{\mathrm{prm}}(t)), \qquad (22.1)$$

where C represents the (overall) calibration function.

Fig. 22.1. Information Flow

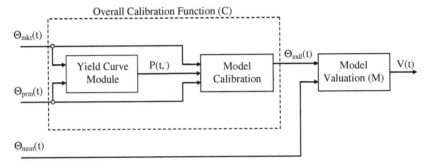

Notes: Basic information flow in derivatives pricing and model calibration.

Given the time t yield curve and a set of model parameters, we can proceed to use the model to price a given portfolio of derivative contracts. This will require us to load contract data for a specified set of securities, and also to read in additional parameters $\Theta_{\mathrm{num}}(t)$ that control the numerical schemes used in the model. Examples of parameters in $\Theta_{\mathrm{num}}(t)$ would include the number of Monte Carlo paths, the size of discretization steps for finite difference grids and for SDE discretization schemes, and so forth. With $V(t) = V_1(t) + \ldots + V_n(t)$ denoting the value of a portfolio of n derivatives, we write (see Figure 22.1),

$$V(t) = M\left(\Theta_{\mathrm{mdl}}(t); \Theta_{\mathrm{num}}(t)\right), \qquad (22.2)$$

for some function M, originating from the expression of arbitrage-free valuation principles through our chosen model. As $\Theta_{\mathrm{mdl}}(t)$ itself originates from $\Theta_{\mathrm{mkt}}(t)$ and $\Theta_{\mathrm{prm}}(t)$, we may, of course, write

---

[3]Recall from Chapter 6 that some yield curve construction algorithms require control parameters (such as tension parameters and precision tolerances), so $\Theta_{\mathrm{prm}}(t)$ may be required for the construction of the discount curve from market inputs.

$$V(t) = H\left(\Theta_{\mathrm{mkt}}(t); \Theta_{\mathrm{prm}}(t), \Theta_{\mathrm{num}}(t)\right), \qquad (22.3)$$

where H is the overall transfer function that translates market data and control parameters into derivatives values.

Finally, let us quickly note that sometimes the calibration function C will be product-specific, i.e. it will depend on the characteristics of the specific security being valued — recall the discussion of "global" versus "local" calibration in Section 14.5.5. For our purposes here, we ignore this additional level of potential inter-connectivity.

22.1.2 Risk: Theory and Practice

According to basic derivatives pricing theory, the function M in (22.2) assigns value based on dynamic hedging and no-arbitrage principles: the price of a derivative security should equal the cost of hedging the security through its lifetime. In doing so, the model will typically rely on idealized assumptions, e.g. that hedging costs are zero, hedging can take place in continuous time, and so forth. These assumptions are, of course, not true in practice, and will require traders to properly charge[4] for the cost of running the hedge, as well as for the fact that the hedge is not truly risk-free. A more subtle issue is the fact that the model will compute value based on an assumption of "infallibility" of the parameter estimates $\Theta_{\mathrm{mdl}}(t)$. In particular, once $\Theta_{\mathrm{mdl}}(t)$ has been established, the underlying model will typically assume that, for any $t' > t$,

$$V(t') = L\left(X(t'); \Theta_{\mathrm{mdl}}(t)\right), \quad t' > t, \qquad (22.4)$$

where $X(t)$ is a random vector of state variables driven by a vector Brownian motion $W(t)$, and L some model-implied map. Our hedging strategy should therefore, as described in Section 1.7, in theory care *only* about neutralizing the effects of movements in X, as caused by W.

In practice, the situation is different. First, actual moves of the yield curve and volatility smiles will inherently deviate from those projected by the model. Second, at time $t' > t$, in reality the model parameter vector $\Theta_{\mathrm{mdl}}(t)$ will be *discarded* and the model calibrated again, followed by an application of (22.2)

---

[4]For simple securities in simple models, it is possible to derive certain analytical results concerning the costs and risks of actual (discrete-time, costly) hedging strategies. A classical paper in this area is Leland [1985], although the approach has recently received some criticism (see, e.g., Kabanov and Safarian [1997]). Other relevant papers include, among many others, Soner et al. [1995], Barles and Soner [1998], and Derman and Kamal [1999]. As all derivatives traders manage risk at the *portfolio* (or *book*) level, the "net" security owned by traders is far more complicated than those covered by most of the literature. In addition, traders tend to rebalance their trading books according to rules much more complex than those assumed in academic papers. As a result, proper charging for transactions normally requires a heavy element of human judgment, and will depend strongly on the portfolio context.

to establish the new value of the portfolio as $V(t') = M(\Theta_{\mathrm{mdl}}(t'); \Theta_{\mathrm{num}}(t'))$. Of course, the recalibrated model parameters $\Theta_{\mathrm{mdl}}(t')$ will rarely, if ever, be consistent with those used at time t, so equation (22.4) will generally fail. As this equation serves as a fundamental assumption of the underlying model, the practical usage of the model is clearly causing quite profound consistency violations. In particular, we constantly change parameters that are assumed by the model to be invariants.

While at first glance the situation outlined above may seem to strike a death blow to the entire foundation of derivatives pricing and hedging, there are several mitigating factors that make the situation less dire than it appears. First, if the model is fundamentally sound, its dynamics will be close to reality most of the time, and (22.2) and (22.4) will consequently be near-identical on average. Second, the trader can employ several strategies to minimize the risk of model mis-specification. One type of strategy involves the use of robust static or super-replicating hedges, as in Sections 16.6.1, 19.4.5 and 20.2.3. As this is not always possible, a more common strategy involves hedging "too much", by neutralizing the portfolio to higher-order sensitivities (gamma hedging), and also by hedging against moves in quantities which are assumed by the model to be non-random. A standard example of the latter is the practice of vega hedging with the Black-Scholes model: despite the fact that the model assumes that the volatility is a constant, the dealer will nevertheless put on a hedge against moves in the volatility parameter. As discussed in Hull [2006] or Taleb [1997], there is empirical evidence that such practices considerably improve the hedge robustness and performance in actual markets.

Vega hedging is an example of the common practice of ignoring the theoretical ideal of (22.4), and instead constructing a hedge around (22.3), with the hedge aiming to neutralize (in a standard Taylor-series sense, see Section 22.1.5) as many of the movements in the entire market data vector $\Theta_{\mathrm{mkt}}(t)$ as possible, irrespective of whether a particular model may suggest that this is reasonable or not. As the dimension of the market data vector N_{mkt} can be very high, often 100 or more, it may be too costly and too onerous to hedge against all components of $\Theta_{\mathrm{mkt}}(t)$ individually, so some type of principal components analysis (as in Section 14.3.1, for instance) may be undertaken to guide the level of granularity required in the hedge. Additionally, one would need to contemplate whether hedges against both first- and second-order (or even higher orders) risk are required. The answer to this question would typically be settled by careful analysis of the convexity properties of the portfolio value V as a function of the market data $\Theta_{\mathrm{mkt}}(t)$: whenever there is significant convexity or concavity with respect to a given parameter, it is reasonable to attempt to put on a second-order hedge. As the (Hessian) matrix of second-order derivatives of V with respect to the components of $\Theta_{\mathrm{mkt}}(t)$ has $N_{\mathrm{mkt}}(N_{\mathrm{mkt}} + 1)/2$ distinct elements (which will often reach thousands), again some selectivity will be required in practice.

The "art of derivatives trading" — that is, the practice of cost-efficiently managing of book of derivatives from market data sensitivity reports — requires considerable and detailed market knowledge and is hard, if not impossible, to describe in purely mathematical terms. Consequently we abstain from attempting to do so, but simply notice that the very foundation of the trading exercise are the market data sensitivities themselves. For readers interested in a description of derivative trading practices, the available material is, unfortunately, rather limited. A common reference is Taleb [1997]; Miron and Swannell [1991] could also be consulted.

22.1.3 Example: the Black-Scholes Model

To make the discussion in Sections 22.1.1 and 22.1.2 more concrete, let us assume that our derivatives portfolio is written on a single underlying asset $X(t)$, the risk-neutral dynamics of which are

$$dX(t)/X(t) = r \, dt + \sigma \, dW(t), \tag{22.5}$$

where r and σ are constants and $W(t)$ is a one-dimensional Brownian motion. We shall shortly (in Section 22.1.4) make the model a bit more realistic by introducing time-dependence into the parameters, but for now we assume that they are constants. In addition, we assume that r and σ are directly observable in the market (again, we relax this shortly), such that $\Theta_{\mathrm{mkt}}(t) = \Theta_{\mathrm{mdl}}(t) = (X(t), r, \sigma)^{\top}$.

The theoretical hedge for a derivative (or a derivative portfolio) V written on X will take the form of a pure delta hedge, where a position $(-\partial V/\partial X)$ in X is maintained at all times. In practice, a trader will not only be concerned with neutralizing against first-order movements in $X(t)$, but will also manage many of the instantaneous sensitivities listed in Table 22.1.

Common Name	Definition
theta	$\partial V/\partial t$
rho	$\partial V/\partial r$
delta	$\partial V/\partial X$
gamma	$\partial^2 V/\partial X^2$
vega	$\partial V/\partial \sigma$
volga (or vomma)	$\partial^2 V/\partial \sigma^2$
vanna	$\partial^2 V/\partial \sigma \partial X$

Table 22.1. Common greeks in the Black-Scholes model.

Neutralization of rho will typically involve taking positions in interest rate swaps, whereas gamma, vega, volga, vanna can only be eliminated by trading derivative securities with non-linear payoffs in X and volatility exposure — typically liquid European options.

We should note that in Table 22.1, the theta (or time decay) has a special status, as the passage of time is both unavoidable and non-random — as such, it makes no sense to try to hedge this "exposure". We also note that the theta, in a sense, emerges as a combination of other greeks, as can be confirmed from the basic Black-Scholes-Merton valuation PDE from Section 1.9:

$$\frac{\partial V}{\partial t} + rX\frac{\partial V}{\partial X} + \frac{1}{2}\sigma^2 X^2 \frac{\partial^2 V}{\partial X^2} = rV \tag{22.6}$$

or, equivalently,

$$\text{theta} + rX \times \text{delta} + \frac{1}{2}\sigma^2 X^2 \times \text{gamma} = r \times \text{value}.$$

Notice, in particular, that a delta hedged position (delta $= 0$), will identify theta $= rV - \frac{1}{2}\sigma^2 X^2 \times$ gamma, so for a delta hedger the time decay of his position originates in part from a pure discount effect due to interest accruing on the net present value of the portfolio (the term rV), and in part from a convexity term (the term $-\frac{1}{2}\sigma^2 X^2 \times$ gamma). The latter, of course, represents optionality leaking away over time, and not surprisingly scales with σ^2.

While traders tend to monitor all entries in Table 22.1, particular attention is typically paid to the delta, gamma and vega exposures. Hedging the vega ensures that inevitable changes to implied volatility will not lead to large moves in the portfolio value, and hedging the gamma helps prevent "slippage" in between dates where the delta hedge is rebalanced. As discussed already in Section 19.7.1, most traders strongly prefer being long gamma exposure (i.e. the net gamma of their portfolio, including hedges, is positive), as a negative gamma position can lose a very substantial amount of money in periods of financial turmoil when the trader cannot adjust his delta hedge quickly enough to track the market.

One would intuitively expect that a portfolio that has low gamma exposure would also have a low vega, a result that in fact can be formalized as follows:

Lemma 22.1.1. *Consider a European-style claim with maturity T and payout function $g(X(T))$, where $X(t)$ satisfies (22.5). With $V(t) = V(t, X(t))$ denoting time t value of this security, we have (employing somewhat loose notation)*

$$\frac{\partial V(t)}{\partial \sigma} = (T-t)X(t)^2\sigma\frac{\partial^2 V(t)}{\partial X(t)^2}, \quad \text{or} \quad \text{vega} = (T-t)X(t)^2\sigma \times \text{gamma}. \tag{22.7}$$

Proof. An elementary, if rather inelegant, proof proceeds as follows. First, we let E_t denote time t risk-neutral expectation, such that

$$V(t) = e^{-r(T-t)}E_t\left(g\left(X(T)\right)\right),$$

where

$$X(T) = X(t) \exp\left(\left(r - \frac{1}{2}\sigma^2\right)(T - t) + \sigma\sqrt{T - t}Z\right),$$

with $Z \sim \mathcal{N}(0, 1)$. Notice that

$$\frac{\partial V(t)}{\partial X(t)} = e^{-r(T-t)}\mathrm{E}_t\left(\frac{\partial g\left(X(T)\right)}{\partial X(T)} \cdot \frac{X(T)}{X(t)}\right), \tag{22.8}$$

and

$$\frac{\partial X(T)}{\partial t} = -X(T)\left(r - \frac{1}{2}\sigma^2 + \frac{\sigma}{2\sqrt{T - t}}Z\right),$$

such that

$$\frac{\partial V(t)}{\partial t} = re^{-r(T-t)}\mathrm{E}_t\left(g\left(X(T)\right)\right) + e^{-r(T-t)}\mathrm{E}_t\left(\frac{\partial g\left(X(T)\right)}{\partial X(T)} \cdot \frac{\partial X(T)}{\partial t}\right)$$

$$= rV(t) - e^{-r(T-t)}\mathrm{E}_t\left(\frac{\partial g\left(X(T)\right)}{\partial X(T)}X(T)\left(r - \frac{1}{2}\sigma^2 + \frac{\sigma}{2\sqrt{T - t}}Z\right)\right)$$

$$= rV(t) - X(t)\frac{\partial V(t)}{\partial X(t)}\left(r - \frac{1}{2}\sigma^2\right)$$

$$- \frac{\sigma}{2\sqrt{T - t}}e^{-r(T-t)}\mathrm{E}_t\left(\frac{\partial g\left(X(T)\right)}{\partial X(T)} \cdot X(T)Z\right), \tag{22.9}$$

where we have used (22.8) in the last equality. Using the same principles, we get

$$\frac{\partial V(t)}{\partial \sigma} = e^{-r(T-t)}\mathrm{E}_t\left(\frac{\partial g\left(X(T)\right)}{\partial X(T)} \cdot \frac{\partial X(T)}{\partial \sigma}\right)$$

$$= e^{-r(T-t)}\mathrm{E}_t\left(\frac{\partial g\left(X(T)\right)}{\partial X(T)} \cdot X(T)\left(-\sigma(T - t) + \sqrt{T - t}Z\right)\right)$$

$$= -X(t)\frac{\partial V(t)}{\partial X(t)}\sigma(T - t) + e^{-r(T-t)}\sqrt{T - t}\mathrm{E}_t\left(\frac{\partial g\left(X(T)\right)}{\partial X(T)}X(T)Z\right). \tag{22.10}$$

Combining (22.9) and (22.10), we get

$$\frac{\partial V(t)}{\partial \sigma} = -X(t)\frac{\partial V(t)}{\partial X(t)}\sigma(T - t)$$

$$+ \frac{2(T - t)}{\sigma}\left(rV(t) - \frac{\partial V(t)}{\partial t} - X(t)\frac{\partial V(t)}{\partial X(t)}\left(r - \frac{1}{2}\sigma^2\right)\right)$$

$$= \frac{2(T - t)}{\sigma}\left(rV(t) - \frac{\partial V(t)}{\partial t} - rX(t)\frac{\partial V(t)}{\partial X(t)}\right).$$

The result (22.7) follows after insertion of the pricing PDE (22.6) into the expression above. □

Remark 22.1.2. A more elegant way to prove Lemma 22.1.1 relies on operator calculus, as in Carr [2000]. As this technique is sometimes quite handy, we demonstrate it in Appendix 22.A, where it is also shown that, for a general European-style claim,

$$\frac{\partial V(t)}{\partial r} = (T - t)\left(X(t)\frac{\partial V(t)}{\partial X(t)} - V(t)\right).$$

While Lemma 22.1.1 only holds for European-style claims in the Black-Scholes model, the observation that there is a close link between vega and gamma holds in general. We shall examine the importance of vega and gamma hedging in more detail shortly, but first let us make the model setting a bit more realistic.

22.1.4 Example: Black-Scholes Model with Time-Dependent Parameters

In actual usage of the Black-Scholes model, one would always allow the short rate and volatility to be time-dependent, in order to match observed term structures of discount bonds and (term) option volatilities. While technically an obvious extension to Section 22.1.3, let us go over the mechanics nevertheless to better illustrate the concepts from Sections 22.1.1 and 22.1.2. We first change (22.5) to

$$dX(t)/X(t) = r(t)\,dt + \sigma(t)\,dW(t),$$

where $r(t)$ and $\sigma(t)$ are deterministic functions of time, to be calibrated to the term structure of discount bonds and implied at-the-money (ATM) volatilities observed in the market. We assume that the yield curve is computed from a vector of swap yields $S(t) = (S_1(t), \ldots, S_J(t))^\top$, where it is understood that $S_i(t)$ represents the time-t par yield of a swap that matures on some date $T_i > t$, $i = 1, \ldots, J$. For simplicity, let us assume, as in Section 6.2.1, that the constructed yield curve is bootstrapped as a piecewise flat curve in forward rate space, with breakpoints located at the swaption maturities $\{T_i\}$. Defining $T_0 \triangleq t$, the resulting time t forward curve may therefore be represented as

$$f(t, u) = -\frac{\partial \ln P(t, u)}{\partial u} = \sum_{i=0}^{J-1} \gamma_i(t)\mathbf{1}_{\{u \in [T_i, T_{i+1})\}},$$

where we use the vector $\gamma(t) = (\gamma_1(t), \ldots, \gamma_J(t))^\top$ to store the resulting J forward curve levels. As r is assumed deterministic, we set $r(u) = f(t, u)$, $u \geq t$, and have then completed the interest rate calibration.

To construct $\sigma(t)$, assume that a vector $v(t) = (v_1(t), \ldots, v_D(t))^\top$ is observed in the market, where each $v_i(t)$ represents a term volatility to t_i on a specified maturity grid $\{t_i\}_{i=1}^D$, with $t_1 > t$. That is,

$$v_i(t) = \sigma_{\text{ATM}}(t, t_i),$$

where $\sigma_{\text{ATM}}(t, t_i) \triangleq \sigma_{\text{BS}}(t, X(t); t_i, X(t))$ is the ATM implied (Black-Scholes) volatility to time t_i, seen from time t (see Section 7.1.2). If we assume that $\sigma(u)$ is piecewise flat on the $\{t_i\}$-grid, we can construct $\sigma(u)$ by bootstrapping from the basic relation (see Section 1.9.3)

$$\sigma_{\text{ATM}}(t, u)^2 = (u - t)^{-1} \int_t^u \sigma(s)^2 \, ds. \tag{22.11}$$

The result of this exercise is a vector $\varsigma(t) = (\varsigma_1(t), \ldots, \varsigma_D(t))^\top$ of flat volatility levels, such that

$$\sigma(u) = \sum_{i=0}^{D-1} \varsigma_i(t) 1_{\{u \in [t_i, t_{i+1})\}},$$

where we set $t_0 \triangleq t$.

The calibration procedure described above turns $\Theta_{\text{mkt}}(t) = (S(t), v(t))^\top$ into $\Theta_{\text{mdl}}(t) = (\gamma(t), \varsigma(t))^\top$, with both vectors having dimension $J + D$, i.e. $N_{\text{mkt}} = N_{\text{prm}} = J + D$. The vector $\Theta_{\text{prm}}(t)$ is here empty, but would have contained precision parameters if, say, the more elaborate yield curve construction algorithm of Section 6.3 had been used. The contents of the vector $\Theta_{\text{num}}(t)$ would depend on what numerical method the calibrated model would implement for the purpose of pricing a specific derivative. For instance, if we were using a finite difference grid, $\Theta_{\text{num}}(t)$ might contain a confidence level multiplier (see Section 2.1) to dimension the grid, a θ-parameter to determine the level of implicitness in the solver (see Section 2.2.3), various flags (e.g. whether to use upwinding or not, see Section 2.6.1) and, of course, information to determine the number of grid points in the time (t) and space (X) directions.

As r and σ are now vector-valued, the quantities rho, vega, volga, and vanna in Table 22.1 are no longer scalars[5] but must be represented as vectors. Also, one issue is how we wish to present this risk information in the first place: for vega, say, do we want to report sensitivities with respect to the market volatilities v (so-called *market vegas*) or the model volatilities ς (so-called *model vegas*)? The former is the most common, but we can here freely translate between the two. Specifically, applying the chain rule to the relationship between v_i's and ς_j's given by (22.11) we have that

$$\frac{\partial}{\partial v_i} = \sum_{j=1}^D \frac{\partial}{\partial \varsigma_j} \left(\frac{\partial \varsigma_j}{\partial v_i} \right),$$

---

[5]If, say, a single vega is nevertheless required, it is often most reasonable to report the sensitivity to a parallel shift of the function $\sigma(t)$ at all values of t. We can think of this as roughly representing the sensitivity with respect to the first principal component of volatility curve moves. A similar principle can be applied to compute a single rho.

where the matrix of partial derivatives $\partial \varsigma / \partial v$ can be obtained by inverting the Jacobian matrix $\partial v / \partial \varsigma$ which can be obtained in closed form from the relation (22.11). A similar translation between sensitivities to swap yields and to forward rate buckets was discussed already, in Section 6.4.

22.1.5 Actual Risk Computations

Getting back to the general representation (22.3), assume that we perturb the market data by a vector-valued amount $\delta = (\delta_1, \ldots, \delta_{N_{\mathrm{mkt}}})^{\top}$. Let us use a Taylor expansion to write (dropping the argument t for clarity)

$$V(\delta) = H\left(\Theta_{\mathrm{mkt}} + \delta\right) \approx H\left(\Theta_{\mathrm{mkt}}\right) + \nabla^{H} \cdot \delta + \frac{1}{2}\delta^{\top} \cdot A^{H} \cdot \delta, \qquad (22.12)$$

where ∇^{H} is an N_{mkt}-dimensional row vector and A^{H} an $N_{\mathrm{mkt}} \times N_{\mathrm{mkt}}$ matrix, with elements

$$\nabla^{H}_{i} = \left.\frac{\partial H(x)}{\partial x_i}\right|_{x=\Theta_{\mathrm{mkt}}} \quad, \quad A^{H}_{i,j} = \left.\frac{\partial^2 H(x)}{\partial x_i \partial x_j}\right|_{x=\Theta_{\mathrm{mkt}}} \quad, \quad i,j = 1, \ldots, N_{\mathrm{mkt}}.$$

While the situation in interest rate modeling is obviously a little more complicated than for the single-asset Black-Scholes setup in Sections 22.1.3 and 22.1.4, loosely speaking the gradient vector ∇^{H} will contain deltas (first-order sensitivities with respect to swap yields) and vegas (first-order sensitivities with respect to swaption volatilities), while the Hessian matrix A^{H} will contain gammas, volgas, and vannas. Notice that the risk measure rho is not used for interest rate derivatives (where, in a sense, delta and rho coincide).

Although not all elements of A are always requested, in a nutshell the main role of an interest rates derivatives risk system is to report[6] ∇^{H} and A^{H} for consumption by the trading desk, risk management, and the middle office. The trading desk will, as we discussed earlier in Section 22.1.2, use the sensitivities to evaluate how much it should rebalance the portfolio to keep it broadly market-neutral and robust to market shocks; the "ideal" configuration for a pure[7] hedger will obviously be to arrange the portfolio such that $\nabla^{H} = 0$ and $A^{H} = 0$. Risk management will use the sensitivities to ensure that the exposures to individual market data components are within given sensitivity *limits*. In addition, risk management will typically

---

[6]Sometimes the risk system is asked to perform the Taylor expansion around a series of different (perturbed) market data scenarios, not just the current market data. The resulting collections of reports are known as *ladders*.

[7]Most derivatives traders have views on the future market evolution and are allowed to express their views in proprietary ("prop") positions, meant to make money if the trader's views turn out to be correct. In this situation, the trader will purposely leave the portfolio open to certain market risk exposures. Banks typically enforce strict limits on the size of these exposures.

compute an overall measure of the portfolio risk, based on a statistical (or historical) model for the perturbation vector δ over a given time horizon, typically one day. We discuss this computation briefly in Section 22.3. In the middle office of a bank, the sensitivities are used for P&L *analysis*, i.e. the process of reconciling observed moves in portfolio value with changes in market data. We review this process in Section 22.2.

Recall that market data input used to compute the value of a derivative security typically goes through a two-step procedure, where first a calibration turns the market data vector Θ_{mkt} into a model data vector Θ_{mdl}, which in turn is used to compute the derivatives price, $V = M(\Theta_{\mathrm{mdl}})$. It is often natural to first compute sensitivities with respect to the model parameters — a process that is independent of the chosen calibration procedure — and then combine these sensitivities with calibration-specific sensitivity information to compute the market data input sensitivities. For instance, we would write, for $j = 1, \ldots, N_{\mathrm{mkt}}$,

$$
\nabla_j^H = \sum_{i=1}^{N_{\mathrm{mdl}}} \frac{\partial M(y)}{\partial y_i} \frac{\partial C_i(x)}{\partial x_j}\bigg|_{y=\Theta_{\mathrm{mdl}}, x=\Theta_{\mathrm{mkt}}},
$$

where C_i is the i-th component of the N_{mdl}-dimensional calibration function C in (22.1). The $N_{\mathrm{mdl}} \times N_{\mathrm{mkt}}$ matrix J with elements

$$
J_{i,j} = \frac{\partial C_i}{\partial x_j}
$$

is known as the *Jacobian* for the map from market to model parameters. The Jacobian can normally be computed in the calibration module, as part of the calibration procedure itself. We saw a simple example of Jacobian matrix usage in Section 22.1.4 above, and will consider the idea in a more realistic setting in Section 26.3.3.

22.1.6 What about Θ_{prm} and Θ_{num}?

The reader will have noticed that the portfolio value depends not only on market data, but also on various technical parameters that control numerics and the calibration, as well as certain unobservable model parameters. This type of data is fairly static, and sensitivity information is rarely reported on a running basis. As the numbers do affect the official profit-and-loss (P&L) produced by the model, the elements of Θ_{prm} and Θ_{num} are typically supervised by control groups that may impose standards on numerical parameters (e.g. require that the numerical error be within a certain tolerances) and may request that monetary buffers — so-called *reserves* — be set aside to cover the uncertainty of unobservable model parameters. The latter will require some estimates of the uncertainty associated with a given parameter, as well as a computation of the portfolio value sensitivity to the parameters.

While the reserves need to be dynamically updated to reflect changes in the portfolio and in the parameter sensitivities, this is normally done relatively infrequently, e.g. every month or quarter. Given this, from a computational perspective sensitivity generation with respect to market data Θ_{mkt} — which is often done on an inter-day basis — is, by far, the more challenging task. As such, our emphasis in the rest of this book is solely on computation of market data sensitivities.

22.1.7 A Note on Trading P&L and the Computation of Implied Volatility

Before proceeding to discuss applications of the sensitivity analysis of Section 22.1.5, we insert a brief interlude to demonstrate an important result (sometimes known as the *fundamental theorem of derivatives trading*) that provides a link between a portfolio's gamma and expected hedging P&L over a given horizon. The setup is as follows. At time 0 a trader buys a contingent claim on a single non-dividend paying asset X, and chooses to value his position by using a Black-Scholes model with fixed volatility σ_{BS}. Let the trader's mark for his portfolio be $V_{\text{BS}}(t)$ and assume that σ_{BS} is such that the value $V_{\text{BS}}(0)$ coincides with the time 0 market value. We assume that the contingent claim expires at time T with value $g(X(T))$, and pays no cash flows before then. The trader is actively hedging his position, but commits two "sins": i) he does not gamma or vega hedge his position, but only delta hedges; and ii) he never re-calibrates the model but assumes that σ_{BS} is the correct volatility to use when computing hedge information, even if the volatility of X is observed to change over time.

In analyzing the performance of the hedger's strategy, let us assume that the volatility of X is a random process $\sigma(t)$, i.e. that dynamics in the real-life measure P are of the form

$$dX(t)/X(t) = O(dt) + \sigma(t)\,dW^P(t),$$

where $W^P(t)$ is a P-Brownian motion. Now, to hedge his long position in the contingent claim, the trader sets up a short position in a portfolio Π with $n_X(t)$ units of X held at time t, along with a cash position $N(t)$. As described above, the trader delta hedges according to the Black-Scholes model, so

$$n_X(t) = \frac{\partial V_{\text{BS}}(t)}{\partial X(t)}.$$

In other words, we have

$$\Pi(t) = n_X(t)X(t) + N(t),$$

where, by construction, $\Pi(0) = V_{\text{BS}}(0)$. Assuming that $N(t)$ is rolled over at the short-term interest rate r (assumed constant for convenience), we therefore get

$$d\Pi(t) = n_X(t)\,dX(t) + N(t)r\,dt, \tag{22.13}$$

where the self-financing condition (see (1.10)) justifies ignoring the change in n_X at $t + dt$. The following important result now holds.

Proposition 22.1.3. *The time T terminal value of the delta hedging account in (22.13) is*

$$\Pi(T) = g(X(T)) + \frac{1}{2}e^{rT}\int_0^T e^{-rt}\left(\sigma_{\mathrm{BS}}^2 - \sigma(t)^2\right)X(t)^2\frac{\partial^2 V_{\mathrm{BS}}(t)}{\partial X(t)^2}\,dt,$$

where $g(x)$ is the terminal payout function.

Proof. By Ito's lemma, observe that

$$dV_{\mathrm{BS}}(t) = \frac{\partial V_{\mathrm{BS}}(t)}{\partial t}\,dt + \frac{\partial V_{\mathrm{BS}}(t)}{\partial X(t)}\,dX(t) + \frac{1}{2}\sigma(t)^2 X(t)^2\frac{\partial^2 V_{\mathrm{BS}}(t)}{\partial X(t)^2}\,dt, \tag{22.14}$$

so combining (22.13) and (22.14) yields

$$\begin{aligned}
d(V_{\mathrm{BS}}(t) - \Pi(t)) &= \frac{\partial V_{\mathrm{BS}}(t)}{\partial t}\,dt + \frac{1}{2}\sigma(t)^2 X(t)^2\frac{\partial^2 V_{\mathrm{BS}}(t)}{\partial X(t)^2}\,dt - N(t)r\,dt \\
&= \frac{\partial V_{\mathrm{BS}}(t)}{\partial t}\,dt + \frac{1}{2}\sigma(t)^2 X(t)^2\frac{\partial^2 V_{\mathrm{BS}}(t)}{\partial X(t)^2}\,dt \\
&\quad - r\left(\Pi(t) - \frac{\partial V_{\mathrm{BS}}(t)}{\partial X(t)}X(t)\right)dt. \tag{22.15}
\end{aligned}$$

We now recall that $V_{\mathrm{BS}}(t)$ satisfies the Black-Scholes PDE with constant volatility σ_{BS}, wherefore

$$\frac{\partial V_{\mathrm{BS}}(t)}{\partial t} + rX(t)\frac{\partial V_{\mathrm{BS}}(t)}{\partial X(t)} + \frac{1}{2}\sigma_{\mathrm{BS}}^2 X(t)^2\frac{\partial^2 V_{\mathrm{BS}}(t)}{\partial X(t)^2} = rV_{\mathrm{BS}}(t).$$

Inserting this expression into (22.15) yields, after a little algebra,

$$\begin{aligned}
d(V_{\mathrm{BS}}(t) &- \Pi(t)) \\
&= \frac{1}{2}(\sigma(t)^2 - \sigma_{\mathrm{BS}}^2)X(t)^2\frac{\partial^2 V_{\mathrm{BS}}(t)}{\partial X(t)^2}\,dt + r(V_{\mathrm{BS}}(t) - \Pi(t))\,dt.
\end{aligned}$$

We can integrate this equation to yield

$$\begin{aligned}
V_{\mathrm{BS}}(T) - \Pi(T) &= e^{rT}(V_{\mathrm{BS}}(0) - \Pi(0)) \\
&\quad + e^{rT}\int_0^T e^{-rt}\frac{1}{2}\left(\sigma(t)^2 - \sigma_{\mathrm{BS}}^2\right)X(t)^2\frac{\partial^2 V_{\mathrm{BS}}(t)}{\partial X(t)^2}\,dt,
\end{aligned}$$

and the result of the proposition follows from the observation that $V_{\mathrm{BS}}(T) = g(X(T))$ and $V_{\mathrm{BS}}(0) = \Pi(0)$. $\square$

Proposition 22.1.3 demonstrates that the hedging strategy followed by the trader generally does not work, in the sense that the terminal value $\Pi(T)$ of the self-financing hedging portfolio will fail to equal $g(X(T))$. In certain special cases, however, the hedge will work, e.g. when the hedged claim is gamma-neutral $(\partial^2 V_{\mathrm{BS}}/\partial X^2 = 0)$ or when $\sigma(t)$ is close to σ_{BS} "on average". These observations, while trivial, strongly support the strategy of re-calibrating the model to changing market conditions and to use gamma-hedging to keep portfolio convexity low.

As an aside, we notice that if i) the Black-Scholes gamma is strictly positive, and ii) the realized volatility $\sigma(t)$ is consistently higher than σ_{BS}, we have $\Pi(T) < g(X(T))$ for sure. As the trader is short the hedging portfolio, it follows that the trader keeping a positive gamma benefits from financial turmoil (high volatility), a point we have made several times already.

Finally, let us present an important corollary of Proposition 22.1.3.

Corollary 22.1.4. *Let the claim in Proposition 22.1.3 be a European call or put option with strike K. The time 0 implied volatility $\sigma_{\mathrm{BS}} = \sigma_{\mathrm{BS}}(0, X(0); T, K)$ is given by*

$$\sigma_{\mathrm{BS}}^2 = \frac{\mathrm{E}\left(\int_0^T e^{-rt}\sigma(t)^2 X(t)^2 \frac{\partial^2 V_{\mathrm{BS}}(t)}{\partial X(t)^2}\, dt\right)}{\mathrm{E}\left(\int_0^T e^{-rt} X(t)^2 \frac{\partial^2 V_{\mathrm{BS}}(t)}{\partial X(t)^2}\, dt\right)}, \tag{22.16}$$

where E *denotes expectation in the risk-neutral measure, and*

$$X(t)^2 \frac{\partial^2 V_{\mathrm{BS}}(t)}{\partial X(t)^2} = X(t)\frac{\phi\left(d_+(X(t))\right)}{\sigma_{\mathrm{BS}}\sqrt{T-t}}, \tag{22.17}$$

$$d_+(x) = \frac{\ln(x/K) + \left(r + \frac{1}{2}\sigma_{\mathrm{BS}}^2\right)(T-t)}{\sigma_{\mathrm{BS}}\sqrt{T-t}},$$

where $\phi(x) = (2\pi)^{-1/2}e^{-x^2/2}$ *is the Gaussian density.*

Proof. The hedge portfolio generates no cash flows on $[0, T]$, so its time 0 value must equal

$$\Pi(0) = e^{-rT}\mathrm{E}\left(\Pi(T)\right) = e^{-rT}\mathrm{E}\left(g\left(X(T)\right)\right)$$

$$+ \mathrm{E}\left(\int_0^T e^{-rt}\frac{1}{2}\left(\sigma(t)^2 - \sigma_{\mathrm{BS}}^2\right)X(t)^2\frac{\partial^2 V_{\mathrm{BS}}(t)}{\partial X(t)^2}\, dt\right),$$

where $g(x) = (x - K)^+$ for a call and $g(x) = (K - x)^+$ for a put. Here, the term

$$e^{-rT}\mathrm{E}\left(g\left(X(T)\right)\right)$$

equals the time 0 market value of the put or call being hedged, and equals $\Pi(0)$ by assumption. It follows that

$$\mathrm{E}\left(\int_0^T e^{-rt}\frac{1}{2}\left(\sigma(t)^2 - \sigma_{\mathrm{BS}}^2\right) X(t)^2 \frac{\partial^2 V_{\mathrm{BS}}(t)}{\partial X(t)^2}\, dt\right) = 0,$$

which immediately leads to (22.16). The result (22.17) follows from an explicit evaluation of gamma in the Black-Scholes model.   □

To get some insights into the result of the corollary above, assume that $\sigma(t)$ is of the local volatility type, $\sigma(t) = \sigma(X(t))$, in which case (22.16) can be written

$$\sigma_{\mathrm{BS}}(T, K)^2 = \frac{\int_0^\infty \int_0^T e^{-rt}\sigma(x)^2 w(t, x)\psi(t, x)\, dt\, dx}{\int_0^\infty \int_0^T e^{-rt}w(t, x)\psi(t, x)\, dt\, dx}, \qquad (22.18)$$

where $\psi(t, x)$ is the density of $X(t)$ as seen from time 0, and

$$w(t, x) = x\frac{\phi\left(d_+(x)\right)}{\sigma_{\mathrm{BS}}(T, K)\sqrt{T - t}}.$$

In a sense, (22.18) demonstrates that implied volatility is a weighted average of local volatility, where weights are proportional to the product of gamma and the asset density.

Direct usage of (22.18) is complicated by the fact that the implied volatility figures in both the left- and right-hand sides of the equation, and by the fact that the density of the asset is rarely, if ever, known explicitly. On the other hand, the product $w(t, x)\psi(t, x)$ can be seen to typically form a "ridge" from $x = X(0)$ at time 0 to $x = K$ at time T, a result that holds irrespective of the model specification. This, among other considerations, has inspired some authors to suggest that

$$\sigma_{\mathrm{BS}}(0, X(0); T, K)^2 \approx \frac{1}{T}\int_0^T \mathrm{E}\left(\sigma(t)^2 | X(t) = x^*(t)\right)\, dt,$$

where $x^*(t)$ is to be interpreted as *the most likely path from $X(0)$ to K*. This idea has found applications for both local and stochastic volatility models, see e.g. Gatheral [2006]. While often quite intuitive, approximation techniques based on (22.16) involve a fair amount of heuristics[8], and precision is often neither impressive nor easy to characterize. As a result, the method — which we believe was originally suggested by Bruno Dupire — is often reserved for qualitative analysis. See Lee [2005] and Gatheral [2006] for additional discussion and applications.

22.2 P&L Analysis

Besides being used by traders to manage the exposure of their books, the sensitivity information contained in Taylor expansions such as (22.12) is

---

[8]In a pinch, it is often reasonable to simply assume that $x^*(t) = T^{-1}(X(0)(T - t) + Kt)$.

consumed by various support and control functions in a bank. We discuss two such uses: P&L analysis and, in Section 22.3 below, value-at-risk computation.

22.2.1 P&L Predict

The expansion (22.12) may be used in an accounting analysis to analyze and reconcile the realized P&L from one trading day to the next. Although the analysis is carried out at time $t + h$ (where the market data movement δ is known), it is known as a *P&L prediction analysis*, or just a *P&L predict*. Given, at time t, expansion terms $\nabla^H(t)$ and $A^H(t)$, if the observed market data movements over the period $[t, t + h]$ (with h typically equal to one business day) is δ, then, all things equal, we would expect the time $t + h$ portfolio value to be approximately

$$V(t + h) \approx V(t) + \frac{\partial V(t)}{\partial t} h + \nabla^H(t) \cdot \delta + \frac{1}{2} \delta^\top \cdot A^H(t) \cdot \delta. \qquad (22.19)$$

Notice the inclusion of the term $\partial V / \partial t$ (theta) in this expansion, to account for the passage of time. The right-hand side of this equation is known as the *second-order P&L predict*; if we omit the convexity term (i.e. set $A^H(t) = 0$), the right-hand side is, naturally, the *first-order P&L predict*. The difference between the right- and left-hand sides of (22.19) may be called the *unpredicted* P&L. If systems and models are working properly, the P&L predict should generally be an accurate and unbiased estimated of actual P&L, so monitoring of the unpredicted P&L serves an important control purpose. Unusually large values of unpredicted P&L may, for instance, hint at problems in the computation of risk sensitivities (and therefore in hedges) or suggest that the portfolio is exposed to large unhedged high-order risks.

We should note that when writing down (22.19), we implicitly made several simplifying assumptions, most notably that the portfolio at time t is the same as at time $t + h$. In reality, trades may expire, get canceled or amended, or entirely new trades may be added to the portfolio on the interval $(t, t + h]$. In addition, cash payments (coupons and settlement amounts) may take place on $(t, t+h]$ and must be added to the left-hand side of (22.19). The function V in (22.19) should therefore really be thought of as representing the part of the portfolio trade population that involves no special events over the period $[t, t + h]$; a full P&L predict analysis will additionally require accounting for a number of adjustments due to cash payments, changes to the portfolio, and rate fixings. Getting all details right is often a fairly complex exercise, and as mentioned earlier is normally handled by dedicated personnel in a bank's middle office.

An important issue in P&L predict (and also in P&L explain, see Section 22.2.2 below) is the computation of the theta term in (22.19). In particular, when advancing time forward, what precisely is it that we should hold fixed?

While one might say that, by definition, all elements of $\Theta_{\mathrm{mkt}}(t)$ should stay at their time t values, this generally causes problems. To demonstrate, assume, as is common, that the yield curve is constructed from a series of Eurodollar contracts, and swap quotes. As discussed in Section 5.4, a Eurodollar futures contract will settle at a fixed point in time — i.e. it has a fixed time *of* maturity — so when advancing time from t to $t + h$, its remaining time *to* maturity will shrink by an amount h. On the other hand, a market-quoted swap always is associated with a standardized time *to* maturity (a fixed swap tenor), so no maturity shrinkage occurs when time is advanced. In total, when advancing time forward, the time to maturity of some, but not all, yield curve instruments will undergo a change. This is not compensated for by a change in the market quote (which is held fixed), which results in an effective move in the forward curve that typically will be highly erratic and entirely unsuitable for a perturbation analysis. To avoid problems of this type, and to properly reflect short-term funding costs, it is natural to compute the expected change in market data

$$\Theta^f_{\mathrm{mkt}}(t) = \mathrm{E}^{t+h}\left(\Theta_{\mathrm{mkt}}(t+h)|\mathcal{F}_t\right), \tag{22.20}$$

and then write

$$\frac{\partial V(t)}{\partial t} \approx \frac{V(t+h; \Theta^f_{\mathrm{mkt}}(t)) - V(t)}{h}. \tag{22.21}$$

The term $V(t + h; \Theta^f_{\mathrm{mkt}}(t))$ may be computed by an outright re-valuation of the portfolio after i) advancing calendar time to $t + h$; and ii) moving the market data to $\Theta^f_{\mathrm{mkt}}(t)$. In computation of $\Theta^f_{\mathrm{mkt}}(t)$, the discount curve constructed at time t will simply be "rolled" up to its time $t + h$ forward curve as seen from time t, i.e. for any $T > t + h$ we set

$$P(t+h, T) = \mathrm{E}^{t+h}\left(P(t, T)|\mathcal{F}_t\right) = \frac{P(t, T)}{P(t, t+h)} \approx P(t, T)\left(1 + r(t)h\right),$$

where $r(t)$ is the short rate. Moving the discount curve in this fashion is consistent with the notion that a risk-free portfolio should earn a rate of $r(t)$ over a short holding period and rationally anchors the P&L predict analysis around forward values of discount bonds, i.e. values that can be locked in by a risk-free trading strategy at time t.

With the choice (22.21) for theta, we have, in effect, moved the expansion point of the Taylor series (22.19) from $\Theta_{\mathrm{mkt}}(t)$ to $\Theta^f_{\mathrm{mkt}}(t)$, which suggest a modified (and improved) expression for the P&L predict:

$$V(t+h) \approx V(t+h; \Theta^f_{\mathrm{mkt}}(t)) + \nabla^H(t) \cdot (\delta - \delta^f) + \frac{1}{2}(\delta - \delta^f)^\top \cdot A^H(t) \cdot (\delta - \delta^f), \tag{22.22}$$

with

$$\delta^f \triangleq \Theta^f_{\mathrm{mkt}}(t) - \Theta_{\mathrm{mkt}}(t). \tag{22.23}$$

In (22.22), the term $V(t+h; \Theta^f_{\mathrm{mkt}}(t))$ represents the value that the portfolio will reach if the market data moves "according to expectations", and remaining terms add first- and second-order corrections based on the deviation of the time $t+h$ market data away from its time t expectation,

$$\delta - \delta^f = \Theta_{\mathrm{mkt}}(t+h) - \Theta_{\mathrm{mkt}}(t) - (\Theta^f_{\mathrm{mkt}}(t) - \Theta_{\mathrm{mkt}}(t)) = \Theta_{\mathrm{mkt}}(t+h) - \Theta^f_{\mathrm{mkt}}(t).$$

For the reasons explained earlier, (22.22) is typically preferable to (22.19), yet it is not uncommon for P&L analysis systems to implement both.

Finally, let us note that in our description of the P&L predict process we assumed that interest rate risk was captured as sensitivities with respect to market quotes of yield curve instruments, i.e. we start from a par-point report, in the convention of Section 6.4. As discussed in that section, it is, however, not uncommon to instead capture interest rate risk through sensitivities with respect to buckets of the forward curve itself (a forward rate report). This change in approach is easily accommodated by the methodology above, by simply altering the definitions of $\Theta_{\mathrm{mkt}}(t)$, δ, and δ^f accordingly.

22.2.2 P&L Explain

The objective of a *P&L explain*[9] analysis is to estimate the contribution of each component of the market vector move δ to the overall move in the portfolio value. In a sense, such information is also captured in the P&L predict (through the sensitivities ∇^H and, perhaps, A^H), but the P&L explain analysis does away with Taylor expansions and instead relies on brute-force bumping of market data. As was the case for the P&L predict, the explain analysis is carried out at time $t+h$ when the market data movement δ is known.

22.2.2.1 Waterfall Explain

In one type of P&L explain — a so-called *waterfall explain* — the impact of the i-th component of δ is basically captured as

$$
\begin{aligned}
E_i = \; & V\left(t+h; \Theta_{\mathrm{mkt}}(t) + (\delta_1, \delta_2, \ldots, \delta_i, 0, 0, \ldots, 0)^\top\right) \\
& - V\left(t+h; \Theta_{\mathrm{mkt}}(t) + (\delta_1, \delta_2, \ldots, \delta_{i-1}, 0, 0, \ldots, 0)^\top\right), \quad (22.24)
\end{aligned}
$$

with $i = 1, \ldots, N_{\mathrm{mkt}}$. In other words, the impact of market variable i is recorded as the difference in portfolio values arising from moving the first $i-1$ and i market data variables, respectively, to their time $t+h$ values. The resulting attribution of P&L is often, quite descriptively, termed a "bump-and-do-not-reset" P&L explain.

---

[9] Also known by the more grammatically sensible names *P&L explanation* or *P&L attribution*.

Notice that

$$\sum_{i=1}^{N_{\text{mkt}}} E_i = V\left(t+h; \Theta_{\text{mkt}}(t+h)\right) - V\left(t+h; \Theta_{\text{mkt}}(t)\right) \neq V(t+h) - V(t),$$

since $V(t) = V(t; \Theta_{\text{mkt}}(t)) \neq V(t+h; \Theta_{\text{mkt}}(t))$. A complete P&L explain report must therefore add back a theta-type term that measures time decay, i.e. we write

$$V(t+h) - V(t) = \sum_{i=1}^{N_{\text{mkt}}} E_i + \{V\left(t+h; \Theta_{\text{mkt}}(t)\right) - V\left(t; \Theta_{\text{mkt}}(t)\right)\}, \quad (22.25)$$

where the term in the curly brackets accounts for the effect associated with keeping market data fixed and letting time progress from t to $t+h$.

As argued in Section 22.2.1, the time decay definition used in (22.25) is often problematic. A more meaningful definition is given in (22.21), which leads to the following improved accounting for the P&L explain:

$$V(t+h) - V(t) = \sum_{i=1}^{N_{\text{mkt}}} E_i^f + \left\{V\left(t+h; \Theta_{\text{mkt}}^f(t)\right) - V\left(t; \Theta_{\text{mkt}}(t)\right)\right\}, \quad (22.26)$$

where

$$E_i^f = V\left(t+h; \Theta_{\text{mkt}}^f(t) + (\delta_1^f, \delta_2^f, \ldots, \delta_i^f, 0, 0, \ldots, 0)^\top\right)$$
$$- V\left(t+h; \Theta_{\text{mkt}}^f(t) + (\delta_1^f, \delta_2^f, \ldots, \delta_{i-1}^f, 0, 0, \ldots, 0)^\top\right),$$

with δ^f defined in (22.23). Both (22.25) and (22.26) can be found in actual bank systems, but the latter is typically preferable.

22.2.2.2 Bump-and-Reset Explain

By construction, the waterfall P&L explain procedure in Section 22.2.2.1 is always fully able to explain P&L moves, in the sense that both (22.25) and (22.26) are identities, rather than approximations. While this is convenient, one drawback of the method is that the amount (E_i or E_i^f) of the P&L move that is allocated to an individual market data variable depends on how the vector $\Theta_{\text{mkt}}(t)$ happens to be ordered. This lends a certain amount of arbitrariness to the waterfall method, which sometimes can affect the P&L attribution process fairly substantially. To see this, assume that $\Theta_{\text{mkt}}(t)$ consists of interest rate and volatility data, and that interest rates are listed before the volatilities. Consider a position in an out-of-the-money caplet, and a market scenario where both interest rates and volatilities increase over the interval $[t, t+h]$. Further, assume that the shift in interest rates just happens to make the caplet position move from being out-of-the-money

(OTM) to at-the-money (ATM). In the waterfall P&L explain, since our ordering was such that we move interest rates before we move volatilities, when measuring the contribution from volatilities to the P&L move, we will register a decent amount, since ATM options have high vega. On the other hand, had we arbitrarily listed volatilities before interest rates in $\Theta_{\mathrm{mkt}}(t)$, the contribution from volatility would have been computed on an OTM option with little vega, resulting in a much smaller P&L effect.

To avoid the consistency problems of the waterfall method, an alternative approach is to change (22.24) to

$$E_i = V\left(t + h; \Theta_{\mathrm{mkt}}(t) + (0, 0, \ldots, \delta_i, 0, 0, \ldots, 0)^{\top}\right) - V\left(t + h; \Theta_{\mathrm{mkt}}(t)\right),$$

which is often called *bump-and-reset P&L explain*. With this definition, however, an exact P&L explain such as (22.25) is not possible, but we must instead content ourselves with an expression of the form

$$V(t + h) - V(t) = \sum_{i=1}^{N_{\mathrm{mkt}}} E_i + \{V\left(t + h; \Theta_{\mathrm{mkt}}(t)\right) - V\left(t; \Theta_{\mathrm{mkt}}(t)\right)\} + U,$$

$$(22.27)$$

where U represents the unexplained part of the P&L (the "unexplain"), primarily caused by cross-convexity terms in the Hessian matrix A^H, i.e. terms of the type $\partial^2 V / \partial \delta_i \partial \delta_j$, $i \neq j$. We note that (22.27) can be improved to incorporate the same notion of time decay as in (22.26); we leave this straightforward modification to the reader.

If the term U is consistently large, it may be necessary to explicitly add terms that capture cross-convexity exposure. This can be done using terms from A^H, which makes the overall procedure a bit of a hybrid between true P&L predict and explain. Alternatively, if we, say, identify the interaction of δ_i and δ_j as being considerable, we may do a joint bump-and-reset of these market data perturbations to split out a cross-term contribution of

$$V\left(t + h; \Theta_{\mathrm{mkt}}(t) + (0, 0, \ldots, \delta_i, 0, \ldots, \delta_j, 0, \ldots 0)^{\top}\right)$$
$$- V\left(t + h; \Theta_{\mathrm{mkt}}(t) + (0, 0, \ldots, \delta_i, 0, 0, \ldots, 0)^{\top}\right)$$
$$- V\left(t + h; \Theta_{\mathrm{mkt}}(t) + (0, 0, \ldots, \delta_j, 0, 0, \ldots, 0)^{\top}\right).$$

Carefully supplementing the basic bump-and-reset P&L explain with cross-term contributions will help ensure that the residual amount of unexplained P&L is small.

22.3 Value-at-Risk

While the P&L predict and explain are largely backward-looking accounting exercises, the risk management team in a bank is primarily focused on analyzing the distribution of *future* portfolio values, in order to gauge the overall riskiness of a portfolio. Rather than report the entire P&L distribution, it is common to summarize it in a few summary statistics, known as *risk measures*. Many such risk measures exist, but the so-called *value-at-risk* (VaR) is probably the most commonly used in practice. VaR at level α (denoted Λ_α) is simply the $(1-\alpha)$-percentile of the distribution of the P&L move $V(t+h) - V(t)$ in the real-life measure P:

$$\mathrm{P}\left(V(t+h) - V(t) \le \Lambda_\alpha | \mathcal{F}_t\right) = 1 - \alpha. \tag{22.28}$$

In other words, the probability of losing[10] more than $-\Lambda_\alpha$ over the time interval $[t, t+h]$ is less than $1 - \alpha$. Typically α is set to 99% or 95%, and h to one business day.

Another commonly used risk measure is *conditional value-at-risk* (cVaR), which is defined as the conditional expectation

$$\Xi_\alpha \triangleq \mathrm{E}^{\mathrm{P}}\left(V(t+h) - V(t) | V(t+h) - V(t) \le \Lambda_\alpha\right). \tag{22.29}$$

cVaR has certain theoretical advantages to VaR[11], but VaR is nevertheless the more common in practice.

To compute Λ_α and Ξ_α, we need a statistical description for the market data increment vector δ. One popular choice uses the historical distribution of δ directly, giving rise to the so-called *historical VaR* risk measure. Here, one takes the actual realizations of δ over the last N_{VaR} trading days (e.g. $N_{\mathrm{VaR}} = 500$, roughly corresponding to two years) and applies them to the current market data, thereby generating the empirical distribution of $V(t+h) - V(t)$. The calculation of VaR then amounts to ranking the impact of the last N_{VaR} market moves on the current portfolio, from worst to best, and using the impact of the market data move on the day with the rank $(1 - \alpha)N_{\mathrm{VaR}}$ as the VaR.

Another VaR methodology uses a parameterized, rather than historical, distribution of market moves. As h is typically a short interval, it is, for instance, often justified to assume that each element in δ is Gaussian with zero mean and standard deviation s_i,

$$\delta_i \sim \mathcal{N}(0, s_i^2), \quad i = 1, \ldots, N_{\mathrm{mkt}}, \tag{22.30}$$

where s_i may be estimated from the annualized basis point volatility σ_i of market element i through the relation

$$s_i = \sigma_i \sqrt{h}.$$

---

[10]Notice that Λ_α virtually always is a negative number. Sometimes it is the absolute value of this number that is reported as the VaR.

[11]Specifically, VaR is not a *coherent risk measure*, in the sense defined in Artzner et al. [1999].

We capture co-dependence between the elements of δ in a correlation matrix R, typically estimated from historical time series. Notice that even if market data element i has some non-zero drift, the mean of δ_i would be of order $O(h)$ and typically negligible relative to s_i (order $O(\sqrt{h})$); the assumption of zero mean is therefore an innocuous one.

To compute VaR and cVaR in the Gaussian setup, one option is to perform a brute-force simulation of the portfolio value $V(t+h)$, using a full portfolio revaluation for each simulated value of the vector δ. While this is, in fact, sometimes done, it is far easier to rely on the Taylor expansion (22.19). For VaR/cVaR purposes, we may safely ignore the time decay term in (22.19), hence we can take as our starting point the equation

$$V(t+h) = V(t) + \nabla^H(t) \cdot \delta + \frac{1}{2}\delta^\top \cdot A^H(t) \cdot \delta. \qquad (22.31)$$

With (22.31) and (22.30), the VaR and cVaR computations are analytically tractable. One simple result is the following.

Proposition 22.3.1. *Let $V(t+h)$ be given as in (22.31), with $A^H(t) = 0$. Also, let the elements of the N_{mkt}-dimensional vector δ have correlation matrix R and satisfy (22.30). Setting $s = (s_1, \ldots, s_{N_{\mathrm{mkt}}})^\top$, we have*

$$\Lambda_\alpha = v\Phi^{-1}(1-\alpha), \qquad (22.32)$$

$$\Xi_\alpha = -v(1-\alpha)^{-1}\phi\left(\Phi^{-1}(1-\alpha)\right), \qquad (22.33)$$

where $\phi(x) = (2\pi)^{-1/2}e^{-x^2/2}$ is the Gaussian density, and

$$v^2 = \nabla^H(t)\operatorname{diag}(s) R\operatorname{diag}(s)\nabla^H(t)^\top.$$

Proof. First observe that the covariance matrix C of δ is

$$C = \operatorname{diag}(s) R\operatorname{diag}(s),$$

where $\operatorname{diag}(s)$ is a square matrix with s along the diagonal and zeros elsewhere. Under our assumptions, it is clear that $V(t+h) \sim \mathcal{N}(V(t), v^2)$, where the variance v^2 is given by

$$v^2 = \nabla^H(t) C\nabla^H(t)^\top.$$

Defining $\Delta V = V(t+h) - V(t)$ and writing $\Delta V = vZ$ for $Z \sim \mathcal{N}(0,1)$, we have

$$\mathrm{P}\left(\Delta V \leq \Lambda_\alpha\right) = \mathrm{P}\left(Z \leq \frac{\Lambda_\alpha}{v}\right) = \Phi\left(\frac{\Lambda_\alpha}{v}\right).$$

Equating this expression to $1-\alpha$, per (22.28), results in (22.32). To compute the cVaR, we write

$$E^P\left(\Delta V|\Delta V \le \Lambda_\alpha\right) = P\left(\Delta V \le \Lambda_\alpha\right)^{-1} \times E^P\left(\Delta V 1_{\{\Delta V \le \Lambda_\alpha\}}\right)$$
$$= (1-\alpha)^{-1}E^P\left(vZ1_{\{Z \le m\}}\right)$$
$$= (1-\alpha)^{-1}vE^P\left(Z1_{\{Z \le m\}}\right),$$

where $m = \Lambda_\alpha/v = \Phi^{-1}(1-\alpha)$. Observe that

$$E^P\left(Z1_{\{Z \le m\}}\right) = \int_{-\infty}^m z\phi(z)\,dz = -\phi(m),$$

and (22.33) follows. $\square$

The results in Proposition 22.3.1 are often denoted *delta VaR* and *delta cVaR*, respectively, to reflect the fact that we have ignored the Hessian matrix A^H. If we wish to include A^H — to compute what is known as *delta-gamma VaR/cVaR* — matters get a bit more complicated, as the distribution of $V(t+h) - V(t)$ is no longer simple. One method, described in Rouvinez [1997], shows that $V(t+h) - V(t)$ can be expressed as a sum of independent non-central chi-square random variables. From this representation, the characteristic function of $V(t+h) - V(t)$ can be constructed and, using a numerical technique, turned into a cumulative distribution function from which VaR and cVaR can be computed. As the topic is somewhat tangential to our needs in this book, we omit the details here but just note that at the end of the day the key to a good delta-gamma VaR/cVaR computation is a reliable and accurate estimate for ∇^H and A^H, a topic that shall occupy the remainder of this book.

22.A Appendix: Alternative Proof of Lemma 22.1.1

Consider a contingent claim with terminal value $g(X(T))$, where $X(t)$ satisfies the Black-Scholes SDE (22.5). Let us write the time t value of this claim as $V(t) = h(T-t, X(t))$, where h satisfies the PDE

$$\frac{\partial h}{\partial \tau} = \mathcal{L}h, \quad \mathcal{L} = -r + rX\frac{\partial}{\partial X} + \frac{1}{2}\sigma^2 X^2\frac{\partial^2}{\partial X^2}, \tag{22.34}$$

subject to an initial condition $h(0, X) = g(X)$. Operator calculus treats this equation as an ordinary differential equation in $\tau = T - t$. The solution to (22.34) is then given by

$$h(\tau, X) = \exp\left(\tau\mathcal{L}\right)g(X), \tag{22.35}$$

where the exponential must be interpreted as

$$\exp\left(\tau\mathcal{L}\right) = \sum_{i=0}^\infty \frac{(\tau\mathcal{L})^i}{i!}.$$

Differentiating with respect to τ verifies that (22.35) is, indeed, the solution to the initial value problem (22.34).

To form the derivative $\partial V/\partial\sigma = \partial h/\partial\sigma$, we notice that all dependency on σ in the expression (22.35) is in the operator $\mathcal{L}$. Differentiating with respect to σ, we get

$$\frac{\partial h}{\partial \sigma} = \tau \frac{\partial \mathcal{L}}{\partial \sigma} \exp\left(\tau\mathcal{L}\right) g(X) = \tau\sigma X^2 \frac{\partial^2}{\partial X^2} \left(\exp\left(\tau\mathcal{L}\right) g(X)\right) = \tau\sigma X^2 \frac{\partial^2 h}{\partial X^2},$$

or, equivalently,

$$\frac{\partial V}{\partial \sigma} = (T-t)\sigma X^2 \frac{\partial^2 V}{\partial X^2},$$

which is (22.7).

The operator representation (22.35) makes it easy to compute other parameter derivatives. For instance, we note that

$$\frac{\partial h}{\partial r} = \tau\left(-1 + X\frac{\partial}{\partial X}\right)\left(\exp\left(\tau\mathcal{L}\right) g(X)\right) = \tau\left(X\frac{\partial h}{\partial X} - g\right),$$

such that

$$\frac{\partial V}{\partial r} = (T-t)\left(X\frac{\partial V}{\partial X} - V\right),$$

as mentioned in Remark 22.1.2.

23

Payoff Smoothing and Related Methods

As made clear in the previous chapter, practical risk management of a portfolio of interest rate securities revolves around price sensitivities with respect to various valuation inputs, such as market prices and model parameters. These sensitivities are often[1] calculated by applying small perturbations to market and model parameters, followed by a re-pricing of the securities portfolio in question.

Being derivatives of a model price function, price sensitivities (greeks) are inherently less smooth than the price function itself. For instance, it is well-known (see Section 1.10.3) that while the value of a Bermudan security on an exercise date is continuous across the exercise boundary, its delta is not. This lack of smoothness will often put significant stress on a numerical scheme, which effectively is faced with the problem of resolving an irregular boundary condition. As a result, a careless implementation of a numerical sensitivity computation will often produce poor results, with the resulting greeks being less stable than what is expected theoretically. In this chapter we study this problem, with an emphasis on how to adapt numerical schemes to avoid introducing spurious instabilities into the calculation of greeks. Some of the discussion in this chapter builds on previous material, and we suggest that the reader briefly review Sections 2.5 and 3.3 before proceeding.

23.1 Issues with Discretization Schemes

As we saw in Section 3.3.1.2, fixing the random seed when computing deltas (and other greeks) by Monte Carlo methods significantly reduces the standard error. This is a simple example of the general rule that one should ideally attempt to freeze as many aspects of a numerical calculation (such as a Monte Carlo seed, the geometry of a PDE grid, etc.) as possible when doing perturbation analysis in a numerical scheme. In particular, adhering to this

---

[1]But not always — see Chapter 24.

simple rule often ensures that no additional discontinuities are introduced by the numerical method itself.

23.1.1 Problems with Grid Dimensioning

While straightforward in theory, consistently following the rule above can be quite difficult. Sometimes violations are subtle and unintentional, and sometimes they are unavoidable due to systems limitations or computational precision requirements. Let us look at the former case, using as our first example the problem of valuing a simple option by numerically integrating the payout against a Gaussian density. As is frequently done in practice, suppose that the numerical domain of integration is chosen to be $[-5\sigma\sqrt{T}, 5\sigma\sqrt{T}]$, where T is the expiry of the option and σ is the asset volatility (a model input). In addition, let the number of integration nodes, N, be chosen so that the integration grid is uniform with a pre-specified and fixed step δ, i.e.

$$N = \left\lfloor \frac{10\sigma\sqrt{T}}{\delta} \right\rfloor, \tag{23.1}$$

where $\lfloor \cdot \rfloor$ denotes the integer part of a real number. The resulting option valuation scheme appears reasonable, if slightly unconventional. However, imagine now that we wish to compute the option vega by comparing the base value of the option to a value computed after shocking the volatility to a new level of $\sigma + \Delta\sigma$. Since the integration domain depends on σ, it will be slightly larger in the perturbation scenario. Moreover, as the integration step is kept constant, the number of integration nodes may change between the base and the bumped scenarios. As the number of integration points can only move by an integer amount, the change in the grid geometry would not be continuous, introducing a purely artificial contribution of the order $O((\Delta\sigma)^{-1})$. This contribution explodes to infinity as $\Delta\sigma \to 0$ (as long as the number of steps changes as a result of the volatility perturbation).

The issue that arises in the example above stems from the fact that the number of integration nodes N as given by (23.1) is *not a smooth function* of σ, as the function $x \mapsto \lfloor x \rfloor$ is not differentiable (or even continuous). Since the numerical value of the security is a function of the number of integration nodes N, the value will not be smooth with respect to σ, and the vega, while continuous in the theoretical model, will be discontinuous in the numerical scheme. Of course, the problem is easy to rectify: heed our advice and avoid altering the geometry of the grid when perturbing market inputs.

23.1.2 Grid Shifts Relative to Payout

The example in Section 23.1.1 above is an example of grid geometry changing outright, due to changes in asset moments used for grid position and/or dimensioning. Another problematic case occurs when the grid is frozen in space, but the nature of the perturbation itself will cause an *effective* shift of the grid relative to the payout. To give an example of this, consider a problem of valuing a European option with a payoff $f(x)$ on an underlying $S(T)$ observed at time T. Assuming zero interest rates, the value of this option is equal to the integral of the payoff against the PDF of the underlying. Let the initial (time 0) spot value of the underlying be denoted by S. Assuming for simplicity that the distribution of the increment $S(T) - S$ is independent of S, we denote the density of $S(T) - S$ by

$$Q((S(T) - S) \in dx) = \pi(x)\, dx.$$

The value of the option as a function of the spot S today is then given by two equivalent expressions

$$V(S) = \int_{-\infty}^{\infty} f(x)\pi(x - S)\, dx \tag{23.2}$$

$$= \int_{-\infty}^{\infty} f(S + x)\, \pi(x)\, dx. \tag{23.3}$$

While (23.2) and (23.3) are mathematically equivalent, (23.2) is better suited for numerical computations of sensitivities of V with respect to S. In particular, notice that changes in S here get absorbed into the *density* of S, which in a grid setting simply amounts to changing the weights on individual grid points in a numerical quadrature rule. On the other hand, the formulation (23.3) absorbs changes in S into the *payout*, which effectively causes the grid to move relative to the payout function[2].

To demonstrate the kind of problems that arise when the discretization grid is not fixed relative to the payoff, let us consider a simple example. We recall that a typical non-adaptive quadrature scheme (including rectangle, trapezoidal and Gaussian quadrature rules) specifies a collection of fixed knot points $\{x_n\}_{n=0}^{N}$ and weights $\{w_n\}_{n=0}^{N}$, and approximates $V(S) \approx \tilde{V}(S)$, where

$$\tilde{V}(S) = \sum_{n=0}^{N} w_n f\left(x_n + S\right).$$

Note again that the weights and knots are fixed relative to the density of the process and not the payoff, i.e. contrary to our earlier advice. Let us analyze the behavior of the scheme under shifts of S. For concreteness, we consider a European call option, i.e.

$$f(x) = (x - K)^{+}$$

for a fixed choice of K. Then

---

[2]The observant reader may have noticed a strong connection to material in Chapter 3 on pathwise and likelihood ratio methods for Monte Carlo applications.

$$\widetilde{V}(S) = \sum_{n=0}^{N} w_n \left(x_n + S - K\right)^+.$$

The exact derivative of the numerical value $\widetilde{V}(S)$ with respect to the initial value of the asset, i.e. the delta, is given by

$$\frac{d}{dS}\widetilde{V}(S) = \sum_{n=0}^{N} w_n 1_{\{x_n + S - K > 0\}}.$$

As a function of S, this function has discontinuities of sizes w_n at points $S = S_n$,

$$S_n \triangleq K - x_n,$$

for all $n = 0, \ldots, N$. Thus, as the spot S moves, the delta will jump whenever the spot crosses one of the levels S_n. Moreover, in this scheme, the delta does not change as long as S does not cross one of S_n, which is obviously unrealistic. A typical plot of such a "delta" is shown in Figure 23.1.

Fig. 23.1. Discontinuous Delta

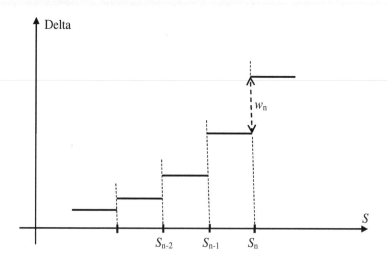

Notes: Delta of a derivative evaluated with an integration scheme that is not fixed relative to the payoff.

The irregular deltas in Figure 23.1 are caused by the call option kink crossing over knot points in the grid as a result of parameter perturbation; similar behavior will occur for all payouts with payout discontinuities, kinks, and the like.

23.1.3 Additional Comments

Problems of the types described in Sections 23.1.1 and 23.1.2 are easy to introduce, but often difficult to track down. This is particularly true for complex models that often use sophisticated numerical methods with complicated dependencies on market and model data. When examining numerical routines for problems, we note that one must obviously include both valuation and calibration algorithms, as the computation of stable sensitivities requires that both types of algorithms behave smoothly with respect to moves in market parameters. Local, bootstrap-type calibrations (such as that developed in Section 13.1.7 for quasi-Gaussian models) generally outperform global, best-fit calibrations (such as that from Section 14.5.7 for LM models). We postpone our treatment of calibration effects to Chapter 26, and in this chapter focus on the (post-calibration) problem of building numerically smooth valuation routines.

Even if one is vigilant and tracks down all cases of non-constant (effective) grid geometry, there may, as mentioned earlier, exist violations that are impractical to resolve. For example, the Monte Carlo "grid" is difficult to keep constant by the nature of how it is generated. Also, by not explicitly tailoring a finite difference grid geometry to the market data used in a perturbed scenario, an unacceptable loss of precision may occur. Finally, we note that the organizational setup in many banks may make it hard for valuation code developers to fully control risk sensitivity computations. For all these reasons, it is worthwhile to develop methods that will produce stable sensitivities, *even* if the grid geometry cannot be guaranteed to stay fixed under market data moves.

As a final comment, let us briefly note that some numerical methods, in principle, give us greeks "for free" as part of valuation. For example, in a finite difference grid solution of a vanilla model PDE, the derivative with respect to the asset can be read off the PDE grid by forming a central finite difference of the solution (at time 0) at grid nodes surrounding today's value of the asset[3]. This avenue is not available for all greeks, however, and even for deltas and gammas the utility of the method is quite limited in term structure models, since we rarely are interested in theoretical derivatives with respect to the abstract model variables, but instead nearly always wish to compute sensitivities to specific perturbations of the yield curve. In addition, as described in Section 16.1.1, we may often be interested in working with joint moves of interest rates and volatilities that are user-prescribed and incompatible with the theoretical model dynamics. Even for PDE-based

---

[3]Sometimes one gets better numerical properties if a spline is fit through all grid values at $t = 0$; deltas and gammas can then be computed by differentiating this spline.

models we therefore typically will need to calculate greeks by applying market data bumps[4].

23.2 Basic Techniques

An obvious remedy to many of the issues identified in earlier sections is simply to increase the number of grid points. As a rule of thumb, the discretization step of a grid should be significantly smaller than a typical shift applied to inputs when calculating greeks. For instance, in Figure 23.1, if the perturbation size for delta calculation covered a few grid intervals, the delta would vary fairly smoothly with the value of the spot. Of course, the smaller the grid size, the more computationally intensive the numerical scheme will be, and we cannot always trade speed for accuracy with impunity, since greek calculations are often time-sensitive. The next few sections cover several more sophisticated, and less computationally costly, alternatives to brute-force grid refinement.

23.2.1 Adaptive Integration

Increasing the density of grid points uniformly is not the best way to spend a computational budget, as adding extra points away from a discontinuity of a payoff does little to improve the numerical properties of greeks. A reasonable yet relatively simple way of improving numerical stability of an algorithm at a modest computational cost is to first identify the region of the state space where the payoff is likely to have singularities[5], and then sample this part of the space at a higher resolution than elsewhere. For instance, suppose we know that the function $f(x)$ that we are integrating has a singularity in a particular interval $[x_m, x_{m+1}]$ of the integration grid. Then we can further subdivide $[x_m, x_{m+1}]$ into 10 (or 100) subintervals and apply the trapezoidal rule for the finer grid. This will insure that the singularity is handled accurately, while no extra effort is wasted in the regions where f is smooth.

For any particular payoff function, it is often relatively simple to identify the regions where a denser grid is beneficial — the type of knowledge that can be incorporated directly into the integration routine. For a more generic setup (or for payouts that are hard to analyze directly), a good

---

[4]In fact, several commonly used risk measures are explicitly meant to be computed by finite-sized shifts. For instance, many swaption traders' definition of gamma is the change in delta, for a 10 basis point move in the yield curve.

[5]Here and elsewhere in this chapter, the term "singularity" refers to a point in the state space at which the payoff function, or one of its derivatives, is discontinuous. Examples include the barrier for a digital option and the strike of a European put or call.

alternative is to rely on adaptive integration routines, often prepackaged in numerical libraries such as IMSL and NAG. In this class of routines, the integral is approximated using ordinary quadrature rules on adaptively refined subintervals of the integration domain until a stopping criterion is met. In effect, the grid points are chosen automatically based directly on properties of the function being integrated. Adaptive algorithms generally work quite well, but care must be taken to keep grid geometry fixed in perturbed scenarios, something that can occasionally be a bit of a challenge if a third-party library routine is used.

23.2.2 Adding Singularities to the Grid

In the situation where perturbations cause a grid shift relative to the payoff, if the position of a payoff singularity is known exactly, numerical properties of the valuation algorithm can be improved substantially by simply adding the singularity to the integration grid. This serves to effectively lock down the grid geometry in the immediate vicinity of the singularity. The method is closely related to the *grid shifting* method used to improve convergence of numerical solutions of PDEs, see Section 2.5.3.

Using the integration problem (23.3) as an example, let us consider a continuous payoff $f(y)$ whose derivative is discontinuous at a single point $y = K$ (multiple singularities can be handled similarly). Let us rewrite the value of the option as

$$V(S) = \int_{-\infty}^{K-S} f(x+S)\,\pi(x)\,dx + \int_{K-S}^{\infty} f(x+S)\,\pi(x)\,dx.$$

Suppose we proceed to apply a simple numerical quadrature to each integral separately. We start with $N+1$ integration knots fixed in x-space, and add one more knot at the singularity, i.e. at $x = K - S$. To characterize the location of the additional knot, let the index $\mu(S)$ be defined by

$$x_{\mu(S)} \le K - S < x_{\mu(S)+1}.$$

Using a trapezoidal integration rule for simplicity, the resulting numerical scheme can formally be written as

$$\widetilde{V}(S) = \widetilde{V}_1(S) + \widetilde{V}_2(S) + \widetilde{V}_3(S) + \widetilde{V}_4(S),$$

$$\widetilde{V}_1(S) = \frac{1}{2} \sum_{n=1}^{\mu(S)} \left(f(x_n + S)\,\pi(x_n) + f(x_{n-1} + S)\,\pi(x_{n-1}) \right) \Delta x_n,$$

$$\widetilde{V}_2(S) = \frac{1}{2} \left(f(K)\pi(K - S) + f(x_{\mu(S)} + S)\,\pi(x_{\mu(S)}) \right) \left(K - S - x_{\mu(S)} \right),$$

$$\tilde{V}_3(S) = \frac{1}{2} \left(f \left(x_{\mu(S)+1} + S \right) \pi \left(x_{\mu(S)+1} \right) + f(K)\pi \left(K - S \right) \right)$$
$$\times \left(x_{\mu(S)+1} - (K - S) \right),$$

$$\tilde{V}_4(S) = \frac{1}{2} \sum_{n=\mu(S)+2}^{N} \left(f \left(x_n + S \right) \pi(x_n) + f \left(x_{n-1} + S \right) \pi \left(x_{n-1} \right) \right) \Delta x_n.$$

The terms $\tilde{V}_1(S)$ and $\tilde{V}_4(S)$ collect the contributions of integration intervals before and after the singularity, respectively, whereas the terms $\tilde{V}_2(S)$ and $\tilde{V}_3(S)$ represent the contributions of the integration interval containing the singularity.

Let us fix m such that $\mu(S) = m$ for the initial (pre-perturbed) value of S. Clearly there are no issues with the smoothness of $d\tilde{V}(S)/dS$ as we move S in such a way that $K - S \in [x_m, x_{m+1})$, so the only potential discontinuity could arise when $K - S$ crosses one of the integration nodes, i.e. when $\mu(S)$ jumps. To show that the scheme implies smooth behavior across grid points, let us investigate what happens when S crosses $S_m \triangleq K - x_m$. For this analysis, we only need to keep track of the terms $\tilde{U}(S)$ that originate from the intervals adjacent to x_m. For $x_{m+1} > K - S \geq x_m$ we have

$$\tilde{U}(S) = \frac{1}{2} \left(f \left(x_m + S \right) \pi(x_m) + f \left(x_{m-1} + S \right) \pi \left(x_{m-1} \right) \right) \Delta x_m$$
$$+ \frac{1}{2} \left(f(K)\pi \left(K - S \right) + f \left(x_m + S \right) \pi(x_m) \right) \left(K - S - x_m \right)$$
$$+ \frac{1}{2} \left(f \left(x_{m+1} + S \right) \pi \left(x_{m+1} \right) + f(K)\pi \left(K - S \right) \right) \left(x_{m+1} - (K - S) \right),$$

and for shifted S such that $x_{m-1} \leq K - S < x_m$, i.e. for $\mu(S) = m - 1$,

$$\tilde{U}(S) = \frac{1}{2} \left(f(K)\pi \left(K - S \right) + f \left(x_{m-1} + S \right) \pi \left(x_{m-1} \right) \right) \left(K - S - x_{m-1} \right)$$
$$+ \frac{1}{2} \left(f \left(x_m + S \right) \pi(x_m) + f(K)\pi \left(K - S \right) \right) \left(x_m - (K - S) \right)$$
$$+ \frac{1}{2} \left(f \left(x_{m+1} + S \right) \pi \left(x_{m+1} \right) + f \left(x_m + S \right) \pi(x_m) \right) \Delta x_{m+1}.$$

Note that $\tilde{U}(S)$ is continuous at $S = S_m$ and

$$\tilde{U}(S_m) = \frac{1}{2} \left(f(K)\pi(x_m) + f \left(K - \Delta x_m \right) \pi \left(x_{m-1} \right) \right) \Delta x_m$$
$$+ \frac{1}{2} \left(f \left(K + \Delta x_{m+1} \right) \pi \left(x_{m+1} \right) + f(K)\pi(x_m) \right) \Delta x_{m+1}.$$

In fact, the *derivative* of $\tilde{U}(S)$ is also continuous across the grid point, as we prove in Appendix 23.A.

As a final observation, we note that to add the singularity to the grid, the location of the singularity obviously needs to be detected first. In many cases, this must be done numerically using, for example, the method of Section 23.3.2.1. The numerical improvements to the greeks, however, are usually well worth the extra cost.

23.2.3 Singularity Removal

Most of the noise in greeks comes from the fact that numerical schemes have difficulty handling payoff singularities. It follows that removing these singularities should restore smoothness. The method based on this idea is quite powerful when it works, but is somewhat limited in its scope.

Fig. 23.2. Singularity Removal

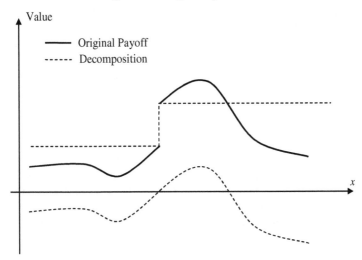

Value

—— Original Payoff
----- Decomposition

x

Notes: A discontinuous payoff could be decomposed into a continuous one and a step function.

Suppose that, as in Figure 23.2, an otherwise-smooth payoff $f(x)$ has a jump discontinuity at one point $x = K$, with the size of the jump equal to a. The payoff can evidently then be decomposed into the sum of two functions, one being a simple step function $a1_{\{x>K\}}$, and the other equal to $g(x) = f(x) - a1_{\{x>K\}}$. Notice that the function $g(x)$ is smooth, unlike $f(x)$ itself. The integration problem

$$V(S) = \int_{-\infty}^{\infty} f(S+x)\pi(x)\,dx$$

may now be split in two,

$$V(S) = a \int_{-\infty}^{\infty} 1_{\{x>K\}} \pi (x - S) \, dx + \int_{-\infty}^{\infty} g(x)\pi (x - S) \, dx.$$

Let $\Psi(x)$ be the cumulative distribution function corresponding to the density $\pi(x)$, then

$$V(S) = a \left(1 - \Psi (K - S)\right) + \int_{-\infty}^{\infty} g(x)\pi (x - S) \, dx.$$

Suppose the CDF $\Psi(x)$ can be computed analytically, or numerically with high precision (and, say, tabulated). As the only function being integrated numerically, g, is smooth by construction, this scheme produces smooth greeks.

Shifted and scaled step functions $a1_{\{x>K\}}$, and combinations thereof, can be used to remove outright discontinuities in the payoff f. To remove discontinuities in the derivative of f, linear combinations of functions $(ax+b)^+$ for various a, b (i.e., call/put type payoffs) can be used instead. It should be clear, however, that the applicability of the singularity removal method will be limited by the availability of accurate (ideally analytic) methods to compute CDF and call/put option values.

23.2.4 Partial Analytical Integration

For the cases where the CDF or call/put option values are not known analytically, suitable approximations could still be used for smoothing the payoff. In the common case where the density $\pi(x)$ corresponds to a diffusive random variable $S(T)$, one approach is to focus on short times to maturity, where the conditional transition density

$$\pi(x, T; y, T - \delta) \, dx \triangleq Q(S(T) \in dx \,|\, S(T - \delta) = y)$$

can often be approximated by, say, a Gaussian or log-normal distribution for small $\delta > 0$. Then the value of a derivative with the payoff $f(x)$ is given by

$$E(f(S(T))) = E\left(E_{T-\delta} \left(f(S(T))\right)\right) = \int_{-\infty}^{\infty} \pi(x, T - \delta; S, 0)V(x, T - \delta) \, dx,$$

$$(23.4)$$

where $V(x, T - \delta)$ is the value of the derivative at time $T - \delta$,

$$V(x, T - \delta) = \int_{-\infty}^{\infty} \pi(y, T; x, T - \delta)f(y) \, dy.$$

With the approximation to $\pi(y, T; x, T - \delta)$ by a Gaussian or log-normal density, this integral can often be calculated analytically. This is certainly true for puts/calls and digitals. More complex payoffs can, for each value of x, often be approximated as combinations of simple payoffs in the vicinity of x, as the width of distribution of $S(T)$ given $S(T - \delta)$ is small for small δ. Moreover, $V(x, T - \delta)$ is often a (much) smoother function of x than $f(x)$; for example the Black-Scholes value of a call option is infinitely differentiable, unlike the payoff itself. Hence, a numerical integration scheme applied to (23.4) should result in smoother and more stable greeks.

The method above (as well as several others reviewed in Section 23.2) is not limited to numerical integration, but can equally well be applied for PDE and Monte Carlo valuation; in fact we already saw a PDE application in Section 2.8. For example, a non-trivial application to TARNs in Monte Carlo is presented in Pietersz and van Regenmortel [2006]. To briefly review this method, let us recall the setup of Section 20.1 and note that

$$V_{\text{tarn}}(0) = \sum_{n=1}^{N-1} V_{\text{cpn},n}(0),$$

where $V_{\text{cpn},n}(0)$ is the value of the n-th (net) coupon conditional on no early redemption,

$$V_{\text{cpn},n}(0) = \text{E}\left(B(T_{n+1})^{-1} X_n 1_{\{Q_n < R\}}\right).$$

Here Q_n is the sum of all structured coupons paid on or before T_n, i.e.

$$Q_n = Q_{n-1} + \tau_{n-1} C_{n-1}.$$

Observing that Q_{n-1} is $\mathcal{F}_{T_{n-2}}$-measurable allows us to write

$$V_{\text{cpn},n}(0) = \text{E}\left(B(T_{n-2})^{-1} V_{\text{cpn},n}(T_{n-2})\right), \tag{23.5}$$

where

$$V_{\text{cpn},n}(T_{n-2}) = P(T_{n-2}, T_{n+1})\text{E}_{T_{n-2}}^{T_{n+1}}\left(X_n 1_{\{C_{n-1} < (R-Q_{n-1})/\tau_{n-1}\}}\right).$$

For Libor-based TARNs, X_n is a function of $L_n(T_n)$ and C_{n-1} is a function of $L_{n-1}(T_{n-1})$, so the calculation of $V_{\text{cpn},n}(T_{n-2})$ involves an integral of a discontinuous function over a joint distribution of $(L_{n-1}(T_{n-1}), L_n(T_n))$, conditioned on $\mathcal{F}_{T_{n-2}}$. As coupon periods are rarely longer than a year, a Gaussian (or log-normal) approximation to this distribution is often accurate enough. Drifts and (co-)variances of Libor rates $(L_{n-1}(T_{n-1}), L_n(T_n))$ can typically be estimated with relative ease from the term structure model used for valuation (see Section 20.1.3 for a relevant discussion), at which point $V_{\text{cpn},n}(T_{n-2})$ would be calculated by an exact or approximate quadrature, perhaps aided by the various methods from Section 17.6. Calculating $V_{\text{cpn},n}(T_{n-2})$ by integration removes the digital discontinuity in the coupon, helping to stabilize Monte Carlo based sensitivity computations for $V_{\text{cpn},n}(0)$ in (23.5).

23.3 Payoff Smoothing For Numerical Integration and PDEs

Upon reflection, it is clear that singularity-removal technique outlined in Section 23.2.4 works by smoothing out an irregular boundary condition by integrating it against a density kernel. A closely related idea involves a direct modification of the payoff, to pre-smooth it before numerical integration or PDE schemes (or even Monte Carlo, as covered in the next section) are applied. We discuss several such payoff smoothing techniques in this section. Let us quickly remind the reader that payoff smoothing has two different, but related benefits. First, payoff smoothing will improve convergence of greeks calculated by PDE or Monte Carlo methods as the number of PDE steps or Monte Carlo paths is increased: the smoother the payoff, the faster the convergence. We have covered this angle in Sections 2.5 and 3.3. Second, payoff smoothing will help alleviate the problems arising if we, for the various reasons mentioned in Section 23.1, are unable to keep discretization grids constant.

23.3.1 Introduction to Payoff Smoothing

In a nutshell, the method of *payoff smoothing* replaces one payoff with a smoother one, to which standard numerical integration or PDE methods are then applied. Payoff smoothing serves to remove points of discontinuity in the payoff and its derivatives which, as we have seen earlier, will help improve the stability and smoothness of greeks.

A simple example of payoff smoothing replaces $f(x)$ with its moving average,

$$f_{\text{smooth}}(x) = \frac{1}{\epsilon} \int_{x-\epsilon/2}^{x+\epsilon/2} f(y)\, dy, \qquad (23.6)$$

for some small $\epsilon > 0$, the choice of which will be discussed later. Payoff smoothing based on (23.6) was already applied in Section 2.5.2 as the *continuity correction* method for improving convergence of numerical solutions of PDEs. An example of moving average payoff smoothing is presented in Figure 23.3.

Continuing with the sample setup and the notations of Section 23.1, we recall that the standard numerical quadrature with knots $\{x_n\}$ and weights $\{w_n\}$ specifies that

$$\widetilde{V}(S) = \sum_{n=0}^{N} w_n f(x_n + S). \qquad (23.7)$$

The payoff smoothing method replaces this with

$$\widetilde{V}(S) = \sum_{n=0}^{N} w_n f_{\text{smooth}}(x_n + S). \qquad (23.8)$$

Fig. 23.3. Payoff Smoothing

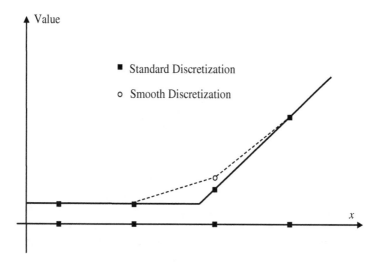

Notes: Smoothing a discretized payoff using moving average (23.6).

Because $f_{\text{smooth}}(x)$ has a higher degree of smoothness than $f(x)$, the numerically computed greeks of $\widetilde{V}(S)$ behave more smoothly with respect to market inputs.

The function $f_{\text{smooth}}(x)$ usually cannot be computed exactly. However, for small ϵ, various approximations can be made. Here, again, the knowledge of singularities of $f(x)$ is important. If $f(x)$ is known to have no singularities on $[x - \epsilon/2, x + \epsilon/2]$, a simple linear approximation to $f(x)$ on this interval will suffice as the corresponding term in (23.7) will be sufficiently smooth:

$$f(y) \approx f(x - \epsilon/2) + (f(x + \epsilon/2) - f(x - \epsilon/2))\frac{y - x + \epsilon/2}{\epsilon}$$

for $y \in [x - \epsilon/2, x + \epsilon/2]$, so that

$$f_{\text{smooth}}(x) = \frac{1}{\epsilon} \int_{x - \epsilon/2}^{x + \epsilon/2} f(y)\, dy \approx f(x). \tag{23.9}$$

If, however, it is known that there is a singularity x^* in $[x - \epsilon/2, x + \epsilon/2]$, then the integral should be handled more carefully by, for example, using separate linear approximations to f in two subintervals, $[x - \epsilon/2, x^*)$ and $(x^*, x + \epsilon/2]$:

$$f(y) \approx \begin{cases} f(x - \epsilon/2) + (f(x^*-) - f(x - \epsilon/2))\frac{y - x + \epsilon/2}{x^* - x + \epsilon/2}, & y \in [x - \epsilon/2, x^*), \\ f(x^*+) + (f(x + \epsilon/2) - f(x^*+))\frac{y - x^*}{x + \epsilon/2 - x^*}, & y \in (x^*, x + \epsilon/2], \end{cases}$$

so that

$$f_{\text{smooth}}(x) = \frac{1}{\epsilon} \int_{x-\epsilon/2}^{x+\epsilon/2} f(y)\, dy \approx \frac{x^* - x + \epsilon/2}{\epsilon} f\left(\frac{x - \epsilon/2 + x^*}{2}\right)$$
$$+ \frac{x + \epsilon/2 - x^*}{\epsilon} f\left(\frac{x + \epsilon/2 + x^*}{2}\right). \quad (23.10)$$

The method is not dissimilar to the singularity-extended grid method, at least in one dimension, but could be more practical to apply in a PDE schemes, say, if multiple singularities at different locations are introduced at different times.

The performance of moving average smoothing will depend on the choice of ϵ. Higher values of ϵ lead to smoother schemes but makes it more difficult to approximate the required integral, since linear approximations may no longer be accurate enough. More importantly, the introduction of smoothing adds bias to the valuation, as the payoff being integrated becomes increasingly different from the actual one when ϵ is increased. In many cases the choice of ϵ is done semi-empirically, with numerical experiments to determines the highest value of the smoothing window that keeps the bias within acceptable limits. In some cases, the discretization of the grid itself drives the size of ϵ, a case that we consider next.

23.3.2 Payoff Smoothing in One Dimension

To develop the method above in more detail, and to link it more directly to grid-based methods, let us introduce a discrete set of x-values, $\{x_n\}_{n=0}^N$. It is helpful to think of (23.7) as a special case of a more general setup where we define the discretized value f_n as the weighted average of $f(x)$ in a neighborhood of x_n,

$$f_n = \int_{-\infty}^{\infty} \kappa_n(x) f(x)\, dx, \quad n = 0, \ldots, N, \quad (23.11)$$

with $\{\kappa_n(x)\}_{n=0}^N$ a collection of averaging weights (e.g. we use $\kappa_n(x) = \epsilon^{-1} 1_{\{x \in [x_n - \epsilon/2, x_n + \epsilon/2]\}}$ in the previous section), such that

$$\int_{-\infty}^{\infty} \kappa_n(x)\, dx = 1, \quad \kappa_n(x) \geq 0, \quad n = 0, \ldots, N.$$

Often the weights are taken to be shifted and scaled versions of a common weight function, in the sense that

$$\kappa_n(x) = \frac{1}{\epsilon_n} \kappa\left(\frac{x - x_n}{\epsilon_n}\right),$$

where

$$\int_{-\infty}^{\infty} \kappa(x)\, dx = 1, \quad \kappa(x) \geq 0.$$

The weights are shifted to center them around the grid points $\{x_n\}_{n=0}^{N}$, and scaling parameters ϵ_n control the dispersion of the weight around x_n. As the scaling parameters tend to zero, the averages of f tend to the values of f on the grid $\{x_n\}$.

23.3.2.1 Box Smoothing

A particularly simple averaging weight is the indicator function on the interval between middle-points of the grid buckets to the either side of x_n,

$$\kappa_n(x) = (c_{n+1} - c_n)^{-1} 1_{\{x \in [c_n, c_{n+1}]\}}, \tag{23.12}$$
$$c_n = (x_n + x_{n-1})/2, \quad n = 1, \dots, N-1.$$

Because of the shape of the weight function, the method is sometimes called the *box smoothing method*. The resulting discretization formula is given by

$$f_n = \frac{1}{c_{n+1} - c_n} \int_{c_n}^{c_{n+1}} f(x)\, dx, \quad c_n = (x_n + x_{n-1})/2, \quad n = 1, \dots, N-1. \tag{23.13}$$

If the function $f(x)$ is known for all x, as is the case for numerical integration, the box smoothing method is easy to apply, using the arguments that lead to (23.9) and (23.10). A more challenging situation arises when only a *discretized* version of the payoff is known, as may happen when f represents a PDE solution rolled back to some intermediate date. While backward induction in a PDE is, in itself, a smoothing operation, singularities may be introduced through the enforcement of jump conditions, as required if the security in question happens to pay a coupon, is exercisable, or has a barrier condition of some kind. Sometimes (e.g., for barrier options) the location of the singularity is known exactly, but often (e.g., for Bermudan style options) it is not. This complicates the application of (23.13), since knowledge of the location of singularities is critical to our ability to compute smooth greeks. We proceed to discuss a scheme to handle the case when the singularity location is not known.

To properly fix our setup, consider a security with terminal payout date T^* whose value is being computed by solving the corresponding PDE numerically, backwards from T^* to time 0. Let $V(t, x)$ represent the true value of the security at time t at state x. As always, let $V(t-, x)$ and $V(t+, x)$ be the value of the security just before and just after time t, respectively. Assuming that a lifecycle event takes place at some intermediate time T (such as an exercise opportunity, a knock-in/knock-out barrier check, a fixing of a structured coupon, and so on), a jump condition will be applied when crossing from $T+$ to $T-$ in the backward recursion scheme, see Section 2.7. Specifically, if $\{V_n^+\}_{n=0}^{N}$ represents the numerical approximation to $V(T+,x)$ on the grid $\{x_n\}_{n=0}^{N}$, the jump condition determines how to compute $\{V_n^{-}\}_{n=0}^{N}$, the grid approximation to $V(T-,x)$. Here we make an important observation: most jump conditions can be represented in the following form,

$$V(T-,x) = 1_{\{g(x) \leq h(x)\}} p(x) + 1_{\{g(x) > h(x)\}} q(x), \qquad (23.14)$$

where the discretized versions of the *smooth* functions $g(x)$, $h(x)$, $p(x)$ and $q(x)$ are known at $t = T+$.[6] Some of these functions could be based on $V(T+,x)$, and others are defined by the specifics of the event. Let us give a few examples.

Example 23.3.1. If the security can be canceled at time T, then

$$V(T-,x) = 1_{\{V(T+,x)>0\}} V(T+,x),$$

i.e. $g(x) = V(T+,x)$, $h(x) = 0$, $p(x) = 0$, $q(x) = V(T+,x)$.

Example 23.3.2. If the security is callable at time T with the exercise value $e(x)$, then

$$V(T-,x) = 1_{\{V(T+,x) \leq e(x)\}} e(x) + 1_{\{V(T+,x) > e(x)\}} V(T+,x),$$

i.e. $g(x) = V(T+,x)$, $h(x) = e(x)$, $p(x) = e(x)$, $q(x) = V(T+,x)$.

Example 23.3.3. Suppose the security knocks out at time T if the rate $r(x)$ is above the barrier b, and the knockout rebate is $a(x)$. Then

$$V(T-,x) = 1_{\{r(x) \leq b\}} V(T+,x) + 1_{\{r(x) > b\}} a(x),$$

i.e. $g(x) = r(x)$, $h(x) = b$, $p(x) = V(T+,x)$, $q(x) = a(x)$.

Example 23.3.4. If a coupon of the form $\max(r(x), s)$ is paid at time T, then

$$V(T-,x) = 1_{\{r(x) \leq s\}} (V(T+,x) + s) + 1_{\{r(x) > s\}} (V(T+,x) + r(x)),$$

i.e. $g(x) = r(x)$, $h(x) = s$, $p(x) = V(T+,x) + s$, $q(x) = V(T+,x) + r(x)$.

Going forward we assume that the event in question can indeed be represented in the form (23.14); let us denote the discretized versions of the smooth functions involved by $\{g_n\}$, $\{h_n\}$, $\{p_n\}$ and $\{q_n\}$.

Turning to the question of localizing singularities in $V(T-,x)$, we notice that the representation of the function in the form (23.14) simplifies our search, as all singularities are given by the solutions to the equation

---

[6]If there is more than one singularity introduced in the event, the decomposition above holds locally around each singularity. We consider a single singularity case only, with trivial extension to multiple ones.

$$g(x) - h(x) = 0.$$

Assume for simplicity that this equation has only one root[7], and denote it by x^*. The problem of finding x^* is complicated somewhat by the fact that the values of functions $g(x)$, $h(x)$ are only known at the grid points $\{x_n\}$. However, since $g(x)$ and $h(x)$ are smooth, linear interpolation on each of the intervals $[x_{n-1}, x_n]$, $n = 1, \ldots, N$, can be used instead. Specifically, we define

$$\widehat{h}(x) = h_n \frac{x - x_{n-1}}{x_n - x_{n-1}} + h_{n-1} \frac{x_n - x}{x_n - x_{n-1}}, \quad x \in [x_{n-1}, x_n],$$

$$\widehat{g}(x) = g_n \frac{x - x_{n-1}}{x_n - x_{n-1}} + g_{n-1} \frac{x_n - x}{x_n - x_{n-1}}, \quad x \in [x_{n-1}, x_n],$$

and use the solution $\widehat{x}^*$ to $\widehat{g}(x) - \widehat{h}(x) = 0$ as an approximation to x^*. To locate x^*, first note that an interval $[x_{n-1}, x_n]$ contains x^* (and $\widehat{x}^*$) provided that

$$(g_{n-1} - h_{n-1})(g_n - h_n) \leq 0.$$

If this inequality is satisfied, we pinpoint $\widehat{x}^*$ by solving a linear equation $\widehat{g}(x) - \widehat{h}(x) = 0$ on the interval $[x_{n-1}, x_n]$, so that $\widehat{x}^*$ is a solution to

$$h_n \frac{x - x_{n-1}}{x_n - x_{n-1}} + h_{n-1} \frac{x_n - x}{x_n - x_{n-1}} = g_n \frac{x - x_{n-1}}{x_n - x_{n-1}} + g_{n-1} \frac{x_n - x}{x_n - x_{n-1}}.$$

After some trivial algebraic manipulations,

$$\widehat{x}^* = \frac{h_n - g_n}{(h_n - g_n) + (g_{n-1} - h_{n-1})} x_{n-1} + \frac{g_{n-1} - h_{n-1}}{(h_n - g_n) + (g_{n-1} - h_{n-1})} x_n.$$

Having established (an approximation of) the singularity in $V(T-, x)$, we can proceed to approximate the integrals

$$\frac{1}{c_{n+1} - c_n} \int_{c_n}^{c_{n+1}} V(T-, x)\, dx,$$

for all n. There are two cases to consider. When the root is not inside $[c_n, c_{n+1}]$, i.e. $\widehat{x}^* \notin [c_n, c_{n+1}]$, (23.9) tells us that we can simply use the value of $V(T-, x)$ at x_n as an approximation to the integral. In other words, for such n we set

$$V_n^- = 1_{\{g_n \leq h_n\}} p_n + 1_{\{g_n > h_n\}} q_n.$$

For the case when $\widehat{x}^* \in [c_n, c_{n+1}]$, we split the integration domain into two, $[c_n, \widehat{x}^*)$ and $(\widehat{x}^*, c_{n+1}]$. By assumption, the function $V(T-, x)$ is smooth on each of the two intervals, so according to (23.10) we can approximate each

---

[7]In other words, we assume that the discretization is fine enough to guarantee that there is only one singularity per interval.

of the two integrals by the value of the function $V(T-, x)$ in the center of each of the two subintervals,

$$V_n^- = \frac{\widehat{x}^* - c_n}{c_{n+1} - c_n} V\left(T-, \frac{\widehat{x}^* + c_n}{2}\right) + \frac{c_{n+1} - \widehat{x}^*}{c_{n+1} - c_n} V\left(T-, \frac{\widehat{x}^* + c_{n+1}}{2}\right).$$

(23.15)

As should be clear from (23.14), on the interval $[c_n, \widehat{x}^*)$ the function $V(T-, x)$ is equal to one of the functions $p(x)$ or $q(x)$, and on the interval $(\widehat{x}^*, c_{n+1}]$ it is equal to the other. More precisely,

$$V(T-, x) = p(x) \quad \text{for } x \in [c_n, \widehat{x}^*), \qquad (23.16)$$
$$V(T-, x) = q(x) \quad \text{for } x \in (\widehat{x}^*, c_{n+1}],$$

if and only if $g_{n-1} < h_{n-1}$. Assume for concreteness that (23.16) in fact holds. Then (23.15) can be rewritten as

$$V_n^- = \frac{\widehat{x}^* - c_n}{c_{n+1} - c_n} p\left(\frac{\widehat{x}^* + c_n}{2}\right) + \frac{c_{n+1} - \widehat{x}^*}{c_{n+1} - c_n} q\left(\frac{\widehat{x}^* + c_{n+1}}{2}\right).$$

To find $p((\widehat{x}^* + c_n)/2)$, $q((\widehat{x}^* + c_{n+1})/2)$, we (once again) use the fact that the functions $p(x)$ and $q(x)$ are smooth, and approximate them linearly. Then, $p((\widehat{x}^* + c_n)/2)$, $q((\widehat{x}^* + c_{n+1})/2)$ can be computed from the known values of $p(x)$ and $q(x)$ on the grid $\{x_n\}$,

$$p\left(\frac{\widehat{x}^* + c_n}{2}\right) \approx \widehat{p}, \quad q\left(\frac{\widehat{x}^* + c_{n+1}}{2}\right) \approx \widehat{q},$$

where

$$\widehat{p} = p_{n-1} \frac{x_n - (\widehat{x}^* + c_n)/2}{x_n - x_{n-1}} + p_n \frac{(\widehat{x}^* + c_n)/2 - x_{n-1}}{x_n - x_{n-1}},$$
$$\widehat{q} = q_n \frac{x_{n+1} - (\widehat{x}^* + c_{n+1})/2}{x_{n+1} - x_n} + q_{n+1} \frac{(\widehat{x}^* + c_{n+1})/2 - x_n}{x_{n+1} - x_n}.$$

Combining various approximations, we finally obtain that for n such that $\widehat{x}^* \in [c_n, c_{n+1}]$,

$$V_n^- = \frac{\widehat{x}^* - c_n}{c_{n+1} - c_n} \widehat{p} + \frac{c_{n+1} - \widehat{x}^*}{c_{n+1} - c_n} \widehat{q},$$

with $\widehat{p}$, $\widehat{q}$ given just above. This concludes the description of the box smoothing method.

23.3.2.2 Other Smoothing Methods

In (23.11), weight functions other than indicator functions could be considered. A fairly popular alternative to box smoothing uses a weight function based on the linear Lagrange (triangular, or "hat") weight functions that we defined in footnote 10 on page 58. The resulting smoothing method is often known as the *hat smoothing method*. Apart from a slightly different functional form of the weight functions, its implementation differs little from the box smoothing method. In particular, for the method to be fully effective, discontinuities still need to be detected (as in the previous section) and incorporated into the calculation of integrals.

Unlike indicator functions, triangular weight functions are continuous, leading to smoother greeks than for box smoothing, especially for higher-order greeks such as gammas. On the other hand, hat smoothing is less "local" than box smoothing, in the sense that f_n computed in hat smoothing will depend on the values of $f(x)$ for $x \in [x_{n-1}, x_{n+1}]$, whereas f_n in box smoothing only depends on the values of $f(x)$ for $x \in [(x_n + x_{n-1})/2, (x_{n+1} + x_n)/2]$. This could be important when, for example, a trade feature (such as the exercise boundary) is close to the initial point of the grid in time and space, where more local functions tend to give better resolution, i.e. lower bias.

Stability of greeks of even higher order can be obtained by using weights that are even more smooth, i.e. Gaussian kernels. This, however, leads to more computationally intensive schemes, as well as schemes of ever more deteriorating locality.

23.3.3 Payoff Smoothing in Multiple Dimensions

The weight-based smoothing methods discussed in the previous sections also apply to multi-dimensional PDEs, although certain challenges quickly become evident, especially when one attempts to split the smoothing integrals into sub-domains around singularities. In one dimension, a singularity is just a single point which splits a single interval into two subintervals. In two dimensions, a singularity is typically a curve which affects multiple rectangles, and splits each affected rectangle into two subdomains of generally irregular geometry. Things get even more complicated in dimensions 3 and higher.

Let us consider the case of box smoothing in two dimensions in more detail, as it presents most of the challenges appearing in higher dimensions. We define $\{x_n\}_{n=0}^N$ and $\{y_m\}_{m=0}^M$ to be grids in x and y dimensions, and denote by (c_n^x, c_m^y) the center point of the rectangle $[x_{n-1}, x_n] \times [y_{n-1}, y_n]$,

$$c_n^x = (x_n + x_{n-1})/2, \quad c_m^y = (y_m + y_{m-1})/2.$$

Furthermore, let us define a rectangle

$$D_{n,m} = \left[c_n^x, c_{n+1}^x\right] \times \left[c_m^y, c_{m+1}^y\right].$$

Then we can introduce a collection of two-dimensional box weights

$$\kappa_{n,m}(x) = |D_{n,m}|^{-1} 1_{\{(x,y) \in D_{n,m}\}}, \quad n = 1, \ldots, N-1, \quad m = 1, \ldots, M-1,$$

where $|D|$ is the area of D. To smooth a function $f(x, y)$, an integral

$$f_{n,m} = \frac{1}{|D_{n,m}|} \int_{D_{n,m}} f(x,y)\, dx\, dy$$

is calculated for each n, m.

Recall that the one-dimensional box smoothing method we presented in Section 23.3.2.1 was based on the representation (23.14). In a similar vein, and using similar notations, we assume that the time $T-$ value of the security is given by

$$V(T-,x,y) = 1_{\{g(x,y) \le h(x,y)\}} p(x,y) + 1_{\{g(x,y) > h(x,y)\}} q(x,y). \quad (23.17)$$

Here the discretized versions of the smooth functions $g(x,y)$, $h(x,y)$, $p(x,y)$ and $q(x,y)$ are assumed known at time $T+$ on the 2-dimensional mesh $\{(x_n, y_m)\}$.

The singularity of $V(T-,x,y)$ is given by the one-dimensional curve

$$s \subset \mathbb{R}^2, \quad s = \{(x,y) : g(x,y) = h(x,y)\}.$$

As in one dimension, the value

$$\frac{1}{|D_{n,m}|} \int_{D_{n,m}} V(T-,x,y)\, dx\, dy \quad (23.18)$$

for $D_{n,m}$ such that $D_{n,m} \cap s = \emptyset$ can be approximated with the value $V(T-,x_n,y_m)$. If, however, $D_{n,m} \cap s \neq \emptyset$, the domain $D_{n,m}$ needs to be split up into two, and the integral of V computed on each of the subdomains separately. We note that, in general, there will be many rectangles $D_{n,m}$ where $D_{n,m} \cap s \neq \emptyset$.

The box smoothing method naturally splits into the following steps.

1. Approximate the value $g(x,y) - h(x,y)$ at the corners of each of $D_{n,m}$ by linear interpolation. We denote

$$\widehat{\xi}_{n,m} \triangleq \widehat{g}(c_n^x, c_m^y) - \widehat{h}(c_n^x, c_m^y),$$

where $\widehat{g}$ and $\widehat{h}$ are approximations to g, h computed using bi-linear interpolation off the grid $\{(x_n, y_m)\}$ (on which the values of g and h are known by assumption).

2. Find those rectangles $D_{n,m}$ for which $D_{n,m} \cap s \neq \emptyset$. The search can be conducted efficiently by looking for those $D_{n,m}$ for which the signs of the difference $g(x,y) - h(x,y)$ are not all the same in the corners of $D_{n,m}$. Specifically, we decide that $D_{n,m}$ contains a singularity if not all of $\widehat{\xi}_{n,m}$, $\widehat{\xi}_{n+1,m}$, $\widehat{\xi}_{n,m+1}$, $\widehat{\xi}_{n+1,m+1}$ have the same sign.

3. For those $D_{n,m}$ that do not contain a singularity, approximate the integral (23.18) with $V(T-,x_n,y_m)$, available from (23.17).

4. For those $D_{n,m}$ that do contain a singularity, approximate the singularity curve with a straight line through those two edges of the rectangle $D_{n,m}$ that are crossed by s. Note that s crosses the edge $(c_n^x, c_m^y) \to (c_{n+1}^x, c_m^y)$ if $\widehat{\xi}_{n,m} \cdot \widehat{\xi}_{n+1,m} \le 0$, and so on. Find approximations to the positions of the points where s crosses the two edges by linearly interpolating the appropriate values of $\widehat{\xi}_{n,m}$, $\widehat{\xi}_{n+1,m}$, $\widehat{\xi}_{n,m+1}$, $\widehat{\xi}_{n+1,m+1}$. For example, if s crosses the edge $(c_n^x, c_m^y) \to (c_{n+1}^x, c_m^y)$, then the point where s crosses the edge is given by $(\widehat{x}_1^*, \widehat{y}_1^*)$, where $\widehat{y}_1^* = c_m^y$, and $\widehat{x}_1^*$ is a solution to

$$\xi_{n,m} + \frac{\xi_{n+1,m} - \xi_{n,m}}{c_{n+1}^x - c_n^x} (x - c_n^x) = 0.$$

Denote the second crossing point by $(\widehat{x}_2^*, \widehat{y}_2^*)$ and approximate the part of s inside $D_{n,m}$ by a straight line[8] segment connecting $(\widehat{x}_1^*, \widehat{y}_1^*)$ and $(\widehat{x}_2^*, \widehat{y}_2^*)$.

5. Compute the integral (23.18) on each of the two subdomains of $D_{n,m}$. For that, split $D_{n,m}$ into two parts, $D_{n,m}^1$ and $D_{n,m}^2$, by the line segment connecting $(\widehat{x}_1^*, \widehat{y}_1^*)$ and $(\widehat{x}_2^*, \widehat{y}_2^*)$. On one subdomain ($D_{n,m}^1$ for concreteness), $V(T-, x, y)$ is equal to $p(x, y)$ and on the other ($D_{n,m}^2$) to $q(x, y)$. Use the fact that p and q are smooth and approximate them with linear functions $\widehat{p}$ and $\widehat{q}$ on $D_{n,m}^i$, $i = 1, 2$. To compute the integrals

$$\int_{D_{n,m}^1} \widehat{p}(x, y)\, dx\, dy, \quad \int_{D_{n,m}^2} \widehat{q}(x, y)\, dx\, dy,$$

use the fact that

$$\int_P l(x, y)\, dx\, dy = l(x_P, y_P) \times |P|$$

for any (integrable) domain P and linear function $l(x, y)$, where (x_P, y_P) is the center of mass of P.

6. Repeat Steps 4 and 5 for all $D_{n,m}$ such that $D_{n,m} \cap s \ne \emptyset$.

In three dimensions, singularities are given by two-dimensional surfaces; within each discretization cube, the singularities can be approximated by planes, and various cases as to how these planes intersect cubes need to be considered. In four dimensions, singularities must be approximated by cubes, and so forth. As the dimensionality of the problem grows, so does the amount of effort required to do smoothing. In a K-dimensional space, a singularity has dimension $K - 1$, so if the K-dimensional grid has an order N discretization for each of the K dimensions, then the number of K-dimensional grid segments (intervals in one-dimension, rectangles in two dimensions, cubes in three dimensions, etc.) that intersect the singularity

---

[8]In line with the footnote 7 we assume that the rectangles are small enough so that we can assume there is only two crossing points.

is of order N^{K-1}. The amount of work required per segment also generally grows with K, so a direct implementations of the smoothness algorithm in dimension 3 and above can already constitute a significant, if not dominant, proportion of calculation time.

In some multi-dimensional problems, the workload can be reduced by using known features of a product and/or model to understand the structure of singularities. For example, in some cases one of the PDE dimensions in a given model can be identified as being "dominant", in the sense that the singularity surface will be mostly orthogonal to this direction. In this case, rather than applying the full multi-dimensional smoothing method, a series of one-dimensional smoothing methods in the dominant dimension can be used instead, often at considerable time savings and with good smoothing results. For example, in SV models singularities will typically be present in the asset (S) dimension, while the payoff will be smooth in the variance (z) dimension. This suggests a scheme where one applies, at each discretized value of the stochastic variance, *one-dimensional* payoff smoothing in the direction of the asset state variable.

23.4 Payoff Smoothing for Monte Carlo

While the methods of Section 23.3.1 can be applied in a Monte Carlo setting, there exist more natural payoff smoothing methods for Monte Carlo applications. Starting from a very simple example in Section 23.4.1, we construct one such method here. The method is designed to be applied when calculating sensitivities by direct perturbation ("bump-and-reprice"). Alternative methods for sensitivity computations by Monte Carlo are described in Section 3.3 and, in particular, in Chapter 24).

23.4.1 Tube Monte Carlo for Digital Options

One situation where payoff smoothing is almost universally applied is in valuing digital options. Consider an option that pays

$$1_{\{S(T)>B\}} \tag{23.19}$$

at time T, where $S(t)$ is the process for the underlying that is simulated using Monte Carlo.

Since the payoff is discontinuous, the Monte Carlo estimate of $V = E(1_{\{S(T)>B\}})$, defined by (ω_j are sample paths, $j = 1, \ldots, J$)

$$V \approx J^{-1} \sum_{j=1}^{J} 1_{\{S(T,\omega_j)>B\}},$$

exhibits poor convergence (see Section 3.3.1.1) and unstable greeks. The standard way to remedy this is to replace the digital payoff with a call spread (or use the likelihood ratio method — but we do not consider it at the moment). Let us choose $K_1 < K_2$ and replace

$$1_{\{x>B\}} \approx f(x), \quad f(x) = \max\left(\min\left(\frac{x - K_1}{K_2 - K_1}, 1\right), 0\right).$$

Since $f(x)$ is smoother than $1_{\{x>B\}}$ (at least it is continuous), the stability of greeks is improved. Various choices of K_1, K_2 are possible. If $K_1 = B$, the call spread with the payoff $f(x)$ is often called an "underhedge" (as $f(x) \leq 1_{\{x>B\}}$ for all x). Conversely, if $K_2 = B$, then the call spread with the payoff $f(x)$ is called an "overhedge" ($f(x) \geq 1_{\{x>B\}}$ for all x). A symmetric payoff with $K_1 = B - \epsilon$, $K_2 = B + \epsilon$, $\epsilon > 0$, is used most often when the goal is to improve greeks stability while minimizing the bias introduced by the smoothness method. In this case,

$$f_{\text{sym}}(x) = \max\left(\min\left(\frac{x + \epsilon - B}{2\epsilon}, 1\right), 0\right). \tag{23.20}$$

The choice of the smoothing window $\epsilon = (K_2 - K_1)/2$ involves a trade-off between a high degree of smoothness (large ϵ) and low bias (low ϵ) and is usually performed experimentally. As we already mentioned, a typical strategy involves formulating a maximum tolerable level of the difference between the values of options with payoffs $1_{\{x>B\}}$ and $f(x)$, and then setting ϵ accordingly.

The method above is (in its symmetric form, at least) a special case of the moving average smoothing approach from Section 23.3.1. As we shall show next, in a Monte Carlo setting the method can also be justified a completely different way. While not particularly useful for the specific case of the digital option, this alternative interpretation allows us to formulate a generic Monte Carlo smoothing strategy applicable to more complicated payoffs.

In the general spirit of payoff smoothing, let us first replace the standard Monte Carlo approximation (23.19) with the following one,

$$V \approx J^{-1} \sum_{j=1}^{J} V_j, \quad V_j = \mathrm{E}\left(1_{\{S(T)>B\}} \middle| A_j\right), \tag{23.21}$$

where E is the expected value operator that corresponds to the pricing measure Q (whose exact nature is unimportant here, but could be taken to be the risk-neutral measure for concreteness), and where A_j is defined as a small interval centered at $S(T, \omega_j)$,

$$A_j = \{\omega : S(T, \omega) \in [S(T, \omega_j) - \epsilon, S(T, \omega_j) + \epsilon]\}, \quad \epsilon > 0.$$

The difference between (23.21) and the standard estimate (23.19) comes from replacing the "point" sample of the payoff $1_{\{x>B\}}$ at $S(T, \omega_j)$ with an "average" estimate of the payoff in a small interval around the sample asset value $S(T, \omega_j)$.

To compute $E(1_{\{S(T)>B\}}|A_j)$, we assume that the distribution of $S(T)$ within the interval $[S(T, \omega_j) - \epsilon, S(T, \omega_j) + \epsilon]$ can be approximated with a uniform distribution. If ϵ is small, the error introduced by this approximation is small. Then, if $B \notin A_j$, we have that $E(1_{\{S(T)>B\}}|A_j) = 1_{\{S(T,\omega_j)>B\}}$. If, however, $B \in A_j$, then (using the uniform distribution approximation as discussed above)

$$
\begin{aligned}
E\left(1_{\{S(T)>B\}}\big|\, A_j\right) &= Q\left(S(T) > B|\, S(T, \omega_j) - \epsilon \le S(T) \le S(T, \omega_j) + \epsilon\right) \\
&= \frac{S(T, \omega_j) + \epsilon - B}{2\epsilon}.
\end{aligned}
$$

Combining the two cases into one formula, we obtain

$$
\begin{aligned}
V_j &= E\left(1_{\{S(T)>B\}}\big|\, A_j\right) \\
&= \frac{S(T, \omega_j) + \epsilon - B}{2\epsilon} \times 1_{\{S(T,\omega_j)-\epsilon \le B < S(T,\omega_j)+\epsilon\}} \\
&\quad + 0 \times 1_{\{B < S(T,\omega_j)-\epsilon\}} + 1 \times 1_{\{S(T,\omega_j)+\epsilon \le B\}} \\
&= f_{\mathrm{sym}}(S(T, \omega_j)).
\end{aligned}
$$

Hence, (23.21) can be rewritten as

$$
V \approx J^{-1} \sum_{j=1}^{J} f_{\mathrm{sym}}\left(S\left(T, \omega_j\right)\right),
$$

and the "call spread" method is motivated from the probabilistic perspective.

The derivation above points to a systematic approach for obtaining Monte Carlo specific payoff smoothing approximations for a wide variety of payoffs. First, we replace point estimates of the payoff along each sample path with averages of the payoff over a suitably defined small neighborhood of the sample path. Then, to compute the required average value over each neighborhood, we use various approximations that can be justified by the fact that each neighborhood is small. We call this method the *tube Monte Carlo* (also sometimes known as *sausage Monte Carlo*, see Piterbarg [2004c]), with the name reflecting the fact that small neighborhoods around sample paths resemble thin, narrow (multi-dimensional) "tubes". In the next section, we apply the tube Monte Carlo method to a more interesting class of payoffs.

23.4.2 Tube Monte Carlo for Barrier Options

Consider a derivative which is a knock-in barrier into a stream of (net) coupons $X_1, \ldots, X_{N-1}$, with the knock-in feature defined by a stopping time index η: the derivative pays coupons X_i at T_{i+1} for $i = \eta, \ldots, N-1$. The value of the security is then given by

$$V_{\mathrm{ki}}(0) = \mathrm{E}\left(\sum_{i=1}^{N-1} B(T_{i+1})^{-1}X_i 1_{\{i\geq\eta\}}\right), \qquad (23.22)$$

where E is the expected value operator for the spot measure Q^B, with $B(t)$ in (23.22) being the discretely compounded money market account. The standard estimate of this value in Monte Carlo simulation is given by

$$V_{\mathrm{ki}}(0) \approx \frac{1}{J}\sum_{j=1}^{J}\left(\sum_{i=1}^{N-1} B(T_{i+1},\omega_j)^{-1}X_i(\omega_j)1_{\{i\geq\eta(\omega_j)\}}\right), \qquad (23.23)$$

where $\omega_1,\ldots,\omega_J$ are Monte Carlo paths and where we use the notation $\xi(\omega)$ for the value of a random variable ξ on path ω.

The indicator functions $1_{\{i\geq\eta(\omega)\}}$ in (23.23) introduce digital discontinuities in the payoff which, as we know, lead to poor stability of risk sensitivities. To improve on this situation, let us consider how to apply the payoff smoothing ideas of Section 23.4.1 here. Let $x(t,\omega)$ be the d-dimensional vector of state variables of the underlying model which we, without practical loss of generality, assume to be Markovian. We further assume that we can write the stopping time index η as the first hitting time of a state-dependent boundary,

$$\eta(\omega) = \min\{n \geq 1 : \psi_n(x(T_n,\omega)) \geq 0\} \wedge N, \qquad (23.24)$$

where $\psi_n(x)$ are some functions, $\psi_n : \mathbb{R}^d \to \mathbb{R}$.

The idea of the tube method is to replace point estimates (23.23) of the payoff with payoff averages over appropriately defined tubes. Let us fix $\epsilon > 0$, the width of the tube. For each j we define the ϵ-tubes in the state space by

$$A_j^\epsilon = \bigcap_{i=1}^{N-1} A_{j,i}^\epsilon, \qquad (23.25)$$

$$A_{j,i}^\epsilon = \{\omega : \|x(T_i,\omega) - x(T_i,\omega_j)\| < \epsilon\},$$

where, essentially, A_j^ϵ denotes the set of all sample paths that come within ϵ-distance of $x(T_i,\omega_j)$'s for all T_i, $i = 1,\ldots,N-1$. Then we replace (23.23) with the following estimator,

$$V_{\mathrm{ki}}(0) \approx J^{-1}\sum_{j=1}^{J}V_j, \quad V_j \triangleq \mathrm{E}\left(\sum_{i=1}^{N-1}\left[B(T_{i+1},\omega)^{-1}X_i(\omega)1_{\{i\geq\eta(\omega)\}}\right]\bigg| A_j^\epsilon\right).$$

$$(23.26)$$

Since $B(T_{i+1},\omega)^{-1}$ and $X_i(\omega)$ are, often, smooth[9] functions of the path ω, we evaluate them just at the sample path,

---

[9]$X_i(\omega)$ can, of course, be discontinuous, but this is not our focus at the moment.

$$V_j \approx \sum_{i=1}^{N-1} B(T_{i+1}, \omega_j)^{-1} X_i(\omega_j) \, \mathrm{E}\left(1_{\{i \geq \eta(\omega)\}} \middle| A_j^\epsilon\right), \qquad (23.27)$$

which approximates (23.26) to order ϵ. To proceed we need the following proposition.

Proposition 23.4.1. *In (23.27) the probabilities*

$$q_i(\omega_j) \triangleq \mathrm{E}\left(1_{\{i \geq \eta(\omega)\}} \middle| A_j^\epsilon\right),$$

can be approximated as follows:

$$1 - q_i(\omega_j) = \prod_{n=1}^{i} (1 - p_n(\omega_j)), \qquad (23.28)$$

$$p_n(\omega_j) \triangleq \begin{cases} 1, & \psi_{n,j} - \delta_{n,j} \geq 0, \\ \frac{\delta_{n,j} + \psi_{n,j}}{2\delta_{n,j}}, & \psi_{n,j} - \delta_{n,j} < 0 < \psi_{n,j} + \delta_{n,j}, \\ 0, & \psi_{n,j} + \delta_{n,j} \leq 0, \end{cases}$$

where

$$\psi_{n,j} \triangleq \psi_n(x(T_n, \omega_j)), \quad \delta_{n,j} \triangleq \epsilon \|\nabla \psi_{n,j}\|, \quad \nabla \psi_{n,j} \triangleq \nabla \psi_n(x)|_{x=x(T_n, \omega_j)}.$$

Proof. By expressing A_j^ϵ in terms of the functions ψ_n that define the knock-in index time in (23.24), we get

$$1 - q_i(\omega_j) = Q^B\left(\bigcap_{n=1}^{i} \{\psi_n(x(T_n, \omega)) \leq 0\} \middle| A_j^\epsilon\right). \qquad (23.29)$$

We claim that, to order ϵ,

$$1 - q_i(\omega_j) = \prod_{n=1}^{i} Q^B\left(\psi_n(x(T_n, \omega)) \leq 0 \middle| A_{j,n}^\epsilon\right). \qquad (23.30)$$

The proof follows by repeated applications of Lemma 23.B.1 from Appendix 23.B, although the intuition behind it is rather simple. Conditioning on $A_j^\epsilon = \bigcap_i A_{j,i}^\epsilon$ is essentially equivalent to pinning down the Markov process at times T_i, $i = 1, \ldots, N-1$, to known values $\{x(T_i, \omega_j)\}$ with ϵ-accuracy. If a Markov process is conditioned on being at a certain state on a given date, past and future events become conditionally independent. Then, the set intersection on the right-hand side of (23.29) can be unwrapped into the product on the right-hand side of (23.30).

Now, define

$$p_n(\omega_j) = Q^B\left(\psi_n(x(T_n), \omega) > 0 \middle| A_{j,n}^\epsilon\right)$$
$$= Q^B\left(\psi_n(x(T_n, \omega)) > 0 \middle| \|x(T_n, \omega) - x(T_n, \omega_j)\| < \epsilon\right),$$

so that (23.30) becomes

$$1 - q_i(\omega_j) = \prod_{n=1}^{i} (1 - p_n(\omega_j)).$$

To compute the p_n, observe that since we assumed that functions ψ_n are smooth, we may write for x such that $||x - x(T_n, \omega_j)|| < \epsilon$,

$$\psi_n(x) \approx \psi_{n,j} + \nabla \psi_{n,j} \times (x - x(T_n, \omega_j)), \qquad (23.31)$$

(here $\nabla \psi$ is the gradient of ψ, a row vector). Define $O_{n,j} \subset \mathbb{R}$ by

$$O_{n,j} = \psi_n \left(\{ z \in \mathbb{R}^d : ||z - x(T_i, \omega_j)|| < \epsilon \} \right),$$

i.e. the image of the ball $||z - x(T_i, \omega_j)|| < \epsilon$ under mapping $\psi_n : \mathbb{R}^d \to \mathbb{R}$. Then, from (23.31),

$$O_{n,j} \approx [\psi_{n,j} - ||\nabla \psi_{n,j}|| \, \epsilon, \psi_{n,j} + ||\nabla \psi_{n,j}|| \, \epsilon],$$

where $||\nabla \psi_{n,j}||$ denotes the norm of the linear operator $\nabla \psi_{n,j}$. Under the approximation

$$\begin{aligned} A_{j,n}^{\epsilon} &= \{ \omega : ||x(T_n, \omega) - x(T_n, \omega_j)|| < \epsilon \} \\ &\approx \{ \omega : \psi_n(x(T_n, \omega)) \in O_{n,j} \} \end{aligned}$$

we get

$$p_n(\omega_j) \approx Q^B \left(\psi_n(x(T_n, \omega)) > 0 \, | \, \psi_n(x(T_n, \omega)) \in O_{n,j} \right).$$

Approximating conditional distribution of $\psi_n(x(T_n, \omega))$ by a uniform distribution on the set $O_{n,j}$ we obtain

$$\begin{aligned} p_n(\omega_j) &\approx \frac{|\{\psi_n > 0\} \cap O_{n,j}|}{|O_{n,j}|} \\ &= \frac{|\{\psi_n > 0\} \cap [\psi_{n,j} - ||\nabla \psi_{n,j}|| \, \epsilon, \psi_{n,j} + ||\nabla \psi_{n,j}|| \, \epsilon]|}{|[\psi_{n,j} - ||\nabla \psi_{n,j}|| \, \epsilon, \psi_{n,j} + ||\nabla \psi_{n,j}|| \, \epsilon]|}. \end{aligned}$$

where we use $|\cdot|$ to denote the length of intervals in $\mathbb{R}$. Denoting

$$\delta_{n,j} = \epsilon \, ||\nabla \psi_{n,j}||,$$

we obtain

$$|[\psi_{n,j} - ||\nabla \psi_{n,j}|| \, \epsilon, \psi_{n,j} + ||\nabla \psi_{n,j}|| \, \epsilon]| = 2\delta_{n,j},$$

and

$$p_n(\omega_j) = \begin{cases} 1, & \psi_{n,j} - \delta_{n,j} \geq 0, \\ \frac{\delta_{n,j} + \psi_{n,j}}{2\delta_{n,j}}, & \psi_{n,j} - \delta_{n,j} < 0 < \psi_{n,j} + \delta_{n,j}, \\ 0, & \psi_{n,j} + \delta_{n,j} \leq 0. \end{cases} \qquad (23.32)$$

This completes the derivation.

□

Combining the results together, the formula for the tube Monte Carlo of a discrete knock-in barrier is then given by

$$V_{\text{ki}}(0) \approx J^{-1} \sum_{j=1}^{J} V_j, \quad V_j = \sum_{i=1}^{N-1} B(T_{i+1}, \omega_j)^{-1} X_i(\omega_j) q_i(\omega_j), \quad (23.33)$$

with q_i's given by Proposition 23.4.1.

Let us analyze this formula in some detail. The quantity $\psi_{n,j}$ in (23.28) measures how far into the knock-in region the state process went, so we call it the "overshoot" function. The quantity $\delta_{n,j}$ is the "window" over which the overshoot function is smoothed out. It is equal to the universal constant ϵ (smoothing window for the state variables $x(\cdot)$) times the size of the gradient of the overshoot function. This provides consistent scaling of smoothing windows across different times/simulated paths. If the overshoot function is high (above $\delta_{n,j}$) then the knock-in barrier is deemed completely breached, and we set $p_n(\omega_j) = 1$. If the overshoot function is low (below $-\delta_{n,j}$), knock-in region is deemed to not have been reached at all. And for cases in between, the knock-in barrier is considered "partially" breached[10], and a weight of $(\delta_{n,j} + \psi_{n,j})/2\delta_{n,j}$ is used to measure the extent of the barrier breach. Another analogy uses the idea of a partial knock-in: if the path ω_j is near the knock-in boundary, relevant coupons get included in the derivative value only partially, with the weights $q_i(\omega_j)$ defining the fractions of the coupons that count. This is in contrast to the standard Monte Carlo formula (23.23) in which coupons get included in the value either completely or not at all.

Critically, the weights $p_n(\omega_j)$ change smoothly with ω_j as do therefore the V_j's in (23.33) (unlike those in (23.23) with digital discontinuities). The tube Monte Carlo formula (23.33) converges to the standard formula (23.23) as ϵ gets small. Clearly, the larger the smoothing window ϵ is, the smoother the payoff becomes, resulting in more stable risk sensitivities. With larger ϵ, however, the bias of the approximation becomes larger. In practice, to balance smoothness versus accuracy, one would start with a small ϵ and then keep increasing it for as long as the observed bias in the price remains within pre-set tolerances. Once the upper acceptable bound on ϵ is established, it can be used in risk sensitivity calculations.

---

[10]The concept of "partial" membership in a set should be familiar to those schooled in *fuzzy logic* (see Zadeh [1965]), and tube Monte Carlo can, in fact, be considered a probabilistically motivated fuzzy logic algorithm. For more discussion of fuzzy logic applications to Monte Carlo sensitivity computations in finance, see Withington and Lucic [2009].

23.4.3 Tube Monte Carlo for Callable Libor Exotics

The method of Section 23.4.2 can be applied directly to callable Libor exotics in Monte Carlo (see Section 18.3) whose valuation often relies on representing them as knock-in discrete barriers with the knock-in defined by an estimate of the exercise index — see e.g. (18.27), (18.28). Interestingly, the exact value of a CLE is a *smooth* function of the underlying path (as we establish later in Chapter 24), yet the representation such as (18.28) introduces digital discontinuities in the payoff. Therefore, for CLEs it is more advisable to use risk calculation methods that are specifically adopted to the CLE structure and its smoothness; such methods are developed in Chapter 24. Tube Monte Carlo, however, still has its place in the arsenal of valuation methods for CLEs as it often integrates better with standard risk system designs, compared to the more specialized methods of Chapter 24. In terms of performance, the effectiveness of the tube method compared to the alternatives depends on many underlying factors, but it is shown in Piterbarg [2005a] that to achieve comparable risk stability, the tube Monte Carlo method typically requires only about $1/4$ of the path count needed for the standard simulation. The pathwise differentiation method of Chapter 24 reduces the required number of paths by another factor of 3 to 4.

Most of the mechanics required to apply the tube Monte Carlo method to CLEs have already been developed in Section 23.4.2. In fact, we only need to describe the functions ψ_n that define knock-in (or, in the context of CLEs, exercise) regions for each exercise date. This is straightforward to do; with (18.27) in mind, we just set

$$\psi_n(x(T_n,\omega)) = \mathcal{C}\left(\widetilde{U}_n(T_n)\right)^\top \zeta(T_n,\omega) - \mathcal{C}\left(\widetilde{H}_n(T_n)\right)^\top \zeta(T_n,\omega),$$

where we treat the right-hand-side — that is, the difference between exercise and hold values as measured by exercise and hold regression polynomials applied to explanatory variables of the regression — as a function of the model state variable vector $x(T_n,\omega)$. The method of Section 23.4.2 now carries over unchanged.

23.4.4 Tube Monte Carlo for TARNs

A TARN (see Section 20.1) can be represented as a derivative that pays a stream of (net) coupons until a knock-out event takes place when a sum of structured coupons exceeds a certain target. As knock-out derivatives are closely related to knock-in's, it is no surprise that a tube Monte Carlo method similar to that of Section 23.4.2 can be developed for TARNs (see Piterbarg [2004c]).

Let us recall the main TARN valuation formula (20.2), which we rewrite in a form similar to (23.22),

$$V_{\text{tarn}}(0) = \mathrm{E}\left(\sum_{i=1}^{N-1} B(T_{i+1})^{-1} X_i 1_{\{i<\eta\}}\right),$$

where

$$\eta(\omega) = \min\{n \geq 1 : Q_n(\omega) - R \geq 0\} \wedge N.$$

Then, in a close analogy to (23.33), we can write the approximation formula for the tube Monte Carlo method,

$$V_{\text{tarn}}(0) \approx J^{-1}\sum_{j=1}^{J} V_j, \quad V_j = \sum_{i=1}^{N-1} B(T_{i+1}, \omega_j)^{-1} X_i(\omega_j)(1 - q_i(\omega_j)),$$

where

$$1 - q_i(\omega_j) = \prod_{n=1}^{i}(1 - p_n(\omega_j)),$$

$$p_n(\omega_j) = \min\left(\max\left(\frac{Q_n(\omega_j) - R + \delta_{n,j}}{2\delta_{n,j}}, 0\right), 1\right),$$

and

$$\delta_{n,j} = \epsilon \|\nabla Q_n(\omega_j)\|, \tag{23.34}$$

with $\nabla Q_n(\omega_j)$ understood to be the gradient of Q_n expressed as a function of the model state vector.

High level of accuracy is not really required when calculating scaling constants $\delta_{n,j}$ in (23.34). In particular, for efficiency reasons we may use a simpler, deterministic scaling

$$\delta_{n,j} = \epsilon_n$$

for a collection $\{\epsilon_n\}$ or even a time-independent deterministic scaling

$$\delta_{n,j} = \epsilon.$$

The same simplifications could, of course, be adopted for knock-in barrier options and callable Libor exotics.

23.A Appendix: Delta Continuity of Singularity-Enlarged Grid Method

To show that the derivative of $\widetilde{U}(S)$ is continuous across the grid point, it is sufficient to show that the left derivative of $\widetilde{U}(s)$ at S_m equals the right derivative (at S_m). To simplify notations, let us assume that $\Delta x_n \equiv \Delta$, $n = 1, \ldots, N$, $\psi(x) \equiv 1$ for x in some neighborhood of x_m, and redefine

$$\overline{U}(S) = \frac{2}{\Delta}\tilde{U}(S).$$

We then have, for $\epsilon > 0$, that

$$\overline{U}\left(S_m + \epsilon\right) = \left(f(K) + f\left(K - \Delta + \epsilon\right)\right)\left(1 - \frac{\epsilon}{\Delta}\right)$$
$$+ \left(f\left(K + \epsilon\right) + f(K)\right)\frac{\epsilon}{\Delta}$$
$$+ \left(f\left(K + \Delta + \epsilon\right) + f\left(K + \epsilon\right)\right),$$
$$\overline{U}\left(S_m\right) = 2f(K) + f\left(K - \Delta\right) + f\left(K + \Delta\right),$$
$$\overline{U}\left(S_m - \epsilon\right) = \left(f\left(K - \epsilon\right) + f\left(K - \Delta - \epsilon\right)\right)$$
$$+ \left(f(K) + f\left(K - \epsilon\right)\right)\frac{\epsilon}{\Delta}$$
$$+ \left(f\left(K + \Delta - \epsilon\right) + f(K)\right)\left(1 - \frac{\epsilon}{\Delta}\right).$$

In particular

$$\overline{U}\left(S_m + \epsilon\right) - \overline{U}\left(S_m\right) = \left(f\left(K + \Delta + \epsilon\right) + f\left(K + \epsilon\right)\right) - \left(f\left(K + \Delta\right) + f(K)\right)$$
$$+ \left(f(K) + f\left(K - \Delta + \epsilon\right)\right) - \left(f(K) + f\left(K - \Delta\right)\right)$$
$$+ \frac{\epsilon}{\Delta}\left(f\left(K + \epsilon\right) - f\left(K - \Delta + \epsilon\right)\right),$$

and

$$D^+\overline{U}\left(S_m\right) \triangleq \lim_{\epsilon \downarrow 0} \epsilon^{-1}\left(\overline{U}\left(S_m + \epsilon\right) - \overline{U}\left(S_m\right)\right)$$
$$= D^+ f(K) + \frac{1}{\Delta}\left(f(K) - f\left(K - \Delta\right)\right)$$
$$+ D^+ f\left(K + \Delta\right) + D^+ f\left(K - \Delta\right). \tag{23.35}$$

Likewise,

$$\overline{U}\left(S_m\right) - \overline{U}\left(S_m - \epsilon\right) = \left(f(K) + f\left(K - \Delta\right)\right) - \left(f\left(K - \epsilon\right) + f\left(K - \Delta - \epsilon\right)\right)$$
$$+ \left(f\left(K + \Delta\right) - f\left(K + \Delta - \epsilon\right)\right)$$
$$+ \frac{\epsilon}{\Delta}\left(f\left(K + \Delta - \epsilon\right) - f\left(K - \epsilon\right)\right),$$

and

$$D^-\overline{U}\left(S_m\right) \triangleq \lim_{\epsilon \uparrow 0} \epsilon^{-1}\left(\overline{U}\left(S_m\right) - \overline{U}\left(S_m - \epsilon\right)\right)$$
$$= D^- f(K) + \frac{1}{\Delta}\left(f\left(K + \Delta\right) - f(K)\right)$$
$$+ D^- f\left(K + \Delta\right) + D^- f\left(K - \Delta\right). \tag{23.36}$$

Combining (23.35), (23.36) together and using the fact that the derivative of $f(x)$ is continuous everywhere except at K, we obtain

$$D^+\overline{U}(S_m) - D^-\overline{U}(S_m) = \left(D^+ f(K) - D^- f(K)\right)$$
$$+ \left(\frac{1}{\Delta}\left(f(K) - f(K - \Delta)\right) - \frac{1}{\Delta}\left(f(K + \Delta) - f(K)\right)\right).$$

We note that, to the second order,

$$\frac{1}{\Delta}\left(f(K) - f(K - \Delta)\right) \approx D^- f(K),$$
$$\frac{1}{\Delta}\left(f(K + \Delta) - f(K)\right) \approx D^+ f(K),$$

and thus, to the second order,

$$D^+\overline{U}(S_m) \approx D^-\overline{U}(S_m).$$

We conclude that the quadrature method produces smooth deltas.

23.B Appendix: Proof of Approximate Conditional Independence for Tube Monte Carlo

Here we prove a lemma needed in the proof of Proposition 23.4.1. Let $x(t, \omega)$, $t \in [0, T]$, be a Markov process with a state space $\mathbb{R}^d$ for some $d \geq 1$. Assume its transition density

$$Q\left(x(t) = z \mid x(s) = y\right)$$

is differentiable in z and y for all t, $s > 0$. Let $T_1 < T_2$ and define, for some $\epsilon > 0$,

$$U_i^\epsilon = \left\{\omega : \|x(T_i, \omega) - x_i\| < \epsilon\right\}, \quad i = 1, 2,$$

for some x_1, x_2.

Lemma 23.B.1. *Let X_1 and X_2 be two subsets of the state space $\mathbb{R}^d$ and define*

$$Z_i = \left\{\omega : x(T_i, \omega) \in X_i\right\}, \quad i = 1, 2.$$

Then

$$Q\left(Z_1 \cap Z_2 \mid U_1^\epsilon \cap U_2^\epsilon\right) = Q\left(Z_1 \mid U_1^\epsilon\right) Q\left(Z_2 \mid U_2^\epsilon\right)\left(1 + O(\epsilon)\right)$$

as $\epsilon \to 0$.

Proof. We have,

$$Q\left(Z_1 \cap Z_2 \mid U_1^\epsilon \cap U_2^\epsilon\right) = \frac{Q\left(Z_1 \cap Z_2 \cap U_1^\epsilon \cap U_2^\epsilon\right)}{Q\left(U_1^\epsilon \cap U_2^\epsilon\right)}.$$

For the expression in the numerator we have

$$Q\left(Z_1 \cap Z_2 \cap U_1^\epsilon \cap U_2^\epsilon\right) = \mathrm{E}\left(1_{\{Z_1\}}1_{\{Z_2\}}1_{\{U_1^\epsilon\}}1_{\{U_2^\epsilon\}}\right)$$
$$= \int dy \int dz\, Q(x(T_1) = y)1_{\{y \in X_1\}}1_{\{\|y - x_1\| < \epsilon\}}$$
$$\times\, Q(x(T_2) = z | x(T_1) = y)1_{\{z \in X_2\}}1_{\{\|z - x_2\| < \epsilon\}}.$$

As the transition density is differentiable, for y such that $1_{\{U_1^\epsilon\}}(y) \neq 0$ we have that

$$Q(x(T_2) = z | x(T_1) = y) = Q(x(T_2) = z | x(T_1) = x_1)(1 + O(\epsilon)),$$

so we can write

$$Q\left((Z_1 \cap Z_2 \cap U_1^\epsilon \cap U_2^\epsilon\right)$$
$$= \int dy \int dz\, Q(x(T_1) = y)1_{\{y \in X_1\}}1_{\{\|y - x_1\| < \epsilon\}}$$
$$\times\, Q(x(T_2) = z | x(T_1) = x_1)1_{\{z \in X_2\}}1_{\{\|z - x_2\| < \epsilon\}}\, (1 + O(\epsilon))$$
$$= \mathrm{E}\left(1_{\{Z_1\}}1_{\{U_1^\epsilon\}}\right)\mathrm{E}\left(1_{\{Z_2\}}1_{\{U_2^\epsilon\}}\, \big|\, x(T_1) = x_1\right)(1 + O(\epsilon)).$$

Now

$$\mathrm{E}\left(1_{\{Z_2\}}1_{\{U_2^\epsilon\}}\, \big|\, x(T_1) = x_1\right)$$
$$= \mathrm{E}\left(1_{\{Z_2\}}\, \big|\, x(T_1) = x_1, U_2^\epsilon\right)\mathrm{E}\left(1_{\{U_2^\epsilon\}}\, \big|\, x(T_1) = x_1\right)$$
$$= \mathrm{E}\left(1_{\{Z_2\}}\, \big|\, x(T_1) = x_1, x(T_2) = x_2\right)(1 + O(\epsilon))\mathrm{E}\left(1_{\{U_2^\epsilon\}}\, \big|\, x(T_1) = x_1\right),$$

where again we used the regularity properties of the transition density. By the Markovian property and the regularity of density,

$$\mathrm{E}\left(1_{\{Z_2\}}\, \big|\, x(T_1) = x_1, x(T_2) = x_2\right) = \mathrm{E}\left(1_{\{Z_2\}}\, \big|\, x(T_2) = x_2\right)$$
$$= \mathrm{E}\left(1_{\{Z_2\}}\, \big|\, U_2^\epsilon\right)(1 + O(\epsilon)).$$

Therefore, up to $O(\epsilon)$,

$$Q\left(Z_1 \cap Z_2 \cap U_1^\epsilon \cap U_2^\epsilon\right) = \mathrm{E}\left(1_{\{Z_1\}}1_{\{U_1^\epsilon\}}\right)$$
$$\times\, \mathrm{E}\left(1_{\{Z_2\}}\, \big|\, x(T_2) = x_2\right)\mathrm{E}\left(1_{\{U_2^\epsilon\}}\, \big|\, x(T_1) = x_1\right)(1 + O(\epsilon)).$$

Hence

$$Q\left(Z_1 \cap Z_2 | U_1^\epsilon \cap U_2^\epsilon\right)$$
$$= \frac{\mathrm{E}\left(1_{\{Z_1\}}1_{\{U_1^\epsilon\}}\right)\mathrm{E}\left(1_{\{Z_2\}}\, \big|\, U_2^\epsilon\right)\mathrm{E}\left(1_{\{U_2^\epsilon\}}\, \big|\, x(T_1) = x_1\right)(1 + O(\epsilon))}{\mathrm{E}\left(1_{\{U_1^\epsilon\}}\right)\mathrm{E}\left(1 | U_2^\epsilon\right)\mathrm{E}\left(1_{\{U_2^\epsilon\}}\, \big|\, x(T_1) = x_1\right)(1 + O(\epsilon))}$$
$$= \frac{\mathrm{E}\left(1_{\{Z_1\}}1_{\{U_1^\epsilon\}}\right)}{\mathrm{E}\left(1_{\{U_1^\epsilon\}}\right)} \times \frac{\mathrm{E}\left(1_{\{Z_2\}}\, \big|\, U_2^\epsilon\right)}{\mathrm{E}\left(1 | U_2^\epsilon\right)}(1 + O(\epsilon))$$
$$= Q\left(Z_1 | U_1^\epsilon\right)Q\left(Z_2 | U_2^\epsilon\right)(1 + O(\epsilon)),$$

as claimed. $\square$

Pathwise Differentiation

The various payoff smoothing methods of Chapter 23 primarily target greeks computed through outright repricing with perturbed market data. However, as we have already seen in Section 3.3, there exist methods for risk calculations that avoid brute-force repricing entirely. In this chapter, we concentrate on the convenient *pathwise differentiation method,* paying particular attention to applications involving securities with barriers or early exercise rights.

24.1 Pathwise Differentiation: Foundations

24.1.1 Callable Libor Exotics

The pathwise differentiation method for European-style derivatives has been considered (in a Monte Carlo setting) in Section 3.3.2. As it turns out, Bermudan-style callable derivatives are also quite amenable to the pathwise differentiation method, as shown in Piterbarg [2004b]. Let us outline the basic ideas.

Using the notations from Chapter 18, we recall that the main valuation recursion for a CLE is given by (see (18.7))

$$H_{n-1}(T_{n-1}) = B(T_{n-1})\mathrm{E}_{T_{n-1}}\left(B(T_n)^{-1}\max\left(H_n(T_n), U_n(T_n)\right)\right), \quad (24.1)$$

for $n = N - 1, \dots, 1$, with the starting condition $H_{N-1} \equiv 0$. Here, E is the expected value operator for the spot measure Q^B, $H_n(t)$ is the n-th hold value, and $U_n(t)$ the n-th exercise value, that is, the value of all future cash flows received upon exercise at time T_n:

$$U_n(t) = B(t)\sum_{i=n}^{N-1}\mathrm{E}_t\left(B(T_{i+1})^{-1}X_i\right).$$

Here X_i are net coupons, $X_i = \tau_i(C_i - L_i(T_i))$, with C_i being the structured coupons and L_i the Libor rates.

Let Δ_α represent a pathwise differentiation operator with respect to a given parameter α. In this section we derive the main representation result that allows us to write a pathwise derivative of a callable Libor exotic as an expectation of a function of the optimal exercise time.

24.1.1.1 CLE Greeks and the Optimal Exercise Time

In order for the pathwise differentiation method to be applicable, we always assume that all coupons X_n, $n = 1, \ldots, N - 1$, and the inverse numeraire $B(t)^{-1}$, are Lipschitz continuous functions of the parameter α. It follows then than the pathwise derivative $\Delta_\alpha X_n$ exists almost surely for each $n = 1, \ldots, N - 1$.

From (24.1), carrying out the differentiation under the expectation operator, we obtain our first result for pathwise derivatives of CLEs.

Proposition 24.1.1. *Provided the coupons and inverse numeraire are Lipschitz continuous, then, for any n, $n = 1, \ldots, N - 1$,*

$$\Delta_\alpha \left(B(T_{n-1})^{-1} H_{n-1}(T_{n-1}) \right)$$
$$= \mathrm{E}_{T_{n-1}} \left(1_{\{U_n(T_n) > H_n(T_n)\}} \Delta_\alpha \left(B(T_n)^{-1} U_n(T_n) \right) \right)$$
$$+ \mathrm{E}_{T_{n-1}} \left(1_{\{H_n(T_n) > U_n(T_n)\}} \Delta_\alpha \left(B(T_n)^{-1} H_n(T_n) \right) \right).$$

Proof. The assumption of Lipschitz continuity of the coupons and the inverse numeraire implies that $U_n(T_n)$ for each n, $n = 1, \ldots, N - 1$, is Lipschitz continuous in α, as is (by assumption) the inverse numeraire $B(t)^{-1}$. Since the function $\max(x, y)$ is Lipschitz continuous in x (and y), it can be shown recursively from (24.1) that $H_n(T_n)$ for each n, $n = 0, \ldots, N - 1$, is Lipschitz continuous in α as well. Hence, Proposition 3.3.1 applies and we have,

$$\Delta_\alpha \left(B(T_{n-1})^{-1} H_{n-1}(T_{n-1}) \right)$$
$$= \mathrm{E}_{T_{n-1}} \left(\Delta_\alpha \left(\max \left(B(T_n)^{-1} H_n(T_n), B(T_n)^{-1} U_n(T_n) \right) \right) \right).$$

The function $x \longmapsto \max(x, c)$ is absolutely continuous with a derivative that is equal to $1_{\{x > c\}}$. Hence, we can differentiate $\max(H_n(T_n), U_n(T_n))$ inside the expected value to obtain

$$\mathrm{E}_{T_{n-1}} \left(\Delta_\alpha \max \left(B(T_n)^{-1} H_n(T_n), B(T_n)^{-1} U_n(T_n) \right) \right)$$
$$= \mathrm{E}_{T_{n-1}} \left(1_{\{U_n(T_n) > H_n(T_n)\}} \Delta_\alpha \left(B(T_n)^{-1} U_n(T_n) \right) \right)$$
$$+ \mathrm{E}_{T_{n-1}} \left(1_{\{H_n(T_n) > U_n(T_n)\}} \Delta_\alpha \left(B(T_n)^{-1} H_n(T_n) \right) \right).$$

Combining equations we obtain the statement of the proposition. $\quad\square$

Proposition 24.1.1 provides us with a recursive relationship (in n, the exercise date index) between $\Delta_\alpha(B(T_{n-1})^{-1} H_{n-1}(T_{n-1}))$ and

$\Delta_\alpha(B(T_n)^{-1}H_n(T_n))$. The next proposition "unwraps" this recursion to give us the formula for $\Delta_\alpha H_0$.

Proposition 24.1.2. *Let η be the optimal exercise time index (see Section 18.2.2). Then*

$$\Delta_\alpha H_0(0) = \mathrm{E}\left(\sum_{n=\eta}^{N-1} \Delta_\alpha\left(B(T_n)^{-1}X_n\right)\right). \tag{24.2}$$

Proof. Unwrapping the recursive statement of Proposition 24.1.1, we find that

$$\Delta_\alpha H_0(0) =$$

$$\sum_{n=1}^{N-1} \mathrm{E}\left(\left(\prod_{i=1}^{n-1} 1_{\{H_i(T_i)>U_i(T_i)\}}\right) 1_{\{U_n(T_n)>H_n(T_n)\}} \Delta_\alpha(B(T_n)^{-1}U_n(T_n))\right).$$

As η is the optimal exercise time index,

$$1_{\{\eta=n\}} = \left(\prod_{i=1}^{n-1} 1_{\{H_i(T_i)>U_i(T_i)\}}\right) 1_{\{U_n(T_n)>H_n(T_n)\}},$$

from which it follows that

$$\Delta_\alpha H_0(0) = \sum_{n=1}^{N-1} \mathrm{E}\left(1_{\{\eta=n\}}\Delta_\alpha\left(B(T_n)^{-1}U_n(T_n)\right)\right).$$

From Proposition 3.3.1 and the fact that

$$B(T_n)^{-1}U_n(T_n) = \sum_{i=n}^{N-1} \mathrm{E}_{T_n}\left(B(T_{i+1})^{-1}X_i\right)$$

we obtain

$$\Delta_\alpha\left(B(T_n)^{-1}U_n(T_n)\right) = \sum_{i=n}^{N-1} \mathrm{E}_{T_n}\left(\Delta_\alpha\left(B(T_{i+1})^{-1}X_i\right)\right),$$

and therefore

$$\Delta_\alpha H_0(0) = \sum_{n=1}^{N-1} \mathrm{E}\left(1_{\{\eta=n\}} \sum_{i=n}^{N-1} \mathrm{E}_{T_n}\left(\Delta_\alpha\left(B(T_{i+1})^{-1}X_i\right)\right)\right).$$

The event $\{\eta = n\}$ is in the sigma-algebra $\mathcal{F}_{T_n}$ because η is a stopping time. Thus we may carry the indicator $1_{\{\eta=n\}}$ inside the expectation E_{T_n}, to get

$$\Delta_\alpha H_0(0) = E\left(\sum_{n=1}^{N-1}\sum_{i=n}^{N-1} 1_{\{\eta=n\}}\left(\Delta_\alpha\left(B(T_{i+1})^{-1}X_i\right)\right)\right).$$

Changing the order of summation we obtain

$$\Delta_\alpha H_0(0) = E\left(\sum_{i=\eta}^{N-1}\left(\Delta_\alpha\left(B(T_{i+1})^{-1}X_i\right)\right)\right),$$

and the proposition follows. $\square$

The result of Proposition 24.1.2 provides the foundation for computing pathwise derivatives of callable Libor exotics, and relates the derivative of a CLE to derivatives of coupons (that can typically be computed easily) and to the optional stopping time, a quantity that is computed during normal CLE valuation anyway. To proceed further, we need to specialize the setup to either PDE or Monte Carlo based models. First, however, we study some important implications, as well as some generalizations, of Proposition 24.1.2.

24.1.1.2 Keeping the Exercise Time Constant

It is instructive to compare the expression for the value of a callable Libor exotic (18.6) with the one for its pathwise derivative in Proposition 24.1.2:

$$H_0(0) = E\left(\sum_{n=\eta}^{N-1} B(T_{n+1})^{-1}X_n\right), \tag{24.3}$$

$$\Delta_\alpha H_0(0) = E\left(\sum_{n=\eta}^{N-1} \Delta_\alpha\left(B(T_{n+1})^{-1}X_n\right)\right).$$

Somewhat surprisingly, it appears that one can compute the derivative Δ_α by differentiating the sum in (24.3) and pretending that the optimal exercise time index η *does not depend* on α. But, paradoxically, in most cases the distribution of η *does* depend on α.

The seeming contradiction above can be resolved with the help of the following argument, known to economists as the *envelope theorem* (see Sydsaeter and Hammond [2008]). For an arbitrary stopping time index ζ, define $V_{ki}(\zeta, X)$ by

$$V_{ki}(\zeta, X) = E\left(\sum_{n=\zeta}^{N-1} B(T_{n+1})^{-1}X_n\right),$$

where, in somewhat loose notation, X in the argument of $V_{ki}(\zeta, X)$ represents all coupons X_n and all numeraire factors $B(T_n)^{-1}$. We can think of $V_{ki}(\zeta, X)$

as the value of a knock-in barrier option with the barrier defined by ζ. Note that $V_{\text{ki}}(\zeta, X)$ is equal to $H_0(0)$ for $\zeta = \eta$. Formally differentiating with respect to α,

$$\Delta_\alpha V_{\text{ki}}(\zeta, X) = \frac{\partial}{\partial \zeta} V_{\text{ki}}(\zeta, X) \Delta_\alpha \zeta + \frac{\partial}{\partial X} V_{\text{ki}}(\zeta, X) \Delta_\alpha X. \qquad (24.4)$$

Substituting $\zeta = \eta$ into the last equation, we make a critical observation that

$$\left. \frac{\partial}{\partial \zeta} V_{\text{ki}}(\zeta, X) \right|_{\zeta = \eta} = 0, \qquad (24.5)$$

because η by definition is the *optimal* stopping time index that maximizes the value of a callable Libor exotics over all stopping times (and (24.5) is the necessary first-order optimality condition). Due to (24.5), the first term in (24.4) drops out and we are left with

$$\Delta_\alpha H_0 = \Delta_\alpha V_{\text{ki}}(\eta, X) = \frac{\partial}{\partial X} V_{\text{ki}}(\eta, X) \times \Delta_\alpha X.$$

The expression on the right hand side can be interpreted as the partial derivative of the sum in (24.3) with η held constant.

The effective insensitivity of the stopping time with respect to parameter changes has some significant practical applications, even in situations where the pathwise differentiation method cannot be used (or is not used for some other reason). Recall that often a valuation of a callable security in Monte Carlo involves two steps (see Section 18.3.6): first, an optimal exercise boundary is estimated; and second, the value of the callable security is computed as a knock-in option, using the estimated exercise boundary as the barrier. In an implementation where the greeks are computed by shocking the inputs and revaluing the security, the result above states that the exercise time from the base scenario (which is, in a Monte Carlo simulation, just an integer index for each path) could be reused in the shocked scenario — i.e. we would force the exercise on a given path in a shocked scenario at exactly the same index as on the same path in the base scenario[1]. Besides obvious savings in computational time (there is now no need to re-estimate the exercise boundary in the shocked scenario), this scheme improves stability of the greeks, as we explain in the next section.

We should note that if the exercise boundary being used in computations is not truly optimal (which is, of course, nearly always the case in practice), freezing the stopping times in the manner described above will change the meaning of the greeks slightly, in a manner described in Section 24.3.4. Unless the exercise rule is truly inaccurate, these differences are typically small enough to ignore. Also, we point out that a theoretically valid alternative technique involves freezing the exercise *boundary*, rather than the exercise *index*. As explained below, the latter has superior numerical properties.

---

[1] Of course, heeding advice from Chapter 23, we should use the same seed and the same number of paths in the base and shocked scenarios.

24.1.1.3 Noise in CLE Greeks

To expand on the discussion above, and to tie it to greeks computations, let us for concreteness consider a Monte Carlo application where we attempt to evaluate CLE greeks by brute-force perturbation methods. From the results above, we have three valid alternatives when deciding how to treat the exercise decision in the perturbed market data scenario: i) re-estimate the exercise boundary (by regression, say); ii) re-use the base scenario exercise boundary; and iii) re-use the base scenario stopping times. While theoretically equivalent in the large-sample limit (due to the envelope theorem), for a realistic number of Monte Carlo paths the numerical properties of these three alternatives will differ substantially. For instance, it should be intuitively clear that re-estimating the exercise boundary will induce a large amount of spurious noise, so most practitioners have traditionally worked with a frozen boundary, as in alternative ii). Even with this approach, however, the derivatives of Bermudan-callable security prices will typically exhibit much higher levels of simulation error than prices of European options. Let us examine why this is the case.

Armed with the estimate of the optimal exercise index $\widetilde{\eta}$, the Monte Carlo estimate of the value of a callable Libor exotic is given by (see Section 18.3.6)

$$\widetilde{H}_0 = J^{-1} \sum_{j=1}^{J} \sum_{i=1}^{N-1} \left[B(T_{i+1}, \omega_j)^{-1} X_i(\omega_j) \, 1_{\{i \geq \widetilde{\eta}(\omega_j)\}} \right],$$

where simulated paths are denoted by ω_j, $j = 1, \ldots, J$. Clearly, the valuation formula involves exercise indicators $1_{\{i \geq \widetilde{\eta}(\omega_j)\}}$. Importantly for our analysis, these indicators are discontinuous functions of the simulated path ω. Figure 24.1 demonstrates the problem visually, for the case where the exercise boundary is frozen (our alternative ii) above).

Notice from Figure 24.1 that if a simulated path passes sufficiently close to the exercise boundary, then a small change in the parameter α can push the path outside of the exercise region for one of the exercise dates, losing a whole coupon as a result. Such a digital-type discontinuity — which is not present in European call/put or other continuous-payoff securities — leads to poorer stability and larger simulation errors for risk sensitivities in Bermudan-callable securities, compared to their European-call counterparts.

One way to improve stability and accuracy of the greeks is to use the payoff smoothing method from Section 23.4.3. However, it is much easier to use alternative iii) above, i.e. to re-use the estimate of the optimal exercise index $\widetilde{\eta}(\omega)$ from the base scenario. In practice it means that for each simulated path, we just force the exercise of a CLE at exactly the same time in calculations with the shocked market data as with the base market data. In this approach, no discontinuities are introduced.

Fig. 24.1. Discontinuity of CLE Value in Monte Carlo

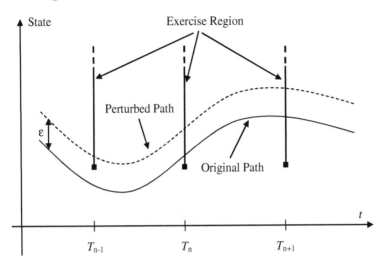

Notes: A whole coupon could be added to/subtracted from the value of a CLE valued in Monte Carlo under small, order-ϵ, perturbations of Monte Carlo paths if the exercise time is not kept constant.

24.1.2 Barrier Options

A CLE can be interpreted as a type of barrier option where the barrier condition is defined by the optimal exercise rule; one might therefore speculate that the pathwise differentiation method could be extended to general barrier options. This is, indeed, the case, although the presence of discontinuities in barrier options requires some additional care. As a warm-up exercise, recall the example of a pathwise derivative of digital option in Section 3.3.2.1 and consider a T-maturity European payoff

$$X = 1_{\{G>h\}}R, \tag{24.6}$$

where G and R are $\mathcal{F}_T$-measurable random variables and h is a particular strike. Differentiating the payoff with respect to α, we obtain

$$\Delta_\alpha \left(1_{\{G>h\}}R\right) = 1_{\{G>h\}}\Delta_\alpha R + \left(\frac{\partial 1_{\{G>h\}}}{\partial G}\right)R\,\Delta_\alpha G.$$

Formally,

$$\frac{\partial 1_{\{G>h\}}}{\partial G} = \delta\left(G - h\right),$$

where $\delta(x)$ is the delta function at zero. Assuming we can exchange the order of differentiation and expectation, we have

$$\Delta_\alpha \mathrm{E}\left(B(T)^{-1}X\right) = \mathrm{E}\left(\Delta_\alpha\left(B(T)^{-1}X\right)\right)$$
$$= \mathrm{E}\left(\Delta_\alpha\left(B(T)^{-1}\right)1_{\{G>h\}}R\right)$$
$$+ \mathrm{E}\left(B(T)^{-1}1_{\{G>h\}}\Delta_\alpha R\right)$$
$$+ \mathrm{E}\left(B(T)^{-1}\delta\left(G-h\right)R\,\Delta_\alpha G\right).$$

Rewriting the second term in the last equality, we find that the sensitivity of this digital option is given by

$$\Delta_\alpha \mathrm{E}\left(B(T)^{-1}1_{\{G>h\}}R\right) = \mathrm{E}\left(\Delta_\alpha\left(B(T)^{-1}\right)1_{\{G>h\}}R\right)$$
$$+ \mathrm{E}\left(B(T)^{-1}1_{\{G>h\}}\Delta_\alpha R\right) + \gamma_G(h)\mathrm{E}\left(B(T)^{-1}R\,\Delta_\alpha G\middle| G=h\right), \quad (24.7)$$

where $\gamma_G(h)$ is the density of G, at $G = h$. While the conditions of Proposition 3.3.1 do not hold for the payoff (24.6) and we cannot rely on Proposition 3.3.1 to justify differentiation inside the expected value operator, the formula (24.7) is, nevertheless, correct, and can be justified by Malliavin calculus, see Fournié et al. [1999].

The expected values on the right-hand side of (24.7) can in general be computed in a numerical scheme such as Monte Carlo, as long as the density of G is known, and the conditional expected value $\mathrm{E}(B(T)^{-1}R\,\Delta_\alpha G|G=h)$ can be evaluated. Both can, in principle, be computed in Monte Carlo using Malliavin calculus techniques (see Fournié et al. [1999]). However, the application of this method is much more practical if both of these quantities can be computed (or approximated) in closed form, which is the case in some models such as, say, Gaussian models.

Proceeding to the case of barrier options, let us introduce a barrier schedule $\{T_n\}_{n=1}^{N-1}$, to which we associate knockout variables G_n and barrier levels h_n, for $n = 1, \ldots, N-1$. We consider an option that pays the value R_n, $n = 1, \ldots, N-1$, on the first T_n where $G_n > h_n$; if this event never takes place, the option pays nothing. Formally, we define η, the knockout index, by

$$\eta = \min\left\{k \geq 1 : G_k > h_k\right\} \wedge N.$$

For notational convenience, set

$$R_N \equiv 0.$$

The time 0 value of our barrier option is then given by

$$V_{\mathrm{ki}}(0) = \mathrm{E}\left(B(T_\eta)^{-1}R_\eta\right).$$

Note that this is the same knock-in option as considered in Section 23.4.2, if we define $R_n = B(T_n)\sum_{i=n}^{N-1}B(T_{i+1})^{-1}X_i$.

More generally, let us denote

$$\eta_n = \min\left\{k \geq n+1 : G_k > h_k\right\} \wedge N$$

and
$$V_{\text{ki},n}(t) = B(t)\mathrm{E}_t\left(B(T_{\eta_n})^{-1}R_{\eta_n}\right),\tag{24.8}$$

with the convention
$$V_{\text{ki},N}(t) \equiv 0.$$

Here, $V_{\text{ki},n}(t)$ can be seen as the value of the option with the barrier condition checked at times $T_{n+1}, \dots, T_{N-1}$ only. In particular,

$$V_{\text{ki}}(0) = V_{\text{ki},0}(0).$$

We denote by $\gamma_n(x)$ the density of G_n at time T_n.

Proposition 24.1.3. *For the barrier option paying R_n on the first T_n where $G_n > h_n$, $n = 1, \dots, N-1$, the pathwise derivative with respect to a parameter α is given by*

$$\Delta_\alpha V_{\text{ki}}(0) = \mathrm{E}\left(B(T_\eta)^{-1}\,\Delta_\alpha R_n|_{n=\eta}\right)$$
$$+ \mathrm{E}\left(B(T_\eta)^{-1}\,\gamma_\eta\,(h_\eta)\,(R_\eta - V_{\text{ki},\eta}\,(T_\eta))\,\Delta_\alpha G_n|_{n=\eta}\,\Big|\,G_\eta = h_\eta\right).\tag{24.9}$$

Proof. (Sketch) The values of the family of knock-in options defined by (24.8) satisfy the following recursive relationship,

$$B(T_n)^{-1}V_{\text{ki},n}(T_n) = \mathrm{E}_{T_n}\left(B(T_{n+1})^{-1}R_{n+1}1_{\{G_{n+1}>h_{n+1}\}}\right)$$
$$+ \mathrm{E}_{T_n}\left(B(T_{n+1})^{-1}V_{\text{ki},n+1}\,(T_{n+1})\,1_{\{G_{n+1}\leq h_{n+1}\}}\right).$$

Differentiating formally,

$$\Delta_\alpha\left(B(T_n)^{-1}V_{\text{ki},n}(T_n)\right) = \mathrm{E}_{T_n}\left(\Delta_\alpha\left(B(T_{n+1})^{-1}R_{n+1}\right)1_{\{G_{n+1}>h_{n+1}\}}\right)$$
$$+ \mathrm{E}_{T_n}\left(\Delta_\alpha\left(B(T_{n+1})^{-1}V_{\text{ki},n+1}\,(T_{n+1})\right)1_{\{G_{n+1}\leq h_{n+1}\}}\right)$$
$$+ \mathrm{E}_{T_n}\left(B(T_{n+1})^{-1}\,(R_{n+1} - V_{\text{ki},n+1}\,(T_{n+1}))\,\delta\,(G_{n+1} - h_{n+1})\right).$$

This defines an expression of $\Delta_\alpha V_{\text{ki},n}(T_n)$ in terms of $\Delta_\alpha V_{\text{ki},n+1}(T_{n+1})$ and other quantities. Unwrapping the recursion, as in Proposition 24.1.2 earlier, proves the proposition. $\square$

Example 24.1.4. Callable Libor exotics are a special case with $R_n = U_n$, $G_n = U_n - H_n$, $h_n = 0$. Proposition 24.1.2 follows from Proposition 24.1.3 once continuity condition for CLEs,

$$(R_n - V_{\text{ki},n}(T_n))|_{G_n=h_n} = 0,$$

is taken into account.

Example 24.1.5. A TARN (See Chapter 20 and Section 23.4.4) can be represented in barrier form. In particular, using the notations of Section 23.4.4, we define

$$G_n \triangleq Q_n, \quad h_n \triangleq R, \quad R_n \triangleq \sum_{k=n}^{N-1} B(T_{k+1})^{-1} X_k(T_k),$$

so that

$$V_{\text{ki},0}(0) + V_{\text{tarn}}(0) = \mathrm{E} \left(\sum_{k=1}^{N-1} B(T_{k+1})^{-1} X_k(T_k) \right),$$

where the right-hand side equals the price of a straight (exotic) swap.

Although Proposition 24.1.3 extends the pathwise differentiation method to barrier options, the complexity of the result limits the practicality of the method. In particular, the transition densities and conditional probabilities in (24.9) are often difficult to compute, and it may ultimately be more fruitful to use methods in Chapters 23 and 25 to smooth or integrate out any discontinuities before applying pathwise differentiation techniques.

24.2 Pathwise Differentiation for PDE Based Models

The pathwise differentiation method can be applied to both PDE and Monte Carlo based models. In this section we consider PDE applications, mostly following Piterbarg [2004a]; we address Monte Carlo applications in Section 24.3.

24.2.1 Model and Setup

The treatment of European-style options in the Section 3.3.2 is rather generic and can easily be implemented in PDE-based numerical schemes. Callable Libor exotics, on the other hand, require more effort, to be undertaken in this section. While the method is developed for, and can be applied to, rather general CLEs, for a number of reasons Bermudan swaptions are probably the most natural target of the techniques described here. Indeed, not only have we shown (in Section 19.2) that low-dimensional, PDE-based Markovian models are appropriate for Bermudan swaptions, but Bermudan swaptions often constitute a dominant part of portfolios of interest rate exotics and are therefore subject to high demands for stable and accurate risk reporting.

To focus on the main features of the method without distraction from minor details, let us consider a Gaussian interest rate model as developed in Section 10.1.2.2, parameterized in terms of the short rate state $x(t)$ as in Proposition 10.1.7. We denote the infinitesimal generator associated with the dynamics of $x(t)$ by $\mathcal{L}$,

$$\mathcal{L} = (y(t) - \varkappa(t)x)\frac{\partial}{\partial x} + \frac{1}{2}\sigma_r(t)^2\frac{\partial^2}{\partial x^2}. \tag{24.10}$$

If $V = V(t,x)$ is the value of a contingent claim at time t given $x(t) = x$, then $V(t,x)$ satisfies the equation (see (10.29))

$$\frac{\partial V}{\partial t}(t,x) + \mathcal{L}V(t,x) = (f(0,t) + x)V(t,x).$$

Note that this valuation expression is associated with the risk-neutral measure Q, induced by the continuous money market account $\beta(t)$. Previous material in this chapter (and in Chapter 18) used a spot Libor measure, but results carry over unchanged to the risk-neutral measure. Recycling notations, we now denote by E the corresponding, i.e. risk-neutral, expected value operator; while E was earlier used as the expected value operator in spot Libor measure, there should be no confusion which measure is used going forward.

Let us consider a CLE with net coupons $\{X_n\}$, as in Section 24.1.1; as we work in a PDE setting, we assume that the value of the net coupon X_n does not depend on the state of the yield curve prior to T_n, $n = 1, \ldots, N-1$. Recall Proposition 24.1.2 which states that the pathwise derivative Δ_α of a CLE is given by

$$\Delta_\alpha H_0(0) = \mathrm{E}\left(\sum_{n=\eta}^{N-1} \Delta_\alpha\left(\beta(T_{n+1})^{-1}X_n\right)\right),$$

where η is the optimal exercise date index. The hold and exercise values are now deterministic functions of x, so we use the notations $H_n(t,x)$, $U_n(t,x)$ where appropriate.

24.2.2 Bucketed Deltas

Arguably, the most important risk measures for an interest rate security are the so-called *bucketed interest rate deltas*, see Section 6.4, that measure the sensitivity of the value of the security to changes in various parts of the yield curve. For the CLE in question, the most natural bucketing[2] of deltas is induced by the tenor structure $\{T_n\}_{n=0}^N$. Specifically, we define the m-th (continuously compounded) forward rate by

$$y_m(0) \triangleq y(0, T_m, T_{m+1}) = -\frac{1}{\tau_m}\ln\left(P\left(0, T_m, T_{m+1}\right)\right), \quad m = 0, \ldots, N-1,$$

and denote by Δ_m the pathwise derivative with respect to $y_m(0)$,

---

[2]Naturally, the sensitivities to these rates can be projected into any other "basis", i.e. a set of rates used to define and aggregate curve sensitivities. For the relevant techniques, see Section 6.4.3.

$$\Delta_m \triangleq \Delta_{y_m}(0), \quad m = 1, \dots, N-1.$$

To establish Δ_m, let us start by rewriting the pathwise derivative in a more convenient form

$$\Delta_m H_0(0) = \sum_{n=1}^{N-1} \mathbb{E}\left(1_{\{\eta \geq n\}} \Delta_m\left(\beta(T_{n+1})^{-1} X_n\right)\right)$$

$$= \sum_{n=1}^{N-1} \mathbb{E}\left(1_{\{\eta \geq n\}} \beta(T_{n+1})^{-1} \frac{\Delta_m\left(\beta(T_{n+1})^{-1}\right)}{\beta(T_{n+1})^{-1}} X_n\right)$$

$$+ \sum_{n=1}^{N-1} \mathbb{E}\left(1_{\{\eta \geq n\}} \beta(T_{n+1})^{-1} \Delta_m\left(X_n\right)\right). \tag{24.11}$$

We shall also need the following lemma.

Lemma 24.2.1. *In the Gaussian model the following holds,*

$$\frac{\Delta_m\left(\beta(T_{n+1})^{-1}\right)}{\beta(T_{n+1})^{-1}} = -1_{\{m \leq n\}} \tau_m.$$

Proof. We have

$$\beta(T_{n+1})^{-1} = \exp\left(-\int_0^{T_{n+1}} r(t)\,dt\right) = \exp\left(-\int_0^{T_{n+1}} (f(0,t) + x(t))\,dt\right),$$

$$\Delta_m\left(\beta(T_{n+1})^{-1}\right) = \beta(T_{n+1})^{-1} \Delta_m\left(-\int_0^{T_{n+1}} (f(0,t) + x(t))\,dt\right).$$

Since the dynamics of $x(t)$ do not depend on the initial yield curve $P(0, \cdot)$,

$$\frac{\Delta_m\left(\beta(T_{n+1})^{-1}\right)}{\beta(T_{n+1})^{-1}} = -\Delta_m\left(\int_0^{T_{n+1}} f(0,t)\,dt\right).$$

Moreover, by definition of y_k's,

$$\int_0^{T_{n+1}} f(0,t)\,dt = \sum_{k=0}^{n} \tau_k y_k(0).$$

Hence,

$$\Delta_m\left(\int_0^{T_{n+1}} f(0,t)\,dt\right) = \Delta_m\left(\sum_{k=0}^{n} \tau_k y_k(0)\right) = 1_{\{m \leq n\}} \tau_m.$$

$\square$

For the next result we need the following definition.

Definition 24.2.2. *The time t survival measure $\Psi(\cdot; t)$ is defined for $\Gamma \subset \mathbb{R}$ by the formula*

$$\Psi(\Gamma; t) = \mathrm{E}\left(\beta(t)^{-1} 1_{\{\eta \geq q(t)\}} 1_{\{x(t) \in \Gamma\}}\right),$$

where the index function $q(t)$ for the tenor structure $\{T_n\}$ is defined in (14.2). The survival density, the density $\psi(x; t)$ of the survival measure with respect to the Lebesgue measure dx, is defined by

$$\Psi(\Gamma; t) = \int_\Gamma \psi(y; t)\, dy,$$

and is assumed to exist.

Combined with this definition of the survival density and the representation (24.11), Lemma 24.2.1 allows us to derive the following representation of bucketed deltas of a CLE.

Proposition 24.2.3. *In the Gaussian model, the m-th bucketed delta of a CLE is given by*

$$\Delta_m H_0(0) = -\sum_{n=m}^{N-1} \tau_m \int_{\mathbb{R}} V_{\mathrm{cpn},n}(x)\, \psi(x; T_n)\, dx$$

$$+ \sum_{n=1}^{m} \int_{\mathbb{R}} D_{\mathrm{cpn},n,m}(x)\psi(x; T_n)\, dx,$$

where $V_{\mathrm{cpn},n}(x)$ and $D_{\mathrm{cpn},n,m}(x)$ are the conditional expectations of the discounted value and the discounted derivative of the n-th coupon,

$$V_{\mathrm{cpn},n}(x) = \mathrm{E}\left(\beta(T_n)\beta(T_{n+1})^{-1} X_n \,\middle|\, x(T_n) = x\right),$$
$$D_{\mathrm{cpn},n,m}(x) = \mathrm{E}\left(\beta(T_n)\beta(T_{n+1})^{-1} \Delta_m X_n \,\middle|\, x(T_n) = x\right).$$

Proof. From (24.11),

$$\Delta_m H_0(0) = -\sum_{n=1}^{N-1} 1_{\{m \leq n\}} \tau_m \mathrm{E}\left(1_{\{\eta \geq n\}} \beta(T_{n+1})^{-1} X_n\right)$$

$$+ \sum_{n=1}^{N-1} \mathrm{E}\left(1_{\{\eta \geq n\}} \beta(T_{n+1})^{-1} \Delta_m X_n\right).$$

By definition of the survival density and from the fact that $q(T_n-) = n$ (see (14.2)) we obtain

$$\mathrm{E}\left(1_{\{\eta \geq n\}} \beta(T_{n+1})^{-1} X_n\right) = \mathrm{E}\left(1_{\{\eta \geq q(T_n-)\}} \beta(T_n)^{-1} V_{\mathrm{cpn},n}\left(x(T_n)\right)\right)$$

$$= \int V_{\mathrm{cpn},n}(x)\, \Psi(dx; T_n)$$

and

$$E\left(1_{\{\eta \geq n\}}\beta(T_{n+1})^{-1}\Delta_m X_n\right) = E\left(1_{\{\eta \geq q(T_n-)\}}\beta(T_n)^{-1}D_{\text{cpn},n,m}\left(x(T_n)\right)\right)$$
$$= \int D_{\text{cpn},n,m}(x)\,\Psi\,(dx;T_n)\,.$$

Since the net coupon X_n depends on the yield curve on or after the fixing time T_n,

$$D_{\text{cpn},n,m}(x) \equiv 0$$

for $m < n$. The result follows.   □

Remark 24.2.4. The functions $V_{\text{cpn},n}(x)$ and $D_{\text{cpn},n,m}(x)$ are usually easy to calculate, as the net coupon X_n is typically a function of discount factors observed at time T_n. The reader may want to consult Section 24.3.2 where related calculations are performed.

Proposition 24.2.3 represents bucketed deltas in terms of the integrals of known (or easily computed) functions $V_{\text{cpn},n}(x)$ and $D_{\text{cpn},n,m}(x)$ against the survival density. Note that the survival density is *universal*, i.e. it does not depend on a particular delta index m. Hence, if we can calculate it efficiently, all pathwise bucketed deltas can be computed quickly, as only simple integrals are required for their calculations. This should be compared to the standard way of computing deltas, where the relevant forward rate is perturbed and the value of a CLE is recomputed by solving a full-blown PDE. We discuss the computation of the survival density in the next section.

24.2.3 Survival Density

As a reference point, let us consider the following family of measures defined on $\mathbb{R}$. We fix time s and position x and define, for $t \geq s$,

$$\Pi_{s,x}(t,\Gamma) \triangleq E_s\left(\beta(t)^{-1}1_{\{x(t)\in\Gamma\}}\,\middle|\,x(s) = x\right)$$
$$= E_s\left(e^{-\int_s^t r(u)\,du}1_{\{x(t)\in\Gamma\}}\,\middle|\,x(s) = x\right), \quad \Gamma \subset \mathbb{R}.$$

For each s,x we can define the density[3] $\pi_{s,x}(t,y)$ by

$$\Pi_{s,x}(t,\Gamma) = \int_\Gamma \pi_{s,x}(t,y)\,dy,$$

where $\pi_{s,x}(t,y)$ can be recognized as values of the Arrow-Debreu securities we introduced in Section 11.3.2.1. For fixed s,x these satisfy an analog to forward Kolmogorov equations, see (11.30), which we rewrite in our notations as

---

[3] Note that this measure density is *not* a probability density, as it does not integrate to 1.

$$\frac{\partial}{\partial t}\pi_{s,x}\left(t, y\right) = \mathcal{L}^* \pi_{s,x}\left(t, y\right) - r(t)\pi_{s,x}\left(t, y\right), \quad (s, x) \text{ fixed}, \qquad (24.12)$$

for $t \geq s$. Here $\mathcal{L}^*$ is the operator adjoint to $\mathcal{L}$ (see (24.10)),

$$\mathcal{L}^* q(t, y) = -\frac{\partial}{\partial y}\left(\left(y(t) - \varkappa(t)x\right)q(t, y)\right) + \frac{\partial^2}{\partial y^2}\left(\frac{1}{2}\sigma_r(t)^2 q(t, y)\right),$$

which is applied to $\pi_{s,x}(t, y)$ as a function of y.

The following proposition outlines an efficient procedure for comput-ing the survival density ψ. The idea of the theorem is that in between the "interesting" times $\{T_n\}$, the density ψ behaves just like the density π in the proposition above. When the time crosses an exercise time T_n, the density ψ gets multiplied by an extra "survival" indicator function $1_{\{H_n(y,T_n)>U_n(y,T_m)\}}$.

Proposition 24.2.5. *For each* n, $0 \leq n \leq N-1$, *the survival density* $\psi(y;t)$ *satisfies the forward PDE*

$$\frac{\partial}{\partial t}\psi\left(y; t\right) = \left(\mathcal{L}^*\psi\right)\left(y; t\right) - r(t)\psi\left(y; t\right),$$

on the time interval

$$t \in (T_n, T_{n+1}),$$

with the boundary condition

$$\psi\left(y; T_n\right) = \psi\left(y; T_n-\right) \times 1_{\{H_n(T_n,y)>U_n(,T_n,y)\}}. \qquad (24.13)$$

The initial condition for the first interval, $(T_0, T_1) = (0, T_1)$, *is given by the delta function*

$$\psi\left(y; 0\right) = \delta\left(y - x(0)\right).$$

Proof. Assume

$$T_n < t < T_{n+1},$$

so that $q(t) = n + 1$. Then

$$\Psi\left(\Gamma; t\right) = \mathrm{E}\left(e^{-\int_0^t r(u)\,du}1_{\{\eta \geq n+1\}}1_{\{x(t)\in\Gamma\}}\right)$$

$$= \mathrm{E}\left(e^{-\int_0^{T_n} r(u)\,du}1_{\{\eta>n\}}\mathrm{E}_{T_n}\left(e^{-\int_{T_n}^t r(u)\,du}1_{\{x(t)\in\Gamma\}}\right)\right)$$

$$= \mathrm{E}\left(e^{-\int_0^{T_n} r(u)\,du}1_{\{\eta>n\}}\Pi_{T_n,x(T_n)}\left(t, \Gamma\right)\right).$$

From this formula we obtain

$$\psi\left(y; t\right) = \mathrm{E}\left(e^{-\int_0^{T_n} r(u)\,du}1_{\{\eta>n\}}\pi_{T_n,x(T_n)}\left(t, y\right)\right).$$

Differentiating this equality with respect to t, exchanging the order of differentiation and taking the expectation, applying (24.12) and exchanging the order of the linear operator $\mathcal{L}^* - r(t)$ and the expectation operator, we obtain that the same equation as (24.12) holds for $\psi(y; t)$,

$$\frac{\partial}{\partial t}\psi(y; t) = \mathcal{L}^*\psi(y; t) - r(t)\psi(y; t)$$

for $t \in (T_n, T_{n+1})$. To derive boundary conditions we notice that

$$\Psi(\Gamma; T_n) = \mathrm{E}\left(\beta(T_n)^{-1}1_{\{\eta \geq n+1\}}1_{\{x(T_n) \in \Gamma\}}\right)$$
$$= \mathrm{E}\left(\beta(T_n)^{-1}1_{\{\eta \geq n\}}1_{\{H_n(T_n, x(T_n)) > U_n(T_n, x(T_n))\}}1_{\{x(T_n) \in \Gamma\}}\right).$$

As $x(t)$ and $\beta(t)$ are continuous at $t = T_n$ we have

$$\Psi(\Gamma; T_n) = \Psi(\Gamma \cap \{H_n(T_n, x(T_n)) > U_n(T_n, x(T_n))\}; T_n-)$$

and, calculating the densities of both sides, we obtain (24.13). For more details see Piterbarg [2004a]. □

Remark 24.2.6. The time-T_n conditions (24.13) require knowledge of the "hold" regions

$$\{x \in \mathbb{R} : H_n(T_n, x) > U_n(T_n, x)\}.$$

These are computed as a by-product of the CLE valuation, since on each exercise date T_n, the hold values $H_n(T_n, x)$ are determined as functions of the state process $x(t)$ evaluated at time $t = T_n$.

Proposition 24.2.5 outlines a procedure for computing the survival density in one forward PDE "sweep", starting at $t = 0$ with a delta function. The solution is computed forward using an appropriate PDE scheme (see e.g. Sections 11.3.2.1, 11.3.2.2) until the first exercise time T_1. At this point, the solution (i.e. the survival density) is multiplied by the indicator function of the no-exercise condition. The density is then rolled forward again until the next exercise date where it is multiplied by another no-exercise indicator function, and so on.

The pathwise differentiation method for calculating deltas handily outperforms the standard approach of re-evaluation of a derivative under shocked scenarios. For the pathwise differentiation method, to calculate all N bucketed deltas, we need to calculate *one* survival density at a cost comparable to one PDE valuation of the derivative, and $2N$ integrals of Proposition 24.2.3, at a combined cost of about twice the PDE valuation of the derivative. In contrast, the perturb-and-revalue method would require N PDE valuations, one for each bucketed delta. As N is typically significantly larger than 3, the cost savings therefore can be quite significant. Nor is the pathwise method limited to deltas only; as shown in Piterbarg [2004a], one can handle vegas and gammas in the same way. As we can reuse much of the calculations (the survival density) among all these greeks, performance improvements are even more dramatic.

24.3 Pathwise Differentiation for Monte Carlo Based Models

Let us now consider applications of pathwise differentiation to Monte Carlo based models. For concreteness, we develop the technique for the LM model (14.13)–(14.14) with separable deterministic local volatility:

$$dL_n(t) = \varphi\left(L_n(t)\right)\lambda_n(t)^\top \left(\mu_n(t)\,dt + dW(t)\right), \qquad (24.14)$$

$$\mu_n(t) = \sum_{j=q(t)}^{n} \frac{\tau_j\varphi\left(L_j(t)\right)\lambda_j(t)}{1 + \tau_j L_j(t)},$$

The basic principles are, however, quite generic and straightforward to apply to other models. With LM models more naturally presented in the spot measure Q^B, we use a setup where the numeraire is chosen to be the rolling money market $B(t)$, see (14.8). For notational convenience, we assume that the LM model and the security to be priced share the same tenor structure $\{T_n\}$; in practice, of course, this need not be the case.

24.3.1 Pathwise Derivatives of Forward Libor Rates

The discrete and spanning nature of forward Libor rates makes the definition of bucketed deltas[4] easy, and we define Δ_m to be the pathwise derivative with respect to $L_m(0)$, $m = 0, \ldots, N-1$,

$$\Delta_m X \triangleq \frac{\partial X}{\partial L_m(0)}$$

for any random variable X. Note that in order to keep notation light, we reuse the definition Δ_m for pathwise derivatives from Section 24.2, but redefine their meaning slightly, as we here calculate derivatives with respect to simply compounded rates, rather than to the continuously compounded rates used in Section 24.2.

As should be clear from the basic discussion in Section 3.3.2, to successfully apply the pathwise differentiation method to a Libor market model, we need to be able to simulate the pathwise derivatives of the forward Libor rates $\Delta_m L_n(t)$, $n, m = 0, \ldots, N-1$. To determine the Q^B-dynamics of $\Delta_m L_n(t)$, we use the standard technique of differentiating the SDEs for $L_n(t)$. From (14.13)–(14.14), differentiating with respect to $L_m(0)$, we get

$$d\left(\Delta_m L_n(t)\right) = \varphi'\left(L_n(t)\right)\lambda_n(t)^\top \Delta_m L_n(t)\left(\mu_n(t)\,dt + dW(t)\right)$$

$$+ \varphi\left(L_n(t)\right)\lambda_n(t)^\top \sum_j \frac{\partial \mu_n(t)}{\partial L_j(t)}\Delta_m L_j(t)\,dt. \quad (24.15)$$

---

[4]As we mentioned before in Section 6.4, once the deltas in a particular "basis" are computed, it is a matter of simple linear algebra to re-express them in any other basis, e.g. the one used by a risk management system.

The initial conditions for these SDEs are found by differentiating the initial conditions for $L_n(t)$'s, resulting in

$$\Delta_m L_n(0) = 1_{\{n=m\}}. \tag{24.16}$$

The system of SDEs given by (24.15) and (14.13)–(14.14) fully specifies the dynamics of the forward Libor rates and their pathwise derivatives through time.

There are N equations in the system (14.13) and N^2 equations in the system (24.15), and simulating all is computationally expensive, even for relatively low values of N. A significant part of the numerical effort originates with drift computations, so it is natural to investigate whether simplifications of the drift term in (24.15) can lighten the computational burden. Glasserman and Zhao [1999] propose to use the following simplified system of SDEs for simulating values and pathwise deltas of forward Libor rates,

$$dL_n(t) = \varphi\left(L_n(t)\right) \lambda_n(t)^\top \left(\mu_n(t)\, dt + dW(t)\right),$$

$$d\left(\Delta_m L_n(t)\right) = \varphi\left(L_n(t)\right) \lambda_n(t)^\top \Delta_m L_n(t) \left(\mu_n(0)\, dt + dW(t)\right)$$

$$+ \varphi'\left(L_n(t)\right) \lambda_n(t)^\top \sum_j \frac{\partial \mu_n(0)}{\partial L_j(0)} \Delta_m L_j(t)\, dt. \tag{24.17}$$

Notice that we here retain the original Libor rate dynamics, but have applied the standard "freezing" technique to the drifts when calculating the dynamics of pathwise derivatives of Libor rates. This allows for a considerable speed-up, as the drifts in the equations for deltas of forward Libor rates can be pre-computed before the simulation. Glasserman and Zhao [1999] show numerically that the loss of accuracy in (24.17) is typically quite small.

The cost of propagating pathwise derivatives of Libor rates often dominates the computations of pathwise deltas, so let us consider computational complexity in more detail. We denote by $\Delta\mathbf{L}(t)$ an $N \times N$ matrix with the (n, m)-th element equal to $\Delta_m L_n(t)$,

$$\Delta\mathbf{L}(t)_{n,m} = \frac{\partial L_n(t)}{\partial L_m(0)}, \quad n, m = 0, \dots, N-1. \tag{24.18}$$

To fix ideas, we assume that we need to propagate $\Delta\mathbf{L}(t)$ for $0 \leq t \leq T = T_k$ for some k, and that we discretize (24.15) using the Euler scheme over the time grid $\{T_i\}_{i=0}^k$ of the LMM tenor structure. Further assume that a path of the Brownian motion $W(t)$ has been drawn, and we have denoted $Z_{i-1} = (W(T_i) - W(T_{i-1}))/\sqrt{\tau_{i-1}}$. Then, for any time step i, $i = 1, \dots, k$, we can rewrite (24.15) in matrix form[5]

$$\Delta\mathbf{L}(T_i) = \mathbf{D}(T_{i-1})\Delta\mathbf{L}(T_{i-1}), \quad i = 1, \dots, k, \tag{24.19}$$

---

[5]We could also use the faster approximation (24.17) here; we leave relevant modifications for the reader to explore.

where the matrix $\mathbf{D}(T_{i-1})$ has elements

$$\mathbf{D}(T_{i-1})_{n,m} = 1_{\{n=m\}} + \varphi(L_n(T_{i-1}))\lambda_n(T_{i-1})^\top \frac{\partial \mu_n(T_{i-1})}{\partial L_m(T_{i-1})}\tau_{i-1}$$
$$+ 1_{\{n=m\}}\varphi'(L_n(T_{i-1}))\lambda_n(T_{i-1})^\top \left(\mu_n(T_{i-1})\tau_{i-1} + Z_{i-1}\sqrt{\tau_{i-1}}\right). \quad (24.20)$$

We see that propagating $\Delta\mathbf{L}(t)$ over one time step requires a matrix-matrix multiplication of order $O(N^3)$, so the calculation of $\Delta\mathbf{L}(T_k)$ (which requires k steps) has total computation effort of order $O(kN^3)$.

It is interesting to compare the computational complexity of this algorithm to a brute-force perturbation method. Let us estimate the cost of calculating N deltas to $L_m(0)$, $m = 0, \ldots, N-1$, by shocking each of these forward Libor rates and re-running the simulation (24.14). Stepping one Libor rate in one perturbation scenario over one time step costs $O(1)$. Hence, stepping all N Libor rates in all N scenarios over all k time steps has complexity $O(kN^2)$, i.e. it is *faster*, by a factor of $O(N)$, than propagating $\Delta\mathbf{L}(t)$. On the other hand, once $\Delta\mathbf{L}(t)$ is simulated, it could be reused for calculating deltas of multiple payoffs, a point we return to in Section 24.3.3. It follows that a naive implementation of the pathwise differentiation method only becomes competitive speed-wise when there are more than N payoffs to differentiate in the same simulation. This seemingly limits the usability of the pathwise method, as simultaneous calculation of risk sensitivities for multiple payoffs is often difficult to achieve in practice, since most risk systems treat each trade as a separate work unit[6]. In Section 24.3.3 below, we show that by suitably rearranging the order of calculations in the pathwise differentiation method, the computational cost can be brought down to $O(kN^2)$, making pathwise differentiation computationally competitive with the revaluation method.

Given that the computational effort is no better than for the significantly simpler perturbation-based methods, the reader may wonder whether the pathwise differentiation method is ultimately worth the effort. The answer to this question is not always entirely obvious. On one hand, pathwise differentiation produces a true derivative estimate without a difference coefficient bias (see Section 3.3.1) and, in a sense, can be seen as the ultimate way of "geometry fixing" for Monte Carlo (or PDEs), since greeks are calculated in exactly the same simulation as the base value. Recalling the analysis in Section 23.1, it is therefore not surprising that pathwise differentiation often produces greeks of superior quality to those produced by run-of-the-mill perturbation methods. On the other hand, by scrutinizing a given product in detail and carefully "locking" all relevant computational details, it is often possible to construct perturbation methods that produce

---

[6] Also, as calibration and time-discretization is normally set up in a product-specific manner, it can be awkward (and even sub-optimal) to attempt to price many securities simultaneously in a single Monte Carlo loop.

greeks of comparable quality to those of the pathwise differentiation — CLEs are good examples of this, as discussed in Section 24.1.1.2.

24.3.2 Pathwise Deltas of European Options

As in Section 14.6.2.1, let $\mathbf{L}(t)$ be the vector of all Libor rates, and consider a European-style option with time T payoff $V(\mathbf{L}(T))$, for a deterministic function $V(\mathbf{x})$, $\mathbf{x} = (x_0, \dots, x_{N-1})$. The option value at time 0 equals

$$V(\mathbf{L}(0)) = \mathrm{E}\left(B(T)^{-1}V(\mathbf{L}(T))\right),$$

where E denotes expectations in the spot measure Q^B. As required by Proposition 3.3.1, we suppose that $V(\mathbf{x})$ is a Lipschitz continuous function of $\mathbf{x}$. Then the pathwise delta Δ_m can be carried under the expectation operator,

$$\Delta_m \mathrm{E}\left(B(T)^{-1}V(\mathbf{L}(T))\right) = \mathrm{E}\left(\Delta_m\left(B(T)^{-1}V(\mathbf{L}(T))\right)\right),$$

so that

$$\Delta_m V(\mathbf{L}(0)) = \mathrm{E}\left(\Delta_m\left(B(T)^{-1}\right)V(\mathbf{L}(T))\right)$$
$$+ \mathrm{E}\left(B(T)^{-1}\sum_{i=0}^{N-1}\frac{\partial V(\mathbf{x})}{\partial x_i}\bigg|_{\mathbf{x}=\mathbf{L}(T)}\Delta_m L_i(T)\right).$$

To compute deltas of the option, we need to be able to compute the deltas of the numeraire $\Delta_m(B(T)^{-1})$, as well as the derivatives of the payoff $\partial V(\mathbf{x})/\partial x_i$. We start with the numeraire.

24.3.2.1 Pathwise Deltas of the Numeraire

Recall that the discrete money market account $B(T)$ is given by

$$B(T) = \left(\prod_{i=0}^{n}(1+\tau_i L_i(T_i))\right)P(T, T_{n+1}),$$

where we have assumed that $T_n \leq T < T_{n+1}$. The pathwise derivative of the stub bond $P(T, T_{n+1})$ will depend (mildly) on the interpolation scheme used in the model, see Section 15.1. To keep the exposition simple, we choose the zero-volatility interpolation for the front stub $P(T, T_{n+1})$ of Section 15.1.4, whereby

$$P(T, T_{n+1}) = P(T_n, T, T_{n+1}).$$

Applying constant interpolation of simply compounded rates, see (15.4), we arrive at

$$P\left(T, T_{n+1}\right) = P\left(T_n, T, T_{n+1}\right) = \frac{1}{1 + \left(T_{n+1} - T\right) L_n(T_n)},$$

so that

$$B(T) = \left(\prod_{i=0}^{n-1} \left(1 + \tau_i L_i(T_i)\right)\right) \frac{1 + \tau_n L_n(T_n)}{1 + \left(T_{n+1} - T\right) L_n(T_n)}.$$

Differentiating, we obtain

$$\Delta_m\left(B(T)^{-1}\right) \tag{24.21}$$

$$= \sum_{j=0}^{n} \frac{\partial \left(B(T)^{-1}\right)}{\partial L_j(T_j)} \Delta_m L_j(T_j) \tag{24.22}$$

$$= -B(T)^{-1} \sum_{j=0}^{n-1} \frac{\tau_j}{1 + \tau_j L_j(T_j)} \Delta_m L_j(T_j)$$

$$- B(T)^{-1} \frac{T - T_n}{\left(1 + \left(T_{n+1} - T\right) L_n(T_n)\right)\left(1 + \tau_n L_n(T_n)\right)} \Delta_m L_n(T_n).$$

24.3.2.2 Pathwise Deltas of the Payoff

A typical (Lipschitz continuous) interest rate payoff $V(\mathbf{x})$ can be represented as an absolutely continuous function, say $f(\cdot)$, of one (or more) Libor or CMS rates. Therefore, the pathwise derivatives of this payoff with respect to initial Libor rates $L_m(0)$, $m = 0, \ldots, N - 1$, can be computed by a chain rule, as long as we know how to differentiate market rates with respect to $L_m(0)$, $m = 0, \ldots, N - 1$. For instance, for some swap rate $S(t)$ (note that t is not necessarily equal to T), we get

$$\Delta_m V(\mathbf{L}(t)) = \Delta_m f(S(t)) = f'(S(t)) \Delta_m S(t).$$

The derivatives $\Delta_m S(t)$ are determined by the way the yield curve at future time t is constructed from simulated primary Libor rates $\mathbf{L}(t)$, as discussed in detail in Section 15.1. In particular, the rate $S(t)$ is always a known function of zero-coupon bonds $P(t, s)$ for various s, so $\partial S(t)/\partial P(t, s)$ is easily computed. Finally, as we have an algorithm to construct all $P(t, s)$ from $\mathbf{L}(t)$ per Section 15.1, we can calculate $\partial P(t, s)/\partial L_m(t)$ along the same lines as in (24.22).

For rates $S(t)$ that are aligned with the tenor structure $\{T_n\}$ of the model (i.e. $S(t) = S_{i,j}(t)$ for some i, j as defined in (4.10)), calculations simplify significantly, and we have already derived relevant derivatives in Section 14.4.2, see (14.31). Other methods from Section 14.4.2 also apply for general rates $S(t)$; in particular, we can recycle ideas from Section 14.4.2 on swap rate volatility approximations used for calibrating LM models. Recall the freezing idea of Proposition 14.4.3,

$$\frac{\partial S(t)}{\partial L_m(t)} \frac{\varphi(L_m(t))}{\varphi(S(t))} \approx \frac{\partial S(0)}{\partial L_m(0)} \frac{\varphi(L_m(0))}{\varphi(S(0))}.$$

By simple algebraic manipulations we obtain that

$$\varDelta_m S(t) = \frac{\partial S(T)}{\partial L_m(T)} \approx \varDelta_m S(0) \frac{\varphi(L_m(0))}{\varphi(S(0))} \frac{\varphi(S(T))}{\varphi(L_m(T))}. \qquad (24.23)$$

Numerical errors arising from the approximation (24.23) are typically small, and performance gains are significant as the quantities $\partial S(0)/\partial L_m(0)$ can now be pre-computed before the simulation starts.

It is worth mentioning at this point that, while a good part of the discussion above was about deltas, other first-order risk sensitivities such as vegas and, even, second-order sensitivities such as gammas could be computed in a pathwise method as well, as briefly discussed in Section 3.3.2.

24.3.3 Adjoint Method For Greeks Calculation

Let us continue contemplating hedge computations for European options paying at time T some function V of the Libor vector $\mathbf{L}(T)$; for notational convenience, define $U \triangleq B(T)^{-1} V(\mathbf{L}(T))$. Once the values of pathwise derivatives of all forward Libor rates $\varDelta_m L_n(T)$, $n, m = 0, \ldots, N-1$, are simulated for a given path (using, for example, (24.15) or (24.17)), then the full set of pathwise deltas $\varDelta U \triangleq (\varDelta_0 U, \ldots, \varDelta_{N-1} U)$ can be calculated at a small cost of multiplying a vector by a matrix (note the row-vector form of left- and right-hand sides),

$$\varDelta U = \frac{\partial U}{\partial \mathbf{L}(T)} \varDelta \mathbf{L}(T), \qquad (24.24)$$

where

$$\frac{\partial U}{\partial \mathbf{L}(T)} = \left(\frac{\partial U}{\partial L_0(T)}, \ldots, \frac{\partial U}{\partial L_{N-1}(T)} \right)$$

is payoff-specific (but often easy to calculate, see Sections 24.3.2.1 and 24.3.2.2), and $\varDelta \mathbf{L}(t)$ is an $N \times N$ matrix with the (n, m)-th element equal to $\varDelta_m L_n(t)$, see (24.18).

As we already mentioned in Section 24.3.1, the representation (24.24) is convenient if we want to calculate pathwise deltas of multiple payoffs simultaneously, as the matrix $\varDelta \mathbf{L}(T)$ can be reused for each payoff. On the other hand, the calculation of the matrix $\varDelta \mathbf{L}(T)$ is computationally costly — as we showed in Section 24.3.1, it is of the order $O(N)$ *slower* than just calculating deltas by revaluation, and if we only need to calculate pathwise deltas of a *single* payoff, it is not clear why one would ever want to use the pathwise differentiation method. However, it turns out that by rearranging the order of calculations in what is known as the *adjoint method* (see Giles and Glasserman [2006]), the speed of calculations in the pathwise method can be significantly improved.

It follows from (24.24), (24.19) and (24.16) that

$$\Delta U = \frac{\partial U}{\partial \mathbf{L}(T_k)} \mathbf{D}(T_{k-1}) \dots \mathbf{D}(T_0), \qquad (24.25)$$

where the matrices $\mathbf{D}(T_i)$ are defined in (24.20). The standard pathwise differentiation method calculates matrices $\mathbf{D}(T_0)$, $\mathbf{D}(T_1)\mathbf{D}(T_0)$, and so on, using a matrix-matrix multiplication on each step and ultimately multiplying the final matrix by the vector $\partial U / \partial \mathbf{L}(T_k)$. We can, however, rearrange the order of calculations so that on each step we have a *vector-matrix* multiplication. To accomplish this, we just need to group the terms in (24.25) "from the left":

$$\Delta U = \left(\cdots \left(\left(\frac{\partial U}{\partial \mathbf{L}(T_k)} \mathbf{D}(T_{k-1}) \right) \mathbf{D}(T_{k-2}) \right) \cdots \right) \mathbf{D}(T_0).$$

In particular, let us define

$$Y^k = \frac{\partial U}{\partial \mathbf{L}(T_k)}$$

and then, recursively,

$$Y^{i-1} = Y^i \mathbf{D}(T_{i-1}), \quad i = k, \dots, 1. \qquad (24.26)$$

Then Y^0 gives the final solution,

$$Y^0 = \Delta U,$$

after applying the recursion (24.26) k times. Each step involves a vector-matrix multiplication and requires only $O(N^2)$ operations — i.e. savings of a factor of N compared to the standard pathwise scheme (see Section 24.3.1) — as is clear from both (24.26) and from the following explicit representation obtained from (24.20):

$$Y_m^{i-1} = Y_m^i + \left(\sum_n Y_n^i \varphi(L_n(T_{i-1})) \lambda_n(T_{i-1})^\top \frac{\partial \mu_n(T_{i-1})}{\partial L_m(T_{i-1})} \right) \tau_{i-1}$$
$$+ Y_m^i \varphi'(L_m(T_{i-1})) \lambda_m(T_{i-1})^\top \left(\mu_m(T_{i-1}) \tau_{i-1} + Z_{i-1} \sqrt{\tau_{i-1}} \right) \quad (24.27)$$

for $m = 0, \dots, N-1$. The computational effort is further reduced by noting that this expression simplifies significantly for some combinations of the indices i, m, n. For instance,

$$\lambda_m(T_{i-1}) = 0, \; \mu_m(T_{i-1}) = 0 \quad \text{for } m \leq i-1$$

in line with our conventions $L_m(t) \equiv L_m(T_m)$ for $t \geq T_m$. Also, in the spot Libor measure, the drift derivatives $\partial \mu_n(T_{i-1})/\partial L_m(T_{i-1})$ are non-zero only for $i \leq m \leq n$, and similar conditions exist for drifts in other measures. All these facts could (and should) be used to obtain an efficient numerical implementation.

The recursion (24.26) proceeds backward in time, but as is clear from (24.27) the i-th step requires the (simulated) value of the Libor vector $\mathbf{L}(T_{i-1})$, which can only be obtained in a forward simulation. This is not much of a problem, however, as we can always save the required values of the Libor rates when calculating the *value* of the option in the (forward) simulation (14.13)–(14.14) and then use these rates in the backward recursion (24.26), (24.27) when calculating deltas. The extra memory requirements are modest as this is done path-by-path.

From the discussion in this section it should be clear that when the pathwise differentiation method is used, there is limited downside to using the adjoint method to arrange the calculation order. An exception occurs if one is able to compute risk on more than $O(N)$ derivatives in the same model, on a time line shared by all products in the same simulation. In this case, the Libor delta matrix $\Delta \mathbf{L}$ should be pre-computed and applied to each payoff via (24.24).

24.3.4 Pathwise Delta Approximation for Callable Libor Exotics

Calculations of pathwise deltas for CLEs can be based on the fundamental result of Proposition 24.1.2 that expresses the pathwise derivative of a CLE in terms of pathwise derivatives of the (net) coupons. Conveniently, as the coupons can be regarded as European options, the results of the previous section can be used to compute pathwise deltas of coupons. Per Proposition 24.1.2, we additionally require an estimate $\widetilde{\eta}$ of the optimal exercise index, which fortunately is almost always found as a by-product of a typical Monte Carlo valuation of a callable security, see Section 18.3. Once $\widetilde{\eta}$ is obtained, the (lower bound) estimate of the value of the CLE is given by

$$\widetilde{H}_0(0) = \mathrm{E}\left(\sum_{n=\widetilde{\eta}}^{N-1} B(T_{n+1})^{-1} X_n \right), \qquad (24.28)$$

as computed in a Monte Carlo simulation.

Replacing the true exercise index η with its estimator $\widetilde{\eta}$ gives an approximation of the value of a callable Libor exotic. In the same vein, replacing in Proposition 24.1.1 the true optimal exercise index η with $\widetilde{\eta}$ gives an approximation of the pathwise delta,

$$\widetilde{\Delta}_m H_0(0) \triangleq \mathrm{E}\left(\sum_{n=\widetilde{\eta}}^{N-1} \Delta_m \left(B(T_{n+1})^{-1} X_n \right) \right), \qquad m = 0, \ldots, N-1. \quad (24.29)$$

It is shown in Piterbarg [2004b] that, as the exercise policy estimate converges to the optimal policy, the estimate in (24.29) approaches the true pathwise delta,

$$\widetilde{\Delta}_m H_0(0) \to \Delta_m H_0(0).$$

This gives rise to an elegant formula for estimating deltas of a callable Libor exotic that is easy to implement in practice. With the estimate of the optimal exercise time, $\widetilde{\eta}$, coming "for free" from the estimation step of the Monte Carlo valuation, we approximate the pathwise delta by

1. Running a forward simulation, for each path ω determining the exercise time index $\widetilde{\eta}(\omega)$.
2. For each path, computing pathwise deltas of all coupons X_n, $n = 1, \ldots, N-1$ (as well as the deltas of the inverse numeraire $B(t)^{-1}$), per Section 24.3.2.
3. Adding up deltas $\Delta_m(B(T_{n+1})^{-1}X_n)$ for those coupons that occur after the exercise index $\widetilde{\eta}(\omega)$.
4. Averaging the result over all paths.

As the pathwise delta of a CLE is given by the sum of pathwise deltas of European-style options, it follows trivially that the adjoint method of Section 24.3.3 could fruitfully be used here as well — an idea discussed at length in Leclerc et al. [2009].

We call the values $\widetilde{\Delta}_m H_0(0)$, $m = 0, \ldots, N-1$, *pathwise delta approximations*. These should not be confused with the true deltas of the lower bound CLE price estimate. To state this more succinctly, recall the definition of $\widetilde{H}_0(0)$ in (24.28), which can be interpreted as the value of a barrier-style Libor exotic that knocks into $U_n(T_n)$ for the first n for which the approximate exercise region (characterized by $\widetilde{\eta}$) is hit. Then, we generally have,

$$\widetilde{\Delta}_m H_0(0) \neq \Delta_m \widetilde{H}_0(0),$$

where on the right-hand side we have the true delta of the lower bound CLE price estimate.

It can easily be shown that under mild regulatory conditions,

$$\Delta_m \widetilde{H}_0(0) \to \Delta_m H_0(0),$$

as the exercise boundary converges to the optimal one. Hence, both approximations $\widetilde{\Delta}_m H_0(0)$ and $\Delta_m \widetilde{H}_0(0)$ provide converging approximations to the true delta $\Delta_m H_0(0)$. We note that it is normally $\Delta_m \widetilde{H}_0(0)$ that is typically computed in the standard perturbation method. Piterbarg [2004b] compares deltas computed by perturbations and by pathwise differentiation, and finds the latter both more stable and significantly faster to compute: in the tests performed, pathwise delta approximations required about 15 times less computational effort than delta computations by direct perturbation methods[7].

As both $\widetilde{\Delta}_m H_0(0)$ and $\Delta_m \widetilde{H}_0(0)$ converge to the same value when the exercise policy approaches optimality, we can use the difference between the two as an informal measure of the quality of our exercise decision approximation (or, equivalently, the gap between the true value and the lower bound value calculated in Monte Carlo). In practice this works best if we aggregate all deltas together, and monitor the difference

$$\sum_{m=0}^{N-1} \widetilde{\Delta}_m H_0(0) - \sum_{m=0}^{N-1} \Delta_m \widetilde{H}_0(0)$$

for significant deviations from 0.

24.4 Notes on Likelihood Ratio and Hybrid Methods

Section 3.3.3 introduced another non-perturbative differentiation method, the likelihood ratio method. The method shifts differentiation from the payoff to the density of the process and is not limited to smooth (Lipschitz continuous) payoffs. Practical applications of likelihood ratio methods in interest rate modeling are typically limited to fairly special situations, so we do not here expand much on our introduction in Section 3.3.3. Still, it is instructive to understand *why* the likelihood ratio method in its basic form is not particularly useful for our purposes[8].

We start by recalling the expression for the log-likelihood ratio (3.80) in the Black-Scholes model. Of particular relevance to our discussion is the presence of $\sqrt{T}$ term in the denominator of the expression for the log-likelihood ratio in (3.80). Clearly, with T approaching zero, the log-likelihood ratio grows to infinity, resulting in exploding variance of the estimate of the likelihood ratio derivative (3.79). In general, it is, in fact, not the time to option expiry that determines how fast the variance of the estimate explodes, but the earliest observation date of the underlying asset process. To demonstrate, we consider a security with payoff $g(Y(T_1), \ldots, Y(T_N))$ for

---

[7]Note that we are here comparing against a brute-force perturbation method where the exercise boundary — rather than the exercise *time* — is kept fixed under perturbations. Had we instead kept the exercise time fixed in perturbed scenarios, we would likely have obtained greeks of quality comparable to those produced by the pathwise method. Recall our comments at the end of Section 24.3.1.

[8]Another potential drawback is the need to know the transition density of the underlying process, although one always has the option of using a Gaussian approximation based on an Euler discretization of the true process. As discussed in Chen and Glasserman [2007b], the limit of this procedure for small time steps is deeply connected to the Malliavin calculus.

some $0 < T_1 < \ldots < T_N$, with the process $Y(t)$ defined in Section 3.3.3.1. Then, clearly

$$\mathrm{E}\left(g\left(Y(T_1), \ldots, Y(T_N)\right)\right) = \mathrm{E}\left(\widetilde{g}\left(Y(T_1)\right)\right),$$

where

$$\widetilde{g}(y) = \mathrm{E}\left(g\left(Y(T_1), \ldots, Y(T_N)\right) \middle| Y(T_1) = y\right).$$

Hence,

$$
\begin{aligned}
\frac{d}{dS_0}\mathrm{E}(g(Y(T_1), \ldots, Y(T_N))) &= \frac{d}{dS_0}\mathrm{E}(\widetilde{g}(Y(T_1))) \\
&= \mathrm{E}(l(Y(T_1))\mathrm{E}(g(Y(T_1), \ldots, Y(T_N))|Y(T_1))) \\
&= \mathrm{E}(l(Y(T_1))g(Y(T_1), \ldots, Y(T_N))) \\
&= \frac{1}{S_0\sigma\sqrt{T_1}}\mathrm{E}(Z_1 g(Y(T_1), \ldots, Y(T_N))),
\end{aligned}
$$

where $Z_1 = W(T_1)/\sqrt{T_1} \sim \mathcal{N}(0, 1)$. The time T_1 here could, for example, be the time to the first coupon fixing date, to the first exercise date of a CLE, or to the first knockout date of a barrier. Because of the regular structure of most interest rate derivatives, the time T_1 will in most cases be rather short, resulting in high variance of the estimate.

The fact that the likelihood ratio method does not work for many interest rate derivatives is unfortunate, since, as described in Section 3.3.3, likelihood ratio methods have the potential of handling irregular (e.g., discontinuous) payouts that are outside the scope of pathwise differentiation and perturbation methods. For such payouts, we will often have to apply one of the payoff smoothing methods from Chapter 23, and *then* apply pathwise differentiation or a perturbation method. An alternative approach involves invoking a *hybrid* method, that attempts to combine features of both the pathwise differentiation and likelihood ratio methods. In a series of works Fries and Kampen [2006], Fries and Joshi [2008a], Fries [2007], the authors have introduced successively more elaborate hybrid schemes that, roughly speaking, attempt to choose the right combination of a pathwise and a likelihood ratio derivative for each Monte Carlo path, depending on the relationship between the path and "interesting" product features such as strikes or barriers. We cannot possibly do justice to all the nuances involved in developing these schemes, so we simply refer interested readers to the source papers. Many of these methods are both fairly involved and rather specialized, so their deployment in generic risk systems will often be challenging. It is fair to say that the jury is still out when it comes to the practicality of these schemes in actual trading systems.

Importance Sampling and Control Variates

Even if sophisticated payoff smoothing and pathwise derivative schemes are employed, obtaining high-quality Monte Carlo greeks will always require the statistical simulation error to be kept low. Several generic variance reduction techniques were already introduced in Chapter 3; here, we expand on certain applications that are of particular relevance in interest rate modeling. As it turns out, some techniques, such as importance sampling techniques of Section 25.2, produce benefits for greeks estimation that go beyond mere variance reduction. On the other hand, other techniques are less impressive for the greeks than for basic value estimation, as described in Section 25.6. Nevertheless, all variance reduction techniques in this chapter are useful to know.

In our discussion here, we first study a number of applications of the importance sampling technique originally introduced in Section 3.4.4, with a particular emphasis on barrier and TARN products. Subsequently, we turn our attention to the control variate method initially considered in Section 3.4.3, discussing a variety of model- and instrument-based strategies for finding good controls.

25.1 Importance Sampling In Short Rate Models

We first look at a classic importance sampling application in simulation of short rate models. For concreteness, let us consider the pricing of a zero-coupon bond maturing at time T in the generic model (11.54); i.e. we are interested in evaluating

$$X(0,T) = \mathrm{E}\left(\exp\left(-\int_0^T x(u)\,du\right)\right) \triangleq \mathrm{E}\left(Y(T)\right) \qquad (25.1)$$

by Monte Carlo methods. Notice that in (25.1) the expectation E is assumed taken under the risk-neutral probability measure Q. We consider now changing probability measure, from Q to some other measure $\widetilde{Q}$, with the measure change characterized by a density $\varsigma(t) = E_t(d\widetilde{Q}/dQ)$ with

$$d\varsigma(t) = -\varsigma(t)q\left(t, x(t)\right) dW(t), \quad \varsigma(0) = 1,$$

for some function $q(t, x(t))$ sufficiently regular for $\varsigma(t)$ to be a Q-martingale. By the Radon-Nikodym theorem

$$X(0, T) = \widetilde{E}\left(Y(T)/\varsigma(T)\right),$$

where $\widetilde{E}$ is the expected value operator for measure $\widetilde{Q}$.

In measure $\widetilde{Q}$, Girsanov's theorem tells us that the joint process for $x(t)$, $1/\varsigma(t)$, and $Y(t)$ becomes

$$d\begin{pmatrix} x(t) \\ 1/\varsigma(t) \\ Y(t) \end{pmatrix} = \begin{pmatrix} \mu_x\left(t, x(t)\right) - q\left(t, x(t)\right)\sigma_x(t, x(t)) \\ 0 \\ -x(t)Y(t) \end{pmatrix} dt$$
$$+ \begin{pmatrix} \sigma_x\left(t, x(t)\right) \\ q\left(t, x(t)\right)/\varsigma(t) \\ 0 \end{pmatrix} d\widetilde{W}(t),$$

where $d\widetilde{W}(t) = dW(t) + q(t, x(t)) dt$ is a $\widetilde{Q}$-Brownian motion.

As shown in Section 3.4.4.3, we can arrange for the random variable $Y(T)/\varsigma(T)$ to have *zero* variance in measure $\widetilde{Q}$, provided that we use (3.96) to set

$$\varsigma(t) = \exp\left(-\int_0^t x(u)\,du\right) X(t, T, x(t))/X(0, T),$$
$$q\left(t, x(t)\right) = -X\left(t, T, x(t)\right)^{-1} \sigma_x\left(t, x(t)\right) \frac{\partial X\left(t, T, x(t)\right)}{\partial x(t)}, \qquad (25.2)$$

where

$$X\left(t, T, x\right) = E_{t,x}\left(\exp\left(-\int_t^T x(u)\,du\right)\right).$$

For the SDE (11.54) we generally do not have an analytical (reconstitution) expression for $X(t, T, x)$, but we are free to provide a guess for it. While doing so will most likely not reduce the variance of $Y(T)/\varsigma(T)$ to zero, if the guess is at all reasonable we can still expect a significant variance reduction effect. One route to an estimate for the function $X(t, T, x)$ is to assume that the SDE for $x(t)$ can be approximated by a simpler SDE for which a closed-form bond reconstitution formula exists; possible candidates would be, say, the affine class of short rate models or the quadratic Gaussian model. For instance suppose that we feel that the SDE for $x(t)$ can be approximated with a mean-reverting Gaussian model

$$dx(t) \approx (m - \varkappa_G x(t))\,dt + \sigma\,dW(t),$$

then we would obtain

$$q(t, x) = \sigma_x (t, x) \frac{1 - e^{-\varkappa_G (T - t)}}{\varkappa_G}. \tag{25.3}$$

In practice, most models will have a linear mean-reverting drift term, so the estimate of the "best" choice of $\varkappa_G$ in (25.3) is often straightforward. If $\mu_x(t, x)$ is non-linear, we could simply linearize it around $x = 0$ for the purpose of estimating $\varkappa_G$.

Andersen [1996] (see also Andersen and Boyle [2000]) tests the efficiency of the choice (25.3) when applied to the problem of computing discount bond prices for the CIR process; the results are far superior to those obtained by traditional variance reduction techniques. Andersen [1996] also notes that the quality of the measure transformation method improves significantly as the number of time steps in the simulation path is increased; this behavior is not surprising given that the method has been designed around the continuous-time limit of the discretized process for $x(t)$. The tendency of the measure transform method to improve with increasing number of discretization steps is quite attractive as it complements the behavior of the bias in the SDE discretization scheme: increasing the number of time steps will lower *both* the systematic bias and the random Monte Carlo error.

Finally, we note that the principles at play in the method above are general and can be applied to more complicated securities than discount bonds; all that is required is some decent estimate of the expectation value as a function of t and x. Often such estimates can be derived — either exactly or at least approximately — in a Gaussian model, for instance. We note that knowledge of an expectation in a closely related model can also form the basis for an application of the control variate method, an idea that we discuss in more detail starting from Section 25.3 below.

25.2 Payoff Smoothing by Importance Sampling in TARNs and General Barrier Options

Let us now take a different tack and demonstrate that the importance sampling method may also be used to produce payoff smoothing, in the vein of Chapter 23.

25.2.1 Binary Options

We first study a simple example that clearly illustrates the connection between importance sampling and payoff smoothing. Let X be a Gaussian random variable with mean μ and variance σ^2, and consider an option that pays $g(X)$ (for some smooth $g(x)$) if X is below a certain barrier b, so that the value of the security is given by

$$V = \mathrm{E}\left(g(X)1_{\{X<b\}}\right), \tag{25.4}$$

where E is an expected value operator for some pricing measure P (note that we do not include discounting in this illustrative example for clarity). Valuing this security by Monte Carlo requires simulating independent Gaussian samples, discarding those that end up above the barrier b, and averaging the payoff values over non-discarded samples. If b is low, then the proportion of paths that contribute to the average is small, which leads to a large simulation error (see related discussion in Section 3.4.4.5). Also, the digital feature in the payoff reduces the accuracy and stability of Monte Carlo estimates of greeks. In light of this, it seems natural to change the probability measure to increase the proportion of "interesting" samples, as we did in Section 3.4.4.5. Alternatively, it is tempting to integrate the digital option analytically. As we shall show, the importance sampling method can be set up to implement *both* strategies.

Let us rewrite the value by conditioning on the survival,

$$V = \mathrm{E}\left(g(X)\mid X < b\right)\mathrm{P}\left(X < b\right). \tag{25.5}$$

The probability of survival in our simple example is known in closed form,

$$\mathrm{P}\left(X < b\right) = \Phi\left(\frac{b-\mu}{\sigma}\right),$$

where $\Phi(z)$ is the standard Gaussian CDF. Calculating the remaining term $\mathrm{E}(g(X)|X<b)$ by Monte Carlo simulation requires us to draw random samples of X, conditioned on the event $\{X<b\}$. In order to do this, let us briefly recall from Section 3.1.1 how Gaussian random variables are typically simulated. If U is a uniform random variable on $[0,1]$, then X is obtained by

$$X = \Phi^{-1}(U). \tag{25.6}$$

Therefore, X conditioned on $\{X<b\}$ can be sampled by simply drawing a random variable U' uniformly distributed on the interval $[0, \Phi(b)]$, followed by an application of the mapping (25.6):

$$X\mid\{X<b\} = \Phi^{-1}\left(U'\right), \quad U' \sim \mathcal{U}\left(0, \Phi(b)\right). \tag{25.7}$$

From (25.5), we may write our option value as

$$V = \mathrm{E}\left(g\left(\Phi^{-1}\left(U'\right)\right)\right)\Phi\left(\frac{b-\mu}{\sigma}\right), \quad U' \sim \mathcal{U}\left(0, \Phi(b)\right). \tag{25.8}$$

Here, the function $g(x)$ is smooth by assumption, and thus a Monte Carlo evaluation of $\mathrm{E}(g(\Phi(U')))$ will have good convergence and exhibit stable greek estimates. By conditioning, we have, in effect, managed to integrate out the discontinuity analytically, and used the Monte Carlo method for the smooth part of the payoff only.

While a close connection of the method above to the payoff smoothing methods of Chapter 23 is obvious, the method can also be interpreted as a particular case of importance sampling, since in (25.8) all drawn samples come from the "interesting" part of the sample space where survival is guaranteed. The measure effectively used for sampling in (25.8) is often known as the *survival measure*. To characterize this measure further, notice first that in (25.4) the variable $1_{\{X<b\}}$ is not strictly positive, which requires some additional considerations before using this variable to define a measure shift. Indeed, starting from Section 1.3, we so far have only considered *equivalent* measures defined by strictly non-zero random variables as Radon-Nikodym derivatives. The definition of measure change can, however, be extended to Radon-Nikodym derivatives which can hit zero, but in this case the new measure $\widetilde{P}$ is not equivalent to the original measure P; instead it is *absolutely continuous* with respect to the original measure:

$$P(A) = 0 \Rightarrow \widetilde{P}(A) = 0,$$

but not necessarily the other way around. Notice that the two measures are equivalent when restricted to the set on which the Radon-Nikodym derivative is strictly positive. This set is what we are interested in here. The specific Radon-Nikodym derivative we need is given by

$$\Lambda = \frac{1_{\{X<b\}}}{P\left(X < b\right)};$$

note the normalization factor so that $E(\Lambda) = 1$. With this definition, we may write

$$V = E\left(g(X)1_{\{X<b\}}\right) = E\left(g(X)\Lambda\right)P\left(X < b\right) = \widetilde{E}\left(g(X)\right)P\left(X < b\right),$$

where $\widetilde{E}$ denotes expectation in the survival measure $\widetilde{P}$, defined by

$$\frac{d\widetilde{P}}{dP} = \Lambda.$$

$\widetilde{P}$ is called a survival measure because it assigns zero probability to all events in the "no-survival" region, i.e. for any event A such that $A \subset \{X \geq b\}$, we have $\widetilde{P}(A) = 0$. The distribution of X under $\widetilde{P}$ coincides with the distribution of X conditioned on $\{X < b\}$, and is given by (25.7).

The example above is simple, but it demonstrates a general approach to smoothing via importance sampling. Even for more complex barrier-style options, conditioning on survival will often allow us to handle discontinuities analytically, and evaluate the smooth part of the payoff by sampling under the survival measure. These ideas are fully developed in Glasserman and Staum [2001], where the authors observe that conditioning on full survival for a general barrier option is usually not analytically tractable, and instead propose to condition on one-step survival, from one barrier observation date to the next; at each time step, the measure is changed locally to allow the process to survive until the next time step. Since the behavior of most processes is much simpler on shorter time scales than on longer ones, this strategy will often lead to analytical tractability and efficient Monte Carlo implementation. Our treatment of TARNs in the next section is based on these ideas.

25.2.2 TARNs

TARNs and their valuation by Monte Carlo have been introduced in Chapter 20. We recall that the main TARN valuation formula (20.2) under the spot measure Q^B reads

$$V_{\text{tarn}}(0) = \text{E}\left(\sum_{n=1}^{N-1} B(T_{n+1})^{-1} X_n(T_n) 1_{\{Q_n < R\}}\right), \qquad (25.9)$$

where $B(t)$ is as the discretely rolled money market numeraire, X_n's are net coupons, $Q_n = \sum_{i=1}^{n-1} \tau_i C_i$ are accumulated structured coupons, and R is the total return. Here E is (re-)defined to be the expected value operator for measure Q^B. As described in detail elsewhere, the net coupon X_n is paid only if the process survives up to time T_n; by analogy to the simple example in Section 25.2.1, we expect that conditioning on survival may reduce variance and improve risk stability. We have already seen a payoff smoothing method applied to TARNs based on partial analytical integration in Section 23.2.4. Here we approach the problem from a different angle.

25.2.3 Removing the First Digital

In many cases, the biggest contributor to the simulation noise in a TARN is the first embedded digital option, i.e. the contract feature that specifies a knock-out event at date T_2 if $C_1(T_1)$ is above a certain barrier[1]. The variance of the estimate can be reduced if we could handle this digital option explicitly, outside of the Monte Carlo simulation.

To develop the idea in detail, let us for concreteness focus on a TARN of the inverse floating type, where the structured coupon is as in (20.1),

$$C_n = (s - g \times L_n(T_n))^+. \qquad (25.10)$$

We also introduce a sequence of random variables

$$b_n = (s - (R - Q_n)/\tau_n)/g, \qquad (25.11)$$

---

[1]Sometimes a TARN is structured so that the first digital is virtually worthless, but the second one is important. The discussion that follows should then be modified accordingly.

with b_n being $\mathcal{F}_{T_{n-1}}$-measurable. The first variable in the sequence,

$$b_1 = (s - R/\tau_1)/g$$

(see (20.4)) is deterministic, and we have

$$\{Q_2 < R\} \Leftrightarrow \{L_1(T_1) > b_1\}.$$

Let us denote by $\mathcal{V}$ the path value of the coupons that depend on the first knockout event (the first coupon $X_1(T_1)$ is paid always and is easy to handle separately),

$$\mathcal{V} = \sum_{n=2}^{N-1} B(T_{n+1})^{-1} X_n(T_n) 1_{\{Q_n < R\}}.$$

Then

$$\begin{aligned}
\mathrm{E}\left(\mathcal{V}\right) = \mathrm{E}\left(\mathcal{V}\middle|\, L_1(T_1) > b_1\right) \mathrm{Q}^B\left(L_1(T_1) > b_1\right) \\
+ \mathrm{E}\left(\mathcal{V}\middle|\, L_1(T_1) \le b_1\right) \mathrm{Q}^B\left(L_1(T_1) \le b_1\right).
\end{aligned}$$

Clearly

$$\mathrm{E}\left(\mathcal{V}\middle|\, L_1(T_1) \le b_1\right) = 0$$

so that

$$\mathrm{E}\left(\mathcal{V}\right) = \mathrm{E}\left(\mathcal{V}\middle|\, L_1(T_1) > b_1\right) \mathrm{Q}^B\left(L_1(T_1) > b_1\right). \tag{25.12}$$

In (25.12), since T_1 is typically small (less than one year), the probability $\mathrm{Q}^B(L_1(T_1) > b_1)$ of not knocking out can nearly always be approximated analytically with a high degree of precision. For instance, since time to expiry is short, the issue of non-deterministic drift of $L_1(T_1)$ under the spot measure can be easily dealt with by, say, freezing the drift along the forward yield curve (see related discussion in Section 23.2.4).

The value $\mathrm{E}(\mathcal{V}|L_1(T_1) > b_1)$ can be interpreted as the value of the TARN under the condition that it does not knock out on the date T_1. This value can be computed in a Monte Carlo simulation by either sampling Gaussian variates that generate simulation steps by a scheme similar to (25.7), or by adjusting the drifts of the forward Libor model in such a way as to move the Libor rate L_1 away from the knockout region. We do not go into details as we will present a more general scheme in the next section.

25.2.4 Smoothing All Digitals by One-Step Survival Conditioning

Removing the first discontinuity from the payoff being calculated by Monte Carlo often reduces the simulation error substantially. However, we can go further. Typically, given the information available on the coupon date T_n, we can evaluate the probability of knockout on the next day (quasi)-analytically. Following Piterbarg [2004c], we can use this information to develop a scheme where *all* discontinuities are integrated outside of Monte Carlo.

Proposition 25.2.1. *A TARN with structured coupon (25.10) can be valued as follows,*

$$V_{\text{tarn}}(0) = \sum_{n=1}^{N-1} \widetilde{E}\left(\psi_n E_{T_{n-1}}\left(B(T_{n+1})^{-1} X_n(T_n)\right)\right), \tag{25.13}$$

$$\psi_n = \prod_{k=1}^{n-1} Q_{T_{k-1}}^B \left(L_k(T_k) > b_k\right). \tag{25.14}$$

Here the measure $\widetilde{Q}^B$ is defined by its Radon-Nikodym derivative with respect to Q^B,

$$\Lambda(t) = E_t \left(\frac{d\widetilde{Q}^B}{dQ^B}\right), \tag{25.15}$$

where $\Lambda(t)$ is a non-negative, normalized Q^B-martingale such that

$$\Lambda(t) = \frac{Q_t^B\left(L_{m+1}\left(T_{m+1}\right) > b_{m+1}\right)}{Q_{T_m}^B\left(L_{m+1}\left(T_{m+1}\right) > b_{m+1}\right)} \prod_{k=1}^{m-1} \frac{1_{\{L_{k+1}(T_{k+1}) > b_{k+1}\}}}{Q_{T_k}^B\left(L_{k+1}\left(T_{k+1}\right) > b_{k+1}\right)}$$

for $t \in [T_m, T_{m+1})$.

Proof. We observe that due to non-negativity of the structured coupon C_{n-1}, the following equality holds Q^B-almost surely,

$$\{Q_n < R\} \Leftrightarrow \left\{Q_{n-1} < R, (s - gL_{n-1}(T_{n-1}))^+ < (R - Q_{n-1})/\tau_{n-1}\right\}$$
$$\Leftrightarrow \{Q_{n-1} < R, L_{n-1}(T_{n-1}) > b_{n-1}\}.$$

Likewise, using non-negativity of all C_i's,

$$\{Q_n < R\} \Leftrightarrow \{Q_1 < R, Q_2 < R, \ldots, Q_n < R\}$$
$$\Leftrightarrow \{L_1(T_1) > b_1, \ldots, L_{n-1}(T_{n-1}) > b_{n-1}\}.$$

Hence

$$1_{\{Q_n < R\}} = \prod_{k=1}^{n-1} 1_{\{L_k(T_k) > b_k\}}.$$

Define

$$\Lambda_n(t) = \begin{cases} \frac{Q_t^B(L_{n+1}(T_{n+1}) > b_{n+1})}{Q_{T_n}^B(L_{n+1}(T_{n+1}) > b_{n+1})}, & t \in [T_n, T_{n+1}), \\ \frac{1_{\{L_{n+1}(T_{n+1}) > b_{n+1}\}}}{Q_{T_n}^B(L_{n+1}(T_{n+1}) > b_{n+1})}, & t \geq T_{n+1}, \\ 1, & t < T_n. \end{cases}$$

We note that $\Lambda_n(t)$ is a non-negative Q^B-martingale. Moreover, $\Lambda_n(t)$ is constant on $[0, T_n]$ and $[T_{n+1}, \infty)$. In addition, $\Lambda_n(t)$ is $\mathcal{F}_{T_{n+1}}$-measurable for $t \geq T_{n+1}$.

We define $\Lambda(t)$ by

$$\Lambda(t) = \prod_{n=0}^{N-2} \Lambda_n(t).$$

It is not hard to show that $\Lambda(t)$ is a Q^B-martingale as well. Let us denote the value of the n-th coupon, contingent on survival to time T_n, by

$$V_{\text{cpn},n}(0) = E\left(B(T_{n+1})^{-1} X_n(T_n) 1_{\{Q_n < R\}}\right).$$

As

$$
\begin{aligned}
\prod_{k=1}^{n-1} 1_{\{L_k(T_k) > b_k\}} &= \left(\prod_{k=1}^{n-1} \frac{1_{\{L_k(T_k) > b_k\}}}{Q^B_{T_{k-1}}(L_k(T_k) > b_k)}\right) \left(\prod_{k=1}^{n-1} Q^B_{T_{k-1}}(L_k(T_k) > b_k)\right) \\
&= \left(\prod_{k=1}^{n-1} \Lambda_{k-1}(T_{n-1})\right) \left(\prod_{k=1}^{n-1} Q^B_{T_{k-1}}(L_k(T_k) > b_k)\right) \\
&= \left(\prod_{k=1}^{N-1} \Lambda_{k-1}(T_{n-1})\right) \left(\prod_{k=1}^{n-1} Q^B_{T_{k-1}}(L_k(T_k) > b_k)\right),
\end{aligned}
$$

we have

$$V_{\text{cpn},n}(0) = E\left(\Lambda(T_{n-1}) B(T_{n+1})^{-1} X_n(T_n) \prod_{k=1}^{n-1} Q^B_{T_{k-1}}(L_k(T_k) > b_k)\right). \tag{25.16}$$

Next, taking the $\mathcal{F}_{T_{n-1}}$-conditional expected value inside the expected value in (25.16) and using the measure $\widetilde{Q}^B$ defined by its Radon-Nikodym derivative with respect to Q^B in (25.15), we obtain

$$V_{\text{cpn},n}(0) = \widetilde{E}\left(\psi_n E_{T_{n-1}}\left(B(T_{n+1})^{-1} X_n(T_n)\right)\right),$$

$$\psi_n = \prod_{k=1}^{n-1} Q^B_{T_{k-1}}(L_k(T_k) > b_k),$$

and the proposition follows. $\square$

Remark 25.2.2. Quantities ψ_n in (25.14) can be calculated or approximated analytically since each term of the form $Q^B_{T_{k-1}}(L_k(T_k) > b_k)$ involves an expected value over a relatively short period $[T_{k-1}, T_k]$, so that short-time approximations (e.g., based on a Gaussian distribution) to the distribution of L_k over $[T_{k-1}, T_k]$ may be applied effectively. Note that the fact that L_k is a Q^B-martingale over time period $[T_{k-1}, T_k]$ would help here. In some simulation schemes, such as those considered in Sections 25.2.5 and 25.2.6 for example, the ψ_n come for free, without any extra work.

Remark 25.2.3. The measure $\widetilde{Q}^B$ is not equivalent to Q^B because $\Lambda(t)$ can be zero. However, since the value of the TARN is zero for those paths for which $\Lambda(t)$ is zero, $\widetilde{Q}^B$ and Q^B are equivalent on the relevant subspace of the sample space.

The formula (25.13) specifies that the value of a TARN can be computed by Monte Carlo simulation under the measure $\widetilde{Q}^B$ by adding values of net coupons X_n scaled by weights ψ_n. This should be contrasted to the original expression (25.9) where the weights on coupons are instead indicator functions $1_{\{Q_n < R\}}$. Obviously, the ψ_n's are much smoother functions of a simulated path than are indicator functions, since in the former the digital discontinuities have been integrated away by computing the probabilities $Q^B_{T_{k-1}}(L_k(T_k) > b_k)$ in (25.14) (quasi)-analytically.

Another feature of note of formula (25.13) is the presence of nested expected values where the inner one is calculated under the original spot measure Q^B, while the outer uses the survival measure $\widetilde{Q}^B$. While $\widetilde{Q}^B$ is the main simulation measure, when computing the value of the n-th coupon we must return to measure Q^B, when stepping from time T_{n-1} to time T_{n+1}.

In measure $\widetilde{Q}^B$, the TARN never knocks out, so a simulation based on the result in Proposition 25.2.1 can be interpreted as a version of the importance sampling method in Section 25.2.1, where the measure is changed from Q^B to $\widetilde{Q}^B$ and the likelihood ratio is partially pre-integrated. Of course, to use the method in practice we need to establish how precisely to simulate model dynamics in measure $\widetilde{Q}^B$; we study this topic in the next three sections.

25.2.5 Simulating Under the Survival Measure Using Conditional Gaussian Draws

We first consider a special case of the Libor market model (see Chapter 14), where we use a single-factor[2] volatility specification with separable deterministic local volatility, see (14.13)–(14.14). Continuing with the inverse floater TARN example and assuming that the tenor structure coincides with the schedule of the TARN, Libor forwards satisfy

$$dL_i(t) = \lambda_i(t)\varphi\left(L_i(t)\right)\left(\mu_i(t)\,dt + dW(t)\right), \qquad (25.17)$$

$i = 1, \ldots, N-1$, with $W(t)$ being a one-dimensional Brownian motion under the spot measure Q^B.

TARN valuation with formula (25.13) requires us to simulate Libor rates under measure $\widetilde{Q}^B$, i.e. in such a way that $L_n(T_n) > b_n$ for each n. To see how this would work, let us consider a simulation time step from T_{n-1} to T_n for a fixed n for all Libor rates. We note that for each n, b_n in (25.11) is $\mathcal{F}_{T_{n-1}}$-measurable, i.e. is known at time T_{n-1}. Employing a simple Euler scheme, we can approximate the Q^B-dynamics as

---

[2]We comment on the multi-factor case later.

$$L_n(T_n) = L_n(T_{n-1}) + \lambda_n (T_{n-1}) \varphi (L_n (T_{n-1})) (\mu_n(T_{n-1})\tau_{n-1} + \sqrt{\tau_{n-1}}Z),$$

where Z is a standard Gaussian random variable. Given that we want to simulate in such a way that $L_n(T_n) > b_n$, we need to make sure that Z satisfies

$$L_n(T_{n-1}) + \lambda_n (T_{n-1}) \varphi (L_n (T_{n-1})) (\mu_n(T_{n-1})\tau_{n-1} + \sqrt{\tau_{n-1}}Z) > b_n, \tag{25.18}$$

which can be solved to yield

$$Z > Z_{\min}, \quad Z_{\min} \triangleq (b_n - m_n)/v_n, \tag{25.19}$$

where we have denoted

$$v_n \triangleq \lambda_n (T_{n-1}) \varphi (L_n (T_{n-1})) \sqrt{\tau_{n-1}}, \quad m_n \triangleq L_n(T_{n-1}) + v_n \mu_n(T_{n-1})\sqrt{\tau_{n-1}}, \tag{25.20}$$

so that

$$L_n(T_n) = m_n + v_n Z. \tag{25.21}$$

The lower bound $Z_{\min}$ in (25.19) is known at time T_{n-1}, and the measure change is expressed by the requirement that the random variable Z should satisfy (25.19). In Section 25.2.1 we have already discussed how to simulate a Gaussian random variable conditioned on it being below (or above) a certain level; all we need to do is to apply the idea behind the scheme (25.7) (with b set to $Z_{\min}$ from (25.19)). In particular we can just set

$$\widetilde{Z} = \Phi^{-1} (\Phi (Z_{\min}) + (1 - \Phi (Z_{\min})) U), \tag{25.22}$$

where U is a uniform draw from $[0, 1]$.

While in this new measure we, by construction, have that $L_n(T_n) > b_n$, it may not be entirely obvious that this is the measure $\widetilde{Q}^B$ as defined by (25.15), since we can satisfy the constraint (25.18) in many different ways. To check, let us denote the measure implicit in the simulation scheme above by $\widehat{Q}^B$ for a moment. We obviously have that for any $l < b_n$,

$$\widehat{Q}^B_{T_{n-1}}(L_n(T_n) > l) = 1, \quad l < b_n. \tag{25.23}$$

For l such that $l > b_n$, we have

$$\widehat{Q}^B_{T_{n-1}}(L_n(T_n) > l) = Q^B_{T_{n-1}}(\widetilde{Z} > (l - m_n)/v_n)$$

and then, from (25.22),

$$Q^B_{T_{n-1}}(\widetilde{Z} > (l - m_n)/v_n) = \frac{1 - \Phi((l - m_n)/v_n)}{1 - \Phi(Z_{\min})}.$$

Now, from (25.20)–(25.21),

$$1 - \Phi((l - m_n)/v_n) = Q^B_{T_{n-1}}(L_n(T_n) > l),$$
$$1 - \Phi(Z_{\min}) = Q^B_{T_{n-1}}(L_n(T_n) > b_n),$$

and we finally obtain

$$\widehat{Q}^B_{T_{n-1}}(L_n(T_n) > l) = \frac{Q^B_{T_{n-1}}(L_n(T_n) > l)}{Q^B_{T_{n-1}}(L_n(T_n) > b_n)}, \quad l \geq b_n$$

which, together with (25.23), demonstrates that $\widehat{Q}^B$ is the same measure as $\widetilde{Q}^B$ defined by (25.15).

To finish the description of the simulation scheme we note that once $\widetilde{Z}$ has been drawn, all Libor rates $L_i(t)$, $i = n, \ldots, N-1$, can be evaluated using

$$L_i(T_n) = L_i(T_{n-1}) + \lambda_i(T_{n-1})\,\varphi\,(L_i(T_{n-1}))\,\left(\mu_i(T_{n-1})\tau_{n-1} + \sqrt{\tau_{n-1}}\widetilde{Z}\right),$$

for $i = n, \ldots, N-1$.

Notice that the algorithm above not only shows how to easily propagate the Libor curve forward in time, it also gives us the weights ψ_n in (25.14) without extra work. Specifically, from (25.19) we see that the one-step survival probability $Q^B_{T_{n-1}}(L_n(T_n) > b_n)$ is simply equal to $1 - \Phi(Z_{\min})$. It should be noted that the simplicity of the algorithm is partly based on the fact that we consider only a single-factor LM model in (25.17), and also by the fact that the payout is such that we can express the survival condition as a simple condition on one of the primary Libor rates over each time period. Both restrictions can, however, be lifted fairly easily. For instance, Pietersz [2005] suggests using a suitable rotation of the local volatility matrix to make sure that the survival (over a given time step) is determined by a single Gaussian draw. A different, and more general, twist is offered in Fries and Joshi [2008b]; we briefly review this approach in the next section.

25.2.6 Generalized Trigger Products in Multi-Factor LM Models

Following Fries and Joshi [2008b], we define a *generalized trigger product* to be a contract that pays a (net) coupon X_n until a knockout event, defined as the first time index n where an $\mathcal{F}_{T_n}$-measurable *trigger variable* G_n exceeds some *trigger level* h_n, i.e. when $G_n > h_n$, with h_n being $\mathcal{F}_{T_{n-1}}$-measurable. In the spot measure, the so-defined security has present value

$$V_{\text{gtp}}(0) = \mathrm{E}\left(\sum_{n=1}^{\eta-1} B(T_{n+1})^{-1}X_n\right), \quad \eta = \min\{n \geq 1 : G_n \geq h_n\} \wedge N.$$

A generalized trigger product is closely linked to barrier options we considered in Sections 23.4.2 and 24.1.2. TARNs are a special case; for a TARN the trigger variable is in fact the n-th structured coupon C_n and the trigger barrier is given by $h_n = (R - Q_{n-1})/\tau_n$, see (20.2). Note that we do not assume any particular form for G_n or h_n at this point.

Leaning on the results in Proposition 25.2.1, the value $V_{\text{gtp}}(0)$ can be rewritten as an expectation in the survival measure $\widetilde{Q}^B$,

$$V_{\text{gtp}}(0) = \sum_{n=1}^{N-1} \widetilde{E}\left(\psi_n E_{T_{n-1}}\left(B(T_{n+1})^{-1} X_n\right)\right), \tag{25.24}$$

$$\psi_n = \prod_{k=1}^{n-1} Q^B_{T_{k-1}}\left(G_k \leq h_k\right). \tag{25.25}$$

Let us now generalize (25.17). We assume that all Libor rates are driven by a d-dimensional Brownian motion,

$$dL_i(t) = \sigma_i(t)^\top \left(\mu_i(t)\,dt + dW(t)\right), \quad i = 1, \ldots, N-1,$$

where $\sigma_i(t)$ is a general process that may depend on (potentially all) Libor rates at time t. Denoting by $\mathbf{L}(t)$ the vector of all forward Libor rates observed at t, we rewrite the dynamics in a vector format

$$d\mathbf{L}(t) = M(t)\,dt + \Sigma(t)^\top dW(t), \tag{25.26}$$

for a suitably defined vector function $M(t)$ and a matrix function $\Sigma(t)$ (both of which are functions of $\mathbf{L}(t)$).

Let us consider a single time step from T_{n-1} to T_n, and assume that the n-th trigger variable G_n is in fact a function $G_n(\mathbf{L}(T_n))$ of the vector of Libor rates observed at T_n. An Euler scheme for (25.26) is given by

$$\mathbf{L}(T_n) = \mathbf{L}(T_{n-1}) + M(T_{n-1})\,\tau_{n-1} + \widehat{\Sigma}^\top Z, \tag{25.27}$$

where Z is a d-dimensional standard Gaussian vector and $\widehat{\Sigma} = \Sigma(T_{n-1})^\top \sqrt{\tau_{n-1}}$. Let us define by $\gamma(z)$ the value of the trigger variable G_n as a function of the realized Gaussian increment in the Euler scheme (25.27),

$$\gamma(z) \triangleq G_n\left(\mathbf{L}(T_{n-1}) + M(T_{n-1})\,\tau_{n-1} + z\right),$$

so that

$$G_n(\mathbf{L}(T_n)) = \gamma(\widehat{\Sigma}^\top Z). \tag{25.28}$$

Next, we define the normalized gradient (a row vector) of the function γ by

$$v = \nabla\gamma(0)/\|\nabla\gamma(0)\|.$$

The survival boundary is given in terms of the function γ as $\gamma(z) = h_n$. Let $Y_{\max}$ be the solution of the linearization of this equation in direction v, i.e. let us set

$$Y_{\max} = (h_n - \gamma(0)) / \|\nabla\gamma(0)\|. \tag{25.29}$$

Then, to first order, as follows from (25.28), the survival condition $G_n(\mathbf{L}(T_n)) < h_n$ is equivalent to the following condition on the Gaussian draw Z,

$$Y < Y_{\max}, \quad Y \triangleq v\widehat{\Sigma}^\top Z. \tag{25.30}$$

In (25.30) the random variable Y is Gaussian, making it straightforward to design a sampling scheme where (25.30) is always satisfied. Drawing on the same idea that lead to (25.22) and (25.7), we define

$$U = \Phi(Y/\sigma_Y), \quad \sigma_Y^2 = v\widehat{\Sigma}^\top \widehat{\Sigma} v^\top,$$

and also set

$$\widetilde{Y} = \sigma_Y \Phi^{-1}(U\Phi(Y_{\max})). \tag{25.31}$$

Clearly $\widetilde{Y} \leq Y_{\max}$ always, and

$$Q^B(Y < K) = \Phi(Y_{\max})Q^B(\widetilde{Y} < K) \tag{25.32}$$

for any $K \in (-\infty, \Phi(Y_{\max})]$. In particular, to first order,

$$\gamma\left(\widehat{\Sigma}^\top Z + v^\top(\widetilde{Y} - Y)\right) \approx \gamma(0) + \|\nabla\gamma(0)\| v\left(\widehat{\Sigma}^\top Z + v^\top(\widetilde{Y} - Y)\right)$$

$$= \gamma(0) + \|\nabla\gamma(0)\| \left(Y + \widetilde{Y} - Y\right)$$

$$= \gamma(0) + \|\nabla\gamma(0)\| \widetilde{Y}$$

and, since $\widetilde{Y} \leq Y_{\max}$, we have that (again to first order)

$$\gamma\left(\widehat{\Sigma}^\top Z + v^\top(\widetilde{Y} - Y)\right) \leq h_n.$$

Therefore, if we replace the stepping scheme (25.27) with

$$\mathbf{L}(T_n) = \mathbf{L}(T_{n-1}) + M(T_{n-1})\tau_{n-1} + \widehat{\Sigma}^\top Z + v^\top(\widetilde{Y} - Y), \tag{25.33}$$

then $\mathbf{L}(T_n)$ will always, to first order, be in the survival region. In particular, to make a time step in the survival measure $\widetilde{Q}^B$, we simply make an adjustment to the Gaussian draw to stay in the survival region and instead of $\widehat{\Sigma}^\top Z$ use $\widehat{\Sigma}^\top Z + v^\top(\widetilde{Y} - Y)$. Moreover, from (25.32) we immediately obtain the weight that we need to apply to a Monte Carlo path with a particular draw Z (for time step $T_{n-1} \to T_n$) — it is simply equal to $\Phi(Y_{\max})$ from (25.29). Putting it all together, we obtain the following result (compare to Proposition 25.2.1).

Proposition 25.2.4. *A generalized trigger product can be valued by (25.24)–(25.25) where an Euler simulation in the survival measure is given by (25.33). The weights ψ_n in (25.25) are given by*

$$\psi_n = \prod_{k=1}^{n-1} Q_{T_{k-1}}^B \left(G_k \left(\mathbf{L}(T_k) \right) < h_k \right) = \prod_{k=1}^{n-1} \varPhi(Y_{\max,k}),$$

where by $Y_{\max,k}$ we denote the value of $Y_{\max}$ in (25.29) for time step $T_{k-1} \rightarrow T_k$.

25.3 Model-Based Control Variates

The method of control variates was first introduced in Section 3.4.3. As we recall from (3.83), the method boils down to replacing the standard Monte Carlo estimate

$$\frac{1}{K} \sum_{j=1}^{K} Y(\omega_j)$$

of $E(Y)$ with

$$\frac{1}{K} \sum_{j=1}^{K} (Y(\omega_j) - \beta^\top (Y^c(\omega_j) - E(Y^c))), \qquad (25.34)$$

where $Y^c(\omega_j)$ are random samples of the potentially multi-dimensional control variate Y^c, chosen such that $E(Y^c)$ is available in closed form. As shown in (3.85), the variance reduction achieved by the method is directly proportional to the correlation between the primary variable Y and its control Y^c. There are multiple ways to select the control variate Y^c. For instance, if Y is the value of a security under a given model, then Y^c may represent the value of the same security under a different, but closely related model; or the value of a different (but related) security under the same model; or the value of an approximate hedging strategy of Y. Of course, we may also select a control variate that is a weighted combination of many different individual control variates, each chosen by a different strategy.

In the next few sections, we shall study several methods to design control variates. We start with the model-based control variate method, which uses the value of a security in a simplified proxy model as a control for the security value in the actual pricing model. To fix ideas, let $\widehat{V}_{\mathrm{orig}}$ be the Monte Carlo estimate for the true security value in the original pricing model, and let $\widehat{V}_{\mathrm{proxy}}$ be the Monte Carlo estimate for the same security in a proxy model. In the proxy model, assume that a highly accurate price estimate V_{PDE} is available, most likely computed by the PDE methods of Chapter 2. As in (25.34), let us introduce a corrected value estimate as

$$\widehat{V}_{\mathrm{corrected}} = \widehat{V}_{\mathrm{orig}} - \beta \left(\widehat{V}_{\mathrm{proxy}} - V_{\mathrm{PDE}} \right),$$

where β is the appropriate regression coefficient. Assuming that $E(\widehat{V}_{\mathrm{proxy}}) = V_{\mathrm{PDE}}$ to high precision, the new estimate will be practically unbiased. If the path values used to compute $\widehat{V}_{\mathrm{orig}}$ are positively correlated with the ones used to obtain $\widehat{V}_{\mathrm{proxy}}$, then the variance of the estimate is reduced.

The computational effort to estimate $\widehat{V}_{\mathrm{corrected}}$ is noticeably higher than that needed to compute $\widehat{V}_{\mathrm{orig}}$, since two additional valuations (one by Monte Carlo and one by PDE methods) are now needed. For the method to lead to an efficiency improvement[3], the achieved variance reduction needs to be high, in turn requiring very high correlation between the original and proxy model path values of the security. This can typically be only achieved if the two models are closely related, and use random numbers in near-identical fashion to generate security path values. For instance, it is unlikely that one could successfully use a short rate model to compute a control variate when the original model is a Libor market (LM) model.

While on the topic of the LM model, we note that this particular model is particularly in need of variance reduction: not only is the LM model always implemented via Monte Carlo methods, it is also more computationally demanding than many other Monte Carlo based models[4] and typically is used for complex, compute-intensive payoffs. To find a suitably faithful proxy model for a full-blown LM model, we note that while PDE methods are generally not available for LM models, PDE *approximations* are possible. While some believe that such approximations may serve as outright substitutes for LM model, in our opinion it is very difficult to make the approximations sufficiently accurate and robust to safely use them for actual security pricing. On the other hand, these approximations are often perfectly adequate for the model-based control variate method, since the requirements on proxy model precision and internal consistency are actually quite low — all that is needed is that the estimator $\widehat{V}_{\mathrm{proxy}}$ is highly correlated to the true model price and has a limit that can be computed accurately by some other scheme. In fact, the proxy model does *not* have to be arbitrage free, which for LM models opens up the possibility of replacing complicated path-dependent drift terms with simpler ones that admit a PDE representation of security values.

25.3.1 Low-Dimensional Markov Approximation for LM models

We recall that an LM model is Markovian only in the full set of all forward Libor rates on the yield curve, plus any additional variables required to model unspanned stochastic volatility. As numerical methods for PDEs start becoming impractical when there are more than 3 or 4 state variables, a fair bit of simplification of the LM model is required to come up with a PDE-friendly model proxy. To show one way of proceeding, let us start with

---

[3]See Section 3.4.1 for a discussion of efficiency measures for variance reduction schemes.

[4]Such as the quasi-Gaussian (qG) model, which normally involves simulation of many fewer state variables than the LM model, see Section 13.1.9.3.

a one-factor model equipped with a deterministic local volatility (14.13)–(14.14), the spot measure dynamics of which we represent as

$$dL_n(t) = \varphi\left(L_n(t)\right)\lambda_n(t)\left(\mu_n\left(t, \mathbf{L}(t)\right) dt + dW(t)\right), \qquad (25.35)$$

$$\mu_n\left(t, \mathbf{L}(t)\right) = \sum_{i=q(t)}^{n} \frac{\tau_i\varphi\left(L_i(t)\right)\lambda_i(t)}{1 + \tau_i L_i(t)},$$

for $n = 1, \ldots, N-1$, with $\mathbf{L}(t)$ being the vector of all forward Libor rates. $W(t)$ is here a one-dimensional Brownian motion under the spot measure Q^B; an extension to two dimensions is studied in Section 25.3.2 below.

In a first step towards a low-dimensional Markovian approximation of (25.35), we look to get rid of the local volatility $\varphi(L_n(t))$. To that end, let us introduce the following transform,

$$f(x) = \int_{x_0}^{x} \frac{d\xi}{\varphi\left(\xi\right)}, \qquad (25.36)$$

where x_0 is an arbitrary but fixed number (see also (2.81)–(2.82)). Defining new variables

$$l_n(t) \triangleq f(L_n(t)), \quad n = 1, \ldots, N-1, \qquad (25.37)$$

we eliminate φ from the diffusion part of the SDE and get, for $n = 1, \ldots, N-1$,

$$dl_n(t) = \lambda_n(t)\left(\left(\mu_n\left(t, \mathbf{L}(t)\right) - \frac{1}{2}\lambda_n(t)\varphi'\left(L_n(t)\right)\right) dt + dW(t)\right). \quad (25.38)$$

For our purposes here, the main issue with (25.38) is the fact that the drift $\mu_n(t, \mathbf{L}(t))$ at each point in time depends on the whole vector of forward Libor rates. An easy way to deal with this is to simply replace in μ_n all Libor forwards with their values at $t = 0$, i.e.,

$$\mu_n(t, \mathbf{L}(t)) \approx \mu_n(t, \mathbf{L}(0)). \qquad (25.39)$$

As the LM model drift terms are generally small, the usage of the first-order approximation (25.39) is certainly justifiable for control variate purposes[5].

With approximation (25.39) we are a long way towards a low-dimensional Markov representation, but still need to simplify the term $\varphi'(L_n(t))$ in (25.38). For local volatility functions $\varphi(x)$ that are close to linear, we can use $\varphi'(L_n(t)) \approx \varphi'(L_n(0))$, an approximation that is exact for the important cases of log-normal and displaced log-normal model specifications. With this, we arrive at the following approximate SDE,

---

[5]Approximations to the drift could be improved by using the Brownian bridge techniques described in Section 14.6.2.5, but the impact of these improvements on the intended control variate applications is negligible.

$$dl_n(t) = \lambda_n(t) \left(\left(\mu_n(t, \mathbf{L}(0)) - \frac{1}{2}\lambda_n(t)\varphi'(L_n(0)) \right) dt + dW(t) \right). \quad (25.40)$$

In (25.40), each $l_n(t)$ is an integral of $\lambda_n(t)$ against a Brownian motion with deterministic drift. To make all the variables functions of the same state variable, we approximate the volatility structure with the following *separable* one,

$$\lambda_n(t) \approx \hat{\lambda}_n(t), \quad \hat{\lambda}_n(t) = \sigma_n \alpha(t), \quad n = 1, \ldots, N-1. \quad (25.41)$$

In a separable volatility structure, each forward Libor volatility function equals a Libor-specific scalar multiplied by a function of time common to all Libor rates. This special structure allows us to define a one-dimensional Markovian state variable by

$$dX(t) = \alpha(t) \, dW(t), \quad (25.42)$$

and all variables $l_n(t)$ are then deterministic functions of $X(t)$:

$$l_n(t) = l_n(0) + d_n(t) + \sigma_n X(t), \quad (25.43)$$

$$d_n(t) = \int_0^t \lambda_n(s) \left(\mu_n(s, \mathbf{L}(0)) - \frac{1}{2}\lambda_n(s)\varphi'(L_n(0)) \right) ds.$$

Translated back to Libor forwards, we arrive at the reconstitution formula

$$L_n(t) = f^{-1}\left(f\left(L_n(0)\right) + d_n(t) + \sigma_n X(t)\right), \quad n = 1, \ldots, N-1, \quad (25.44)$$

where we have made an implicit assumption that $f^{-1}(\cdot)$ exists (which is the case if, for example, $\varphi(\cdot)$ is positive in (25.36)).

With the representation above, at each point in time t, the value of any path-independent derivative V can be expressed as a function of t and $X(t)$,

$$V = V(t, X(t)),$$

where the function $V(t, x)$ satisfies the following PDE,

$$\frac{\partial V(t, x)}{\partial t} + \frac{\alpha(t)^2}{2} \frac{\partial^2 V(t, x)}{\partial x^2} = r(t, x)V(t, x), \quad (25.45)$$

subject to appropriate boundary and jump conditions. In (25.45), $r(t, X(t))$ is the discounting rate applied to any payoff over an instantaneous period of time $[t, t+dt]$, whose specific expression in terms of Libor rates (and, ultimately, $X(t)$) depends on the interpolation method used. For instance, under the assumption that instantaneous forward rates $f(t, u)$ are constant[6] for $u \in [t, T_{q(t)}]$ (see Section 15.1.6 and equation (15.20)), we obtain

$$r(t, X(t)) = \frac{1}{\tau_{q(t)}} \ln\left(1 + \tau_{q(t)} L_{q(t)}(t)\right),$$

where the right-hand side is understood to be a function of $X(t)$ via (25.44).

---

[6]We generally do not recommend this interpolation scheme, but it makes for a good example.

25.3.2 Two-Dimensional Extension

Before turning to the question of how to pick the volatility term structure for the LM proxy model, let us consider the extension to LM models driven by a two-dimensional Brownian motion. To build a two-factor Markovian proxy model, assume that

$$W(t) = \left(W^1(t), W^2(t)\right),$$

and that forward Libor volatilities are two-dimensional processes,

$$\lambda_n(t) = \left(\lambda_n^1(t), \lambda_n^2(t)\right).$$

In the spot measure, the deterministic local volatility LM model then follows

$$dL_n(t) = \sum_{k=1}^{2} \lambda_n^k(t)\varphi\left(L_n(t)\right)\left(\mu_n^k\left(t, \mathbf{L}(t)\right) dt + dW^k(t)\right),$$

$$\mu_n^k\left(t, \mathbf{L}(t)\right) = \sum_{i=q(t)}^{n} \frac{\tau_i \varphi\left(L_i(t)\right)\lambda_i^k(t)}{1 + \tau_i L_i(t)},$$

for $n = 1, \ldots, N - 1$. Following the steps in Section 25.3.1, we eventually come to the point where we need to approximate the volatility structure with a separable one similar to (25.41). A naive generalization would specify

$$\lambda_n^1(t) \approx \widehat{\lambda}_n^1(t), \quad \widehat{\lambda}_n^1(t) = \sigma_n^1 \alpha^1(t), \tag{25.46}$$
$$\lambda_n^2(t) \approx \widehat{\lambda}_n^2(t), \quad \widehat{\lambda}_n^2(t) = \sigma_n^2 \alpha^2(t),$$

for $n = 1, \ldots, N - 1$. However, an extension is possible (and desirable, as will be clear later), as we can use the more general expression

$$\widehat{\lambda}_n^1(t) = \sigma_n^1 \alpha^{11}(t), \tag{25.47}$$
$$\widehat{\lambda}_n^2(t) = \sigma_n^1 \alpha^{21}(t) + \sigma_n^2 \alpha^{22}(t),$$

while keeping the approximation Markovian. In particular we can then define two (correlated) state variables by[7]

$$dX_1(t) = \alpha^{11}(t)\, dW^1(t) + \alpha^{21}(t)\, dW^2(t),$$
$$dX_2(t) = \alpha^{22}(t)\, dW^2(t).$$

The Libor rates can then be computed by (compare to (25.44))

---

[7] We can use the triangular form here because the square root of a variance-covariance matrix can always be written this way by application of the Cholesky decomposition, see Section 3.1.2.1. A more general form, however, could be beneficial for fitting as discussed later, see footnote 9.

$$L_n(t) = f^{-1}\left(f\left(L_n(0)\right) + \tilde{d}_n(t) + \sigma_n^1 X_1(t) + \sigma_n^2 X_2(t)\right), \qquad (25.48)$$

for $n = 1, \ldots, N-1$, where the deterministic part $\tilde{d}_n(t)$ is suitably defined. The (two-dimensional) valuation PDE for $V = V(t,x,y)$ is now given by

$$\frac{\partial V}{\partial t} + \frac{\alpha^{11}(t)^2 + \alpha^{21}(t)^2}{2}\frac{\partial^2 V}{\partial x^2} + \alpha^{21}(t)\alpha^{22}(t)\frac{\partial^2 V}{\partial x \partial y}$$
$$+ \frac{\alpha^{22}(t)^2}{2}\frac{\partial^2 V}{\partial y^2} = r(t,x,y)V, \quad (25.49)$$

and can be solved with, for example, the Craig-Sneyd scheme from Section 2.11.2. The terminal condition $V(T,x,y)$ here is determined from the payoff of the derivative, after expressing the yield curve at time T in terms of the state variables $X_1(T)$, $X_2(T)$ through the reconstitution formula (25.48).

The specification (25.47) is normally substantially more accurate than (25.46) in approximating the volatility structure of the original model. To understand why, recall from the principal components analysis in Section 14.3.1 that the first volatility component of the original model $\lambda_n^1(t)$ normally represents a near-parallel yield curve shift and is positive for all t and n — a shape that can be represented in the form (25.46) quite well. However, the second component $\lambda_n^2(t)$ models a yield curve twist (see Figure 14.1) which, for a fixed value of t, will require that $\{\lambda_n^2(t)\}$ crosses zero for some value of n. With this in mind, consider the approximations in (25.46) and (25.47). In the former, a function of the form $\sigma_n^2\alpha^2(t)$ can cross zero either "vertically" ($\sigma_n^2\alpha^2(t) = 0$ for some $t = t_0$ for all n) or "horizontally" ($\sigma_n^2\alpha^2(t) = 0$ for some $n = n_0$ for all t). In reality, due to an imposed (or desired) time-homogeneity, $\lambda_n^2(t)$ usually crosses zero "diagonally", in the loose sense that the function $n_0(t) \triangleq \{n : \lambda_n^2(t) = 0\}$ grows with t. Figure 25.1 demonstrates the point. On each of the three diagrams of the figure, we plot signs of the second volatility component for all points $(t,n) \in [0, T_{N-1}] \times \{1, \ldots, N-1\}$. The plus symbol indicates a positive value and the minus symbol represents a negative one. Diagrams (A) and (B) represent the only two possibilities for the second volatility component of the form (25.46). The diagram (C) shows how a typical second volatility component really looks like, a behavior that can be replicated by (25.47) but *not* by (25.46).

25.3.3 Approximating Volatility Structure

So far we have glossed over the actual mechanics of approximating the original model volatility $\lambda(t)$ with a separable proxy version, as in (25.41) or (25.47). One approach would be to perform an outright calibration (see Section 14.5) of the Markov proxy model to the same swaption quotes used to calibrate the original model. We generally do not recommend this when we use the Markov LM model to generate a control variate. Instead, recall

Fig. 25.1. Sign of the Second Volatility Component

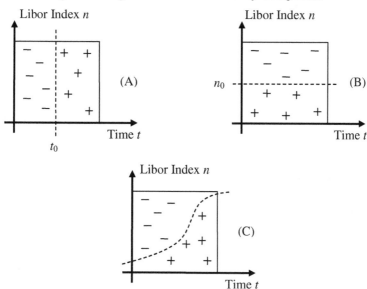

Notes: Signs of the second volatility component for separable (diagrams (A), (B)) parameterization of the form (25.46), and a typical non-separable one (diagram (C)).

from Section 3.4.3 that the variance reduction achieved by a control variate method is strongly correlation dependent, which suggests that we attempt to approximate the *factor* volatilities of the original model directly, without much consideration given to the precision with which the proxy model can price swaptions. In this spirit, calibration of, say, the two-factor separable volatility can be stated as a least-squares problem[8]

$$\sum_{i=0}^{N-2}\sum_{n=1}^{N-1}\left(\lambda_n^1(T_i) - \sigma_n^1\alpha^{11}(T_i)\right)^2$$

$$+ \sum_{i=0}^{N-2}\sum_{n=1}^{N-1}\left(\lambda_n^2(T_i) - \left(\sigma_n^1\alpha^{21}(T_i) + \sigma_n^2\alpha^{22}(T_i)\right)\right)^2 \to \min. \quad (25.50)$$

Here we optimize over all σ's and α's, for a total of $5 \times (N-1)$ variables. Of course, we may extend the norm as we see fit, to include smoothing penalty terms or to use different weights on different terms or factors.

---

[8]Note that, in line with (14.42), we assume that λ's and α's are piecewise constant over $[T_i, T_{i+1})$, $i = 0, \ldots, N-2$.

Denson and Joshi [2009] propose a slightly different fitting algorithm. First, we would set

$$\sigma_n^1 = \lambda_n^1(T_0), \quad \sigma_n^2 = \lambda_n^2(T_0), \quad n = 1, \ldots, N-1$$

and

$$\alpha^{11}(T_0) = \alpha^{22}(T_0) = 1, \quad \alpha^{21}(T_0) = 0.$$

This ensures that $\lambda_n^i(t) = \widehat{\lambda}_n^i(t)$ for $t \in [T_0, T_1)$ for all $n = 1, \ldots, N-1$, and $i = 1, 2$. Then we would solve the minimization problem (notice index i starting at $i = 1$, rather than at $i = 0$)

$$\sum_{i=1}^{N-2} \sum_{n=1}^{N-1} \left(\lambda_n^1(T_i) - \lambda_n^1(T_0)\alpha^{11}(T_i) \right)^2$$

$$+ \sum_{i=1}^{N-2} \sum_{n=1}^{N-1} \left(\lambda_n^2(T_i) - \left(\lambda_n^1(T_0)\alpha^{21}(T_i) + \lambda_n^2(T_0)\alpha^{22}(T_i) \right) \right)^2 \to \min \quad (25.51)$$

for α's only, a quadratic optimization problem that is solved analytically[9], e.g.

$$\alpha^{11}(T_i) = \frac{\sum_{n=1}^{N-1} \lambda_n^1(T_i)\lambda_n^1(T_0)}{\sum_{n=1}^{N-1} \left(\lambda_n^1(T_0) \right)^2}, \quad i = 1, \ldots, N-2,$$

and so on. The direct and analytic linkage of $\widehat{\lambda}$'s to λ's in this approach could lead to better performance of the control variate method when calculating risk sensitivities, especially vegas.

25.3.4 Markov Approximation as a Control Variate

To use the Markov approximation as the control variate, we define $\widehat{V}_{\text{proxy}}$ (see the beginning of Section 25.3 for notations) to be the value of the security in the Markov LM model computed by Monte Carlo, and by V_{PDE} the value computed by the PDE method in the same model. For V_{PDE} to be consistent with $\widehat{V}_{\text{proxy}}$, both the Monte Carlo and the PDE methods should be applied to the same derivative. While this may seem like a trivial point, some cases are fairly subtle and require care. For example, for callable Libor exotics, the $\widehat{V}_{\text{proxy}}$ value would often be a lower bound (Section 18.3) on the actual value of the callable derivative and, effectively, would represent the value of an exotic swap that knocks out at the estimated exercise boundary. In this case the PDE method should be applied not to a callable Libor exotic, but to a knockout swap with a knockout boundary lifted directly from the Monte

---

[9] Denson and Joshi [2009] also extend the specification (25.47) by allowing an extra term for $\widehat{\lambda}_n^1(t)$ with $\widehat{\lambda}_n^1(t) = \sigma_n^1 \alpha^{11}(t) + \sigma_n^2 \alpha^{12}(t)$, which may improve the fit somewhat.

Carlo valuation. For this to work in practice, the LS regression method for the estimation of the exercise boundary in the Markov LM model should use the Markov state variables $X(t)$ as explanatory variables[10] so that relevant regression functions can be easily transferred into the PDE setup.

Achieving a high correlation between the path values of a derivative in the original and proxy models is crucial to the performance of the model-based control variate method, so care must be taken in ensuring that the simulation schemes for the original and Markov models are as similar as possible. This ranges from the obvious requirement of using the same simulation seed for random number generation in the two models, to the more subtle issue of discretization scheme compatibility. For example, suppose we use the Euler discretization scheme on the original LM model SDE (25.35). Then, for the Markov LM model we must also use the Euler scheme on the SDE

$$
\begin{aligned}
dL_n(t) = \widehat{\lambda}_n(t)\varphi\left(L_n(t)\right) \\
\times \left(\mu_n\left(t, \mathbf{L}(0)\right) - \frac{1}{2}\widehat{\lambda}_n(t)\left(\varphi'\left(L_n(0)\right) - \varphi'\left(L_n(t)\right)\right) dt + dW(t),\right)
\end{aligned}
$$

obtained from (25.40) and (25.37). Here $\widehat{\lambda}_n(t) = \sigma_n\alpha(t)$ of course, or the equivalent for the two-dimensional model. In particular, notice that to keep the simulation of the two models in lock-step, we must resist the temptation to be clever, and avoid using special-purpose discretization schemes that take advantage of the simple form of the Markov proxy model dynamics.

The performance of the model-based control variate method can often be quite impressive, with Piterbarg [2003] reporting a reduction in sample standard deviation by a factor of 3 to 10, corresponding to a speed improvement of 10 to 100 times (see Section 3.4.1). Of course, there is extra work involved that includes an extra Monte Carlo simulation for the Markov model, and a (relatively speedy) PDE valuation. The potential downside of the method is the fact that its scope is somewhat limited by the need to perform a PDE valuation. With three dimensions probably being the practical maximum for a reasonably quick PDE scheme, the model-based control variate method is limited to either i) a two-factor Libor market model (as we developed above), ii) a three-factor model (a straightforward extension), iii) a two-factor model with stochastic volatility, or iv) a two-factor model for a path-dependent trade that could be treated in PDE by introducing an extra state variable (e.g., a TARN, see Section 20.1.5). For products and models that do not fit these categories, a proxy model may still be useful for defining dynamic control variates, as demonstrated in Section 25.5. In addition, we always have the option of using instrument-based control variates, as we describe next.

---

[10]This is a good idea for any Markovian model, see Section 18.3.9.1.

25.4 Instrument-Based Control Variates

In the model-based control variate method, we created a control variate by introducing a new model and applying it to the (unchanged) payoff of the security we look to price. In the instrument-based control variate method, in a sense we do the opposite: we keep the model fixed but change the payoff. In fixed income applications, the idea of using proxy securities as control variates is most closely associated with Bermudan swaption pricing, but the basic ideas often extend fairly naturally to more complicated callable Libor exotics.

For concreteness, let us start by considering a Bermudan swaption with $N - 1$ exercise opportunities. We follow the notation of Chapter 19 and, in particular, denote the $N - 1$ exercise values by $U_n(t)$, $n = 1, \ldots, N - 1$ (see (19.1)). The K-path Monte Carlo estimate value of a Bermudan swaption, or indeed a general CLE, is given by

$$H_0(0) \approx \frac{1}{K} \sum_{k=1}^{K} Y(\omega_k), \quad Y(\omega) = \sum_{i=\eta(\omega)}^{N-1} B(T_{i+1}, \omega)^{-1} X_i(\omega), \quad (25.52)$$

where $\{\omega_k\}_{k=1}^{K}$ are the simulated paths and η the (estimate of the) optimal exercise time index.

Naively, we could try to introduce control variates based on the $N - 1$ (deflated) exercise values, as observed at the final expiry time T_N. That is, we could define controls $Y^c = (Y_1^c, \ldots, Y_{N-1}^c)^\top$, where

$$Y_n^c(\omega) \triangleq \sum_{i=n}^{N-1} B(T_{i+1}, \omega)^{-1} X_i(\omega), \quad n = 1, \ldots, N - 1. \quad (25.53)$$

Each control is then a sum of path values of net coupons. Alternatively, for Bermudan swaptions in particular, we can use

$$Y_n^c(\omega) = U_n(T_n, \omega), \quad n = 1, \ldots, N - 1, \quad (25.54)$$

where, of course, each $U_n(t)$ is the value at time t of all net coupons from the n-th one onward (see (18.2)). For Bermudan swaptions the $U_n(t)$ are just swap values and are available bias-free in a closed form expression. Note that (25.54) constitutes a *different* set of control variates than (25.53).

Both of the control variate schemes outlined above are quite simplistic and typically fail to yield good variance reduction. One reason is that both these control variates are, in the case of a standard Bermudan swaption, essentially linear functions of rates, whereas the payoff of a Bermudan swaption is option-like and clearly not well-approximated by a linear function. We can attempt to rectify this issue by using control variates that are non-linear in rates. European swaptions are natural choices for this, but often must be ruled out due to lack of exact valuation formulas[11] or approximations that are

---

[11] The swap market models of Section 15.4 are, however, not affected by this issue.

sufficiently accurate across a wide range of moneyness and expiries. For Libor market models, however, caps (or floors) often have exact pricing formulas, so these instruments may be a good option; we explore this idea in more detail below. First, however, let us note that a more subtle reason for the failure of the schemes (25.53) and (25.54) is due to the fact that the control variates effectively are observed "at the wrong time". To elaborate, notice that the value of a Bermudan swaption (or a CLE) along a path ω in (25.52) involves cash flows fixed at times $T_{\eta(\omega)}, \ldots, T_N$, whereas all control variates in (25.53) always include a deterministic number of cash flows. Similarly, in (25.54) each control variate is sampled at a single time only. So, compared to controls, a path value of a Bermudan swaption will often have an incorrect number of net coupons included, and will likely have low correlation with the controls.

The fact that the timing mismatch contributes significantly to de-correlation between a Bermudan swaption and naive controls was noted by Rasmussen [2005], who also proposed to rectify the issue by sampling the controls *at the Bermudan swaption exercise time*. Here is the technical result that justifies this choice.

Proposition 25.4.1. *With T_n denoting the n-th exercise date, set $U_n = U_n(T_n)$, and let Z_n, $n = 0, \ldots, N$, be a martingale process with respect to $\{\mathcal{F}_n \triangleq \mathcal{F}_{T_n}\}_{n=0}^N$. Let stopping times $\eta, \sigma \in \{1, \ldots, N-1\}$ be given such that $\eta \leq \sigma$. Then*
$$\left(\operatorname{Corr}(U_\eta, Z_\eta) \right)^2 \geq \left(\operatorname{Corr}(U_\eta, Z_\sigma) \right)^2.$$

Proof. The proof follows by the repeated applications of the optional sampling theorem. For the covariance term, we have
$$\begin{aligned}
\operatorname{Cov}(U_\eta, Z_\sigma) &= \operatorname{E}(U_\eta Z_\sigma) - \operatorname{E}(U_\eta) \operatorname{E}(Z_\sigma) \\
&= \operatorname{E}(U_\eta \operatorname{E}(Z_\sigma | \mathcal{F}_\eta)) - \operatorname{E}(U_\eta) \operatorname{E}(\operatorname{E}(Z_\sigma | \mathcal{F}_\eta)) \\
&= \operatorname{E}(U_\eta Z_\eta) - \operatorname{E}(U_\eta) \operatorname{E}(Z_\eta) \\
&= \operatorname{Cov}(U_\eta, Z_\eta).
\end{aligned}$$

For the variance term,
$$\begin{aligned}
\operatorname{Var}(Z_\sigma) &= \operatorname{E}\left(Z_\sigma^2\right) - \left(\operatorname{E}(Z_\sigma)\right)^2 \\
&= \operatorname{E}\left(\operatorname{E}\left(Z_\sigma^2 - Z_\eta^2 + Z_\eta^2 \middle| \mathcal{F}_\eta\right)\right) - \operatorname{E}\left(\operatorname{E}(Z_\sigma | \mathcal{F}_\eta)\right)^2 \\
&= \operatorname{E}\left(\operatorname{E}\left(Z_\sigma^2 - Z_\eta^2 \middle| \mathcal{F}_\eta\right)\right) + \operatorname{E}\left(Z_\eta^2 - \operatorname{E}(Z_\eta)^2\right) \\
&= \operatorname{E}\left(\operatorname{Var}(Z_\sigma | \mathcal{F}_\eta)\right) + \operatorname{Var}(Z_\eta) \\
&\geq \operatorname{Var}(Z_\eta),
\end{aligned}$$

and the result follows. $\square$

To understand the implications of Proposition 25.4.1, consider using a European option maturing at time T_n as a control variate. If for a particular

Monte Carlo path we have that $\eta < n$, then the proposition essentially suggests that we should use the value of the option at time T_η to generate a control variate, rather than wait until the maturity date T_n. Of course, since the result in the proposition deals with martingale controls, a little care is required in creation of the control variates. Specifically, most interest rate derivatives (including the prospective controls in (25.53)) pay coupons, and hence need to be adjusted to become martingales. Fortunately, this is fairly easy to do: all coupons paid to time t *should not* be dropped from the value of the control at time t, but should be rolled up (using the numeraire) to time t. In particular, since our definition of $U_n(t)$'s in (18.2) makes sense even for $t > T_n$, we define new controls $Y^c = (Y_1^c, \ldots, Y_{N-1}^c)^\top$ by

$$Y_n^c \triangleq U_n\left(T_\eta\right) = B(T_\eta) \sum_{i=n}^{\eta-1} B(T_{i+1})^{-1} X_i$$

$$+ B(T_\eta) \mathrm{E}_{T_\eta}\left(\sum_{i=\max(\eta,n)}^{N-1} B(T_{i+1})^{-1} X_i \right), \quad (25.55)$$

for $n = 1, \ldots, N-1$ (where we use the convention that $\sum_a^b = 0$ if $b < a$). While these controls still do not exhibit the non-linear characteristic of options (we turn to this shortly), they do resolve the timing problem. By the optional sampling theorem the exact values of the controls, as required by the control variate method, are known,

$$\mathrm{E}\left(U_n\left(T_\eta\right)\right) = U_n(0), \quad n = 1, \ldots, N-1.$$

In theory, adding more controls never increases the variance of the estimate. However, as explained in Section 3.4.1, the efficiency of the scheme can very well decrease as the additional computational effort of new controls may not be rewarded by sufficiently high decreases in variance. As a consequence, one often is best served by selecting just a few carefully chosen controls, rather than indiscriminately throwing a large set of suboptimal controls at the problem. For the case above, the set of exercise values U_n, $n = 1, \ldots, N-1$, is composed of various subsets of (net) coupons already contained in the "longest" underlying U_1. The efficiency gains from using sums of subsets of coupons as a vector control compared to using just the sum of all coupons as a single control can rightly be questioned, suggesting that the following one-dimensional control may be useful:

$$Y^c = Y_1^c, \qquad (25.56)$$

$$Y_1^c \triangleq U_1(T_\eta) = B(T_\eta) \sum_{i=1}^{\eta-1} B(T_{i+1})^{-1} X_i + B(T_\eta) \mathrm{E}_{T_\eta}\left(\sum_{i=\eta}^{N-1} B(T_{i+1})^{-1} X_i \right).$$

To introduce non-linearity into the controls for a Bermudan swaption, we can consider using caplets and caps, as these can be valued exactly in the majority of LM models. For concreteness, let us focus on a payer Bermudan swaption with strike k, the exercise values of which are given by

$$U_n(t) = B(t) \sum_{i=n}^{N-1} \mathrm{E}_t \left(B(T_{i+1})^{-1} \left(L_i(T_i) - k \right) \tau_i \right),$$

where $L_i(T_i)$ is a Libor rate observed at time T_i for the period $[T_i, T_{i+1}]$. To construct a suitably non-linear control for the Bermudan swaption, consider using the set of caps[12],

$$V_{\mathrm{cap},n}(t) \triangleq B(t) \sum_{i=n}^{N-1} \mathrm{E}_t \left(B(T_{i+1})^{-1} \left(L_i(T_i) - k \right)^+ \tau_i \right), \quad n = 1, \ldots, N - 1.$$

With these caps we can construct a few possible controls. For example, in direct analogy to (25.55), we can use a collection of all caps (observed at the Bermudan exercise time) as an $(N - 1)$-dimensional control,

$$Y^c = \left(Y_1^c, \ldots, Y_{N-1}^c \right)^\top, \tag{25.57}$$

$$Y_n^c \triangleq V_{\mathrm{cap},n}(T_\eta) = B(T_\eta) \sum_{i=n}^{\eta-1} B(T_{i+1})^{-1} \left(L_i(T_i) - k \right)^+ \tau_i$$

$$+ B(T_\eta) \mathrm{E}_{T_\eta} \left(\sum_{i=\max(\eta,n)}^{N-1} B(T_{i+1})^{-1} \left(L_i(T_i) - k \right)^+ \tau_i \right),$$

for $n = 1, \ldots, N - 1$. Alternatively, since each cap is just a sum of different caplets $(L_i(T_i) - k)^+$, a closely related, but simpler, control can be constructed by using all caplets (instead of all caps), again sampled at the exercise time,

$$Y^c = \left(Y_1^c, \ldots, Y_{N-1}^c \right)^\top,$$

$$Y_n^c \triangleq B(T_\eta) \mathrm{E}_{T_\eta} \left(B(T_{n+1})^{-1} \left(L_n(T_n) - k \right)^+ \tau_n \right),$$

for $n = 1, \ldots, N - 1$. Furthermore, in direct analogy to (25.56), we can use only one control, the longest cap, stopped at the exercise time,

---

[12]While we use the fixed rate of the swap as a strike for a cap, a potentially better control variate could be constructed by using a strike at or near the exercise boundary of the Bermudan swaption.

$$Y^c = Y_1^c, \tag{25.58}$$

$$Y_1^c \triangleq V_{\text{cap},1}(T_\eta) \equiv B(T_\eta) \sum_{i=1}^{\eta-1} B(T_{i+1})^{-1} \left(L_i(T_i) - k \right)^+ \tau_i$$

$$+ B(T_\eta) \sum_{i=\eta}^{N-1} \mathrm{E}_{T_\eta} \left(B(T_{i+1})^{-1} \left(L_i(T_i) - k \right)^+ \tau_i \right).$$

We note that using just the longest cap as a control variate can be interpreted as using all caplets as controls, but enforcing the same regression coefficient β for all of them. Finally, linear (e.g. (25.55)) and non-linear (e.g. (25.57)) controls can be combined together to improve variance reduction further over a wide range of strikes and maturities.

Various strategies for constructing Bermudan swaption control are tested in a LM model setup by Jensen and Svenstrup [2003]; their main conclusion is that the combination of caplets (25.57) and linear controls (25.55) performs well for a diverse set of Bermudan swaptions. Using the longest cap in combination with the longest swap resulted in only a slightly worse control. The achieved reduction in sample standard deviation is of the order 3 to 5 (i.e. speed ups of up to order 9 to 25 times).

For securities more complex than standard Bermudan swaptions, such as general callable Libor exotics, the idea of sampling the controls at the exercise time still applies. The choice of controls, however, becomes non-trivial. If the underlying can be valued in closed form, such as for callable inverse floaters or callable capped floaters on Libor rates, it should be used. The extra non-linearity in the payoff can be handled with caps of different strikes. However, when pricing CLEs we will often find that the underlying exotic swap does not permit closed-form valuation, preventing us from using the underlying as a control. For these securities, a more general, dynamic type of control variate may be an alternative, as we describe next.

25.5 Dynamic Control Variates

As demonstrated in the previous section, even for relatively simple securities such as Bermudan swaptions, finding good control variates is often challenging. For more complicated CLEs, the search becomes increasingly involved and, what is probably worse, usually has to be done on a case-by-case basis: what works for callable range accruals will probably not work for callable CMS spread options.

In contrast, *dynamic*, or delta-based, control variates are always available — at least in theory. As discussed in Section 3.4.3.2, the main idea behind this method is to select as a control variate the value of a self-financing hedging strategy for the security to be priced. Constructing the exact hedging strategy requires knowledge of the deltas of the security, at each point in time and for each realization of the Monte Carlo paths. These are, of course, rarely available, but often we can use deltas constructed using *approximate* risk sensitivities instead.

Approximate deltas can be constructed in a number of ways. One idea, suggested by Clewlow and Carverhill [1994] and mentioned already in Section 3.4.3.2, uses deltas from a tractable proxy model. For the Libor market model, these deltas could originate from, say, an approximate Markov proxy model, as described in Section 25.3. Without resorting to proxy models, a general technique suitable for CLEs is suggested in Moni [2005], who proposes to extract approximate deltas from regressed values of CLE prices, as computed by the LS method (see Section 18.3). The method capitalizes on the fact that the regressed values of the CLE are designed to be good approximations to the actual CLE values under various market scenarios. Let us describe this method in a bit more detail.

We use the basic scheme of Section 18.3.1 as an example, with regression variables defined to be polynomials of explanatory variables $x(t) = (x_1(t), \ldots, x_d(t))$ as described in Section 18.3.9.2. The regression approximation (18.11) to the hold value can be written as (using index n instead of $n-1$ to simplify notations)

$$\widetilde{H}_n(T_n) = p_n(x(T_n)), \quad n = 0, \ldots, N-1,$$

where $p_n(x)$'s are polynomials in d variables, obtained "for free" as part of the LS algorithm. With this representation, we can compute approximate sensitivities with respect to explanatory variables

$$\frac{\Delta H_n(T_n)}{\Delta x_m(T_n)} \approx \frac{\Delta \widetilde{H}_n(T_n)}{\Delta x_m(T_n)} = \frac{\partial p_n}{\partial x_m}(x(T_n)), \quad m = 1, \ldots, d,$$

to which corresponds the approximate hedging strategy

$$V_{\text{hs}}(T_n) = H_0(0) + \sum_{j=0}^{n-1} \left(\sum_{m=0}^{d} \frac{\partial p_n}{\partial x_m}(x(T_j))(x_m(T_{j+1}) - x_m(T_j)) \right)$$

for $n = 1, \ldots, N-1$. The expected value of the hedging strategy at time 0 under some pricing measure (such as the often-used spot measure) is given by

$$E(V_{\text{hs}}(T_n)) = \sum_{j=0}^{n-1} E \left(\sum_{m=0}^{d} \frac{\partial p_n}{\partial x_m}(x(T_j)) \left(E(x_m(T_{j+1}) | \mathcal{F}_{T_j}) - x_m(T_j) \right) \right),$$

(25.59)

which needs to be known bias-free for the control variate method to work. The easiest way to calculate (25.59) is to select all explanatory variables to be martingales, so that

$$E(x(T_{j+1}) | \mathcal{F}_{T_j}) = x(T_j)$$

(25.60)

for each j, and
$$\mathrm{E}\left(V_{\mathrm{hs}}(T_n)\right) = H_0(0).$$

Moreover, in this case the hedging strategy is itself a martingale,
$$\mathrm{E}\left(V_{\mathrm{hs}}(T_n)|\,\mathcal{F}_{T_k}\right) = V_{\mathrm{hs}}(T_k)$$

for $k \leq n$.

The martingale requirement (25.60) on the explanatory variables is a restriction on the set of all possible explanatory variables used in the LS method, but not a very severe one. Recall (Section 18.3.9.2) that we typically advocate using financially meaningful quantities as explanatory variables. Numeraire-deflated values of traded securities are both financially-meaningful *and* are martingales, hence they can be used for construct a dynamic control variate. For explanatory variables that themselves cannot be represented as prices of traded securities prices, slight modifications in variable selection can often be used to make them so. For example, while a swap rate is not a martingale, the closely related (deflated) swap *value* is.

In some cases, linking a required explanatory variable to a particular security price may be difficult, such as for the stochastic variance factor in a stochastic volatility Libor market model (see Section 18.3.9.3). In such cases, we can always construct a martingale from the explanatory variable by simply subtracting out its mean over each simulation time step, as already done in Section 3.5.5 for martingale construction (see e.g. (3.118)).

Once we have constructed an approximate hedging strategy, we can define a control variate as the hedging strategy stopped at the exercise time (employing a key insight from Section 25.4),

$$Y^c = Y_1^c, \quad Y_1^c \triangleq V_{\mathrm{hs}}(T_\eta).$$

Tests in Moni [2005] show that this approach typically yield reductions in the standard error by a factor of two to three.

The quality of the hedging strategy produced by regressions will depend on the quality of the estimated future hold values implied by the regression functions. While the bias in the lower bound value of the CLE itself also depends on the quality of the regression approximation, there is an important difference: for the basic lower bound price to be of good quality, the approximations to the hold (and exercise) values need only be accurate around the exercise boundary. When using the regression to construct a dynamic control variate, however, the approximations need to be accurate over the whole range of possible values of explanatory variables. The former is obviously much easier to achieve than the letter. When using regression to produce a control variate, Moni [2005] recommends to use polynomials of a lower degree than for the basic lower-bound valuation routine (which, incidentally, means that the regression results no longer come for free from the basic valuation, although the extra cost is modest).

Dynamic hedging ideas are studied in Jensen and Svenstrup [2003], in the context of the approach described in Section 25.4. Here, the authors use hedging strategies to represent values of core European swaptions in a Bermudan swaption, as European option values (and not caps) appear to be the more natural option-based controls. As closed-form values of these controls in Libor market models are unavailable, they resort to approximate hedging strategies based on deltas generated by the swaption approximation formulas that *are* available for LM models. The authors report good results for this method as well.

Finally, let us note that a number of additional twists on the theme of dynamic control variates have emerged in the literature, some of which are based on information extracted from the upper bound methods of Sections 3.5.5 and 18.3.8. Representative papers on the application of these techniques to CLEs include Beveridge and Joshi [2009] and Bender et al. [2006]. In Juneja and Kalra [2009], the authors additionally suggest to use a measure change arising from multiplicative duality[13] for importance sampling.

25.6 Control Variates and Risk Stability

We finish this chapter with a caveat. The various control variate methods discussed in this chapter often show impressive reductions in simulation error on the basic security price, but are not always equally effective in reducing simulation errors of *risk sensitivities*. This observation stems from the fact that the sources of simulation errors when calculating risk sensitivities often bear little relationship to the sources of simulation error of the value itself. The following simple example should clarify this point. Consider the problem of pricing a digital option on the underlying X with the payoff

$$1_{\{X>b\}}.$$

As X is positively correlated with $1_{\{X>b\}}$, we can use X itself as a control variate to reduce the variance. Then the value of the security using such control is effectively equal to the value of a new security with the payoff

$$1_{\{X>b\}} - \beta(X - \mathrm{E}(X)),$$

with β the regression coefficient. Clearly, however, the risk sensitivities of this new security would exhibit the same level of simulation error as the original one, as both payoffs have the same jump discontinuity at $X = b$ which, as

---

[13]Multiplicative duality is developed in Jamshidian [1995] and is closely related to the (additive) duality of Section 1.10.2. A comparison of multiplicative and additive duality for upper bound simulations can be found in Chen and Glasserman [2007a].

discussed in Chapter 23, is the dominant factor affecting the simulation error and the stability of risk sensitivities here. This problem is fairly typical of variance reduction techniques in general, and control variate methods in particular. For irregular payoffs, we find that we typically get more "bang for the buck" out of techniques that focus specifically on improving risk stability, rather than on general variance reduction. Sample techniques include the smoothing methods of Chapter 23, the non-perturbation methods of Chapter 24 or, perhaps, payoff smoothing through importance sampling (Section 25.2). Of course, nothing prevents one from combining general variance reduction techniques with payoff smoothing methods, a strategy that often works very well.

Vegas in Libor Market Models

We recall that vega measures the sensitivity of a security price to moves in volatility. In interest rate models with rich volatility structures, calculating (and even defining, for that matter) vega can be surprisingly difficult, especially in a Monte Carlo setting where vega computations bring about a new layer of complexity beyond the standard challenges discussed in recent chapters. Since, as explained in Chapter 22, vega is of fundamental importance in risk management, the ability to robustly and accurately compute vega is a key requirement for any actual model implementation. This final chapter of the book is dedicated to the challenging topic of vega computations, mostly using the Libor market (LM) model as a convenient, and highly relevant, example.

26.1 Basic Problem of Vega Computations

As discussed in Section 4.4, any diffusive (HJM) model of interest rates is uniquely defined by its volatility structure $\sigma_f(t, T)$ (see Lemma 4.4.1). At the most fundamental level, vega calculations involve the computation of interest rate derivative price sensitivities to changes to this fundamental volatility structure. For a general model, $\sigma_f(t, T)$ is two-dimensional[1], depending on both calendar time t and time to maturity T. For a given interest rate derivative we, in principle, are faced with the problem of quantifying the impact on the derivative security value of *all* possible two-dimensional shocks to this volatility structure. While the space of all possible shocks to a two-dimensional surface is quite rich, in theory we can decompose each shock into a linear combination of "Dirac delta" shocks to individual points (t, T), $0 \leq t \leq T < T_{\max}$, and measure vegas to those shocks only. This is sufficient,

---

[1]In multi-factor models, for each t and T, $\sigma_f(t, T)$ is obviously a multi-dimensional vector, but it is not this dimensionality that we are interested in here.

as vega is a first-order sensitivity and must therefore be linear with respect to linear combination of shocks.

While interest rate vegas are fundamentally two-dimensional, simpler types of interest rate models often reduce this dimensionality for tractability. For instance, one-factor Gaussian and quasi-Gaussian models reduce the two-dimensional structure of a generic HJM volatility structure to a separable factor form

$$\sigma_f(t, T) = g(t)h(T) \tag{26.1}$$

for some $g(t)$, $h(T)$ (see (4.44) and (13.2)). A shock to the volatility structure that preserves the form (26.1) obviously cannot be two-dimensional, and a two-dimensional "Dirac delta" shock will not preserve the factor form. Instead, if we wish to measure volatility sensitivities in models satisfying (26.1), we would have to either bump the volatility structure for a fixed t and all T (a shock to function $g(\cdot)$) or to bump it for all t but a fixed T (a shock to function $h(\cdot)$). In other words, the set of volatility shocks that preserve the volatility structure factor form (26.1) is significantly reduced relative to the general case, and we are consequently prevented from measuring the impact of many types of potentially relevant volatility shocks. Of course, if the factor decomposition is refined relative to (26.1) by using additional state variables (see Section 12.1.5, for instance, or our discussion of multi-factor quasi-Gaussian models in Section 13.3.2), then more complicated shock shapes may be approximated to arbitrarily high precision.

While the discussion above concern model vegas, i.e. sensitivities with respect to perturbations of the model volatility structure, it is often the market vegas, i.e. vegas with respect to volatilities of market-observed vanilla options (see Section 22.1.4), that are ultimately of most practical interest. The dimensionality issue touched upon earlier is equally present in market vegas, since the set of European swaptions is two-dimensional, indexed by option expiry and swap tenor[2]. The dimensionality reduction that is implicit in models such as a one-factor Gaussian model means that we are sometimes unable to quantify the sensitivities of a given derivative to *all* market instruments. Instead, we are forced to choose, often somewhat arbitrarily, which (much reduced) set of European swaptions we wish to use for model calibration and, ultimately, for vega reporting. While we can make a reasonably informed choice for some derivatives (e.g., vanilla Bermudan swaptions where we would use coterminal European swaptions and, possibly, caplets), many securities will not allow us to easily locate the dominant vega exposure locations, should these even exist in the first place. In contrast, models that either use a high number of Markov state variables (e.g., a multi-factor quasi-Gaussian model) or do not rely on volatility factorization at all (e.g., an LM model) will better preserve the full dimensionality of the volatility structure and hence, at least in theory, could be used to *tell* us, in

---

[2]We are ignoring the strike dimension for now.

an unambiguous way, which points of the volatility structure have impact on the value of a given security.

The discussion above is clearly intimately related to our earlier analysis of the debate surrounding local versus global calibration, see Section 14.5.5. With models that require product-specific volatility calibration, the choice of the calibration option set effectively decides in which buckets the vega will be reported, sensibly or not. On the other hand, with globally calibrated models with a fully flexible volatility structure, the model itself will ultimately determine the vega bucketing, in a manner that relies little on (possibly flawed) user intuition. The distinction between model types is fundamental, and irrespective of whether we ultimately choose to use product-specific or global calibration, the ability to discover, in a largely automated way, the set of European swaption volatilities that drive the value of any given derivative can often be essential to robust risk management of interest rate product portfolios. Of course, information uncovered this way could also be used to guide more robust and accurate product-specific calibrations for the local projection method (see Sections 18.4, 20.1.3, 20.2.1).

Our focus in this chapter is squarely on globally calibrated models, with the LM model being our primary example. The same techniques could, however, be applied to any globally-calibrated model underpinned with either a genuinely two-dimensional volatility structure, or one that approximates it closely, such as a multi-factor quasi-Gaussian model.

26.2 Review of Calibration

Let us start the technical discussion by recalling some notations from Section 14.5, and also introducing some new ones. We start with G (see Section 14.5.2 and in particular (14.41)), a subset of discretized instantaneous Libor volatilities which we regard as primary model parameters to be calibrated to market data. The $(N_t \times N_x)$-dimensional matrix G is defined by a rectangular grid of times and tenors $\{t_i\} \times \{x_j\}$, $i = 1, \ldots, N_t$, $j = 1, \ldots, N_x$. For the purposes of this chapter, we denote by G^{full} the full grid of instantaneous Libor volatilities $||\lambda_{n,k}||$ for all n, k (see Section 14.5.3 and (14.42)). The matrix G is obtained from matrix G^{full} by selecting rows and columns that correspond to times $\{t_i\}$ and tenors $\{x_j\}$. As in Section 14.5.8, we assume that the calibration, or *benchmark*, set consists of all swaptions with expiries t_i and tenors x_j, $i = 1, \ldots, N_t$, $j = 1, \ldots, N_x$. On the other hand, we call the set of *all* at-the-money swaptions (i.e. swaptions with expiries T_i and tenors $T_j - T_i$ for all $i = 1, \ldots, N-1$, $j = i+1, \ldots, N-1$) the *full swaption set*.

Recall now the sample calibration algorithm of Section 14.5.7. Given a guess of G, we interpolate it to obtain G^{full}, which is then used to calculate model volatilities of swaptions in the benchmark set that we arrange in a matrix $\Lambda(G)$ with entries

$$(\Lambda(G))_{i,j}, \quad i = 1, \ldots, N_t, \quad j = 1, \ldots, N_x. \tag{26.2}$$

Given $\Lambda(G)$, an objective function $\mathcal{I}(G; \widehat{\Lambda})$ may then be constructed, typically involving a sum of precision and smoothness terms (see e.g. (14.51) and (14.54)), where the precision targets measure the distance between the model and market volatilities of swaptions in the benchmark set. Here we explicitly highlight the dependence of the objective function on market volatilities of swaptions in the benchmark set; these market volatilities are here assumed arranged in an $N_t \times N_x$ matrix $\widehat{\Lambda}_{i,j}$, $i = 1, \ldots, N_t$, $j = 1, \ldots, N_x$. The model calibration minimizes the objective function, resulting in a calibrated grid G^* of Libor volatilities, given by

$$G^*(\widehat{\Lambda}) = \operatorname*{argmin}_{G} \mathcal{I}(G; \widehat{\Lambda}). \tag{26.3}$$

Once the model is calibrated, the value of a given derivative security, $V = V(G^*(\widehat{\Lambda}))$, may be calculated.

26.3 Vega Calculation Methods

Having formalized in Section 26.2 above the basic dependency structure of derivative security values on market volatilities, the key question is now how to establish sensitivities with respect to these volatilities. The next few sections outline several potential methods.

26.3.1 Direct Vega Calculations

26.3.1.1 Definition and Analysis

In the *direct method* for vega calculations, we simply apply a shock to the matrix of market swaption volatilities $\widehat{\Lambda}$, redo the model calibration, and reprice our security position. Let δ be an $N_t \times N_x$ matrix characterizing the shape of the chosen shock; then, the set of shocked market volatilities is given by the matrix

$$\widehat{\Lambda} + \epsilon\delta,$$

for some small $\epsilon > 0$. We proceed to calibrate the new grid of model volatilities $G^*(\widehat{\Lambda} + \epsilon\delta)$ by solving (26.3), i.e.

$$G^*(\widehat{\Lambda} + \epsilon\delta) = \operatorname*{argmin}_{G} \mathcal{I}(G; \widehat{\Lambda} + \epsilon\delta),$$

and then estimate the (market) vega $\nu_{\mathrm{mkt}}(\delta)$ in direction δ by the finite difference[3]

---

[3] See footnote 16 in Chapter 6 for a similar definition of sensitivity to a shock to a yield curve.

$$\nu_{\text{mkt}}(\delta) = \epsilon^{-1} \left(V(G^*(\widehat{\Lambda} + \epsilon\delta)) - V(G^*(\widehat{\Lambda})) \right) \approx \frac{d}{du} V(G^*(\widehat{\Lambda} + u\delta)) \Big|_{u=0}.$$

$$(26.4)$$

We note that here, and throughout, we think of vegas as pure derivatives, while in reality for reporting purposes the vega is often normalized to represent a change in value of a derivative that corresponds to, say, 1% change in the quoted market volatility.

The shock δ could take many forms, starting with the most basic *flat shock* (or "parallel shift"), where $\delta = \delta_{\text{flat}}$, $(\delta_{\text{flat}})_{i,j} = 1$ for all i, j. For more granular sensitivities, e.g. to measure sensitivities to individual market volatilities, we could use *bucketed shocks* $\delta = \delta_{n,m}$,

$$(\delta_{n,m})_{i,j} = 1_{\{i=n\}} 1_{\{j=m\}}. \tag{26.5}$$

A full collection of bucketed shocks — one for each swaption in the benchmark set — gives rise to a total of $N_t \cdot N_x$ so-called *bucketed vegas*[4].

While it is often the goal to calculate vegas to all swaptions in the benchmark set — i.e. calculate sensitivities in directions $\delta_{n,m}$ in (26.5) for all n, m — it is not necessary to use the directions $\delta_{n,m}$ directly. As we have already seen in the context of interest rate deltas in Section 6.4, as long as we have the same number of directions $(N_t N_x)$ that span the set $\{\delta_{n,m}\}$, then we can always express vegas in one basis from vegas in another basis by simple linear algebra. To give an example, consider the set of *running cumulative shocks* $\delta'_{n,m}$ given by

$$(\delta'_{n,m})_{i,j} = \begin{cases} 1, & i < n \text{ or } i = n, j \le m, \\ 0, & \text{otherwise.} \end{cases} \tag{26.6}$$

Then, since

$$\delta'_{n,m} = \delta'_{n,m-1} + \delta_{n,m}$$

(with obvious modifications for $m = 1$), we have

$$\nu_{\text{mkt}}(\delta_{n,m}) = \nu_{\text{mkt}}(\delta'_{n,m}) - \nu_{\text{mkt}}(\delta'_{n,m-1}). \tag{26.7}$$

Hence, we can calculate $\nu_{\text{mkt}}(\delta'_{n,m})$ for all n, m using the algorithm described above, and then calculate all $\nu_{\text{mkt}}(\delta_{n,m})$ using (26.7).

Another sometimes used choice for the vega calculation basis is the set of all *cumulative shocks* $\delta''_{n,m}$ defined by

$$(\delta''_{n,m})_{i,j} = 1_{\{i\le n\}} 1_{\{j\le m\}} \tag{26.8}$$

for all n, m. Again, we can easily express $\delta_{n,m}$ in terms of the $\delta''_{n,m}$. The motivation for introducing alternative bases is similar to that for introducing

---

[4]Between the two extremes of the flat and bucketed shocks lie *row shocks* δ_n of the form $(\delta_n)_{i,j} = 1_{\{i=n\}}$, $n = 1, \ldots, N_t$.

alternative ways of bumping the yield curve in Section 6.4.4: cumulative shocks as a rule lead to less distortion in the internal, model-specific volatility representation. We elaborate on this point later in the chapter.

While many variations of the direct vega method are possible, ultimately the accuracy and stability of vegas obtained by direct perturbation are rarely entirely satisfactory. The main reason is the fact that the calibration (26.3) is not exact, in the sense that the model does not exactly replicate all market volatilities of the swaptions in the benchmark set,

$$\Lambda(G^*(\widehat{\Lambda})) \neq \widehat{\Lambda}.$$

This imprecision is typically caused by the presence of regularization (smoothness) terms in the objective function, by usage of low-dimensional parametric forms for the volatility structure, or by other smoothness measures introduced to prevent overfitting of the model. For a well-designed calibration procedure, the resulting calibration errors are typically within bid-ask tolerances of market data[5] and, consequently, are of little concern in securities pricing. However, the accuracy is often insufficient for vega calculations, since the typical size of the shock applied to market data (i.e., the magnitude of ϵ in (26.4)) is usually of the same order as the calibration errors. As a result, when calculating the vega calibration errors (the "noise") might easily be of the same order of magnitude as the sensitivities themselves (the "signal"), making vegas too noisy to be useful. To improve on this, one can try increasing the size of ϵ in (26.4), but as described in Section 3.3.1 this leads to a bias relative to the true infinitesimal volatility sensitivity. More worryingly, applying shocks of a large magnitude to small subsets of the swaption volatility surface may result in an unrealistically choppy market data scenario to which the model can no longer calibrate properly.

The noise problems described above are less severe for global shocks than for local ones. For example, calculating the flat shift vega with $\delta = \delta_{\text{flat}}$ by direct methods is often possible, and in fact can serve as a benchmark and a reality check for the more advanced methods that we introduce later. The relatively good performance for global shocks is easy to understand, as they tend to preserve the distribution of calibration error among swaptions in the base and bumped scenario; i.e. we roughly have

$$\Lambda(G^*(\widehat{\Lambda})) - \widehat{\Lambda} \approx \Lambda(G^*(\widehat{\Lambda} + \delta)) - (\widehat{\Lambda} + \delta). \tag{26.9}$$

In other words, such shocks do not affect (too much) the calibration error for individual market volatilities; when calculating vegas by (26.4), the calibration errors therefore tend to cancel out. In fact, the introduction of cumulative shocks such as (26.6) and (26.8) can, in part, be motivated by the notion of keeping calibration errors relatively constant to ensure that

---

[5]These are typically in the order of 0.1% in implied volatility terms (with typical market swaption volatilities being in the 10–50% range).

(26.9) holds. While usage of a cumulative shock basis can, in fact, improve the vega noise somewhat, it still rarely produces satisfactory results.

Below, we elaborate a bit more on the noise issues plaguing the direct vega method. We should note that even if one were to find a remedy for the noise problem, the direct vega method may still be unattractive due to the need to repeatedly run a computationally intensive calibration algorithm for each shocked scenario. While the calibration algorithm of Section 14.5 is often relatively fast, if multiple scenarios are required, the total computation time per security can easily become impractically large.

26.3.1.2 Numerical Example

To demonstrate how the (basic) direct vega method performs on a simple example, we set up a 20 year one-factor LM model with 6 month Libor tenors, using a relatively coarse calibration grid: $\{t_i\} = \{x_j\} = \{1y, 5y, 10y, 15y\}$. For simplicity, the LM model is log-normal with flat Libor volatilities at 20%, i.e. $\lambda_{n,m} = 20\%$ for all n, m. The yield curve is also assumed flat, at a level of 5% continuously compounded.

In our model calibration, the swaption benchmark set consists of swaptions with expiry/tenor matching Libor volatilities in the matrix G, i.e. on the grid $\{t_i\} \times \{x_j\}$. We use a global calibration as outlined in Section 26.2 with smoothing weights (in expiry and tenor direction) set high enough to remove unwanted oscillations in the model volatility surface. Additionally, the vega shocks $\delta_{n,m}$, $n, m = 1, \ldots, 4$, are assumed to be the bucketed shocks (26.5) applied to the 4×4 swaption matrix $\widehat{\Lambda}$. In this and subsequent examples we consider three instruments:

1. 5y5y European swaption: a European payer swaption with strike 5%, expiry 5y and tenor 5y. Note that this swaption belongs to the benchmark set.
2. 3y7y European swaption: a European payer swaption with strike 5%, expiry 3y and tenor 7y. Note that this swaption does *not* belong to the benchmark set.
3. 10nc1 Bermudan swaption: a 10 no-call 1 (see Section 5.12) Bermudan payer swaption with annual exercise rights and a 5% strike.

All three derivatives have a notional of 1, and in all examples their values are calculated by Monte Carlo with 16,384 paths.

Table 26.1 shows the vegas obtained by the direct method for the 5y5y European swaption. As the 5y5y swaption is part of the benchmark set, we would expect a non-zero bucket vega number in only the 5y5y expiry/tenor bucket. While the largest vega exposure indeed does show up in this bucket, the table results are noisy and there are non-zero vegas in most other buckets as well.

Consider now our second test instrument, the 3y7y European swaption which is not in the benchmark set; its vegas are given in Table 26.2. As this

	1y	5y	10y	15y
1y	-0.1	0.2	0.0	0.0
5y	-0.8	16.5	0.3	0.1
10y	-0.1	0.8	-1.0	
15y	-0.6	2.5		

Table 26.1. Vegas by the direct method for the 5y5y European swaption as defined in the text, in basis points $(1bp = 10^{-4})$ per 1% shift in volatility of each swaption in the benchmark set. Rows are expiries and columns are tenors of swaptions in the benchmark set.

swaption is not in the benchmark set, we here do not expect only a single non-zero bucket vega; instead, a well-behaved algorithm should produce non-zero numbers only in the four buckets that immediately surround the 3y7y point (the 1y5y, 1y10y, 5y5y, and 5y10y swaptions). As is evident from the table, however, the direct vega method assigns non-zero vegas to many other buckets as well, with some of the vegas being substantially negative.

	1y	5y	10y	15y
1y	-3.0	6.6	7.9	-0.7
5y	-0.2	7.0	2.3	-0.2
10y	0.0	0.3	-0.4	
15y	-0.2	0.9		

Table 26.2. Vegas by the direct method for the 3y7y European swaption as defined in the text, in basis points $(1bp = 10^{-4})$ per 1% shift in volatility of each swaption in the benchmark set. Rows are expiries and columns are tenors of swaptions in the benchmark set.

Finally, let us look at vegas of the 10-nocall-1 Bermudan swaption. While results in Tables 26.1 and 26.2 primarily served to show the deficiencies of the direct vega method, the Bermudan swaptions vegas will be used as a useful benchmark for better methods we shall develop later in the chapter. Table 26.3 lists the relevant results; we notice that there are non-zero vega numbers in buckets corresponding to swaptions with total maturity (expiry + tenor) exceeding the 10 year life of the Bermudan swaption, so again we must conclude that the vega report in the table is affected by a significant amount of noise.

26.3.2 What is a Good Vega?

In the previous section, we pointed out some obvious deficiencies of vegas computed by direct perturbation methods. Before developing other methods

	1y	5y	10y	15y
1y	1.9	3.8	1.1	0.4
5y	6.0	6.8	-0.2	-0.1
10y	3.0	-0.6	-0.2	
15y	-0.2	1.1		

Table 26.3. Vegas by the direct method for the 10nc1 Bermudan swaption as defined in the text, in basis points ($1bp = 10^{-4}$) per 1% shift in volatility of each swaption in the benchmark set. Rows are expiries and columns are tenors of swaptions in the benchmark set.

for calculating vegas, it is useful to first define what characteristics a method for computing vegas should ideally have. For starters, we obviously require that all computed risk sensitivities, including vegas, to be both stable and accurate. While stability can (and should, on a regular basis) be tested empirically by observing calculated risk measure over long periods of time, accuracy is often more difficult to measure. One relevant metric for accuracy could be the performance of the P&L predict from Section 22.2.1, since accurately calculated risk measures typically imply low unpredicted P&L. While a P&L predict analysis is always useful, tests of vega accuracy in this manner may be inconclusive, as the P&L predict measures aggregate quality of all risk sensitivities and could be thrown off by inaccuracy of greeks other than the vega. As a consequence, it is often helpful to have more tailored measures of vega "goodness"; the list below contains several relevant measures. All the equalities below should be understood as "equality within tolerance", where tolerances are typically determined by the requirements of the trading desk that uses the LM model.

1. Additivity. We normally would expect that

$$\nu_{\mathrm{mkt}}(\delta_1 + \delta_2) = \nu_{\mathrm{mkt}}(\delta_1) + \nu_{\mathrm{mkt}}(\delta_2),$$

   i.e. that applying two shocks together gives a vega that is a sum of vegas that correspond to the two individual shocks. As a particularly important case, the flat-shift vega should be reproduced as a sum of bucketed vegas:

$$\sum_{n,m} \nu_{\mathrm{mkt}}(\delta_{n,m}) = \nu_{\mathrm{mkt}}(\delta_{\mathrm{flat}}).$$

2. Scaling. Scaling the size of a shock should scale the vega accordingly,

$$\nu_{\mathrm{mkt}}(c\delta) = c\nu_{\mathrm{mkt}}(\delta)$$

   for a reasonable range of values of c, e.g. $c \in [0.5, 2]$. It is often also natural to require that the vega is invariant with respect to the sign of the bump, i.e. the equality holds with $c = -1$,

$$\nu_{\mathrm{mkt}}(-\delta) = -\nu_{\mathrm{mkt}}(\delta).$$

3. Locality. Our notion of vega locality is similar to the one used for yield curve perturbations in Chapter 6, and loosely requires that vega exposure "lives" where we expect it to. In this requirement, we can distinguish between a few variations:

a) Benchmark set locality. The bucketed vegas calculated for a European swaption in the benchmark set are equal to zero everywhere except in the bucket that corresponds to the swaption itself. In other words, for a swaption with expiry t_i and tenor x_j,

$$\nu_{\mathrm{mkt}}(\delta_{n,m}) = 0 \text{ for } n \neq i, m \neq j$$

and $\nu_{\mathrm{mkt}}(\delta_{i,j}) = \partial V_{\mathrm{swaption},i,j}/\partial \widehat{\lambda}_{i,j}$, where the right-hand side, the vega of the European swaption, is calculated in a vanilla model compatible with the LM model used. As we saw from Table 26.1, the direct method does not have benchmark set locality.

b) Full swaption set locality. This is a stronger version of the previous point. For standard European swaptions that are *not* part of the benchmark set, we expect the vega to be non-zero only in the four buckets that surround the swaption in question. In particular, for a European swaption with expiry T_l and final swap maturity of T_k, we expect that

$$\nu_{\mathrm{mkt}}(\delta_{n,m}) = 0 \text{ for } n \notin \{i-1, i\} \text{ and } m \notin \{j-1, j\},$$

where

$$i = \min\{a : t_a \geq T_l\}, \quad j = \min\{b : x_b \geq T_k - T_l\}.$$

Moreover we require that

$$\nu_{\mathrm{mkt}}(\delta_{i-1,j-1}) + \nu_{\mathrm{mkt}}(\delta_{i-1,j}) + \nu_{\mathrm{mkt}}(\delta_{i,j-1}) + \nu_{\mathrm{mkt}}(\delta_{i,j})$$

equals the vega of the European swaption in the compatible vanilla model. Again, the direct method does not satisfy this property as clear from Table 26.2.

c) Exotic locality. For many exotics derivatives such as, for example, Bermudan swaptions, we know *a-priori* in which buckets the vega is supposed to reside (and often what sign it is supposed to have). For example, for a Bermudan swaption with final maturity T_k, we would expect no vega below the coterminal diagonal[6]:

$$\nu_{\mathrm{mkt}}(\delta_{n,m}) = 0 \text{ if } t_n + x_m > T_k.$$

Given that the direct vega method fails simpler tests of locality, it is unlikely that it will respect the theoretical location (or sign) of Bermudan swaption vegas, an observation confirmed by Table 26.3.

---

[6]If the benchmark set does not include all coterminal swaptions for a given Bermudan swaption, non-zero vegas are still possible immediately below the coterminal diagonal due to interpolation effects.

4. Convergence. As with all quantities calculated by Monte Carlo methods, we expect vegas to converge to some value as we increase the number of paths. In particular, for the number of Monte Carlo paths N_{MC} used, the vegas calculated with N_{MC} paths should be within required tolerances compared to vegas calculated with $2N_{\mathrm{MC}}$ paths, and vegas calculated with two different Monte Carlo seeds should be identical to within given tolerance.

5. Stability. Again, as a general requirement on values calculated by numerical methods, we expect vegas to vary smoothly with changing market inputs.

26.3.3 Indirect Vega Calculations

26.3.3.1 Definition and Analysis

While the mapping (26.3) of market volatilities to model volatilities involves non-linear optimization that adds noise, the reverse mapping of model volatilities to market volatilities (26.2) is typically done by direct application of swaption volatility approximation formulas and is, consequently, noiseless. Hence, it is natural to think that Jacobian techniques — which we have already encountered in Sections 6.4.3 and 22.1.4 — could be fruitfully applied here, with the exact mapping (26.2) used to define the transformation from model vegas to market vegas. To motivate the method we write, informally,

$$\frac{\partial V}{\partial G} = \frac{\partial V}{\partial \Lambda}\frac{\partial \Lambda}{\partial G}, \qquad (26.10)$$

where on the left hand side we have (a vector of) model vegas, i.e. sensitivities with respect to changes in model volatilities, and on the right a product of (a vector of) market vegas and (a matrix of) sensitivities of swaption volatilities with respect to model parameters. As it is the market vegas we are interested in, we solve this linear system to obtain

$$\frac{\partial V}{\partial \Lambda} = \left(\frac{\partial \Lambda}{\partial G}\right)^{-1}\frac{\partial V}{\partial G}. \qquad (26.11)$$

In this equation, $\partial \Lambda / \partial G$ can be computed analytically, whereas the term $\partial V / \partial G$ (the model vegas) normally must be computed by Monte Carlo methods.

Let us develop the ideas above a bit more carefully. For a given $N_t \times N_x$ perturbation matrix δ, we define the model vega $\nu_{\mathrm{mdl}}(\delta)$ in direction δ by

$$\nu_{\mathrm{mdl}}(\delta) = \epsilon^{-1}\left(V(G^* + \epsilon\delta) - V(G^*)\right) \approx \frac{d}{du}V(G^* + u\delta)\Big|_{u=0},$$

where $\epsilon > 0$ is a small number, and $G^* = G^*(\widehat{\Lambda})$ as before is the matrix of model volatilities calibrated to market.

Let us consider applying, to the model volatilities, a set of $N_t \cdot N_x$ shocks denoted by $\delta_{n,m}$ for $n = 1, \ldots, N_t$, $m = 1, \ldots, N_x$. These could be the unit shocks of (26.5), or any of the other families we introduced in Section 26.3.1. It often helps to think of these shocks as market data scenarios, with vega hedging being the exercise of finding weights for the hedging instruments (swaptions in the benchmark set) to neutralize as much as possible the sensitivity of a given security to the chosen scenarios. The sensitivity of the volatility of the (i,j)-th swaption in the benchmark set to "scenario" $\delta_{n,m}$ is given by

$$\frac{\partial \Lambda_{i,j}}{\partial \delta_{n,m}} \triangleq \epsilon^{-1} \left(\Lambda_{i,j}(G^* + \epsilon \delta_{n,m}) - \Lambda_{i,j}(G^*) \right) \approx \frac{d}{du} \Lambda_{i,j}(G^* + u \delta_{n,m}) \bigg|_{u=0},$$

a quantity that can easily be calculated by differentiating the formula for approximation swaption volatility in the LM model with respect to model volatilities[7]. Hence, in its most basic form, the market vega matrix $(\nu_{\mathrm{mkt}})_{i,j}$ can be introduced as the solution to the following least-squares minimization problem

$$\sum_{m=1}^{N_x} \sum_{n=1}^{N_t} \left(\nu_{\mathrm{mdl}}(\delta_{n,m}) - \sum_{j=1}^{N_x} \sum_{i=1}^{N_t} (\nu_{\mathrm{mkt}})_{i,j} \frac{\partial \Lambda_{i,j}}{\partial \delta_{n,m}} \right)^2 \rightarrow \min. \qquad (26.12)$$

The definition (26.12) can be extended in a number of ways, along the same lines as in Section 6.4.3. We could, for instance, use a different number of scenarios than hedging instruments, either by supplying more scenarios or by utilizing only a subset of the benchmark set for hedging purposes. We could also use different weights for different scenarios, with higher weights applied to the scenarios we care more about. In addition, we could introduce regularization weights to, say, penalize hedging positions of excessive size. A reasonably general definition of market vegas in the *indirect method* for vega calculations is then given by the solution to the following least-squares minimization problem (compare to (6.29))

$$\sum_{n,m} W_{n,m}^2 \left(\nu_{\mathrm{mdl}}(\delta_{n,m}) - \sum_{i,j} (\nu_{\mathrm{mkt}})_{i,j} \frac{\partial \Lambda_{i,j}}{\partial \delta_{n,m}} \right)^2$$
$$+ \sum_{i,j} U_{i,j}^2 \left((\nu_{\mathrm{mkt}})_{i,j} \right)^2 \rightarrow \min, \qquad (26.13)$$

where $W_{n,m}$ are weights applied to different scenarios and $U_{i,j}$ are penalty weights for different hedges. This problem can be formulated in matrix form

---

[7] See Section 14.4.2 for examples of such formulas. Notice that the Jacobian $\{\partial \Lambda_{i,j} / \partial \delta_{n,m}\}$ is often available for free as part of the initial calibration of the LM model, especially if calibration relies on a gradient-based optimization method.

(see e.g. (6.30)) and could be solved by standard methods of linear algebra, as in (6.31).

The indirect method for computing vegas avoids noisy (and costly) model recalibration, and often results in a marked improvement over the direct vega method of Section 26.3.1. Still, the results are not perfect, as the vegas calculated by the indirect method will often violate several of the criteria for good vegas listed in Section 26.3.2. In particular, while the indirect vega method tends to satisfy the additivity and scalability properties, it is often quite noisy and exhibits unsatisfactory convergence and stability.

Stability and convergence issues could in principle be addressed by modifying (26.12) (or (26.13)). Specifically, we can add penalty terms that would promote smoothness of market vegas in expiry and tenor dimensions, in the same spirit as we smooth model volatilities during calibration, see Section 14.5.6 and in particular equation (14.51). For example, to promote first-order smoothness we can change (26.12) to

$$
\sum_{n,m} \left(\nu_{\text{mdl}}(\delta_{n,m}) - \sum_{i,j} (\nu_{\text{mkt}})_{i,j} \frac{\partial \Lambda_{i,j}}{\partial \delta_{n,m}} \right)^2
$$
$$
+ w_{\partial t} \sum_{i,j} \left((\nu_{\text{mkt}})_{i,j} - (\nu_{\text{mkt}})_{i-1,j} \right)^2
$$
$$
+ w_{\partial x} \sum_{i,j} \left((\nu_{\text{mkt}})_{i,j} - (\nu_{\text{mkt}})_{i,j-1} \right)^2 \to \min .
$$

We can also add second-order smoothing terms along the same lines as in (26.12). These modifications do not make the minimization problem any harder to solve, as it remains quadratic.

As one would expect, the addition of smoothing terms often significantly improves the stability and convergence characteristics of the indirect method. Unfortunately, however, extra smoothing destroys the locality of vegas: if we apply the indirect method with smoothing to a European swaption in the calibration set, then its vega will "leak out" from its native bucket to other nearby buckets. Despite this issue, we believe that the indirect method with smoothing (and its variants) is widely used in industry for vega calculations in LM models. The locality problems of the method are either ignored on pragmatic grounds (in effect choosing the lesser of two evils, non-locality over instability), or justified by the fact that vegas for actual trading books tend to be spread out over all buckets anyway. Such arguments are obviously not entirely convincing, and assigning vega to buckets where there should be none has strong negative implications for hedging and P&L explain.

26.3.3.2 Numerical Example and Performance Analysis

In order to later improve on the indirect vega method, let us first gain some understanding of the actual performance of the method. For concreteness, we continue with the LM model example from Section 26.3.1.2 using Monte Carlo with 16,384 paths (the same as in the examples of Section 26.3.1.2), and apply shocks $\delta_{n,m}$, $n, m = 1, \ldots, 4$, (assumed to be the bucketed shocks (26.5)) to the 4×4 Libor volatility matrix G. In all tests, we do not use smoothing, i.e. we compute vegas by applying the basic equation (26.12).

Considering first the 5y5y European swaption defined in Section 26.3.1.2, vegas computed by the indirect method are given in Table 26.4. Comparison with Table 26.1 shows that there is a marked improvement over the direct method, but a fair amount of noise is still apparent. For example, the vega of -0.9bp in the 5y1y bucket is clearly incorrect.

	1y	5y	10y	15y
1y	0.0	0.0	0.1	0.0
5y	-0.9	17.0	0.2	0.0
10y	0.0	0.0	0.0	
15y	0.0	0.0		

Table 26.4. Vegas by the indirect method for the 5y5y European swaption as defined in the text, in basis points (1bp $= 10^{-4}$) per 1% shift in volatility of each swaption in the benchmark set. Rows are expiries and columns are tenors of swaptions in the benchmark set.

Table 26.5 lists vegas for a 3y7y European swaption. Again, we see an improvement over the vegas calculated by the direct method in Table 26.2, but non-zero values in the 1 year tenor column again indicate that the method is not completely satisfactory.

	1y	5y	10y	15y
1y	-2.9	6.4	8.2	-0.9
5y	-0.2	7.3	1.9	0.0
10y	0.0	0.0	0.0	
15y	0.0	0.0		

Table 26.5. Vegas by the indirect method for the 3y7y European swaption as defined in the text, in basis points (1bp $= 10^{-4}$) per 1% shift in volatility of each swaption in the benchmark set. Rows are expiries and columns are tenors of swaptions in the benchmark set.

Finally, we consider the 10nc1 Bermudan swaption, the indirect method vegas of which are shown in Table 26.6. While overall somewhat cleaner than the vegas in Table 26.3, negative values in the 15 year expiry row indicate the presence of noise.

	1y	5y	10y	15y
1y	1.9	4.3	1.9	0.2
5y	6.2	5.1	1.4	0.5
10y	2.4	0.2	-0.3	
15y	-0.4	-0.2		

Table 26.6. Vegas by the indirect method for the 10nc1 Bermudan swaption as defined in the text, in basis points ($1\mathrm{bp} = 10^{-4}$) per 1% shift in volatility of each swaption in the benchmark set. Rows are expiries and columns are tenors of swaptions in the benchmark set.

To understand why the performance of the indirect vega method is unimpressive, let us first note that the solution to (26.12) is given by (26.11), where $(\nu_{\mathrm{mdl}})_{i,j}$ is arranged into a vector[8] $\partial V/\partial G$, $(\nu_{\mathrm{mkt}})_{i,j}$ is arranged into a vector $\partial V/\partial \Lambda$, and the matrix $\partial \Lambda/\partial G$ is an appropriately arranged matrix of sensitivities of swaption volatilities with respect to Libor volatilities. Some of the buckets — namely 10y15y, 15y10y and 15y15y — are outside of the 20 year model horizon and can be discarded, so the dimension of the matrix $\partial \Lambda/\partial G$ is 13×13.

The vector of model vegas $\partial V/\partial G$ is calculated numerically in Monte Carlo, by perturbing individual entries in the matrix G and repricing the derivative[9]. This procedure induces noise in the model vegas which will be transmitted into market vegas through (26.11), by multiplication with the inverse of the matrix $\partial \Lambda/\partial G$. Let us look at this matrix in more detail, as clearly its properties will influence the propagation of noise. Using our test setup above, Figure 26.1 represents the matrix $(\partial \Lambda/\partial G)^{-1}$ graphically, with a few selected columns plotted as separate lines, showing how a shock to a particular swaption volatility in the benchmark set affects all Libor volatilities in G. Each market vega is obtained by adding up all model vegas weighted by the values of a corresponding column (line in the figure).

Two things are apparent in Figure 26.1. First, each column is rather "wiggly", with positive and negative values alternating in a ringing pattern. This behavior — which is not due to numerical noise, since the calculation of the matrix $(\partial \Lambda/\partial G)^{-1}$ is exact — is likely to exacerbate any noise in model vegas. Second, the (absolute values of) values in the matrix are quite high, reaching values of 10 to 15. This is significant, as any noise in the market vegas is then essentially multiplied by a factor of 10 to 15. This noise-amplifying effect is confirmed by looking at the eigenvalues of the matrix $(\partial \Lambda/\partial G)^{-1}$: the lowest and highest eigenvalues equal 1 and 18, respectively. If we make the reasonable assumption that the level of

---

[8]This could be done arbitrarily, but for concreteness we do it in row-major order, i.e. rows of the matrix are stacked end-to-end to come up with a vector.

[9]Of course, the pathwise or likelihood differentiation methods of Chapter 24 could have been used here as well.

Fig. 26.1. Inverse Jacobian for Indirect Vega Method

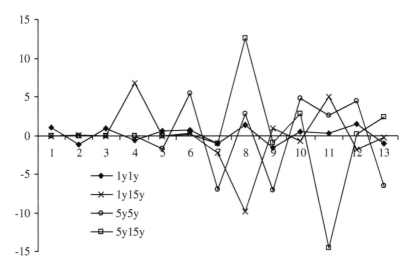

Notes: Each line represents how a shock to a given swaption volatility affects forward Libor (i.e. model) volatilities in matrix G arranged in a row-major order. The lines are graphic representations of (a few selected) columns from the inverse Jacobian $(\partial \Lambda/\partial G)^{-1}$, see text for details.

accuracy in calculating model vegas is roughly the same as for deltas, then the accuracy in *market* vegas could be 10 or even 20 times worse. And this is only for our simplistic example — in real applications with models of longer tenors and larger benchmark sets, the noise amplification factor could easily be in the hundreds, making the indirect vega method perform poorly.

Incidentally, looking at the matrix $(\partial \Lambda/\partial G)^{-1}$ in Figure 26.1 also sheds more light on the poor performance of the *direct* method of calculating vegas, as described earlier in Section 26.3.1. When a particular swaption volatility is shocked and the model is recalibrated, Figure 26.1 shows that the resulting model will effectively have its Libor rate volatilities severely distorted by a large shock with irregular shape. For example, a perturbation of 1% to a 5y5y swaption volatility would move some of the Libor volatilities by almost 15% (and others by -15%). Clearly, with shocks of this size we cannot hope to accurately capture only first-order sensitivity, as second- and higher-order effects will pollute the vega we are trying to calculate.

26.3.4 Hybrid Vega Calculations

26.3.4.1 Definition and Analysis

In Section 26.3.3 we identified poor numerical invertibility (also known as *stiffness*) of the matrix $\partial \Lambda / \partial G$ as the main reason for poor performance of the basic indirect method for vega calculations. This stiffness primarily arises from the usage of shocks to model volatilities that do not adequately take into consideration the dependence of swaption volatilities on Libor volatilities. To improve the indirect vega method, it is therefore natural to change our set of simple bucketed shocks in Libor volatilities to a set of shaped shocks that will result in a better Jacobian, with less ringing and smaller noise amplification factors than in Figure 26.1.

One good choice for the Jacobian would be a unit matrix, which is both perfectly smooth and involves no amplification of noise. A unit Jacobian matrix will arise only if we use Libor volatility shocks that correspond to shocks of individual swaption volatilities in the benchmark set. It may appear that this line of reasoning simply leads us back to the direct method of vega calculations, but we here make a subtle but critical distinction: instead of outright shocking swaption volatilities and recalibrating the model, we instead construct Libor volatility shocks that are approximately equivalent to bucketed swaption volatility shocks, and then apply these shocks through the Jacobian technique outlined earlier. Avoiding recalibration and carefully controlling the shape of shocks to the Libor rate volatility surface not only leads to better computational performance, it also ultimately will lead to better vega quality[10], in the sense defined in Section 26.3.2.

In light of the discussion above, the key problem we have to deal with is how to construct, in a noise-free manner, a shaped shock to Libor volatilities that approximates a shock to a particular European swaption. Here, the bootstrap LM model calibration presented in Section 14.5.8 turns out to be useful. Recall that the idea of bootstrap (or cascade) calibration is to find the instantaneous volatility of each Libor rate over each time period one at a time by solving a quadratic equation, a procedure that is enabled by doing the calculations in a certain (row-major) order. As pointed out in Section 14.5.8, bootstrap calibration is normally not suitable for a full calibration to market data, as market volatilities of swaptions typically come with some amount of noise in them or, at any rate, are not guaranteed to change smoothly across expiries and tenors. As exemplified by Figure 26.1, this leads to rapid accumulation of noise in Libor volatilities during the bootstrap and almost unavoidable calibration failure (where quadratic equations fail to have real roots). On the other hand, *if* the input volatilities happened to be smooth and "compatible" with a Libor market model, there would be no reason why the bootstrap calibration would not work. This suggests

---

[10]An observation also made by Pietersz and Pelsser [2004], although in a somewhat different context.

that we should apply swaption volatility shocks not to market values of swaption volatilities, but to the implied swaption volatilities returned by the calibrated model. The latter are fundamentally compatible with an LM model and, assuming that a reasonable amount of smoothing was enforced in the calibration norm, smooth enough for the bootstrap method to work.

These ideas lead us to the following *hybrid method* for calculating vegas, combining features of both the direct and indirect methods.

1. Calibrate the LM model to market data, i.e. obtain G^* from $\widehat{\Lambda}$, using our global calibration method.
2. Calculate Λ, the model-implied swaption volatilities.
3. Fix expiry t_n and tenor x_m.
4. Apply a unit shock $\alpha_{n,m}\delta_{n,m}$ with $\delta_{n,m}$ shaped as in (26.5), for sufficiently small[11] $\alpha_{n,m} > 0$ to Λ.
5. Bootstrap calibrate an LM model to swaption volatilities $\Lambda + \alpha_{n,m}\delta_{n,m}$, to obtain a matrix of shocked Libor volatilities $G^{*,n,m}$.
6. Calculate a Libor shock $\delta_{\text{Libor},n,m}$ by $\delta_{\text{Libor},n,m} = \beta_{n,m}(G^{*,n,m} - G^*)$, where the scaling constant $\beta_{n,m} \neq 0$ is chosen so that $\max_{i,j} |(\delta_{\text{Libor},n,m})_{i,j}| \leq \epsilon$ for a small $\epsilon > 0$.
7. Repeat Steps 3–6 for all expiries and tenors, and save all shocks $\{\delta_{\text{Libor},n,m}\}$.
8. Apply the indirect method of Section 26.3.3 with the collection of Libor volatility shocks $\{\delta_{\text{Libor},n,m}\}$.

The Jacobian matrix $\partial \Lambda_{i,j}/\partial \delta_{\text{Libor},n,m}$ will here be exactly diagonal, with the element $\alpha_{n,m}\beta_{n,m}$ on the diagonal in the position determined by the ordering of swaptions in the benchmark set; the inverse transformation (26.11) will amount to an appropriate scaling of each model vega. Of course, the choice of $\delta_{n,m}$'s in Step 4 is not unique, and instead of (26.5) we could have used other families such as (26.6) or (26.8). In this case the Jacobian would no longer be diagonal, but otherwise the method would still work the same way.

Let us discuss now the choice of various constants that appear in the algorithm. In Step 4 we need a choice for the positive constant $\alpha_{n,m} > 0$. The idea here is to apply a constant small enough that the bootstrap calibration of Step 5 works. Clearly for $\alpha_{n,m} = 0$ this is the case, so there exists a small enough $\alpha_{n,m} > 0$ that satisfies this criteria. On the other hand, we should not choose $\alpha_{n,m}$ too small as it may adversely affect the Monte Carlo simulation error when computing relevant finite differences (see Section 23.2). In practice, we may start with some reasonable large value of $\alpha_{n,m}$, say 1%, and attempt the bootstrap. If this fails, we reduce $\alpha_{n,m}$ by half and try again — and so on until we find the value of $\alpha_{n,m}$ that allows the bootstrap to succeed.

---

[11] We comment below on the choice of $\alpha_{n,m}$ as well as other required constants.

As for the constant $\epsilon > 0$ required in Step 6, we should choose it in a way that ensures that shocks to Libor volatilities are small enough to prevent significant second-order effects to show up in vegas, yet big enough to control the level of Monte Carlo error in the numerically calculated sensitivity. A reasonable choice here is to set ϵ somewhere between 0.1% to 1%.

26.3.4.2 Numerical Example

To present test results for the hybrid vega method, we continue the numerical example of Sections 26.3.1.2 and 26.3.3.2. Looking first at the 5y5y European swaption, the hybrid method vegas are listed in Table 26.7. As we can see, results are much improved compared to the direct and indirect methods (see Tables 26.1 and 26.4, respectively) with very little noise and the only significant vega correctly showing up in the 5y5y bucket (as expected).

	1y	5y	10y	15y
1y	0.0	0.0	0.1	0.0
5y	-0.1	16.3	0.0	0.0
10y	0.0	0.0	0.0	
15y	0.0	0.0		

Table 26.7. Vegas by the hybrid method for the 5y5y European swaption as defined in the text, in basis points (1bp $= 10^{-4}$) per 1% shift in volatility of each swaption in the benchmark set. Rows are expiries and columns are tenors of swaptions in the benchmark set.

Similar good results are obtained for the 3y7y European swaption, as shown in Table 26.8. Unlike the results in Tables 26.2 and 26.5, there is here hardly any noise visible outside of the four neighboring buckets where we expect the vega to be located.

	1y	5y	10y	15y
1y	0.0	5.3	4.9	0.0
5y	0.0	5.2	4.5	0.0
10y	0.0	0.0	0.0	
15y	0.0	0.0		

Table 26.8. Vegas by the hybrid method for the 3y7y European swaption as defined in the text, in basis points (1bp $= 10^{-4}$) per 1% shift in volatility of each swaption in the benchmark set. Rows are expiries and columns are tenors of swaptions in the benchmark set.

Finally, Table 26.9 lists hybrid method vegas for a Bermudan swaption. Once again, we only see vega where it is expected to be, in contrast to Tables 26.3 and 26.6.

	1y	5y	10y	15y
1y	3.3	3.4	2.0	0.0
5y	6.5	4.6	0.6	0.0
10y	2.8	0.5	0.0	
15y	0.0	0.0		

Table 26.9. Vegas by the hybrid method for the 10nc1 Bermudan swaption as defined in the text, in basis points ($1bp = 10^{-4}$) per 1% shift in volatility of each swaption in the benchmark set. Rows are expiries and columns are tenors of swaptions in the benchmark set.

26.4 Skew and Smile Vegas

So far, we have focused our discussion of vega on the computation of sensitivities to at-the-money swaption volatilities. Together with the correlation sensitivities that we touch upon in Section 26.5, this comprises the full set of volatility sensitivities in models that parametrize each swaption volatility smile with a single number representing the overall level of volatilities across all strikes (such as a log-normal LM model). In models with richer volatility smile parameterizations, such vegas also have a meaningful interpretation as sensitivities to parallel shifts of volatility smiles of each swaption in the benchmark set. For such models, however, there are other volatility sensitivities that so far have been left out of the discussion, namely the sensitivities to changes in *shapes* of swaption volatility smiles. Sensitivities to changes in the volatility smile slope and curvature are often denoted *skew vegas* and *smile vegas*, respectively.

As with ATM vegas, in principle there are skew and smile vegas for each swaption in the benchmark set. However, it is rarely a requirement that one be able to calculate them all individually, as ATM vegas capture the majority of volatility sensitivity. More often, what is required are aggregated measures of skew and smile risk, such as a single number that corresponds to a change in slope or curvature of *all* volatility smiles together. For such aggregated measures of risk, brute-force recalibration and recomputation along the lines of the direct vega method is often sufficient. Moreover, it is typically more useful to use a scenario-based approach with large slope or curvature shocks, rather than true first-order differentiation. For example, in a displaced log-normal LM model (see Table 14.1 in Section 14.2.4) one can switch the skew parameter from 1 (log-normal) to 0 (Gaussian) to get a good idea of the impact of the slope of volatility smile.

In the off-chance that bucketed skew/smile exposure is required, the indirect (Section 26.3.3) method or the hybrid (Section 26.3.4) approach that we developed for the ATM vegas could often be reused. In some cases skew and smile sensitivities are even easier to calculate than ATM vegas, due to a simpler connection between the model and market parameters. For example, the term swaption skew in a displaced log-normal LM model is a linear function of instantaneous Libor skews, see Section 15.2, making the Jacobian-type methods particularly easy to apply. We do not go into further detail here, as the mechanics of these calculations should be clear to the reader by now.

26.5 Vegas and Correlations

Earlier in the chapter we used a one-factor version of the LM model in our numerical examples, but in practice we are often more interested in calculating vegas in multi-factor LM models. For a q-factor LM model, yield curve dynamics are characterized by factor volatilities, i.e. q-dimensional vectors $\lambda_k(T_n)$ associated with each Libor rate $L_k(t)$ and each time period $(T_{n-1}, T_n]$. As we recall from Section 14.5.4, these are constructed from the volatility norm $||\lambda_{n,k}||$ (which we denoted by G^{full} in Section 26.2) and instantaneous correlations of Libor rates. In the indirect and hybrid methods of Sections 26.3.3 and 26.3.4 we shocked the elements of G (and, ultimately, G^{full}), with the understanding that instantaneous correlations of Libor rates remained fixed in all scenarios. In the direct method of Section 26.3.1, we referenced the sample calibration algorithm of Section 14.5.7, which implicitly assumed that the instantaneous correlations of Libor rates were untouched while perturbing swaption volatilities. So, in all three methods we so far have calculated interest rate vegas under the assumption that instantaneous Libor correlations are kept constant when forming the derivative with respect to volatility. While not unreasonable, this choice is not unique and several viable alternatives exist. We discuss some of these in this section.

26.5.1 Term Correlation Effects

While the correlation structure in the LM model is typically captured through a parameterization of instantaneous Libor correlations, the prices of traded correlation-sensitive instruments — CMS spread options, in particular — depend more directly on *term* correlations of *swap* rates (see Section 14.4.3.1). Importantly, when Libor volatilities are changed with instantaneous Libor correlations kept constant, term correlations of swap rates will generally change quite significantly. This effect should be intuitively clear and is a consequence of the dependence of the formula for term correlation in Section 14.4.3.1 on Libor volatilities.

To demonstrate the magnitude of the vega effect on term correlations, we continue the numerical example of Sections 26.3.1.2, 26.3.3.2 and 26.3.4.2, but now extend our setup to a 10-factor LM model with the instantaneous Libor correlations parameterized by a function of the form (14.19) (with $\rho_\infty = 0.5$, $a_0 = 0.42$, $a_\infty = 0$, $\kappa = 0.08$) and instantaneous Libor volatilities fixed at 20%. For concreteness, let us study the sensitivity of the term correlation between the 10 and 1 year swap rates over a 10 year horizon ($\rho_{term}(0, 10)$ in the notation of Section 14.4.3.1). The base value of this correlation in our setup is about 83%. As demonstrated in Table 26.10, $\rho_{term}(0, 10)$ is quite sensitive to shocks to some of the volatilities. For example, a shock of 1% to the volatility of a 10y10y swaption would change this term correlation by -0.85%, which is highly significant.

	1y	5y	10y	15y
1y	0.1	-6.2	-3.1	2.0
5y	-5.0	-5.6	7.0	4.7
10y	14.1	45.7	-85.3	
15y	0.0	0.0		

Table 26.10. Sensitivity for the 10y1y term (10 years) swap rate correlation, in basis points (1bp = 10^{-4}) per 1% shift in volatility of each swaption in the benchmark set. All numbers are computed using the hybrid method in Section 26.3.4.2. Rows are expiries and columns are tenors of swaptions in the benchmark set.

26.5.2 What Correlations should be Kept Constant?

Since term swap rate correlations change under volatility shocks when instantaneous Libor correlations are fixed, we could instead decide to keep term swap rate correlations constant (while allowing instantaneous Libor correlations to move) under volatility shocks; this choice would lead to different vegas, of course. As we discussed in Chapter 22, this ambivalence is not unique to the problem of calculating vegas, and we often need to decide which quantities to keep constant and which to let float when calculating risk sensitivities. Ultimately, such decisions are often driven by traders' preferences for risk representation, or by the types of hedging strategies that they want to pursue. In making these decisions, traders generally (and reasonably) tend to emphasize the issue of *consistency* across different products.

A typical interest rate exotics trading desk will trade correlation-sensitive exotics (e.g., CMS spread TARNs, see Section 5.13.3), as well as vanilla spread options. The exotic derivatives will often be risk managed in an LM model, while for spread options the desk may use a simpler vanilla model, as discussed in Chapter 17. These two models will typically have different (internal) correlation parameters: the LM model will use instantaneous Libor correlations, while a vanilla model (based, say, on a Gaussian copula) will use a term correlation between swap rates as an input. While it is natural for each model to keep its internal correlation parameters constant when calculating vegas, doing so would lead to inconsistency in the definition of vegas between the exotic derivatives and their vanilla hedges. Such inconsistency is typically quite dangerous as it could lead to a position that is deemed hedged, but in fact has an outright exposure.

In the example above, as well as many other similar situations, arguably the easiest way to maintain consistency is to use the more general model (here, the LM model) as the risk "engine" for all products in the book, exotic or not. In an LM model setting, we would then need to compute the volatility sensitivity of both exotics and vanilla securities assuming fixed instantaneous Libor correlations. For the vanilla securities, computation of this sensitivity could be done by either outright valuation of vanilla spread options in a LM model or, perhaps more pragmatically, by calculating the volatility shock impacts on the relevant term correlations of swap rates and combining the results (Jacobian-style) with known correlation sensitivities of the vanilla model. To complement the resulting vega report, it would be natural to also report correlation sensitivities, by calculating sensitivities of the portfolio to instantaneous Libor correlations[12].

While not entirely without merit, the approach outlined above has its limitations. Of course, if the portfolio is fully hedged to both vegas (under the assumption of constant Libor correlations) and to instantaneous Libor correlations, then it has no volatility risk, irrespective of how we define vega. However, a fully hedged position is rarely, if ever, achieved, in which case reported volatility and correlation sensitivities are used as a monitoring tool. As we have commented before, it is often much easier for traders to understand sensitivities expressed in terms of traded quantities, rather than in terms of non-traded quantities such as instantaneous Libor correlations. Traders therefore typically have a strong preference for seeing their correlation risk expressed in terms of market-implied swap rate correlations, which, for consistency reasons, dictates that vegas should be calculated under the assumption that market (and not model) correlations are kept constant. In addition, we should note that the LM model typically uses a fairly parsimonious correlation parameterization, often comprised of just a handful of numbers (see Section 14.3.2). Hence, correlation risk produced by the LM model would tend to be insufficiently granular for risk-managing vanilla spread options which often are quite liquid for a range of expiries

---

[12]But *not* sensitivities to term correlations of swap rates, which would lead to another inconsistency, with double-counting of risk.

and a reasonably large number of swap rate pairs. This issue also favors using term swap rate correlations for risk management purposes.

26.5.3 Vegas with Fixed Term Correlations

In the last section we made the case for holding term correlations of swap rates fixed when computing vegas. Let us discuss how to turn this idea into practice, by suitably modifying the various computational methods discussed earlier in this chapter. The direct method of Section 26.3.1 is the easiest to modify: all we need to do is to add the relevant[13] term swap rate correlations as targets in the basic model calibration, with the calibration algorithm extended along the lines of Section 14.5.9. Note that it is spread option *correlations* that should be the calibration targets and not spread option *values* — we certainly expect the values of spread options to change under different volatility scenarios, even as we keep correlations constant. While it is easy to extend the direct method, its limitations with regards to the quality of vegas produced remain (or are amplified, most likely), and consequently this approach is not recommended.

Extending the indirect (or hybrid, as the procedure is more or less the same) method to control spread option correlations is somewhat more difficult. One naive choice would involve appending a correlation calibration to every shock of Libor volatilities, to ensure that term swap correlations stay fixed after each perturbation of volatilities. In other words, after applying a shock to G, we would then proceed to solve for new parameters to the instantaneous Libor correlation function in order to remain in calibration with term swap rate correlations. In this approach, we would need to run a separate optimization problem for each model vega shock, which most likely would make the method prohibitively slow and introduce extra noise due to non-exact nature of the solution of the optimization problem.

Our preferred method for extending the indirect vega computation is, once again, based on Jacobian methods. In this approach, we would i) apply shocks to model volatilities *and* to model correlations, ii) calculate the value of a derivative as well as changes to swaption volatilities and swap rate correlations, and iii) manipulate these quantities to obtain the vegas. Let us present the blueprint of the scheme using the stylized notations of (26.10) — we trust that the reader can expand our presentation into a workable algorithm.

Let ξ be the vector of instantaneous Libor correlations, and ρ the vector of term swap rate correlations (see footnote 13). We recognize security value and market data dependence on model data through the notations

---

[13]Note that it is impractical to include *all* swap rate correlations in the calibration set. Instead, one would typically choose a set (or perhaps a few sets) of correlations of two specific swap rate tenors, such as 10 year and 2 year, over a collection of time periods.

$$V = V(G, \xi), \quad \Lambda = \Lambda(G, \xi), \quad \rho = \rho(G, \xi).$$

Our goal is to compute the vector

$$\left. \frac{\partial V}{\partial \Lambda} \right|_{\rho=\text{const}}, \tag{26.14}$$

i.e. sensitivities to market swaption volatility shocks, keeping market correlations constant. Implicitly, G and ξ are functions of Λ and ρ,

$$G = G(\Lambda, \rho), \quad \xi = \xi(\Lambda, \rho),$$

and so is V,

$$V(\Lambda, \rho) = V(G(\Lambda, \rho), \xi(\Lambda, \rho)). \tag{26.15}$$

By an application of the chain rule to (26.15), we get

$$\left. \frac{\partial V}{\partial \Lambda} \right|_{\rho=\text{const}} = \frac{\partial V}{\partial G} \frac{\partial G}{\partial \Lambda} + \frac{\partial V}{\partial \xi} \frac{\partial \xi}{\partial \Lambda}. \tag{26.16}$$

Here, the sensitivities $\partial V/\partial G$ and $\partial V/\partial \xi$ may be obtained by application of model parameter shocks to the valuation of the derivative. $\partial G/\partial \Lambda$ and $\partial \xi/\partial \Lambda$ can be found by inverting the (full) Jacobian (inversion should be understood in the generalized least-squares sense as in (26.12) vs. (26.11) as the matrices involved may not even be square),

$$\begin{pmatrix} \partial G/\partial \Lambda & \partial G/\partial \rho \\ \partial \xi/\partial \Lambda & \partial \xi/\partial \rho \end{pmatrix} = \begin{pmatrix} \partial \Lambda/\partial G & \partial \Lambda/\partial \xi \\ \partial \rho/\partial G & \partial \rho/\partial \xi \end{pmatrix}^{-1}. \tag{26.17}$$

The matrix on the right-hand side of (26.17) is obtained by applying shocks to model volatilities and correlations, following the same approach as outlined for the indirect and hybrid vega methods. As a by-product of this calculation we also conveniently obtain risk sensitivities with respect to market *correlations*, since

$$\left. \frac{\partial V}{\partial \rho} \right|_{\Lambda=\text{const}} = \frac{\partial V}{\partial G} \frac{\partial G}{\partial \rho} + \frac{\partial V}{\partial \xi} \frac{\partial \xi}{\partial \rho}, \tag{26.18}$$

where $\partial G/\partial \rho$ and $\partial \xi/\partial \rho$ are obtained in (26.17).

26.5.4 Numerical Example

To demonstrate the difference between various definitions of vegas, we look at a simple, single 10 year option on the spread between 10 year and 1 year swap rates. We calculate its vegas in the 10-factor LM model used in Section 26.5.1, under the assumption of constant instantaneous Libor correlations (Table 26.11) and constant term swap rate correlations (Table

26.12). The second method puts all vega (apart from some minor noise) into the 10y1y and 10y10y buckets, unlike the first method which assigns significant vega to, for example, the 10y5y bucket. Arguably, most traders would consider the vega in Table 26.12 more intuitive, as it is exactly the shape of the vega profile that one would obtain for this spread option from a typical vanilla model.

	1y	5y	10y	15y
1y	0.00	0.10	0.05	-0.03
5y	0.08	0.10	-0.12	-0.08
10y	2.10	-0.77	2.00	
15y	0.00	0.00		

Table 26.11. Vegas by the hybrid method for the 10y option on the spread between 10y and 1y swap rates while keeping instantaneous Libor correlations constant, in basis points ($1bp = 10^{-4}$) per 1% shift in volatility of each swaption in the benchmark set. Rows are expiries and columns are tenors of swaptions in the benchmark set.

	1y	5y	10y	15y
1y	0.00	-0.01	0.00	0.00
5y	-0.01	-0.01	0.01	0.01
10y	2.36	0.05	0.47	
15y	0.00	0.00		

Table 26.12. Vegas by the hybrid method for the 10y option on the spread between 10y and 1y swap rate while keeping 10y term swap rate correlation between 10y and 1y swap rates constant, in basis points ($1bp = 10^{-4}$) per 1% shift in volatility of each swaption in the benchmark set. Rows are expiries and columns are tenors of swaptions in the benchmark set.

26.6 Deltas with Backbone

As we saw in Section 26.5, the need for consistency between exotic and vanilla models often drives the definitions of risk sensitivities. Such consistency requirements, it turns out, also affect calculations of deltas (and, of course, gammas) in the LM and other term structure models. To describe this effect in more detail, we first recall the discussion of Section 16.1.2 and, in particular, the fact that vanilla models are sometimes set up to attribute some user-specified amount of the vega to delta. If such a procedure is used, it would be useful to ensure that the deltas computed in models for more exotic derivatives have the same meaning as in the vanilla model. In essence, this would require a link in the exotic model between the volatility smile and the level of rates.

Sometimes vanilla-exotic delta consistency is ensured automatically, as a consequence of the choice of the models in use. For example, the vanilla SV model (16.8)–(16.9) is naturally consistent with the SV LM model (14.15)–(14.16). On the other hand, if we start adjusting the backbone of the vanilla model as in Section 16.1.2, the consistency would often be lost. For example, were we to use a vanilla model of the type (16.5), we would need to modify the volatility terms for SDEs for Libor rates under the LM model to have the same form, e.g.

$$dL_n(t) = O(dt) + \Big(b_n(t)L_n(t) + (m - b(t))\,L_n(0) + (1 - m)L\Big)$$
$$\times \lambda_n(t)^\top\,dW(t), \quad n = 1,\ldots,N-1,$$

for some mixing m and level L.

A more complicated situation would arise were we to use a vanilla delta convention without a natural exotic counterpart, such as the SABR model of Section 8.6 or the SVI interpolation rule of Section 16.1.5. In these cases, it would be difficult to "internalize" the same smile move logic in the LM model dynamics. Fortunately, we can use an external brute-force approach that, in principle, works for any combination of the vanilla and exotic models.

The method we have in mind is quite straightforward, and we describe it with a log-normal LM model representing the exotic model. With f denoting the yield curve, let $\widehat{\Lambda}(f)$ be the ATM Black volatilities of the swaptions in the benchmark set, given the yield curve f. Suppose the delta is calculated by shifting the yield curve from f to f', which causes a move in swaptions to $\widehat{\Lambda}(f')$, as dictated by our vanilla rule for smile moves. How we proceed depends on the vega calculation method in use. In the direct method, we simply recalibrate the LM model to the new set of swaption volatilities $\widehat{\Lambda}(f')$, and then proceed to use the resulting LM model parameterization together with the shifted yield curve f' to calculate the shocked value of the security in question.

In the indirect vega method, we would do conceptually the same calculation as for the direct method, except we would obtain the LM model parameterization for the shocked yield curve scenario by applying the inverse Jacobian $(\partial \Lambda/\partial G)^{-1}$ to the shifted swaption volatilities $\widehat{\Lambda}(f')$. The inverse Jacobian would automatically be available as part of the basic vega calculation. In the hybrid method, we would first apply the shift in market swaption volatilities arising from the shift in the yield curve, i.e. $\widehat{\Lambda}(f') - \widehat{\Lambda}(f)$, to the base model swaption volatilities, by setting

$$\Lambda' = \Lambda + \Big(\widehat{\Lambda}(f') - \widehat{\Lambda}(f)\Big).$$

Subsequently, we would bootstrap-calibrate the LM model to Λ' and, once again, use this model to calculate the shocked value of the security in question.

26.7 Vega Projections

After our short detour into delta computations, we return to LM vegas. Clearly, the reporting of vega depends on the benchmark set of swaptions used in the vega calculation method: simply put, the vega is reported only to those swaptions that are in the benchmark set. Note that, while we so far assumed that this benchmark set is the same as used for calibrating the LM model in the first place, it actually need not be. Indeed, it is often a good idea to use different sets for calibration and vega calculation. For calibration we often seek to include as many European swaptions as possible to capture the maximum amount of market information in the model calibration, but for vegas it may be preferable to choose a smaller set of benchmarks. There are several reasons for this, starting with the fact that liquidity in different European swaptions is not the same, and the desk may want to express vegas in only the most liquid swaptions[14]. In addition, both the computation time (which is linear in the number of shocks applied) and the numerical properties of all vega calculation methods (properties such as the stiffness of the matrix $\partial\Lambda/\partial G$ in Section 26.3.3 or the shape of the Libor volatility bumps in Section 26.3.4) tend to deteriorate as the number of benchmark swaptions grows. For crisper and quicker vegas, it is therefore often useful to cut down on the number of benchmark swaptions.

To understand the issues that a reduced benchmark set of swaptions may lead to, let us revert to the setup used for numerical results in earlier sections (see Section 26.3.1.2, for instance) and imagine that we use an LM model calibrated to European swaptions with expiries and maturities of 1y, 2y, ..., 19y (the "full swaption set" discussed in Section 26.2), yet the vega is to be calculated with the 4×4 benchmark set of swaptions used in previous numerical results (see e.g. Table 26.3). Then the vega for the 10nc1 Bermudan swaption would be reported in the 5y5y bucket but not, say, in 4y5y buckets (as the 4y5y swaption is not in the benchmark set). This, of course, is slightly misleading — it is not that the Bermudan swaption has zero vega in the 4y5y swaption bucket, but that the choice of our benchmark set effectively aggregates that sensitivity and reports it in the 5y5y and, less pronounced, 1y5y buckets. As a trading desk may want to use a fairly granular grid for keeping track of its vega exposure, we should think about how to rationally "project" our coarser vegas onto a finer grid. The idea of

---

[14]This situation would often also be reflected in the usage of different swaption weights in the calibration norm (14.51), with precision weights on illiquid swaptions set lower than on liquid ones.

just assigning our computed 5y5y LMM vega to the 5y5y bucket of the full grid is clearly suboptimal; instead we should somehow spread some of the vega around to buckets surrounding the 5y5y grid point.

There are various methods for projecting vegas from small to full grids, and they all suffer from a degree of arbitrariness as, ultimately, we are trying to create information where there is none. Perhaps the simplest method here is to interpolate (bi)linearly between the points of the small grid to get the values for all points on the full grid, and then rescale to make sure the total vega (i.e. the sum of all vegas in the grid) is the same for the full and reduced-size grids. A slightly more advanced — but nevertheless still somewhat arbitrary — method utilizes the LM model itself to come up with the interpolation scheme. To elaborate on this, let ν^{ex} be the LM vegas for some exotic derivative on the small grid $N_t \times N_x$ (which we are trying to project on a full grid). Furthermore, let the matrix $\nu^{i,j}$ be the matrix (of size $N_t \times N_x$) of vegas for (i,j)-th swaption in the full swaption set; see the start of Section 26.2 for more detail. Then we find the matrix Υ^{ex} of vegas on the full grid by solving the minimization problem

$$\left\| \nu^{\mathrm{ex}} - \sum_{i,j} (\Upsilon^{\mathrm{ex}})_{i,j} \, \nu^{i,j} \right\|^2 + \mathcal{I}_{\mathrm{smooth}} (\Upsilon^{\mathrm{ex}}) \to \min, \qquad (26.19)$$

where the norm $\|\cdot\|$ is some suitable matrix norm (such as the Frobenius norm used in Section 3.1.3) and $\mathcal{I}_{\mathrm{smooth}}(\Upsilon^{\mathrm{ex}})$ is a smoothing objective along the lines of the definition (14.51); for instance, for first-order smoothness in expiry and tenor directions we would specify

$$\mathcal{I}_{\mathrm{smooth}} (\Upsilon^{\mathrm{ex}}) = w_{\partial t} \sum_{i,j} \left((\Upsilon^{\mathrm{ex}})_{i,j} - (\Upsilon^{\mathrm{ex}})_{i-1,j} \right)^2$$

$$+ w_{\partial x} \sum_{i,j} \left((\Upsilon^{\mathrm{ex}})_{i,j} - (\Upsilon^{\mathrm{ex}})_{i,j-1} \right)^2.$$

The problem (26.19) is quadratic and easily solved with linear algebra methods.

Without a smoothing term, the problem (26.19) is under-specified as there are more free variables that constraints. The smoothing term is essential to pick a unique solution, yet it may lead to undesirable effects like affecting the locality of the vega. Another issue to keep in mind here is that it is not clear how these smoothing weights should be estimated, yet they would impact strongly the allocation of vega. Ultimately, however, one has to live with such issues since, as pointed out, we are filling information "gaps" using fairly arbitrary rules.

On the positive side, the method of projecting LM vegas on a full grid allows for consistent risk representation across a whole portfolio that a trading desk normally trades, including European swaptions, other vanilla products, and interest rate exotics. This method also allows benchmark sets to be tailored to the features of each derivative thus, potentially, getting better risk resolution and saving computational time by minimizing the number of shocks applied to each derivative. On the other hand, it obviously also introduces a certain level of arbitrariness into the vega calculation and aggregation, and even a danger that the vega for a particular derivative will be reported in inappropriate buckets. To guard against errors, it is often advisable to also calculate LM vegas for all products on the same — and relatively large — set of benchmark swaptions. This could be done relatively infrequently, for instance as part of weekly or monthly control calculations, while leaving daily vegas to be calculated with smaller, product-specific benchmark sets.

Besides the problem of projecting benchmark vegas "up" to a large common grid, we could also contemplate the possibility of projecting vegas "down", to a smaller grid of potentially different benchmark swaptions. While this capability may seem rather esoteric, some traders find it useful to be able to express their vegas in terms of different sets of European swaptions. It could also be useful for other functions within a bank, such as the risk management department who may use a volatility grid of different shape for calculating risk numbers such as the VaR (see Section 22.3). As the "down" projection compresses information rather than creates it, it is easy to imagine a reasonable algorithm — one just needs to decide how to aggregate "old" buckets into "new" ones. This could be done by, for example, adding up all vegas in the old buckets that are within a certain distance (in expiry/tenor space) from a given new bucket.

26.8 Some Notes on Computing Model Vegas

In all of the vega calculation methods covered in this section, at some point we still need to compute a sensitivity, most typically by Monte Carlo. The standard advice from chapters in Part V of this book for calculating these risk sensitivities apply; however, let us emphasize a few salient points.

- Just like for deltas, the main source of noise for model vegas of *callable* securities is the jumps in the exercise indicators, so the exercise boundary should be kept constant when calculating model vegas, see Section 24.1.1.2.
- More generally, pathwise differentiation of Section 24.3 could be used for model vegas. SDEs for model vegas can be derived by differentiating the SDEs for primary Libor rates with respect to volatility.
- The likelihood ratio or hybrid methods that we just touched upon in Section 24.4 actually work rather better for vegas than for deltas, and could be a viable alternative. Intuitively, a shock to the initial value of a forward Libor rate to compute a delta affects a Monte Carlo path

only up until the first event time (such as an option exercise or a barrier check), whereas the bump to a vega affects the whole path. As the time to the first event goes to zero, the likelihood weight for the delta then explodes, but the one for the vega does not.

- Smoothing of the payoffs by tube Monte Carlo (see Section 23.4) or by importance sampling (see Section 25.2) benefits vegas as well as deltas.
- The variance reduction method in Section 25.3 based on a Markovian approximation could be applied to vegas, but a direct linkage between the original volatility structure and the Markovian approximation is needed, so we should use (25.51) instead of (25.50).

Appendix

A

Markovian Projection

Markovian projection is a powerful method for simplifying complex process dynamics to a form that enables rapid calibration of model parameters to quoted option prices. We use the method several times in this book, see for instance Chapters 13 and 15. The usefulness of Markovian projection, however, extends beyond interest rate modeling applications. In this appendix, we develop the relevant theory behind the method and present additional examples and applications.

A.1 Marginal Distributions of Ito Processes

Models used in quantitative finance generally serve to define dynamics of market observables. Some models impose such dynamics directly on the observables; this is, for instance, the case for vanilla models (see Chapters 7, 8, 16 and 17) where the evolution of swap rates is modeled explicitly. Outside of interest rate modeling, equity and FX models typically fall in this category as well. In other cases, the dynamics of market observables are specified indirectly, through modeling of abstract Markovian state variables that drive the market observables through functional relationships (i.e., reconstitution formulas). This style of modeling is common in term structure models for commodities and interest rates (see Chapter 13 for a typical model).

Regardless of type, all models ultimately need to be calibrated to liquidly traded options, most often European call/put options on market observables. To facilitate efficient model calibration, it is generally helpful if exact or approximate analytical expressions exist for European options. Often, the derivation of such results is significantly aided by an initial simplification of the underlying dynamic processes, either because these processes are outright too complex to handle analytically, or because non-linear reconstitution formulas translate simple state variable dynamics into intractable dynamics for market observables. The viability of such simplifications stems from the simple structure of European options, which only depend on the one-dimensional marginal distributions of the market observable process. As it turns out, irrespective of how complicated a process for a particular market observable is, it is often possible to find much simpler process that preserves the marginal distributions.

A systematic way of finding process simplifications is based on the following fundamental result, see Theorem 4.6 in Gyöngy [1986].

Theorem A.1.1 (Gyöngy). *Let $X(t)$ be given by an SDE*

$$dX(t) = \lambda(t)\, dW(t), \tag{A.1}$$

where $W(t)$ is a one-dimensional Brownian motion under some probability measure P. Assume that the process $\lambda(t)$ is adapted, bounded, and uniformly bounded away from 0, such that (A.1) admits a unique strong solution. Define $b(t, x)$ by

$$b(t, x)^2 = \mathrm{E}\left(\lambda(t)^2 \big| X(t) = x\right), \tag{A.2}$$

where E is the expected value operator for the pricing measure P. Then the SDE

$$dY(t) = b(t, Y(t))\, dW(t), \quad Y(0) = X(0), \tag{A.3}$$

admits a weak solution Y that has the same one-dimensional distributions as X.

Remark A.1.2. The original result by Gyöngy also includes a drift in the dynamics of X, considering

$$dX(t) = \mu(t)\, dt + \lambda(t)\, dW(t)$$

instead of (A.1). The theorem then still holds with (A.3) replaced by

$$dY(t) = a(t, Y(t))\, dt + b(t, Y(t))\, dW(t), \quad Y(0) = X(0),$$

where

$$a(t, x) = \mathrm{E}\left(\mu(t) \big| X(t) = x\right).$$

In financial applications we nearly always have $\mu(t) = 0$ as we tend to consider the dynamics of X in its own martingale measure. For this reason we do not consider drifts in what follows.

Proof. The original proof in Gyöngy [1986] is fairly involved. A rigorous proof under much weaker assumptions than we stated (see Proposition A.1.4 below) is given in Brunick [2008] and is also highly technical. We do not reproduce either of these proofs, but instead we present a somewhat informal argument[1] originally due to Dupire, see Dupire [1994], Dupire [1997], who independently discovered essentially the same results as in Gyöngy [1986].

---

[1] A version of which we have already seen in Proposition 7.4.2.

The function $b(t, x)$ is often called the *Dupire local volatility* function for the process X.

Let us denote

$$c(t, K) \triangleq c(0, X(0); t, K) = \mathrm{E}\left((X(t) - K)^+\right)$$

to be the values of European call options on X for expiries t and strikes K. It follows from a time-dependent extension of Proposition 7.4.2 that, if we define Y by

$$dY(t) = b(t, Y(t)) \, dW(t) \tag{A.4}$$

with

$$b(t, K)^2 = \frac{2\frac{\partial}{\partial t} c(t, K)}{\frac{\partial^2}{\partial K^2} c(t, K)}, \tag{A.5}$$

then the values of European call options in the model (A.4) will be equal to $c(t, K)$ for all expiries t and strikes K, i.e. will be the same as in the model (A.1). To compute the right-hand side, we first write (the use of Dirac delta functions in the integrands can be justified by Tanaka's formula, see Section 1.9.2 and Karatzas and Shreve [1991])

$$d\left(X(t) - K\right)^+ = 1_{\{X(t) > K\}} dX(t) + \frac{1}{2}\delta(X(t) - K)\lambda(t)^2 \, dt$$

and, since $X(t)$ is a P-martingale,

$$\mathrm{E}\left(X(t) - K\right)^+ - (X(0) - K)^+ = \frac{1}{2}\int_0^t \mathrm{E}\left(\delta(X(t) - K)\lambda(t)^2\right) \, dt.$$

Clearly

$$\mathrm{E}\left(\delta(X(t) - K)\lambda(t)^2\right) = \mathrm{E}\left(\delta(X(t) - K)\right) \times \mathrm{E}\left(\lambda(t)^2 \middle| X(t) = K\right)$$

and

$$\mathrm{E}\left(\delta(X(t) - K)\right) = \frac{\partial^2}{\partial K^2}\mathrm{E}\left(X(t) - K\right)^+ = \frac{\partial^2}{\partial K^2} c(t, K).$$

In particular,

$$\frac{\partial}{\partial t} c(t, K) = \frac{\partial}{\partial t}\left(\mathrm{E}\left(X(t) - K\right)^+ - (X(0) - K)^+\right)$$

$$= \frac{1}{2}\frac{\partial^2}{\partial K^2} c(t, K) \times \mathrm{E}\left(\lambda(t)^2 \middle| X(t) = K\right).$$

Substituting this equality into (A.5) we obtain

$$b(t, K)^2 = \mathrm{E}\left(\lambda(t)^2 \middle| X(t) = K\right),$$

consistent with (A.2). □

Since X and Y have the same one-dimensional marginal distributions, the prices of European options on X and Y for all strikes K and expiries T will be identical (a result that is also implicit in our proof of the theorem). Thus, for the purposes of European option valuation, a potentially complicated process X can be replaced with a simpler Markov process Y; we call Y the *Markovian projection* of X. Notice that the process Y conveniently is of the local volatility type considered in Chapter 7, for which we have developed many exact or approximate methods for valuation of European options.

Theorem A.1.1 can be extended in a number of ways. Possibly the simplest extension involves relaxing the assumption that the Brownian motion $W(t)$ in (A.1) is one-dimensional, in which case the following trivial corollary holds.

Corollary A.1.3. *Suppose that X follows multi-factor dynamics*

$$dX(t) = \lambda(t)^\top dW(t) \tag{A.6}$$

with $W(t)$ a d-dimensional Brownian motion and $\lambda(t)$ a d-dimensional adapted process whose norm is bounded and uniformly bounded away from 0. Define the SDE

$$dY(t) = b(t, Y(t))\, d\widetilde{W}(t), \quad Y(0) = X(0), \tag{A.7}$$

where $\widetilde{W}(t)$ a one-dimensional Brownian motion, and

$$b(t,x)^2 = \mathrm{E}\left(\lambda(t)^\top \lambda(t) \,\middle|\, X(t) = x \right).$$

Then (A.7) admits a weak solution $Y(t)$ that has the same one-dimensional distributions as $X(t)$.

Proof. Clearly X can be written in one-dimensional form

$$dX(t) = \left(\lambda(t)^\top \lambda(t)\right)^{1/2} d\widetilde{W}(t),$$

where

$$d\widetilde{W}(t) = \frac{1}{(\lambda(t)^\top \lambda(t))^{1/2}} \lambda(t)^\top dW(t).$$

Simple quadratic variance calculations show that $\widetilde{W}(t)$ is a one-dimensional Brownian motion, and the corollary then is a direct consequence of Theorem A.1.1. $\square$

The original result by Gyöngy required that the variance process $\lambda(t)^\top \lambda(t)$ in (A.6) be both bounded and uniformly bounded away from 0. These are rather severe limitations that are violated in some standard models of mathematical finance, including the Heston model of Chapter 8. Brunick [2008] has proved the same result under much milder regulatory conditions and also extended it to the case where the asset process X itself is multi-dimensional.

Proposition A.1.4. *Let $X(t)$ be a p-dimensional stochastic process given by the strong solution of the SDE*

$$dX(t) = \lambda(t)^{\top} dW(t),$$

where now $\lambda(t)$ is a $(d \times p)$-matrix-valued adapted process and $W(t)$ is a d-dimensional Brownian motion. We assume that

$$\mathrm{E}\left(\int_0^t \left\| \lambda(s)^{\top} \lambda(s) \right\| ds \right) < \infty$$

for all $t \geq 0$. Then there exists a $(d \times p)$-matrix-valued function b such that

$$b(t,x)^{\top} b(t,x) = \mathrm{E}\left(\lambda(t)^{\top} \lambda(t) \middle| X(t) = x \right)$$

for any $t \geq 0$, the SDE

$$dY(t) = b(t, Y(t))^{\top} dW(t), \quad Y(0) = X(0),$$

admits a weak solution Y, and the random vector $Y(t)$ has the same distribution as the random vector $X(t)$ for any $t \geq 0$.

More generally, Brunick [2008] also proves that we can construct a "mimicking" process Y such that marginal distributions of some *functional* of X, rather than of X itself, are matched by a suitable functional of Y. The definition of functionals for which this result works is fairly technical, but the allowed set includes such financially relevant cases as functions of the running average of X, functions of the running maximum (or minimum) of X, and many others. To avoid technicalities, let us consider only the case of a running maximum of a one-dimensional asset process. Specifically, let us define $M(t, \xi)$ to be the running maximum for a given process ξ, i.e.

$$M(t, \xi) = \sup_{0 \leq s \leq t} \xi(s), \quad t \geq 0.$$

Proposition A.1.5. *Let X follow a one-dimensional diffusion*

$$dX(t) = \lambda(t) dW(t),$$

where $W(t)$ is a one-dimensional Brownian motion, and $\lambda(t)$ is a scalar adapted process that satisfies

$$\mathrm{E}\left(\int_0^t \lambda(s)^2 ds \right) < \infty, \quad t \geq 0.$$

Then i) there exists a deterministic function $b(t, x, m)$ such that

$$b(t, x, m)^2 = \mathrm{E}\left(\lambda(t)^2 \middle| X(t) = x, M(t, X) = m \right)$$

holds for all $t \geq 0$; ii) the SDE

$$dY(t) = b(t, Y(t), M(t, Y)) dW(t), \quad Y(0) = X(0),$$

admits a weak solution Y; and iii) the pair $(Y(t), M(t, Y))$ has the same distribution as the pair $(X(t), M(t, X))$ for any $t \geq 0$.

Example A.1.6. A European up-and-out barrier call option (see Section 2.1) with expiry T, strike K and barrier level B is an option with the payoff

$$1_{\{M(T,X)<B\}}(X(T) - K)^+.$$

The values of such options on X for all expiries, strikes and barrier levels match those on the mimicking process Y defined by Proposition A.1.5.

While the projection defined by Proposition A.1.5 matches all (up-and-out) barrier option prices, the standard projection result in Theorem A.1.1 does not, as it does not preserve joint distributions of the process observed at multiple times. This is an important point that should be kept in mind in a calibration setting: the Markovian projection of Theorem A.1.1 should be used solely as the means to calibrate to *European* option prices and not to more complicated derivatives[2].

A.2 Approximations for Conditional Expected Values

According to Theorem A.1.1, the coefficients for the SDE of the Markovian projection are obtained by calculating conditional expected values as in (A.2). This, in the majority of interesting cases, is a non-trivial task. Below, we consider several possible approximations.

A.2.1 Gaussian Approximation

Of the few probability distributions that allow us to calculate conditional expected values in closed form, the most important is, of course, the Gaussian distribution (a fact we use extensively in many places of the book, see for instance Chapter 17). Not surprisingly, we can get good mileage out of the idea of approximating the original distributions of $X(t)$ and $\lambda(t)^2$ with Gaussian distributions, in order to calculate the conditional expected value in (A.6). Many variations are possible here; we present one approach in Proposition A.2.1 below. To fix our setup, we assume that X follows the SDE (A.1) with a process $\lambda(t)$ given by the SDE

$$d\lambda(t)^2 = \nu(t)\,dt + \varepsilon(t)\,dZ(t),$$

with two adapted stochastic processes $\nu(t)$ and $\varepsilon(t)$, and a Brownian motion $Z(t)$.

Proposition A.2.1. *The conditional expected value in (A.2) can be approximated by*

---

[2] Although some creative approximations for such securities can occasionally be derived from Markovian projection, see Section 13.1.9.4.

$$b(t, x)^2 \approx \overline{\lambda(t)^2} + s(t) \left(x - X(0) \right), \qquad (A.8)$$

$$s(t) = \frac{\int_0^t \overline{\varepsilon}(s) \overline{\lambda}(s) \overline{\rho}(s) \, ds}{\int_0^t \overline{\lambda(s)^2} \, ds},$$

where $\overline{\varepsilon}(t)$, $\overline{\lambda(t)^2}$, $\overline{\rho}(t)$, $\overline{\lambda}(t)$ are deterministic approximations to $\varepsilon(t)$, $\lambda(t)^2$, $\rho(t) \triangleq \langle dW(t), dZ(t) \rangle / dt$, and $\lambda(t)$, respectively. In particular, we can take

$$\overline{\varepsilon}(t) = \mathrm{E}\left(\varepsilon(t) \right), \quad \overline{\lambda(t)^2} = \mathrm{E}\left(\lambda(t)^2 \right) = \mathrm{E}\left(\int_0^t \nu(s) \, ds \right),$$

$$\overline{\rho}(t) = \mathrm{E}\left(\langle dW(t), dB(t) \rangle / dt \right), \quad \overline{\lambda}(t) = \sqrt{\overline{\lambda(t)^2}}.$$

Proof. First, we approximate the dynamics of $(X(t), \lambda(t)^2)$ with the Gaussian processes

$$dX(t) \approx \overline{\lambda}(t) \, dW(t),$$
$$d\lambda(t)^2 \approx \overline{\nu}(t) \, dt + \overline{\varepsilon}(t) \, d\overline{Z}(t),$$

where $\overline{\nu}(t) = \mathrm{E}(\nu(t))$ and $\overline{Z}(t)$ is a Brownian motion such that $\langle dW(t), d\overline{Z}(t) \rangle = \overline{\rho}(t) \, dt$. The result then follows from the standard conditioning formula for Gaussian variables U, V:

$$\mathrm{E}\left(U \mid V \right) = \mathrm{E}(U) + \frac{\mathrm{Cov}\left(U, V \right)}{\mathrm{Var}(V)} \left(V - \mathrm{E}(V) \right). \qquad (A.9)$$

□

Rather than approximating $X(t)$ and $\lambda(t)^2$ directly by Gaussian processes, we can instead use deterministic functions of Gaussian processes. For instance, if we use exponential "mapping functions", we would then arrive at a log-normal (rather than Gaussian) approximation. Furthermore, instead of approximating the drift of $\lambda(t)^2$ as a deterministic function, we could instead approximate it by a linear function of $\lambda(t)^2$ itself, which would retain a Gaussian distribution for $\lambda(t)^2$. Ultimately, the original form of the SDEs for the asset and variance in a given model would typically suggest the most proper usage of the Gaussian approximation principle.

As evident from (A.8), the local variance function that emerges from the Gaussian approximation method is linear, whereby the approximating model (A.3) will always generate a monotonic *implied* volatility smile. On the other hand, we may know *a-priori* that the true volatility smile of the original model (A.1) is not close to linear — it could be U-shaped, say. In such cases, a wholesale replacement of the original dynamics by the approximating linear local variance model is unlikely to be satisfactory. As it turns out, it is possible to apply the Gaussian approximation in a more sophisticated manner, leading to a better approximating model. We discuss this in Section A.3.1 below, but first we outline an alternative approach to estimating conditional expected values.

A.2.2 Least-Squares Projection

Section 16.6.2 develops the least-squares projection method for conditional expected values, using the insight that a conditional expected value can be defined as a projection onto a suitable functional space. If we project onto some subspace of the full functional space, we obtain an approximation to the conditional expected value, as stated formally in Proposition 16.6.2. For our purposes here, we focus only on the particularly tractable case of linear subspaces, utilized in Section 16.6.4 to produce a linear least-squares projection. Restating the linear projection in the notations of this appendix, we obtain the following result.

Proposition A.2.2. *The linear least-squares approximation to the conditional expected value in (A.2) is given by*

$$b(t, x)^2 \approx \overline{\lambda(t)^2} + s(t) (x - X(0)), \qquad (A.10)$$

where

$$s(t) = \frac{\mathrm{E}\left(\lambda(t)^2 (X(t) - X(0))\right)}{\mathrm{E}\left((X(t) - X(0))^2\right)}, \quad \overline{\lambda(t)^2} = \mathrm{E}\left(\lambda(t)^2\right) \qquad (A.11)$$

or, more compactly,

$$s(t) = \frac{\mathrm{Cov}\left(\lambda(t)^2, X(t)\right)}{\mathrm{Var}\left(X(t)\right)}.$$

We notice a strong similarity between the expressions for the Dupire local volatility approximations in Propositions A.2.1 and A.2.2. By design (as we projected the variance only on linear functions of $X(t)$), the local variance function approximation in (A.10) is still linear in x, and the expression (A.11) reduces to the formulas of Proposition A.2.1 if we apply a Gaussian approximation. This is not surprising, as the linear least-squares projection is known to produce an exact result for conditional expectations of Gaussian variables.

As Propositions A.2.1 and A.2.2 approximate local *variance* with a linear function of spot, both suggest that the SDE for the approximating process is of the displaced square-root type:

$$dY(t) = \left(\overline{\lambda(t)^2} + s(t)(Y(t) - Y(0))\right)^{1/2} dW(t).$$

While such processes are analytically tractable (see Sections 7.2.4 and 10.2), it is often more convenient to work with processes of the displaced log-normal type (see Proposition 7.2.12) where the local *volatility* function is linear:

$$dY(t) = \sigma(t) \left(1 + b(t)(Y(t) - Y(0))\right) dW(t). \qquad (A.12)$$

To obtain an approximating process of this type, we can expand the square root function to the first order around $x = Y(0)$, yielding the following result.

Proposition A.2.3. *The displaced log-normal approximation to the process* $X(t)$ *in (A.1) is given by (A.12), where*

$$\sigma(t) = \sqrt{\overline{\lambda(t)^2}} = \sqrt{\mathrm{E}\left(\lambda(t)^2\right)},$$

$$b(t) = \frac{s(t)}{2\sigma(t)^2} = \frac{\mathrm{Cov}\left(\lambda(t)^2, X(t)\right)}{2\mathrm{E}\left(\lambda(t)^2\right)\mathrm{Var}\left(X(t)\right)}.$$

The same result was obtained in Antonov and Misirpashaev [2009a] by a direct application of the least-squares method, i.e. by solving the minimization problem

$$\mathrm{E}\left(\left(\lambda(t)^2 - \sigma(t)^2\left(1 + b(t)(X(t) - X(0))\right)^2\right)^2\right) \to \min$$

in $\sigma(t)$ and $b(t)$ and keeping only the leading $O(\lambda(t)^2)$ terms.

A.3 Applications to Local Stochastic Volatility Models

A.3.1 Markovian Projection onto a Stochastic Volatility Model

In applying the Markovian projection method, we are limited by the accuracy of approximations to conditional expected values. As reviewed in Section A.2, the methods that are generally available approximate local volatility or variance functions with linear[3] functions which, as discussed in Section A.2.1, is insufficient for the case where $X(t)$ has complex dynamics. Fortunately, Theorem A.1.1 provides us with means to approximate a given model by a model of essentially *any* type, not just local volatility. The following, borderline trivial, corollary to Theorem A.1.1 is the key[4].

Corollary A.3.1. *If two processes* X_1 *and* X_2 *have the same Markovian projections (i.e., they imply identical Dupire local volatility functions), then European put/call option prices on* X_1 *and* X_2 *both are identical for all strikes and expiries.*

Let us demonstrate the usefulness of Corollary A.3.1 by applying it to a stochastic volatility model. Let $X_1(t)$ follow a stochastic volatility SDE (recall from Section 8.1 that for non-linear functions $b(t, x)$ such models are sometimes called local stochastic volatility, or LSV, models)

$$dX_1(t) = b_1\left(t, X_1(t)\right)\sqrt{z_1(t)}\,dW(t),$$

---

[3]Markovian projection on processes with *quadratic* local volatility (see Section 7.3) was developed in Antonov and Misirpashaev [2009b], using Wiener chaos expansion techniques. Predictably it outperforms projections on linear local volatility processes but is significantly more complicated.

[4]Dupire [1997] dubs this corollary the *universal law of volatility*.

where $z_1(t)$ is some variance process. Suppose we would like to derive approximations for European options on X_1. One possibility is to approximate X_1 with a local volatility model, using Theorem A.1.1 directly. As we discussed, this is unlikely to work well, at least if we compute conditional expected values with the approximations developed in Section A.2. Instead, we can use Corollary A.3.1 and approximate X_1 with a *stochastic* volatility process that employs a more tractable process for stochastic variance. Let us call this variance process z_2, and consider a model of the form

$$dX_2(t) = b_2(t, X_2(t))\sqrt{z_2(t)}\,dW(t). \tag{A.13}$$

Then Corollary A.3.1 and Theorem A.1.1 imply that to match European option prices in the two models for all strikes and expiries, we need to set $b_2(t,x)$ such that

$$b_2(t,x)^2 = b_1(t,x)^2 \frac{E(z_1(t)|\,X_1(t) = x)}{E(z_2(t)|\,X_2(t) = x)}. \tag{A.14}$$

While we still need to apply formulas from Section A.2 to approximate conditional expected values in (A.14), the fact that we calculate the *ratio* of two expected values gives us some hope for error cancellation — i.e. even if each individual approximation is not particularly accurate, they are inaccurate "in the same way" and the overall error diminishes when the ratio is formed. To maximize the error cancellation effect, it is obviously beneficial to choose z_2 as similar to z_1 as possible, while still retaining analytical tractability.

Using the SV model (A.13) as the target model for Markovian projection exercise will benefit from the fact that the model (A.13) is quite rich, even for linear local volatility functions $b_2(t,x)$. If z_2 is a square-root process and $b_2(t,x)$ is linear (in x), then the resulting model reduces to the displaced Heston model (8.3)–(8.4) which, as we saw in Chapter 8, is both tractable and capable of generating a wide variety of implied volatility smiles.

In general, limitations of available conditional expected value approximations impose certain restrictions on designing approximating models. In particular, as we want the "output" local volatility $b_2(t,x)$ to be as close to linear as possible — so that the inevitable linear approximation is not far off — we should choose the stochastic variance process z_2 in such a way that the characteristics of this process, and not the shape of the local volatility, explain as much of the *curvature* of the implied volatility smile of the model for X_1 as possible (note that this also holds true for X_1 processes of the non-SV type).

Let us turn our attention to another common application of Corollary A.3.1. Suppose that X_1 follows the SDE

$$dX_1(t) = \lambda(t)\sqrt{z(t)}\,dW(t), \tag{A.15}$$

where $z(t)$ is a stochastic variance process and $\lambda(t)$ is now a stochastic process in its own right. For instance, $\lambda(t)$ could be a complicated function of state variables in a term structure model of interest rates, which is a relevant example when $S(t)$ represents a swap rate. We would like to replace the SDE (A.15) with a local stochastic volatility model,

$$dX_2(t) = b\left(t, X_2(t)\right) \sqrt{z(t)}\, dW(t),$$

where we use *the same* stochastic variance process $z(t)$ as in (A.15). Then, according to Corollary A.3.1 we need to set

$$b\left(t,x\right)^2 = \frac{\mathrm{E}\left(\lambda(t)^2 z(t)\middle| X_1(t) = x\right)}{\mathrm{E}\left(z(t)\middle| X_2(t) = x\right)}. \tag{A.16}$$

This formula can be simplified when $\lambda(t)$ and $z(t)$ are (approximately) conditionally independent given $X_1(t)$, in which case we get

$$b\left(t,x\right)^2 \approx \mathrm{E}\left(\lambda(t)^2\middle| X_1(t) = x\right) \frac{\mathrm{E}\left(z(t)\middle| X_1(t) = x\right)}{\mathrm{E}\left(z(t)\middle| X_2(t) = x\right)}. \tag{A.17}$$

In many situations, it can be safely assumed that

$$\mathrm{E}\left(z(t)\middle| X_1(t) = x\right) \approx \mathrm{E}\left(z(t)\middle| X_2(t) = x\right),$$

in which case the formula simplifies further,

$$b\left(t,x\right)^2 \approx \mathrm{E}\left(\lambda(t)^2\middle| X_1(t) = x\right). \tag{A.18}$$

This formula forms the basis for European swaption approximations in term structure models with stochastic volatility; we use it for both quasi-Gaussian models (see Section 13.3.3) and Libor market models (see Proposition 15.2.1).

A.3.2 Fitting the Market with a Local Stochastic Volatility Model

While the focus of this appendix is on approximating more complicated models with simpler ones, direct calibration of local stochastic volatility (LSV) models to the market is another possible application of the techniques we consider.

Let $S(t)$ be the value of a given market variable (for example, a swap rate or an equity price), with initial value $S(0) = S_0$. Suppose market prices of European call or put options are known for all expiries T and strikes K. These can be easily converted (see (A.5)) into a market-implied Dupire local volatility $b_{\mathrm{mkt}}(t, x)$, such that the market European option prices are reproduced by the model

$$dY(t) = b_{\mathrm{mkt}}\left(t, Y(t)\right) dW(t), \quad Y(0) = S_0.$$

Suppose we postulate a stochastic variance process $z(t)$ (such as the square-root process of the Heston model, or whatever multi-factor variance process is in favor in equity modeling circles at the moment), and aim to construct a stochastic volatility model

$$dS(t) = b\left(t, S(t)\right) \sqrt{z(t)}\, dW(t), \quad S(0) = S_0, \tag{A.19}$$

consistent with market European option prices. As follows from Theorem A.1.1, the local volatility function $b(t, x)$ is then given by

$$b\left(t, x\right)^2 = \frac{b_{\mathrm{mkt}}\left(t, x\right)^2}{\mathrm{E}\left(z(t)|\, S(t) = x\right)}. \tag{A.20}$$

As mentioned before, the challenge of computing $\mathrm{E}(z(t)|S(t) = x)$ makes the method difficult to implement in practice. One choice is to apply finite difference methods to compute the conditional expected value numerically in a forward Kolmogorov PDE for $(S(t), z(t))$, see Ren et al. [2007]. Here, we instead look for analytic approximations.

Linear projections of the type explored in Section A.2 are possible, but are likely to be inaccurate in this case if $b_{\mathrm{mkt}}(t, x)$ has a high degree of convexity. We instead wish to explore methods based on comparing (A.19) to another stochastic volatility model in which European option prices can be cheaply computed. Suppose we have identified such a "proxy" model, defined through a known local volatility function $\widetilde{b}(t, x)$,

$$dX(t) = \widetilde{b}\left(t, X(t)\right) \sqrt{z(t)}\, dW(t), \quad X(0) = S_0, \tag{A.21}$$

where $z(t)$ is the same process as in (A.19). For tractability, we often choose $\widetilde{b}(t, x)$ to be a linear function of x. We will have more to say about the parameterization of the proxy model in Section A.3.3, but for now it suffices to assume that this model allows us to quickly and efficiently calculate European call (or put) option prices. These prices can be turned into a "proxy" Dupire local volatility function $b_{\mathrm{proxy}}(t, x)$ by means of (A.5). Rewriting (A.20), we then have

$$\mathrm{E}\left(z(t)|\, X(t) = x\right) = \frac{b_{\mathrm{proxy}}\left(t, x\right)^2}{\widetilde{b}\left(t, x\right)^2}. \tag{A.22}$$

In other words, from a proxy stochastic volatility model which easily computed European option prices, we can efficiently compute the conditional expected values $\mathrm{E}(z(t)|X(t) = x)$. One way to take advantage of this observation is to combine (A.20) and (A.22) as follows.

Proposition A.3.2. *The local volatility function $b(t, x)$ that makes the model (A.19) consistent with the market is given by*

$$b\left(t, x\right)^2 = \widetilde{b}\left(t, x\right)^2 \frac{b_{\mathrm{mkt}}\left(t, x\right)^2}{b_{\mathrm{proxy}}\left(t, x\right)^2} \frac{\mathrm{E}\left(z(t)|\, X(t) = x\right)}{\mathrm{E}\left(z(t)|\, S(t) = x\right)},$$

where $X(t)$ follows the "proxy" model (A.21) with a known local volatility function $\widetilde{b}(t, x)$.

The ratio $b_{\mathrm{mkt}}(t, x)/b_{\mathrm{proxy}}(t, x)$ can, as discussed, usually be computed efficiently. Approximating

$$\frac{\mathrm{E}\left(z(t)|\,X(t) = x\right)}{\mathrm{E}\left(z(t)|\,S(t) = x\right)} \approx 1, \tag{A.23}$$

we obtain the following useful corollary.

Corollary A.3.3. *Under the approximation (A.23) we have that*

$$b\left(t, x\right) \approx \widetilde{b}\left(t, x\right) \frac{b_{\mathrm{mkt}}\left(t, x\right)}{b_{\mathrm{proxy}}\left(t, x\right)}.$$

To obtain a more sophisticated approximation than that of Corollary A.3.3, we can attempt to improve on (A.23) by looking for an (approximate) functional relationship between $X(t)$ and $S(t)$. Denote

$$h\left(t, x\right) = \int_{x_0}^{x} \frac{dy}{b\left(t, y\right)}, \quad \widetilde{h}\left(t, x\right) = \int_{x_0}^{x} \frac{dy}{\widetilde{b}\left(t, y\right)}, \tag{A.24}$$

$$H\left(t, x\right) = \widetilde{h}^{-1}\left(t, h\left(t, x\right)\right),$$

where $\widetilde{h}^{-1}(t, x)$ is the inverse of $\widetilde{h}(t, x)$ in the second (i.e., x) argument. Furthermore, denote

$$\widetilde{X}(t) = H\left(t, S(t)\right).$$

Then

$$d\widetilde{X}(t) = \frac{\partial}{\partial x} H\left(t, x\right)\bigg|_{x=S(t)} dS(t) + O(dt)$$

$$= \widetilde{b}\left(t, \widetilde{X}(t)\right) \sqrt{z(t)}\, dW(t) + O(dt).$$

We see that $\widetilde{X}(t)$ and $X(t)$ have identical diffusion coefficients, which suggests the approximation

$$X(t) \approx H\left(t, S(t)\right). \tag{A.25}$$

This leads to the following result.

Proposition A.3.4. *The local volatility function $b(t, x)$ that makes the model (A.19) consistent with the market is approximately given by*

$$b\left(t, x\right) \approx \widetilde{b}\left(t, H\left(t, x\right)\right) \frac{b_{\mathrm{mkt}}\left(t, x\right)}{b_{\mathrm{proxy}}\left(t, H\left(t, x\right)\right)}, \tag{A.26}$$

with $H(t, x)$ given by (A.24).

Proof. By (A.25),

$$E\left(z(t)|\,S(t)=x\right)=E\left(z(t)|\,H\left(t,S(t)\right)=H\left(t,x\right)\right)$$
$$\approx E\left(z(t)|\,X(t)=H\left(t,x\right)\right).$$

From (A.22),

$$E\left(z(t)|\,X(t)=H\left(t,x\right)\right)=\frac{b_{\mathrm{proxy}}\left(t,H\left(t,x\right)\right)^2}{\widetilde{b}\left(t,H\left(t,x\right)\right)^2},$$

and the result follows from (A.20).   □

We emphasize that $H(t,x)$ depends on the (unknown) function $b(t,x)$, hence (A.26) is, in fact, an equation for $b(t,x)$. This equation can be solved. Let us first denote

$$h_{\mathrm{proxy}}\left(t,x\right)=\int_{x_0}^{x}\frac{dy}{b_{\mathrm{proxy}}\left(t,y\right)},\qquad h_{\mathrm{mkt}}\left(t,x\right)=\int_{x_0}^{x}\frac{dy}{h_{\mathrm{mkt}}\left(t,y\right)}. \qquad (A.27)$$

The following then holds (see Henry-Labordère [2009]).

Proposition A.3.5. *The mapping function $H(t,x)$ is given by*

$$H(t,x)=h_{\mathrm{proxy}}^{-1}\left(t,h_{\mathrm{mkt}}\left(t,x\right)\right), \qquad (A.28)$$

where $h_{\mathrm{proxy}}^{-1}(t,x)$ is the inverse of $h_{\mathrm{proxy}}(t,x)$ defined by (A.27) in the second (x) argument. Furthermore,

$$b\left(t,x\right)\approx\widetilde{b}\left(t,h_{\mathrm{proxy}}^{-1}\left(t,h_{\mathrm{mkt}}\left(t,x\right)\right)\right)\frac{b_{\mathrm{mkt}}\left(t,x\right)}{b_{\mathrm{proxy}}\left(t,h_{\mathrm{proxy}}^{-1}\left(t,h_{\mathrm{mkt}}\left(t,x\right)\right)\right)}. \qquad (A.29)$$

Proof. Differentiating $H(t,x)$ in (A.24) with respect to x we obtain

$$\frac{\partial H(t,x)}{\partial x}=\frac{\partial h(t,x)}{\partial x}\bigg/\frac{\partial\widetilde{h}(t,f)}{\partial f}\bigg|_{f=H(t,x)}=\frac{\widetilde{b}\left(t,H(t,x)\right)}{b\left(t,x\right)}.$$

Therefore, we can rewrite (A.26) as an (approximate) equation

$$\frac{\partial H(t,x)}{\partial x}=\frac{b_{\mathrm{proxy}}\left(t,H\left(t,x\right)\right)}{b_{\mathrm{mkt}}\left(t,x\right)}. \qquad (A.30)$$

Treating this as an ODE in x for fixed t, we solve it to find

$$\int_{x_0}^{H(t,x)}\frac{dy}{b_{\mathrm{proxy}}\left(t,y\right)}=\int_{x_0}^{x}\frac{dy}{b_{\mathrm{mkt}}\left(t,y\right)}, \qquad (A.31)$$

resulting in (A.28). Then (A.29) follows from (A.26) and (A.28).   □

Remark A.3.6. Henry-Labordère [2009] notices that in fact (A.30) implies a condition more general than (A.31), namely that

$$\int_{H_0}^{H(t,x)} \frac{dy}{b_{\text{proxy}}(t,y)} = \int_{x_0}^{x} \frac{dy}{b_{\text{mkt}}(t,y)}$$

for any H_0. He proposes to choose H_0 so that the difference in drifts of $S(t)$ and $X(t)$ is minimized (our approximation above matches diffusion terms only).

A.3.3 On Calculating Proxy Local Volatility

The techniques in Section A.3.2 above all hinge on the critical assumption that we can pick a proxy stochastic volatility model that allows for efficient computation of the Dupire local volatility function $b_{\text{proxy}}(t,x)$. Provided that $z(t)$ follows the standard mean-reverting square-root process (as in the Heston model), then an obvious choice for the proxy model is a displaced Heston model, with

$$\widetilde{b}(t,x) = \widetilde{b}_1 x + \widetilde{b}_2. \tag{A.32}$$

Special cases include the $\widetilde{b}(t,x) = \widetilde{b}_1 x$ (the original Heston model) and $\widetilde{b}(t,x) = \widetilde{b}_2$ (the "Gaussian" Heston model), but any choice of constants $\widetilde{b}_1$ and $\widetilde{b}_2$ will allow for quick pricing of European put and call options; see Chapter 8 for details. Using the averaging techniques of Chapter 9, we can, in fact, extend (A.32) to linear local volatility function with time-dependent coefficients,

$$\widetilde{b}(t,x) = \widetilde{b}_1(t)x + \widetilde{b}_2(t). \tag{A.33}$$

The time-dependent coefficients in (A.33) can be chosen to make the proxy model resemble as much as possible the true model for S, thereby improving the quality of various approximations made. Ideally we should use

$$\widetilde{b}_2(t) = b(t,S_0), \quad \widetilde{b}_1(t) = \frac{\partial}{\partial x} b(t,S_0),$$

which is the first-order approximation to the Dupire local volatility $b(t,x)$ along the forward value of $S(t)$. Of course, the value and derivative of $b(t,x)$ are unknown *a-priori*, but one can easily envision various approximations or, perhaps, an iterative procedure where an approximation for the Dupire local volatility in step n is used to define the proxy local volatility $\widetilde{b}$ for step $n+1$.

The choice of a mean-reverting square-root process for $z(t)$ for variance leads to a specification that is quite amenable to the methods of Section A.3.2 and is often sufficient. However, more advanced applications such as those considered in Section 15.7 (and some popular models in equity modeling, see e.g. Bergomi [2009]) involve multi-factor stochastic variance dynamics and require extra effort as European options are then not always easy to calculate or approximate. As should be clear from Section A.3.2, we actually do not need to be able to calculate European option prices in the proxy model; all we really need is the Dupire local volatility for the proxy model, $b_{\text{proxy}}(t, x)$. This function is defined by (see (A.22))

$$b_{\text{proxy}}(t, x)^2 = \tilde{b}(t, x)^2 \, \mathrm{E}\left(z(t) \middle|\, X(t) = x\right). \tag{A.34}$$

It turns out that there exists a reasonably efficient algorithm for calculating the right-hand side of this question for a large selection of proxy models. To make this statement precise, let us define the proxy stochastic volatility model by

$$dX(t)/X(t) = \lambda\sqrt{z(t)} \left(\rho \, dW(t) + \left(1 - \rho^2\right)^{1/2} dW_X(t)\right), \tag{A.35}$$

$$dz(t) = \nu(t) \, dt + \varepsilon_1(t) \, dW(t) + \varepsilon_2(t) \, dW_z(t), \tag{A.36}$$

for a deterministic $\lambda > 0$ (in the notation of (A.21) therefore $\tilde{b}(t, x) = \lambda x$). Here $W(t)$, $W_X(t)$ and $W_z(t)$ are independent Brownian motions and $\nu(t)$, $\varepsilon_1(t)$ and $\varepsilon_2(t)$ are sufficiently regular adapted processes. The algorithm is based on the following result (see Romano and Touzi [1997] and Lee [2001]).

Proposition A.3.7. *For the model (A.35)–(A.36), we have*

$$\mathrm{E}\left(z(t) \middle|\, X(t) = x\right) = \frac{\mathrm{E}\left(z(t)\xi(t, x)\right)}{\mathrm{E}\left(\xi(t, x)\right)}, \quad \xi(t, x) = \frac{1}{D(t)^{1/2}} \phi\left(\frac{\ln(x) - m(t)}{D(t)^{1/2}}\right), \tag{A.37}$$

where

$$m(t) = \ln(X(0)) + \lambda\rho \int_0^t \sqrt{z(s)} \, dW(s) - \frac{\lambda^2}{2} \int_0^t z(s) \, ds,$$

$$D(t) = \lambda^2 \left(1 - \rho^2\right) \int_0^t z(s) \, ds,$$

and $\phi(z) = (2\pi)^{-1/2} e^{-z^2/2}$ is the standard Gaussian PDF.

Proof. Proceeding informally, we observe that

$$\mathrm{E}\left(z(t) \middle|\, X(t) = x\right) = \frac{\mathrm{E}\left(z(t)\delta(X(t) - x)\right)}{\mathrm{E}\left(\delta(X(t) - x)\right)}, \tag{A.38}$$

where $\delta(x)$ is the Dirac delta function. Let $\mathcal{F}_t^z$ be the filtration generated by $W(s)$ and $W_z(s)$, $0 \leq s \leq t$. Clearly $z(t)$ is adapted to $\mathcal{F}_t^z$. We can write

$$X(t) = X(0) \exp\left(\lambda\rho \int_0^t \sqrt{z(s)} \, dW(s) - \frac{\lambda^2}{2} \int_0^t z(s) \, ds\right)$$

$$\times \exp\left(\lambda \left(1 - \rho^2\right)^{1/2} \int_0^t \sqrt{z(s)} \, dW_X(s)\right),$$

where the first exponential is adapted to $\mathcal{F}_t^z$ and the second is driven by a Brownian motion W_X that is independent of this filtration. Conditioned on the filtration $\mathcal{F}_t^z$, $\ln(X(t))$ is Gaussian with known moments,

$$\ln(X(t))|\,\mathcal{F}_t^z \sim \mathcal{N}\,(m(t), D(t))\,. \tag{A.39}$$

Notice that

$$\mathrm{E}\,(\delta(X(t) - x)) = \mathrm{E}\,(\mathrm{E}\,(\,\delta(X(t) - x)|\,\mathcal{F}_t^z))$$

where $\mathrm{E}(\delta(X(t) - x)|\mathcal{F}_t^z)$ is the log-normal density for the conditional distribution of $X(t)$ defined by (A.39). We therefore have

$$\mathrm{E}\,(\mathrm{E}\,(\,\delta(X(t) - x)|\,\mathcal{F}_t^z)) = \mathrm{E}\,\left(\frac{1}{xD(t)^{1/2}}\phi\left(\frac{\ln(x) - m(t)}{D(t)^{1/2}}\right)\right),$$

and

$$\mathrm{E}\,(z(t)\delta(X(t) - x)) = \mathrm{E}\,(z(t)\mathrm{E}\,(\,\delta(X(t) - x)|\,\mathcal{F}_t^z))$$
$$= \mathrm{E}\,\left(\frac{z(t)}{xD(t)^{1/2}}\phi\left(\frac{\ln(x) - m(t)}{D(t)^{1/2}}\right)\right),$$

The result of the proposition then follows from (A.38).   □

To compute $b_{\mathrm{proxy}}(t, x)$ in a proxy model of the type (A.35)–(A.36), Henry-Labordère [2009] suggests simulating paths of the variance process $z(t)$ and then tabulating the values of $\mathrm{E}(z(t)|X(t) = x)$ by calculating the formula (A.37) for a selection of values of x and t. To obtain the values of $b_{\mathrm{proxy}}(t, x)$ for all t and x, we would then use (A.34) with some sort of interpolation to fill in values of $\mathrm{E}(z(t)|X(t) = x)$ between the tabulated ones.

A.4 Basket Options in Local Volatility Models

So far, our primary application for Markovian projection has been the pricing of European options on scalar processes. However, Markovian projection can easily be extended to options on multiple underlyings, e.g. basket options. Such options appear naturally in interest rate modeling — for example, we can often think of swap rates as baskets of Libor rates — a representation that we use in Section 15.2 together with a Markovian projection to approximate the swap rate process in a Libor market model. CMS spread options could also be thought of as options on baskets of two assets with weights ± 1. Furthermore, baskets serve as a good example of the practical usage of the Markovian projection method and the various "tricks" that go along with it.

Let us consider a collection of N assets $\mathbf{S}(t) = (S_1(t), \dots, S_N(t))^\top$, each driven by its own local volatility model

$$dS_n(t) = \sigma_n(t)\varphi_n\,(t, S_n(t))\,dW_n(t),\quad n = 1, \dots, N. \tag{A.40}$$

We assume from now on that each local volatility can be well-approximated by a linear function, i.e. that

$$\varphi_n(t, x) = 1 + b_n(t)(x - S_n(0)). \tag{A.41}$$

The Brownian motions $(W_1(t), \ldots, W_N(t))$ are assumed to be correlated with correlations

$$\langle dW_i(t), dW_j(t) \rangle = \rho_{i,j} \, dt, \quad i, j = 1, \ldots, N.$$

We define by $S(t)$ the value of the basket (also sometimes called the *index*)

$$S(t) = \sum_{n=1}^{N} w_n S_n(t). \tag{A.42}$$

We wish to calculate values of European options on $S(t)$, a problem first considered by Avellaneda et al. [2002] (but using quite different methods).

Applying Ito's lemma to $S(t)$, we see that

$$dS(t) = \sum_{n=1}^{N} w_n \sigma_n(t) \varphi_n(t, S_n(t)) \, dW_n(t). \tag{A.43}$$

If we define

$$\lambda(t)^2 \triangleq \sum_{n,m=1}^{N} w_n w_m \sigma_n(t) \varphi_n(t, S_n(t)) \, \sigma_m(t) \varphi_m(t, S_m(t)) \, \rho_{n,m}, \tag{A.44}$$

$$dW(t) \triangleq \frac{1}{\lambda(t)} \sum_{n=1}^{N} w_n \sigma_n(t) \varphi_n(t, S_n(t)) \, dW_n(t), \tag{A.45}$$

then

$$dS(t) = \lambda(t) \, dW(t),$$

where $W(t)$ is easily seen to be a standard Brownian motion. The process $\lambda(t)$ is here a complicated function of the vector of asset prices $\mathbf{S}(t)$.

As in Proposition A.2.3, let us approximate the dynamics of S with a displaced log-normal process. We then have $S \approx \widetilde{S}$, where

$$d\widetilde{S}(t) = \sigma(t) \left(1 + b(t)(\widetilde{S}(t) - S(0)) \right) dW(t), \tag{A.46}$$

with

$$\sigma(t)^2 = \mathrm{E}\left(\lambda(t)^2 \right), \quad b(t) = \frac{\mathrm{Cov}\left(\lambda(t)^2, S(t) \right)}{2\sigma(t)^2 \mathrm{Var}\left(S(t) \right)}. \tag{A.47}$$

Let us denote by $\overline{\lambda}(t)$ the approximation to $\lambda(t)$ obtained by "freezing" the $S_n(t)$'s to their initial values in (A.44),

$$\overline{\lambda}(t)^2 \triangleq \sum_{n,m=1}^{N} w_n w_m \sigma_n(t) \sigma_m(t) \rho_{n,m}. \tag{A.48}$$

In the same spirit, we approximate $W(t)$ in (A.45) by $\overline{W}(t)$, where

$$d\overline{W}(t) \triangleq \frac{1}{\overline{\lambda}(t)} \sum_{n=1}^{N} w_n \sigma_n(t) \, dW_n(t),$$

is also a Brownian motion. Finally, let us denote

$$\rho_n(t) \triangleq \langle dW_n(t), d\overline{W}(t) \rangle / dt = \frac{1}{\overline{\lambda}(t)} \sum_{m=1}^{N} w_m \sigma_m(t) \rho_{n,m}. \tag{A.49}$$

With all this notation in place, we then can state the following proposition.

Proposition A.4.1. *Let the skew functions φ_n be as stated in (A.41). A displaced log-normal approximation to the dynamics of the basket $S(t)$ in (A.43) is then given by (A.46) where*

$$\sigma(t)^2 \approx \sum_{n=1}^{N} \sum_{m=1}^{N} w_n w_m \sigma_n(t) \sigma_m(t) \rho_{n,m}, \tag{A.50}$$

$$b(t) \approx \frac{\sum_{n=1}^{N} \overline{\lambda}(t) \sigma_n(t) \rho_n(t) \left(\int_0^t \overline{\lambda}(s) \sigma_n(s) \rho_n(s) \, ds \right) w_n b_n(t)}{\overline{\lambda}(t)^2 \int_0^t \overline{\lambda}(s)^2 \, ds}, \tag{A.51}$$

with the ρ_n's defined in (A.49) and $\overline{\lambda}(t)$ defined in (A.48).

Proof. The approximation (A.50) follows from

$$\sigma(t)^2 = \mathrm{E}\left(\lambda(t)^2\right) \approx \mathrm{E}\left(\overline{\lambda}(t)^2\right) = \overline{\lambda}(t)^2$$

and (A.48). To find the covariance term in (A.47), we recall (A.41) and write

$$\mathrm{Cov}\left(\lambda(t)^2, S(t)\right)$$

$$= \sum_{n,m=1}^{N} w_n w_m \sigma_n(t) \sigma_m(t) \rho_{n,m}$$

$$\times \mathrm{Cov}\left((1 + b_n(t)(S_n(t) - S_n(0)))(1 + b_m(t)(S_m(t) - S_m(0))), S(t)\right)$$

$$\approx 2\overline{\lambda}(t) \sum_{n=1}^{N} w_n \sigma_n(t) \rho_n(t) b_n(t) \mathrm{Cov}\left(S_n(t), S(t)\right),$$

where we disregarded the terms $\mathrm{Cov}((S_n(t) - S_n(0))(S_m(t) - S_m(0)), S(t))$ as being of higher order in volatility. Furthermore,

$$\text{Cov}\,(S_n(t), S(t))$$

$$= \sum_{m=1}^{N} w_m \text{Cov}\,(S_n(t), S_m(t)) \tag{A.52}$$

$$= \sum_{m=1}^{N} \int_0^t w_m \rho_{n,m} \sigma_n(s) \sigma_m(s) \text{E}\Big(\varphi_n\,(s, S_n(s))\,\varphi_m\,(s, S_m(s))\Big)\,ds$$

$$\approx \int_0^t \sigma_n(s)\overline{\lambda}(s)\rho_n(s)\,ds.$$

To summarize,

$$\text{Cov}\,\big(\lambda(t)^2, S(t)\big) \approx 2\overline{\lambda}(t) \sum_{n=1}^{N} w_n \sigma_n(t)\rho_n(t)b_n(t) \int_0^t \sigma_n(s)\overline{\lambda}(s)\rho_n(s)\,ds.$$

In the same spirit, from (A.52),

$$\text{Var}\,(S(t)) = \sum_{n=1}^{N} w_n \text{Var}\,(S_n(t), S(t))$$

$$\approx \sum_{n=1}^{N} w_n \int_0^t \sigma_n(s)\overline{\lambda}(s)\rho_n(s)\,ds = \int_0^t \overline{\lambda}(s)^2\,ds,$$

and the proposition follows from (A.47). □

Remark A.4.2. Proposition A.4.1 depends on numerous ad-hoc approximations, the nature of which we did not characterize rigorously. A more detailed analysis can be found in Antonov and Misirpashaev [2009a] where it is shown that our expressions for $\sigma(t)$ and $b(t)$ are leading order terms in the small-volatility limit.

The parameter $b(t)$ in (A.46) represents the slope of the local volatility function for the basket S, and the expression (A.51) relates this slope to a weighted average of the slopes $b_n(t)$ of the individual volatility functions for the basket components. This approximation works best when the skews $b_n(t)$ are of the same sign and the weights w_n are all positive. For skews b_n (or weights w_n) of mixed signs, the approximation is not very accurate, however. This should be intuitively clear, since the difference of two processes with positive skews (say) can easily have a U-shaped smile, which is obviously not well-approximated by a projection of the difference onto a displaced log-normal process. To handle this case, one possible solution is to use a projection on a stochastic volatility process (even though the components of the basket are local volatility processes), as described in Section A.5 below. Alternatively, we can use a projection on a local volatility process with a quadratic local volatility as mentioned in footnote 3. Finally, we can always fall back on the copula-based methods of Chapter 17.

A.5 Basket Options in Stochastic Volatility Models

We continue investigating basket options, but now augment the model (A.40), (A.41) with stochastic volatility. Specifically, we replace (A.40) with

$$dS_n(t) = \sigma_n(t)\varphi_n\left(t, S_n(t)\right)\sqrt{z_n(t)}\,dW_n(t), \quad n = 1, \dots, N,$$

where (A.41) still holds, and where individual stochastic variance processes $z_n(t)$ are defined by

$$dz_n(t) = \theta_n(t)(1 - z_n(t))\,dt + \eta_n(t)\sqrt{z_n(t)}\,dW_{N+n}(t), \quad z_n(0) = 1, \quad (A.53)$$

$n = 1, \dots, N$. Here, $(W_1(t), \dots, W_{2N})$ is a $2N$-dimensional Brownian motion with correlations

$$\langle dW_i, dW_j \rangle = \rho_{i,j}\,dt, \quad i, j = 1, \dots, 2N.$$

To simplify the already cumbersome notation, let us absorb the basket weights w_n into a redefinition of the asset processes: $S_n \leftarrow w_n S_n$. The basket value $S(t)$ is now

$$S(t) = \sum_{n=1}^{N} S_n(t),$$

where we now have

$$dS(t) = \sum_{n=1}^{N} \sigma_n(t)\varphi_n\left(t, S_n(t)\right)\sqrt{z_n(t)}\,dW_n(t) = \lambda(t)\,dW(t),$$

with

$$\lambda(t)^2 = \sum_{n,m=1}^{N} \sigma_n(t)\varphi_n\left(t, S_n(t)\right)\sigma_m(t)\varphi_m\left(t, S_m(t)\right)\rho_{n,m}\sqrt{z_n(t)z_m(t)},$$

$$(A.54)$$

$$dW(t) = \frac{1}{\lambda(t)}\sum_{n=1}^{N} \sigma_n(t)\varphi_n\left(t, S_n(t)\right)\sqrt{z_n(t)}\,dW_n(t).$$

S above is driven by a multi-dimensional stochastic volatility process, and, as we discussed previously in Section A.3.1, projecting S on a displaced log-normal local volatility process is unlikely to lead to accurate option approximations. Following Antonov et al. [2009], we instead investigate projections on a displaced Heston process.

Let us first assume that the skew parameter $b(t)$ of the target approximation is given exogenously (we will discuss its computation later in this section). With $b(t)$ given, we rewrite the dynamics of the process $S(t)$ in a way more suitable for approximations:

$$dS(t) = \sigma(t)\left(1 + b(t)(S(t) - S(0))\right)\sqrt{z(t)}\,dW(t), \tag{A.55}$$

where

$$z(t) \triangleq \frac{\Lambda(t)^2}{\sigma(t)^2}, \quad \Lambda(t)^2 \triangleq \frac{\lambda(t)^2}{\left(1 + b(t)(S(t) - S(0))\right)^2}, \tag{A.56}$$

and

$$\sigma(t)^2 \triangleq \mathrm{E}(\Lambda(t)^2). \tag{A.57}$$

For future reference, we apply Ito's lemma to $z(t)$ and write

$$dz(t) = \nu(t)\,dt + \varepsilon(t)\sqrt{z(t)}\,dZ(t), \tag{A.58}$$

where $Z(t)$ is a Brownian motion such that $\langle dW(t), dZ(t)\rangle = \chi(t)\,dt$, where the exact form of stochastic processes $\nu(t)$, $\varepsilon(t)$, $\chi(t)$ is not important for the moment.

By the multi-dimensional extension of Gyöngy's theorem in Proposition A.1.4, we replicate the exact distribution of the pair $(S(t), z(t))$ for each $t \geq 0$ with $(\widetilde{S}(t), \widetilde{z}(t))$, where

$$d\widetilde{S}(t) = \sigma(t)\left(1 + b(t)(\widetilde{S}(t) - \widetilde{S}(0))\right)\sqrt{\widetilde{z}(t)}\,dW(t),$$

$$d\widetilde{z}(t) = \mathrm{E}\left(\nu(t)\,\Big|\,S(t) = \widetilde{S}(t), z(t) = \widetilde{z}(t)\right)dt$$

$$+ \left(\mathrm{E}\left(\varepsilon(t)^2\,\Big|\,S(t) = \widetilde{S}(t), z(t) = \widetilde{z}(t)\right)\right)^{1/2}dZ(t). \tag{A.59}$$

We cannot calculate the conditional expectations in (A.59) exactly, so we proceed to assume a particular parametric form for the process $(\widetilde{S}(t), \widetilde{z}(t))$,

$$d\widetilde{S}(t) \approx \sigma(t)\left(1 + b(t)(\widetilde{S}(t) - \widetilde{S}(0))\right)\sqrt{\widetilde{z}(t)}\,dW(t), \tag{A.60}$$

$$d\widetilde{z}(t) \approx \theta(t)(1 - \widetilde{z}(t))\,dt + \eta(t)\sqrt{\widetilde{z}(t)}\,dZ(t), \tag{A.61}$$

with $\langle dW(t), dZ(t)\rangle = \rho(t)\,dt$. The unknown parameters $\theta(t)$, $\eta(t)$ and $\rho(t)$ emerge as solutions to the following optimization problems,

$$\mathrm{E}\left(\left(\varepsilon(t)^2 - \eta(t)^2 z(t)\right)^2\right) \to \min,$$

$$\mathrm{E}\left(\left(\varepsilon(t)\chi(t)z(t) - \eta(t)\rho(t)z(t)\right)^2\right) \to \min,$$

$$\mathrm{E}\left(\left(\nu(t) - \theta(t)(1 - z(t))\right)^2\right) \to \min. \tag{A.62}$$

From the usual first-order optimality conditions, we then find

$$\eta(t)^2 = \frac{\mathrm{E}\left(\varepsilon(t)^2 z(t)\right)}{\mathrm{E}\left(z(t)^2\right)},$$

$$\rho(t) = \frac{\mathrm{E}\left(\varepsilon(t)\chi(t)z(t)^2\right)}{\eta(t)\mathrm{E}\left(z(t)^2\right)},$$

$$\theta(t) = \frac{\mathrm{E}\left(\nu(t)(1 - z(t))\right)}{\mathrm{E}\left((1 - z(t))^2\right)}. \tag{A.63}$$

The expectations required in these equations can be approximated from the definition of $z(t)$ in (A.56) and the coefficients $\nu(t)$, $\varepsilon(t)$, $\chi(t)$ in (A.58). The (laborious) calculations are performed in Antonov et al. [2009], and we do not reproduce them here.

We have not yet specified how to set the skew function $b(t)$ in (A.60), originally appearing in (A.55) and (A.56). One idea here is to use the results from the local volatility approximation in Section A.4 and simply use (A.51). Another alternative suggested in Antonov et al. [2009] finds $b(t)$ by minimizing the *defect* $D_\theta(t)$ for the solution (A.63) of the problem (A.62),

$$D_\theta(t) = \mathrm{E}\left(\nu(t)^2\right) - \frac{\left(\mathrm{E}\left(\nu(t)(1 - z(t))\right)\right)^2}{\mathrm{E}\left((1 - z(t))^2\right)}. \qquad (A.64)$$

The defect $D_\theta(t)$ measures the error in the objective function in (A.62) for the solution (A.63), and clearly is a function of $b(t)$. By minimizing the defect, an ODE for $b(t)$ arises that can be solved in closed form. Again, we refer the interested reader to Antonov et al. [2009] for details.

Our final topic in this section is the calculation of the effective volatility $\sigma(t)$ in (A.57). Among all parameters of the model (A.60)–(A.61), $\sigma(t)$ typically is the one that has the biggest impact on the quality of approximations. From (A.54) and (A.56), the standard "freezing" (where $S_n(t) = S_n(0)$ for all $n = 1, \ldots, N$) approximation gives us

$$\sigma(t)^2 = \sum_{n,m=1}^{N} \sigma_n(t)\sigma_m(t)\rho_{n,m}\mathrm{E}\left(\sqrt{z_n(t)z_m(t)}\right). \qquad (A.65)$$

In principle, the same freezing idea could be applied to z_n's, leading us to an approximation that is identical to the one we already derived for the local volatility case, see (A.50). This is, indeed, the leading term of the expansion of $\mathrm{E}(\Lambda(t)^2)$ in small volatilities (see Antonov et al. [2009]), but for typical parameter settings encountered in interest rate modeling the quality of approximations based on this choice for $\sigma(t)$ is rather poor — volatility of variance parameters η_n are often simply too far from being "small" (larger than 1 is typical).

To improve the accuracy of the overall Markovian projection onto a displaced Heston model, we need to find a way to calculate $\sigma(t)$ in (A.65) without the assumption of small variances of z_n's. Clearly, our ability to do so hinges on accurate approximations to $\mathrm{E}(\sqrt{z_n(t)z_m(t)})$ in (A.65), which could be interesting for other purposes as well. We discuss such a "non-perturbative" approximation in Appendix A.A (see Proposition A.A.1) where we also consider a related problem of approximating $\mathrm{E}(\sqrt{z_n(t)})$ (Lemma A.A.2).

With these approximations in place, Antonov et al. [2009] demonstrate excellent performance of Markovian projection onto a displaced Heston for basket and spread options.

To conclude, let us quickly summarize the entire algorithm. First, we calculate a non-perturbative approximation to $\sigma(t)$ given by (A.65), with the square roots obtained by the methods of Appendix A.A. Second, we calculate the optimal skew function $b(t)$ by minimizing the function $D_\theta(t)$ in (A.64). As it turns out, this minimization problem (or, rather, a collection of minimization problems indexed by t) leads to a first-order ODE on $b(t)$ with coefficients that only depend on $\sigma(t)$ (in the small volatility limit), as derived in Antonov et al. [2009]. Finally, having established $\sigma(t)$ and $b(t)$, we can now solve for optimal coefficients for the stochastic variance process (and its correlation with the asset process) by solving (A.62) for each $t > 0$.

A.A Appendix: Approximations for $E(\sqrt{z_n(t)z_m(t)})$ and $E(\sqrt{z_n(t)})$

As discussed in Section A.5, European option approximations in multi-dimensional SV models require calculations of certain expected values of square-root processes, a subject we consider in this section. Let us simplify the notations of (A.53) a little, and consider a two-dimensional square-root process

$$dz_i(t) = \theta_i\left(1 - z_i(t)\right)dt + \eta_i\sqrt{z_i(t)}\,dW_i(t), \quad z_i(0) = 1, \qquad \text{(A.66)}$$

where $i = 1, 2$, and $\langle dW_1(t), dW_2(t)\rangle = \rho\,dt$. We first consider the problem of approximating $E(\sqrt{z_1(t)z_2(t)})$.

Proposition A.A.1. *For a two-dimensional square-root process (A.66), let us define*

$$E(\theta_1, \theta_2, \eta_1, \eta_2, \rho) = \left(1 + \rho\eta_1\eta_2 q_1 q_2 \frac{1 - e^{-(\theta_1 + \theta_2 - \rho\eta_1\eta_2 q_1 q_2)t}}{\theta_1 + \theta_2 - \rho\eta_1\eta_2 q_1 q_2}\right)$$
$$\times E\left(\sqrt{z_1(t)}\right)E\left(\sqrt{z_2(t)}\right), \qquad \text{(A.67)}$$

where q_i, $i = 1, 2$, are obtained by solving

$$1 = \frac{\left(E\left(\sqrt{z_i(t)}\right)\right)^2}{2\theta_i - \eta_i^2 q_i^2}\left(2\theta_i - \eta_i^2 q_i^2 e^{-(2\theta_i - \eta_i^2 q_i^2)t}\right).$$

This function gives an approximation to $E(\sqrt{z_1(t)z_2(t)})$, which is

1. *Exact in the limit $\rho = 0$.*
2. *Exact in the limit $\rho = 1$, $\theta_1 = \theta_2$, $\eta_1 = \eta_2$.*
3. *Has correct leading behavior in the expansion in powers of η_1 and η_2.*

The proposition is proved in Section A.A.1. We note that enforcement of the first two non-perturbative conditions substantially improves the accuracy of the approximation to $\sigma(t)$ in (A.65) when $\eta_1, \eta_2 \geq 1$.

The approximation (A.67) relies on our ability to calculate $E(\sqrt{z_i(t)})$, the calculation that is also of independent interest sometimes. To state the result, we assume that $z(t)$ follows the square-root process (8.4):

$$dz(t) = \theta\left(z_0 - z(t)\right) dt + \eta\sqrt{z(t)}\, dZ(t), \quad z(0) = z_0, \qquad (A.68)$$

and recall the definitions (8.6):

$$d = 4\theta z_0/\eta^2, \quad n(t, T) = \frac{4\theta e^{-\theta(T-t)}}{\eta^2\left(1 - e^{-\theta(T-t)}\right)}.$$

The following result, proven in Section A.A.2, derives the required representation.

Lemma A.A.2. *In the model (A.68) we have*

$$E\left(\sqrt{z(t)}\right) = \left(\frac{2e^{-\theta t}}{n(0,t)}\right)^{1/2} e^{-n(0,t)/2} \sum_{j=0}^{\infty} \frac{\left(n(0,t)/2\right)^j}{j!} \frac{\Gamma(d/2+j+1/2)}{\Gamma(d/2+j)},$$
$$(A.69)$$

where $\Gamma(z)$ is the Gamma function.

Remark A.A.3. The series in (A.69) converges rapidly, so only the first few terms need to be computed. In each term, the ratio of Gamma functions can be evaluated by standard algorithms present in most numerical software packages. We find that the following approximation is sufficient for our purposes:

$$\Gamma(x+1/2)/\Gamma(x) = \begin{cases} x\left(0.5619x^2 - 1.3353x + 1.6651\right), & x \in [0, 0.9], \\ \sqrt{x}\left(1 - \frac{1}{8x} + \frac{1}{128x^2}\right), & x > 0.9, \end{cases}$$

where the first line is obtained by fitting $\Gamma(x+1/2)/\Gamma(x)$ with a third-degree polynomial over the interval $[0, 1]$, and the second line is the truncation of the series expansion of $\Gamma(x+1/2)/\Gamma(x)$ valid for $x > 1$, see Cevher et al. [2007]. The cut-off point of 0.9 is chosen to make the function continuous and (nearly) C^1.

A.A.1 Proof of Proposition A.A.1

A.A.1.1 Step 1. Reduction to Covariance

Taking expected values of both sides of (A.66) we see that $E(z_i(t)) = 1$, $i = 1, 2$. Furthermore, we note that

$$d\left(z_1(t)z_2(t)\right) = z_1(t)\left(\theta_2\left(1 - z_2(t)\right) dt + \eta_2\sqrt{z_2(t)}\, dW_2(t)\right)$$
$$+ z_2(t)\left(\theta_1\left(1 - z_1(t)\right) dt + \eta_1\sqrt{z_1(t)}\, dW_1(t)\right)$$
$$+ \eta_1\eta_2\rho\sqrt{z_1(t)z_2(t)}\, dt,$$

so that

$$\frac{d}{dt}E(z_1(t)z_2(t)) = (\theta_1 + \theta_2)\left(1 - E(z_1(t)z_2(t))\right) + \eta_1\eta_2\rho E\left(\sqrt{z_1(t)z_2(t)}\right).$$
(A.70)

Hence, calculating $E(\sqrt{z_1(t)z_2(t)})$ is equivalent to calculating $E(z_1(t)z_2(t))$.

A.A.1.2 Step 2. Linear Approximation to Volatility

To proceed, we consider a linear approximation to (A.66) obtained by substituting

$$\sqrt{z_i} \rightarrow p_i + q_i z_i,$$
(A.71)

with the coefficients to be found. Define by $c(t) = c(t; p_1, q_1, p_2, q_2)$ the value of $E(\tilde{z}_1(t)\tilde{z}_2(t))$ in the approximate model,

$$d\tilde{z}_i(t) = \theta_i(1 - \tilde{z}_i(t))\,dt + \eta_i(p_i + q_i\tilde{z}_i(t))\,dW_i(t), \quad \tilde{z}_i(0) = 1.$$
(A.72)

Simple calculations yield an ODE on $c(t)$,

$$\frac{d}{dt}c(t) = -(\theta_1 + \theta_2)\left(c(t) - 1\right)$$
$$+ \rho\eta_1\eta_2\,(p_1 + q_1)\,(p_2 + q_2) + \rho\eta_1\eta_2 q_1 q_2\,(c(t) - 1),$$

which can be solved,

$$c(t) = 1 + \rho\eta_1\eta_2\,(p_1 + q_1)\,(p_2 + q_2)\,\frac{1 - e^{-(\theta_1 + \theta_2 - \rho\eta_1\eta_2 q_1 q_2)t}}{\theta_1 + \theta_2 - \rho\eta_1\eta_2 q_1 q_2}.$$
(A.73)

Then, from (A.70) and the equality (A.73) that we use to approximate $E((z_1(t)z_2(t))$ in the original model (A.66), we get

$$E\left(\sqrt{z_1(t)z_2(t)}\right)$$
(A.74)
$$\approx \frac{(p_1 + q_1)\,(p_2 + q_2)}{\theta_1 + \theta_2 - \rho\eta_1\eta_2 q_1 q_2}$$
$$\times \left(\theta_1 + \theta_2 - \rho\eta_1\eta_2 q_1 q_2 e^{-(\theta_1 + \theta_2 - \rho\eta_1\eta_2 q_1 q_2)t}\right)$$
$$= (p_1 + q_1)\,(p_2 + q_2)\left(1 + \rho\eta_1\eta_2 q_1 q_2\frac{1 - e^{-(\theta_1 + \theta_2 - \rho\eta_1\eta_2 q_1 q_2)t}}{\theta_1 + \theta_2 - \rho\eta_1\eta_2 q_1 q_2}\right).$$

A.A.1.3 Step 3. Coefficients of the Linear Approximation

Applying (A.74) to $z_2(t) = z_1(t)$ we obtain

$$1 = E((z_1(t)) = \frac{(p_1 + q_1)^2}{2\theta_1 - \eta_1^2 q_1^2}\left(2\theta_1 - \eta_1^2 q_1^2 e^{-(2\theta_1 - \eta_1^2 q_1^2)t}\right),$$
(A.75)

which gives one equation on p_1, q_1. The other one is obtained by using as $z_2(t)$ an independent copy of $z_1(t)$,

$$\left(E\left(\sqrt{z_1(t)} \right) \right)^2 = (p_1 + q_1)^2 . \tag{A.76}$$

The coefficients p_1 and q_1 are determined as a solution to the system (A.75), (A.76), provided $E(\sqrt{z_1(t)})$ is known. Similar system holds for p_2, q_2. This completes the derivation of the approximating function (A.67).

A.A.1.4 Step 4. Order of Approximation

The first two features of the approximation listed in the Proposition A.A.1 are valid by construction. The validity of the last feature is obvious because the usage of a linear approximation (A.71) is consistent with the first order of perturbative expansion in volatilities. This is also easy to verify directly. Indeed, a straightforward perturbative calculation gives[5]

$$E\left(\sqrt{z_i(t)} \right) = 1 - \frac{\eta_i^2 \left(1 - e^{-2\theta_i t} \right)}{16\theta_i} + O(\eta_i^3),$$

$$\frac{E\left(\sqrt{z_1(t)z_2(t)} \right)}{E\left(\sqrt{z_1(t)} \right) E\left(\sqrt{z_2(t)} \right)} = 1 + \frac{\rho\eta_1\eta_2 \left(1 - e^{-(\theta_1+\theta_2)t} \right)}{4(\theta_1 + \theta_2)} + O((\max(\eta_1, \eta_2))^3),$$

which is in agreement with the leading order of expansion of (A.67) in η_1 and η_2. Note that we used the zero order expansion for coefficients $q_i = 1/2 + O(\eta_i)$ to prove the agreement.

A.A.2 Proof of Lemma A.A.2

By Proposition 8.3.2, $z(t)$ is distributed as $e^{-\theta t}/n(0,t)$ times a non-central chi-square distributed random variable ξ with $\nu = d$ degrees of freedom and non-centrality parameter $\gamma = n(0,t)$. Thus we obtain that

$$E\left(\sqrt{z(t)} \right) = \left(\frac{e^{-\theta t}}{n(0,t)} \right)^{1/2} E\left(\sqrt{\xi} \right),$$

where ξ has the density (see (8.5))

$$\frac{\partial}{\partial z}\Upsilon(z;\nu,\gamma) = e^{-\gamma/2} \sum_{j=0}^{\infty} \frac{(\gamma/2)^j}{j!2^{\nu/2+j}\Gamma\left(\nu/2 + j \right)} z^{\nu/2+j-1} e^{-z/2} .$$

---

[5]Due to the reflecting symmetry of the Brownian motion, an arbitrary average of the variables $z_1(t)$ and $z_2(t)$ is an even function of η_1 and η_2. Thus, the order of the approximations below is effectively higher.

Then

$$\mathrm{E}\left(\sqrt{\xi}\right) = e^{-\gamma/2} \sum_{j=0}^{\infty} \frac{(\gamma/2)^j}{j! 2^{\nu/2+j} \Gamma\left(\nu/2+j\right)} \int_0^{\infty} z^{1/2+\nu/2+j-1} e^{-z/2} \, dz$$

$$\tag{A.77}$$

$$= e^{-\gamma/2} \sum_{j=0}^{\infty} \frac{(\gamma/2)^j}{j! 2^{\nu/2+j} \Gamma\left(\nu/2+j\right)} 2^{1/2+\nu/2+j} \Gamma\left(\nu/2+j+1/2\right)$$

$$= \sqrt{2} e^{-\gamma/2} \sum_{j=0}^{\infty} \frac{(\gamma/2)^j}{j!} \frac{\Gamma\left(\nu/2+j+1/2\right)}{\Gamma\left(\nu/2+j\right)}.$$

References

M. Abramowitz and I. A. Stegun, editors. *Handbook of Mathematical Functions.* Dover, 1965.

P. J. Acklam. An algorithm for computing the inverse Normal cumulative distribution function. Technical report, Unaffiliated, 2003. URL `http:// home.online.no/ p̂jacklam/ notes/ invnorm`.

K. J. Adams and D. R. van Deventer. Fitting yield curves and forward rate curves with maximum smoothness. *Journal of Fixed Income*, 4:52–62, 1994.

D.-H. Ahn, R. F. Dittmar, and A. R. Gallant. Quadratic term structure models: Theory and evidence. *Review of Financial Studies*, 15(1):243–288, 2002.

J. Ahrens and U. Dieter. Computer methods for sampling from the gamma, beta, Poisson, and binomial distributions. *Computing*, 12(3):223–246, 1974.

Y. Aït-Sahalia. Testing continuous-time models of the spot interest rate. *Review of Financial Studies*, 9(2):385–426, 1996.

F. Åkesson and J. P. Lehoczky. Discrete eigenfunction-expansion of multi-dimensional Brownian motion and the Ornstein-Uhlenbeck process. Carnegie Mellon University working paper, 1998.

H. Albrecher, P. Mayer, W. Schoutens, and J. Tistaert. The little Heston trap. *Wilmott*, 1(1):83–92, 2007.

L. B. Andersen. Simulation and calibration of the HJM model. General Re Financial Products working paper, 1995.

L. B. Andersen. *Five Essays on Contingent Claims Pricing.* PhD thesis, Aarhus Business School, 1996.

L. B. Andersen. Monte Carlo simulation of options on joint minima and maxima. In B. Dupire, editor, *Monte Carlo: Methodologies and Applications for Pricing and Risk Management.* Risk Books, 1998.

L. B. Andersen. A simple approach to the pricing of Bermudan swaptions in the multi-factor Libor market model. *Journal of Computational Finance*, 3(2): 5–32, 2000a.

L. B. Andersen. Separable Libor market models. General Re Financial Products working paper, 2000b.

L. B. Andersen. Yield curve construction with tension splines. *Review of Derivatives Research*, 10(3):227–267, 2005.

L. B. Andersen. Simple and efficient simulation of the Heston stochastic volatility model. *Journal of Computational Finance*, 11:1–42, 2008.

L. B. Andersen. Option pricing with quadratic volatility: A revisit. *Finance and Stochastics*, 2010. Forthcoming.

L. B. Andersen and J. Andreasen. Jump-diffusion processes: Volatility smile fitting and numerical methods for option pricing. *Review of Derivatives Research*, 4 (3):231–262, 2000a.

L. B. Andersen and J. Andreasen. Volatility skews and extensions of the Libor market model. *Applied Mathematical Finance*, 7(1):1–32, 2000b.

L. B. Andersen and J. Andreasen. Factor dependence of Bermudan swaption prices: Fact or fiction? *Journal of Financial Economics*, 62(1):3–37, 2001.

L. B. Andersen and J. Andreasen. Volatile volatilities. *Risk*, 15(12):163–168, 2002.

L. B. Andersen and P. Boyle. Monte Carlo methods for interest rate derivatives. In N. Jegadeesh and B. Tuckman, editors, *Advanced Fixed-Income Valuation Tools*. Wiley, 2000.

L. B. Andersen and M. Broadie. A primal-dual simulation algorithm for pricing of multi-dimensional American options. *Management Science*, 50(9):1222–1234, 2004.

L. B. Andersen and R. Brotherton-Ratcliffe. Exact exotics. *Risk*, 9(10):85–89, 1996.

L. B. Andersen and R. Brotherton-Ratcliffe. The equity option volatility smile: An implicit finite difference approach. *Journal of Computational Finance*, 1 (2):5–38, 1998.

L. B. Andersen and R. Brotherton-Ratcliffe. Extended Libor market models with stochastic volatility. *Journal of Computational Finance*, 9(1):1–40, 2005.

L. B. Andersen and D. Buffum. Implementation and calibration of convertible bond models. *Journal of Computational Finance*, 7(2):1–34, 2003.

L. B. Andersen and N. Hutchings. Parameter averaging of quadratic SDEs with stochastic volatility. *Journal of Computational Finance*, 2010. Forthcoming.

L. B. Andersen and V. V. Piterbarg. Moment explosions in stochastic volatility models. *Finance and Stochastics*, 11(1):29–50, 2007.

L. B. Andersen, J. Andreasen, and D. Eliezer. Static replication of barrier options: Some general results. *Journal of Computational Finance*, 5(4):1–25, 2002.

L. B. Andersen, J. Sidenius, and S. Basu. All your hedges in one basket. *Risk*, 11: 67–72, 2003.

T. G. Andersen and J. Lund. Estimating continuous-time stochastic volatility models of the short-term interest rate. *Journal of Econometrics*, 77(2):343–377, 1997.

S. Andradóttir. A stochastic approximation algorithm with varying bounds. *Operations Research*, 43(6):1037–1048, 1995.

S. Andradóttir. A scaled stochastic approximation algorithm. *Management Science*, 42(4):475–498, 1996.

J. Andreasen. The pricing of discretely sampled Asian and lookback options: A change of numeraire approach. *Journal of Computational Finance*, 1(1):5–30, 1998.

J. Andreasen. Turbo-charging the Cheyette model. Bank of America working paper, 2001.

J. Andreasen. Pricing simple exotics under stochastic volatility. Bank of America working paper, 2002.

J. Andreasen. Back to the future. *Risk*, 18(9):104–109, 2005.

J. Andreasen, B. Jensen, and R. Poulsen. Eight valuation methods in financial mathematics: The Black-Scholes formula as an example. *Mathematical Scientist*, 23:18–40, 1998.

A. Antonov and M. Arneguy. Analytical formulas for pricing CMS products in the LIBOR market model with the stochastic volatility. *SSRN eLibrary*, 2009.

A. Antonov and T. Misirpashaev. Markovian projection onto a displaced diffusion: Generic formulas with applications. *International Journal of Theoretical and Applied Finance*, 12(4):507–522, 2009a.

A. Antonov and T. Misirpashaev. Projection on a quadratic model by asymptotic expansion with an application to LMM swaption. *SSRN eLibrary*, 2009b.

A. Antonov, T. Misirpashaev, and V. Piterbarg. Markovian projection onto a Heston model. *Journal of Computational Finance*, 13(1):23–47, 2009.

L. Arnold. *Stochastic Differential Equations: Theory and Practice*. Wiley, 1974.

P. Artzner, F. Delbaen, J.-M. Eber, and D. Heath. Coherent measures of risk. *Mathematical Finance*, 9(3):203–228, 1999.

S. Assefa. Calibration and pricing in a multi-factor quadratic Gaussian model. Research Paper Series 197, Quantitative Finance Research Centre, University of Technology, Sydney, 2007. URL http://ideas.repec.org/p/uts/rpaper/197.html.

M. Attari. Option pricing using Fourier transforms: A numerically efficient simplification. *SSRN eLibrary*, 2004.

P. Austing. Valuing multi-asset options on foreign exchange: A joint density repricing the cross smile. Submitted to Risk, 2010.

M. Avellaneda, R. Buff, C. Friedman, N. Grandechamp, L. Kruk, and J. Newman. Weighted Monte Carlo: A new technique for calibrating asset-pricing models. *International Journal of Theoretical and Applied Finance*, 4(1):91 – 119, 2001.

M. Avellaneda, D. Boyer-Olson, J. Busca, and P. Friz. Reconstructing volatility. *Risk*, 15(10), 2002.

O. Axelsson and V. Barker. *Finite Element Solution of Boundary Value Problems: Theory and Computation*, volume 35 of *Classics in applied mathematics*. SIAM, 1991.

S. Babbs. *The Term Structure of Interest Rates: Stochastic Processes and Contingent Claims*. PhD thesis, University of London, 1990.

Y. Balasanov. A gentle introduction to the BEEMIR model. NationsBanc working paper, 1996.

D. Bang. Numerical computation of Fourier transforms in Heston model. Bank of America working paper, 2009.

G. Barles and H. M. Soner. Option pricing with transaction costs and a nonlinear Black-Scholes equation. *Finance and Stochastics*, 2(4):369–397, 1998.

J. D. Beasley and S. G. Springer. Algorithm AS 111: The percentage points of the Normal distribution. *Applied Statistics*, 26(1):118–121, 1977.

D. Belomestny, C. Bender, and J. Schoenmakers. True upper bounds for Bermudan products via non-nested Monte Carlo. *Mathematical Finance*, 19(1):53–71, 2007.

C. Bender, A. Kolodko, and J. Schoenmakers. Iterating cancellable snowballs and related exotics. *Risk*, 9:126–130, 2006.

H. Berestycki, J. Busca, and I. Florent. Computing the implied volatility in stochastic volatility models. *Communications on Pure and Applied Mathematics*, 57 (10):1352–1373, 2004.

L. Bergomi. Smile dynamics III. *Risk*, 5:90–96, 2009.

M. Berrahoui. Pricing CMS spread options and digital CMS spread options with smile. In *The Best of Wilmott 2*. Wiley, 2005.

C. Beveridge and M. S. Joshi. Juggling snowballs. *Risk*, 12:100–104, 2008.

C. Beveridge and M. S. Joshi. Practical policy iteration: Generic methods for obtaining rapid and tight bounds for Bermudan exotic derivatives using Monte Carlo simulation. *SSRN eLibrary*, 2009.

C. Beveridge, N. Denson, and M. S. Joshi. Comparing discretization of the Libor market model in the spot measure. *SSRN eLibrary*, 2008.

P. Billingsley. *Probability and Measure*. Wiley Series in Probability and Mathematical Statistics. Wiley, 3d edition, 1995.

T. Björk. A geometric view of interest rate theory. In E. Jouini, J. Cvitanic, and M. Musiela, editors, *Option Pricing, Interest Rates and Risk Management*, pages 241–277. Cambridge University Press, 2001.

F. Black. The pricing of commodity contracts. *Journal of Financial Economics*, 3: 167–179, 1976.

F. Black. Interest rates as options. *Journal of Finance*, 50(5):1371–1376, 1995.

F. Black and P. Karasinski. Bond and option pricing when short rates are lognormal. *Financial Analysis Journal*, 47(4):52–59, 1991.

F. Black and M. Scholes. The pricing of options and corporate liabilities. *Journal of Political Economy*, 81(3):637–654, 1973.

F. Black, E. Derman, and W. Toy. A one-factor model of interest rates and its application to Treasury bond options. *Financial Analysts Journal*, 46(1):33–39, 1990.

R. R. Bliss and D. C. Smith. The stability of interest rate processes. Federal Reserve Bank of Atlanta working paper, 1997.

A. N. Borodin and P. Salminen. *Handbook of Brownian Motion — Facts and Formulae*. Probability and Its Applications. Birkhäuser, Basel, 1996.

N. Boyarchenko and S. Z. Levendorski. On errors and bias of Fourier transform methods in quadratic term structure models. *International Journal of Theoretical and Applied Finance*, 10(2):273–306, 2007.

P. Boyle, M. Broadie, and P. Glasserman. Monte Carlo methods for security pricing. *Journal of Economic Dynamics and Control*, 21(8-9):1267–1321, 1997.

A. Brace, D. Gatarek, and M. Musiela. The market model of interest rate dynamics. *Mathematical Finance*, 7:127–154, 1997.

D. T. Breeden and R. H. Litzenberger. Price of state-contingent claims implicit in option prices. *Journal of Business*, 51(4):621–651, 1978.

K. Brekke and B. Øksendal. The high contact principle as a sufficiency condition for optimal stopping. In D. Lund and B. Øksendal, editors, *Stochastic Models and Option Values: Applications to Resources, Environment and Investment Problems*. North-Holland, 1991.

M. J. Brennan and E. S. Schwartz. Analyzing convertible bonds. *Journal of Financial and Quantitative Analysis*, 15(4):907–929, 1980.

D. Brigo and F. Mercurio. *Interest-Rate Models – Theory and Practice*. Springer-Verlag, 2001.

D. Brigo and M. Morini. Efficient analytical cascade calibration of the Libor market model with endogenous interpolation. *Journal of Derivatives*, 14(1): 40–60, 2006.

M. Broadie and M. Cao. Improved lower and upper bound algorithms for pricing American options by simulation. *Quantitative Finance*, 8(8):845–861, 2008.

M. Broadie and J. Detemple. American capped call options on dividend-paying assets. *Review of Financial Studies*, 8(1):161–91, 1995.

M. Broadie and J. Detemple. Valuation of American options on multiple assets. *Review of Financial Studies*, 7(3):241–286, 1997.

M. Broadie and P. Glasserman. Estimating security price derivatives using simulation. *Management Science*, 42(2):269–285, 1996.

M. Broadie and P. Glasserman. Pricing American style securities using simulation. *Journal of Economic Dynamics and Control*, 21(8-9):1323–1352, 1997.

M. Broadie and P. Glasserman. A stochastic mesh method for pricing high-dimensional American options. *Journal of Computational Finance*, 7(4):35–72, 2004.

M. Broadie and O. Kaya. Exact simulation of stochastic volatility and other affine jump diffusion processes. *Operations Research*, 54(2):217–231, 2006.

M. Broadie, P. Glasserman, and S. Kou. A continuity correction for discrete barrier options. *Mathematical Finance*, 7(4):325–349, 1997.

M. Broadie, P. Glasserman, and S. Kou. Connecting discrete and continuous path-dependent options. *Finance and Stochastics*, 3:55–82, 1999.

M. Broadie, D. M. Cicek, and A. Zeevi. General bounds and finite-time improvement for stochastic approximation algorithms. Columbia University working paper, 2009.

R. Brotherton-Ratcliffe. Monte Carlo motoring. *Risk*, 7(12):53–57, 1994.

R. Brotherton-Ratcliffe. The BGM model for path-dependent swaps. General Re Financial Products working paper, 1997.

G. Brunick. *A Weak Existence Result with Application to the Financial Engineer's Calibration Problem*. PhD thesis, Carnegie Mellon University, 2008.

G. Burghardt. *The Treasury Bond Basis*. McGraw-Hill, New York, 2005.

J. Campos, N. R. Ericsson, and D. F. Hendry, editors. *General-to-Specific Modelling*. International Library of Critical Writings in Econometrics. Edward Elgar Publishing Ltd, 2005.

L. Capriotti. Least squares importance sampling for Libor market models. *Wilmott*, 7:100–107, 2007.

P. Carr. Deriving derivatives of derivative securities. *Journal of Computational Finance*, 4(2):5–29, 2000.

P. Carr and R. Jarrow. The stop-loss start-gain paradox: A new decomposition into intrinsic and time value. *The Review of Financial Studies*, 3(3):469–492, 1990.

P. Carr and R. Lee. Put-call symmetry: Extensions and applications. *Mathematical Finance*, 19(4):523–560, 2009a.

P. Carr and R. Lee. Volatility derivatives. *Annual Review Financial Economics*, 1: 1–21, 2009b.

P. Carr and D. Madan. Option valuation using the fast Fourier transform. *Journal of Computational Finance*, 2:61–73, 1999.

P. Carr and L. Wu. What type of process underlies options? A simple robust test. *Journal of Finance*, 58:2581–2610, 2003.

P. Carr, R. Jarrow, and R. Myneni. Alternative characterizations of American puts. *Mathematical Finance*, 2:87–106, 1992.

J. Carrière. Valuation of early-exercise price of options using simulations and nonparametric regression. *Insurance: Mathematics and Economics*, 19(12): 19–30, 1996.

A. P. Carverhill. When is the short rate Markovian? *Mathematical Finance*, 4(4): 305–312, 1994.

A. P. Carverhill and L. J. Clewlow. American options: Theory and numerical analysis. In S. Hodges, editor, *Options: Recent Advances in Theory and Practice*. Manchester University Press, 1990.

J. Casassus, P. Collin-Dufresne, and B. Goldstein. Unspanned stochastic volatility and fixed income derivatives pricing. *Journal of Banking & Finance*, 29(11): 2723–2749, 2005.

E. Catmull and R. Rom. A class of local interpolating splines. In R. Barnhill and R. Reisenfeld, editors, *Computer Aided Geometric Design*, pages 317–326. Academic Press, 1974.

V. Cevher, R. Chellappa, and J. H. McClellan. Gaussian approximations for energy-based detection and localization in sensor networks. In *IEEE Statistical Signal Processing Workshop*, Madison, 2007. URL http:// www.umiacs.umd.edu/ users/ volkan/ SSP07.pdf.

K. C. Chan, G. A. Karolyi, F. A. Longstaff, and A. Sanders. An empirical comparison of alternative models of the short-term interest rate. *Journal of Finance*, 47:1209–1227, 1992.

D. Chapman and N. Pearson. Is the short rate drift actually nonlinear? *Journal of Finance*, 55:355–388, 2000.

L. Chen, D. Filipović, and H. V. Poor. Quadratic term structure models for risk-free and defaultible rates. *Mathematical Finance*, 14(4):515–536, 2004.

N. Chen and P. Glasserman. Additive and multiplicative duals for American option pricing. *Finance and Stochastics*, 11:153–179, 2007a.

N. Chen and P. Glasserman. Malliavin greeks without Malliavin calculus. *Stochastic Processes and their Applications*, 117(11):1689–1723, 2007b.

R.-R. Chen and L. Scott. Pricing interest rate options in a two-factor Cox-Ingersoll-Ross model of the term structure. *Review of Financial Studies*, 5:613–636, 1992.

R.-R. Chen and L. Scott. Stochastic volatility and jumps in interest rates: An empirical analysis. Rutgers University working paper, 2001.

Z. Chen and P. A. Forsyth. A semi-Lagrangian approach for natural gas storage valuation and optimal operation. *SIAM Journal of Scientific Computing*, 30 (1):339–368, 2007.

R. C. H. Cheng and G. M. Feast. Gamma variate generators with increased shape parameter range. *Communications of the ACM*, 23(7):389–395, 1980.

T. H. F. Cheuk and T. C. F. Vorst. Complex barrier options. *Journal of Derivatives*, 4:8–22, 1996.

O. Cheyette. Markov representation of the Heath-Jarrow-Morton model. BARRA working paper, 1991.

C. Chiarella, A. Kucera, and A. Ziogas. A survey of the integral representation of American option prices. University of Technology Sydney working paper, 2004.

B. J. Christensen, R. Poulsen, and M. Sørensen. Optimal inference in diffusion models of the short rate of interest. *Centre for Analytical Finance Aarhus* working paper, 2001.

L. Clewlow and A. P. Carverhill. On the simulation of contingent claims. *Journal of Derivatives*, 999:66–74, 1994.

L. Clewlow and C. Strickland. A note on parameter estimation in the two-factor Longstaff and Schwartz interest rate model. *Journal of Fixed Income*, 3:95–100, 1994.

P. Collin-Dufresne and B. Goldstein. Do credit spreads reflect stationary leverage ratios. *Journal of Finance*, 56:2177–2208, 2001.

P. Collin-Dufresne and R. Goldstein. Pricing swaptions within an affine framework. *Journal of Derivatives*, 10:9–26, 2002a.

P. Collin-Dufresne and R. S. Goldstein. Do bonds span the fixed income markets? Theory and evidence for unspanned stochastic volatility. *The Journal of Finance*, 57(4):1685–1730, 2002b.

G. Courtadon. The pricing of options on default-free bonds. *Journal of Financial and Quantitative Analysis*, 17:75–100, 1982.

T. M. Cover and J. A. Thomas. *Elements of Information Theory*. Wiley, 2006.

J. C. Cox. The constant elasticity of variance option pricing model. *Journal of Portfolio Management*, 22:15–17, 1996.

J. C. Cox, J. E. Ingersoll, and S. A. Ross. The relationship between forward prices and futures prices. *Journal of Financial Economics*, 9:321–346, 1981.

J. C. Cox, J. E. Ingersoll, and S. A. Ross. A theory of the term structure of interest rates. *Econometrica*, 53(2):385–407, 1985.

J. J. D. Craig and A. D. Sneyd. An alternating-direction implicit scheme for parabolic equations with mixed derivatives. *Computers and Mathematics with Applications*, 16(4):341–350, 1988.

P. Craven and G. Wahba. Smoothing noisy data with spline functions: Estimating the correct degree of smoothing by the method of generalized crossvalidation. *Numerische Matematik*, 31:377–403, 1979.

D. Davydov and V. Linetsky. Pricing and hedging path-dependent options under the CEV process. *Management Science*, 47(7):949–965, 2001.

F. Delbaen and W. Schachermayer. A general version of the fundamental theorem of asset pricing. *Mathematische Annalen*, 300:463–520, 1994.

N. Denson and M. S. Joshi. Vega control. *SSRN eLibrary*, 2009.

E. Derman and M. Kamal. When you cannot hedge continuously: The corrections of Black-Scholes. *Risk*, 12:82–85, 1999.

E. Derman and I. Kani. Riding on a smile. *Risk*, 2:32–39, 1994.

J. Dhaene and M. J. Goovaerts. Dependency of risks and stop-loss order. *Astin Bulletin*, 26(2):201–212, 1996.

Y. d'Halluin, P. A. Forsyth, and G. Labahn. A numerical PDE approach for pricing callable bonds. *Applied Mathematical Finance*, 8:49–77, 2001.

Y. d'Halluin, P. A. Forsyth, and K. R. Vetzal. Robust numerical methods for contingent claims under jump diffusion processes. *IMA Journal on Numerical Analysis*, 25:87–112, 2005.

C. G. Ding. Algorithm AS 275: Computing the non-central chi-squared distribution function. *Applied Statistics*, 41:478–482, 1992.

J. Dollard and C. Friedman. *Product Integration with Applications to Differential Equations*. Addison-Wesley, 1979.

L. U. Dothan. On the term structure of interest rates. *Journal of Financial Economics*, 6(1):59–69, 1978.

D. Duffie. *Dynamic Asset Pricing Theory*. Princeton University Press, 2001.

D. Duffie and P. Glynn. Efficient Monte Carlo simulation of security prices. *The Annals of Applied Probability*, 5(4):897–905, 1995.

D. Duffie and M. Huang. Swap rates and credit quality. *Journal of Finance*, 51: 921–949, 1996.

D. Duffie and R. Kan. A yield-factor model of interest rates. *Mathematical Finance*, 6(4):379–406, 1996.

D. Duffie, J. Pan, and K. Singleton. Transform analysis and asset pricing for affine jump-diffusions. *Econometrica*, 68:1343–1376, 2000.

D. Duffie, D. Filipovic, and W. Schachermayer. Affne processes and applications in Finance. *The Annals of Applied Probability*, 13:984–1053, 2003.

D. J. Duffy. Robust and accurate finite difference methods in option pricing: One factor models. DataSim Financial working paper, 2000.

D. Dufresne. The integrated square-root process. University of Montreal working paper, 2001.

B. Dupire. Pricing with a smile. *Risk*, 7(1), 1994.

B. Dupire. A unified theory of volatility. Banque Paribas working paper, 1997.

B. Dupire. Modelling volatility skews. ICBI Conference, Paris, 2006.

P. H. Dybvig. Bond and bond option pricing based on the current term structure. In M. A. H. Dempster and S. Pliska, editors, *Mathematics of Derivative Securities*. Cambridge University Press, 1997.

D. Egloff, M. Kohler, and N. Todorovic. A dynamic look-ahead Monte Carlo algorithm for pricing Bermudan options. *The Annals of Applied Probability*, 17(4):1138–1171, 2007.

E. Ekström and J. Tysk. Existence and uniqueness theory for the term structure equation. Uppsala University working paper, 2008.

S. N. Ethier and T. G. Kurtz. *Markov Processes: Characterization and Convergence*. Wiley, 1986.

I. Evers and F. Jamshidian. Replication of flexi-swaps. *Risk*, 18(3):67–70, 2005.

C.-O. Ewald, R. Poulsen, and K. Schenk-Hopp. Stochastic volatility: Risk minimization and model risk. University of Copenhagen working paper, 2007.

F. J. Fabozzi. *The Handbook of Mortgage-backed Securities*. Wiley, 1985.

F. J. Fabozzi. *Fixed Income Securities*. Wiley, 2nd edition, 2001.

F. J. Fabozzi and T. D. Fabozzi. *Bond Markets, Analysis and Strategies*. Prentice Hall, 1989.

F. J. Fabozzi and F. Modigliani. *Capital Markets: Institutions and Instruments*. Wiley, 1996.

D. Filipovic and J. Teichmann. On the geometry of the term structure of interest rates. In *Proceedings of The Royal Society of London. Series A. Mathematical, Physical and Engineering Sciences*, volume 460, pages 129–167, 2004.

G. S. Fishman. Sampling from the gamma distribution on a computer. *Communications of the ACM*, 19(7):407–409, 1976.

B. Flesaker. Arbitrage free pricing of interest rate futures and forward contracts. *The Journal of Futures Markets*, 13(1):77–91, 1993.

H. Föllmer and M. Schweizer. Hedging of contingent claims under incomplete information. In M. Davis and R. Elliott, editors, *Applied Stochastic Analysis*, pages 389–414. Gordon and Breach, 1990.

H. Fong and O. Vasicek. Fixed income volatility management. *The Journal of Portfolio Management*, 17:41–46, 1991.

P. Forsyth and K. Vetzal. Quadratic convergence of a penalty method for valuing American options. *SIAM Journal of Scientific Computing*, 23:2096–2123, 2002.

E. Fournié, J.-M. Lasry, J. Lebuchoux, P.-L. Lions, and N. Touzi. Application of Malliavin calculus to Monte Carlo methods in finance. *Finance and Stochastics*, 3:391–412, 1999.

S. Frankau, D. Spinellis, N. Nassuphis, and C. Burgard. Going functional on exotic trades. *J. Functional Programming*, 19(1):27–45, 2009.

C. P. Fries. Localized proxy simulation schemes for generic and robust Monte-Carlo Greeks. *SSRN eLibrary*, 2007.

C. P. Fries and M. S. Joshi. Partial proxy simulation schemes for generic and robust Monte-Carlo Greeks. *Journal of Computational Finance*, 11(3):79–106, 2008a.

C. P. Fries and M. S. Joshi. Conditional analytic Monte Carlo pricing scheme for auto-callable products. *SSRN eLibrary*, 2008b.

C. P. Fries and J. Kampen. Proxy simulation schemes for generic robust Monte-Carlo sensitivities, process oriented importance sampling and high accuracy drift approximation. *Journal of Computational Finance*, 10(2):97–128, 2006.

M. Fujii, Y. Shimada, and A. Takahashi. A note on construction of multiple swap curves with and without collateral. *SSRN eLibrary*, 2010.

S. Galluccio and C. Hunter. The co-initial swap market model. *SSRN eLibrary*, 2003.

S. Galluccio, Z. Huang, O. Scaillet, and J.-M. Ly. Theory and calibration of swap market models. *SSRN eLibrary*, 2005.

D. Gatarek. Constant maturity swaps, forward measure and LIBOR market model. *SSRN eLibrary*, 2003.

R. Gâteaux. Sur les fonctionnelles continues et les fonctionnelles analytiques. *Comptes rendus de l'academie des sciences (Paris)*, 157:325–327, 1913.

J. Gatheral. Lecture 2: Fitting the volatility skew. Case Studies in Financial Modelling course notes, Courant Institute, 2001.

J. Gatheral. A parsimonious arbitrage-free implied volatility parameterization with application to the valuation of volatility derivatives. ICBI Conference, Madrid, 2004.

J. Gatheral. *The Volatility Surface: A Practitioner's Guide*. Wiley, 2006.

J. Gatheral and A. Jacquier. Convergence of Heston to SVI. *SSRN eLibrary*, 2010.

J. Gatheral, E. P. Hsu, P. M. Laurence, C. Ouyang, and T.-H. Wang. Asymptotics of implied volatility in local volatility models. *SSRN eLibrary*, 2009.

H. Geman, N. Karou, and J. Rochet. Changes of numeraire, changes of probability measure and option pricing. *Journal of Applied Probability*, 32(2):443–458, 1995.

M. R. Gibbons and K. Ramaswamy. A test of the Cox, Ingersoll, and Ross model of the term structure. *Review of Financial Studies*, 6:619–658, 1993.

M. Giles and R. Carter. Convergence analysis of Rannacher time marching. *Journal of Computational Finance*, 9:89–112, 2006.

M. Giles and P. Glasserman. Smoking adjoints: Fast Monte Carlo Greeks. *Risk*, 19(1):89–112, 2006.

P. Glasserman. *Monte Carlo Methods in Financial Engineering*. Springer-Verlag, New York, 2004.

P. Glasserman and K.-K. Kim. Gamma expansion of the Heston stochastic volatility model. Columbia Business School working paper, 2008.

P. Glasserman and N. Merener. Cap and swaption approximations in LIBOR market models with jumps. *SSRN eLibrary*, 2001.

P. Glasserman and J. Staum. Conditioning on one-step survival in barrier option simulations. *Operations Research*, 49:923–937, 2001.

P. Glasserman and B. Yu. Simulation for American options: Regression now or regression later? In H. Niederreiter, editor, *Monte Carlo and Quasi-Monte Carlo Methods 2002*, pages 213–226, Berlin, 2002. Springer-Verlag.

P. Glasserman and B. Yu. Number of paths versus number of basis functions in American option pricing. *Annals of Applied Probability*, 14(4):2090–2119, 2004.

P. Glasserman and B. Yu. Large sample properties of weighted Monte Carlo estimators. *Operations Research*, 53:298–312, 2005.

P. Glasserman and X. Zhao. Fast greeks in forward Libor models. *Journal of Computational Finance*, 3:5–39, 1999.

P. Glasserman and X. Zhao. Arbitrage-free discretization of lognormal forward Libor and swap rate models. *Finance and Stochastics*, 4:35–68, 2000.

P. Glasserman, P. Heidelberger, and P. Shahabuddin. Asymptotically optimal importance sampling and stratification for pricing path dependent options. *Journal of Mathematical Finance*, 9(2):117–152, 1999.

P. W. Glynn and W. Whitt. Indirect estimation via L = w. *Operations Research*, 37(1):82–103, 1989.

R. Goldstein and W. P. Keirstead. On the term structure of interest rates in the presence of reflecting and absorbing boundaries. Ohio State University working paper, 1997.

G. H. Golub and C. F. van Loan. *Matrix Computations*. The John Hopkins University Press, 1989.

V. Gorovoi and V. Linetsky. Black's model of interest rates as options, eigenfunction expansions and Japanese interest rates. *Mathematical Finance*, 14:49–78, 2004.

J. Gregory. *Counterparty Credit Risk: The New Challenge for Global Financial Markets*. Wiley, 2009.

I. Gyöngy. Mimicking the one-dimensional distributions of processes having an Ito differential. *Probability Theory and Related Fields*, 71:501 – 516, 1986.

P. S. Hagan. Adjusters: Turning good prices into great prices. *Wilmott*, 2:56–59, 2002.

P. S. Hagan and G. West. Interpolation methods for yield curve construction. Working paper, 2004.

P. S. Hagan and D. E. Woodward. Markov interest rate models. *Applied Mathematical Finance*, 6(4):233–260, 1999a.

P. S. Hagan and D. E. Woodward. Equivalent Black volatilities. *Applied Mathematical Finance*, 6:147–157, 1999b.

P. S. Hagan, D. Kumar, A. S. Lesniewski, and D. E. Woodward. Managing smile risk. *Wilmott*, 11:84–108, 2002.

J. M. Hammersley and D. C. Handscomb. *Monte Carlo Methods*. Methuen & Co., 1965.

A. T. Hansen and P. L. Jørgensen. Exact analytical valuation of bonds when spot interest rates are log-normal. *SSRN eLibrary*, 1998.

P. Hansen. Analysis of discrete ill-posed problems by means of the l-curve. *SIAM Review*, 34:561–580, 1992.

M. Harrison and D. Kreps. Martingales and arbitrage in multiperiod securities markets. *Journal of Economic Theory*, 20:381–408, 1979.

M. Haugh and L. Kogan. Pricing American options: A duality approach. *Operations Research*, 52:258–270, 2004.

H. He, W. Keirstead, and J. Rebholz. Double lookbacks. *Mathematical Finance*, 8:201–228, 1998.

D. Heath, R. Jarrow, and A. Morton. Bond pricing and the term structure of interest rates: A new methodology for contingent claims valuation. *Econometrica*, 60(1):77–105, 1992.

V. Henderson and D. Hobson. Local time, coupling, and the passport option. *Finance and Stochastics*, 4:69–80, 2000.

P. Henry-Labordère. A general asymptotic implied volatility for stochastic volatility models. ArXiv working paper, 2005.

P. Henry-Labordère. *Analysis, Geometry and Modeling in Finance*. Chapman and Hall/CRC, 2008.

P. Henry-Labordère. Calibration of local stochastic volatility models to market smiles. *Risk*, 9:112–117, 2009.

S. Heston, M. Loewenstein, and G. Willard. Options and bubbles. *Review of Financial Studies*, 20:359–390, 2007.

S. L. Heston. A closed-form solution for options with stochastic volatility with applications to bond and currency options. *Review of Financial Studies*, 6: 327–343, 1993.

N. Higham. Computing the nearest correlation matrix - a problem from finance. *IMA Journal of Numerical Analysis*, 22:329–343, 2002.

T. Ho and S. Lee. Term structure movements and pricing interest rate contingent claims. *Journal of Finance*, 41:1011–1029, 1986.

M. Hogan and J. Weintraub. The lognormal interest rate model and Eurodollar futures. Citibank working paper, 1993.

P. Honore. Maturity induced bias in estimating spot-rate diffusion models. Aarhus School of Business working paper, 1998a.

P. Honore. *Five Essays on Financial Econometrics in Continuous-Time Models*. PhD thesis, Aarhus School of Business, 1998b.

Z. Hu, J. Kerkhof, P. McCloud, and J. Wackertap. Cutting edges using domain integration. *Risk*, 19(11):95–99, 2006.

J. C. Hull. *Options, futures and other derivatives*. Prentice Hall, 2006.

J. C. Hull and A. White. The pricing of options on assets with stochastic volatilities. *Journal of Finance*, 42:281–300, 1987.

J. C. Hull and A. White. Numerical procedures for implementing term structure models I: Single-factor models. *Journal of Derivatives*, 2:7–16, 1994a.

J. C. Hull and A. White. Numerical procedures for implementing term structure models II: Two-factor models. *Journal of Derivatives*, 2:37–48, 1994b.

P. Hunt and J. Kennedy. *Financial Derivatives in Theory and Practice*. Wiley, 2000.

C. Hunter, P. Jäckel, and M. S. Joshi. Drift approximations in a forward-rate based Libor market model. Quarc working paper, 2001.

T. Hyer. *Derivatives Algorithms*. World Scientific, 2010.

J. E. Ingersoll. Valuing foreign exchange rate derivatives with a bounded exchange process. *Review of Derivatives Research*, 1:159–181, 1997.

ISDA. 2005 ISDA collateral guidelines. Technical report, International Swaps and Derivatives Association, Inc., 2005. http://www.isda.org/ publications/ pdf/ 2005isdacollateralguidelines.pdf.

ISDA. ISDA margin survey 2009. Technical report, International Swaps and Derivatives Association, Inc., 2009. http://www.isda.org/ c_and_a/ pdf/ ISDA-Margin-Survey-2009.pdf.

S. D. Jacka. Optimal stopping and the American put. *Mathematical Finance*, 1: 1–14, 1991.

P. Jäckel. *Monte Carlo Methods in Finance*. Wiley, Chichester, U.K., 2002.

P. Jäckel. Weighted sampling for variance reduction. OTC Analytics working paper, 2004.

F. Jamshidian. An exact bond option pricing formula. *Journal of Finance*, 44: 205–209, 1989.

F. Jamshidian. Forward induction and construction of yield curve diffusion models. *Journal of Fixed Income*, 1:62–74, 1991a.

F. Jamshidian. Bond and option evaluation in the Gaussian interest rate model. *Research in Finance*, 9:131–170, 1991b.

F. Jamshidian. An analysis of American options. *Review of Futures Markets*, 11: 72–80, 1992.

F. Jamshidian. Option and futures evaluation with deterministic volatilities. *Mathematical Finance*, 3(2):149–159, 1993.

F. Jamshidian. The duality of optimal exercise and domineering claims: A Doob-Meyer decomposition approach to the Snell envelope. *SSRN eLibrary*, 1995.

F. Jamshidian. Libor and swap market models and measures. *Finance and Stochastics*, 1(4):293–330, 1997.

C. Jeffery. Reverse cliquets: End of the road? *Risk*, 17(2):20–22, 2004.

M. Jensen and M. Svenstrup. Efficient control variates and strategies for Bermudan swaptions in a Libor market model. *SSRN eLibrary*, 2003.

N. L. Johnson, S. Kotz, and N. Balakrishnan. *Continuous Univariate Distributions, Volume 2*. Wiley, 1995.

S. P. Jones, J.-M. Eber, and J. Seward. Composing contracts: An adventure in financial engineering. ICFP, 2000. URL http:// www.lexifi.com/ downloads/ MLFiPaper.pdf.

F. D. Jong, J. Driessen, and A. Pelsser. Libor market models versus swap market models for pricing interest rate derivatives: An empirical analysis. *European Finance Review*, 5(3):201–237, 2001.

M. S. Joshi. *C++ Design Patterns and Derivatives Pricing*. Cambridge University Press, 2004.

M. S. Joshi and A. Stacey. New and robust drift approximations for the Libor market model. *Quantitative Finance*, 8(4):427–434, 2008.

C. Joy, P. P. Boyle, and K. S. Tan. Quasi-Monte Carlo methods in numerical finance. *Management Science*, 42:926–938, 1996.

N. Ju. Pricing an American option by approximating its early exercise boundary as a multi-piece exponential function. *Review of Financial Studies*, 11:627–646, 1998.

S. Juneja and H. Kalra. Variance reduction techniques for pricing American options using function approximations. *Journal of Computational Finance*, 12 (3):79–102, 2009.

Y. M. Kabanov and M. M. Safarian. On Lelands strategy of option pricing with transaction costs. *Finance and Stochastics*, 1:239–250, 1997.

C. Kahl and P. Jäckel. Not-so-complex logarithms in the Heston model. *Wilmott*, 19(9), 2005.

C. Kahl and P. Jäckel. Fast strong approximation Monte-Carlo schemes for stochastic volatility. *Quantitative Finance*, 6:513–536, 2006.

D. Kainth and N. Saravanamuttu. Multifactor stochastic volatility models. ICBI Conference, Paris, 2007.

R. Kangro and R. Nicolaides. Far field boundary conditions for Black-Scholes equations. *SIAM Journal on Numerical Analysis*, 38:1357–1368, 2000.

I. Karatzas and S. E. Shreve. *Brownian Motion and Stochastic Calculus*. Springer-Verlag, 2nd edition, 1991.

S. Karlin and H. Taylor. *A Second Course in Stochastic Processes*. Academic Press, 1981.

J. Kennedy, P. Hunt, and A. Pelsser. Markov-functional interest rate models. *Finance and Stochastics*, 4:391–408, 2000.

F. Kilin. Accelerating the calibration of stochastic volatility models. Frankfurt School of Finance & Management working paper, 2007.

I. J. Kim. The analytical valuation of American options. *The Review of Financial Studies*, 3:547–572, 1990.

P. E. Kloeden and E. Platen. *Numerical Solution of Stochastic Differential Equations (Stochastic Modelling and Applied Probability)*. Springer-Verlag, 2000.

A. Kolodko and J. Schoenmakers. Iterative construction of the optimal Bermudan stopping time. *Finance and Stochastics*, 10:27–49, 2006.

S. Kotz, N. Balakrishnan, and N. L. Johnson. *Continuous Multivariate Distributions. Volume 1: Models and Applications*. Wiley, 2000.

J. F. B. M. Kraaijevanger, H. W. J. Lenferink, and M. N. Spijker. Stepsize restrictions for stability in the numerical solution of ordinary and partial differential equations. *Journal of Computational and Applied Mathematics*, 20: 67–81, 1987.

M. Kramin. A multi-factor Markovian HJM model for pricing exotic interest rate derivatives. ICBI Conference, Paris, 2008.

H. Kreiss, V. Thomee, and O. Widlund. Smoothing of initial data and rates of convergence for parabolic difference equations. *Communications on Pure and Applied Mathematics*, 23:241–259, 1970.

H. Kunita. *Stochastic Flows and Stochastic Differential Equations*. Cambridge University Press, 1990.

B. Kvasov. *Methods of Shape-Preserving Spline Approximation*. World Scientific, 2000.

P. K. Kythe and M. R. Schäferkotter. *Handbook of Computational Methods for Integration*. Routledge, USA, 2004.

M. Leclerc, Q. Liang, and I. Schneider. Fast Monte Carlo Bermudan greeks. *Risk*, 22(7):84–88, 2009.

P. L'Ecuyer. Uniform random number generation. *Annals of Operations Research*, 53:77–120, 1994.

R. W. Lee. Implied and local volatilities under stochastic volatility. *International Journal of Theoretical and Applied Finance*, 4(1):45–89, 2001.

R. W. Lee. Option pricing by transform methods: Extensions, unification, and error control. *Journal of Computational Finance*, 7(3):51–86, 2004.

R. W. Lee. Implied volatility: Statics, dynamics, and probabilistic interpretation. In R. Baeza-Yates, J. Glaz, H. Gzyl, J. Husler, and J. L. Palacios, editors, *Recent Advances in Applied Probability*, Berlin, 2005. Springer-Verlag.

M. Leippold and L. Wu. Asset pricing under the quadratic class. *Journal of Financial and Quantitative Analysis*, 37:271–295, 2002.

H. Leland. Option pricing and replication with transaction costs. *Journal of Finance*, 40(5):1283–1301, 1985.

H. W. Lenferink and M. N. Spijker. On the use of stability regions in the numerical analysis of initial value problems. *Mathematics of Computation*, 57:221–237, 1991.

A. L. Lewis. *Option Valuation under Stochastic Volatility: with Mathematica Code*. Finance Press, 2000.

A. L. Lewis. A simple option formula for general jump-diffusion and other exponential Levy processes. *SSRN eLibrary*, 2001.

Q. Li and H. Qi. A sequential semismooth Newton method for the nearest low-rank correlation matrix problem. Working paper, 2009.

E. Liebscher. Construction of asymmetric multivariate copulas. *Journal of Multivariate Analysis*, 99(10):2234–2250, 2008.

A. Lipton. *Mathematical Methods For Foreign Exchange: A Financial Engineer's Approach*. World Scientific, 2001.

A. Lipton. The vol smile problem. *Risk*, 15(2):61–65, 2002.

F. A. Longstaff and E. S. Schwartz. Interest rate volatility and the term structure: A two-factor general equilibrium model. *The Journal of Finance*, 4:1259–1282, 1992.

F. A. Longstaff and E. S. Schwartz. Implementation of the Longstaff-Schwartz interest rate model. *Journal of Fixed Income*, 3:7–14, 1993.

F. A. Longstaff and E. S. Schwartz. Valuing American options by simulation: A simple least-squares approach. *The Review of Financial Studies*, 14(1):113–147, 2001.

R. Lord and C. Kahl. Optimal Fourier inversion in semi-analytical option pricing. *SSRN eLibrary*, 2007.

R. Lord, R. Koekkoek, and D. van Dijk. A comparison of biased simulation schemes for stochastic volatility models. Tinbergen Institute working paper, 2006.

V. Lucic. On singularities in the Heston model. *SSRN eLibrary*, 2007.

V. Lucic. Boundary conditions for computing densities in hybrid models via PDE methods. *SSRN eLibrary*, 2008.

R. W. Lynch. A method for choosing a tension factor for spline under tension interpolation. University of Texas working paper, 1982.

W. Margrabe. The value of an option to exchange one asset for another. *Journal of Finance*, 33:177–186, 1978.

M. Matsumoto and T. Nishimura. Mersenne twister: A 623-dimensionally equidistributed uniform pseudo-random number generator. *ACM Transactions on Modeling and Computer Simulation*, 8:3–30, 1998.

J. Mayle. *Standard Securities Calculation Methods: Fixed Income Securities Formulas for Price, Yield and Accrued Interest*. SIFMA, 1993.

R. Merton. The theory of rational option pricing. *Bell Journal of Economics and Management Science*, 4:141–183, 1973.

S. Meyers. *Effective C++: 55 Specific Ways to Improve Your Programs and Designs*. Addison-Wesley, 2005.

K. Miltersen, K. Sandmann, and D. Sondermann. Closed form solutions for term structure derivatives with lognormal interest rates. *Journal of Finance*, 52(1): 409–430, 1997.

P. Miron and P. Swannell. *Pricing and hedging swaps*. Euromoney Institutional Investor PLC, 1991.

A. Mitchell and D. Griffiths. *The Finite Difference Method in Partial Differential Equations*. Wiley, New York, 1980.

C. Moni. Fast American Monte Carlo. *SSRN eLibrary*, 2005.

B. Moro. The full monte. *Risk*, 8(2):57–58, 1995.

V. Morozov. On the solution of functional equations by the method of regularization. *Soviet Mathematics Doklady*, 7:414–417, 1966.

A. Morton. Arbitrage and martingales. Working paper, 1988.

B. Moskowitz and R. Caflisch. General framework for pricing derivative securities. *Mathematical and Computer Modeling*, 23:37–54, 1996.

M. Musiela and M. Rutkowski. *Martingale Methods in Financial Modeling*. Applications of Mathematics. Springer-Verlag, 1997.

R. B. Nelsen. *An Introduction to Copulas (Lecture Notes in Statistics)*. Springer-Verlag, 2nd edition, 2006.

B. L. Nelson. Control variate remedies. *Operations Research*, 38(6):974–992, 1990.

C. R. Nelson and A. F. Siegel. Parsimonious modeling of yield curves. *Journal of Business*, 60:473–489, 1987.

I. E. Nesterov, A. Nemirovskii, and Y. Nesterov. *Interior-Point Polynomial Algorithms in Convex Programming*, volume 13 of *Siam Studies in Applied Mathematics*. Society for Industrial & Applied Mathematics, 1994.

J. Obloj. Fine-tune your smile. Imperial College London working paper, 2008.

M. K. Ochi. *Applied Probability and Stochastic Processes*. Wiley, 1990.

B. Øksendal. *Stochastic differential equations: an introduction with applications*. Springer-Verlag, New York, 1992.

Y. Osajima. General asymptotics of Wiener functionals and application to mathematical finance. Mitsubishi UFJ Securities working paper, 2007.

S. Paskov and J. Traub. Faster valuation of financial derivatives. *Journal of Portfolio Management*, 22:113–120, 1995.

M. B. Pedersen. Calibrating Libor market models. *SSRN eLibrary*, 1998.

M. B. Pedersen. Bermuda swaptions in the LIBOR market model. *SSRN eLibrary*, 1999.

J. Perello, J. Masoliver, and J.-P. Bouchaud. Multiple time scales in volatility and leverage correlations: A stochastic volatility model. *Applied Mathematical Finance*, 11:1–24, 2004.

R. Pietersz. Importance sampling in market models for targeted accrual redemption notes (TARNs). ABN AMRO working paper, 2005.

R. Pietersz and P. Groenen. Rank reduction of correlation matrices by majorization. *Quantitative Finance*, 4:649–662, 2004.

R. Pietersz and A. Pelsser. Swap vega in BGM: Pitfalls and alternatives. *Risk*, 17 (3):91–93, 2004.

R. Pietersz and M. van Regenmortel. Importance sampling for stable Greeks in the LIBOR market model for targeted accrual redemption notes (TARNs). 3d WBS Fixed Income Conference, Amsterdam, 2006.

R. Pietersz, A. Pelsser, and M. van Regenmortel. Fast drift approximated pricing in the BGM model. *Journal of Computational Finance*, 8(1):93–124, 2004.

V. V. Piterbarg. A note on pricing weakly-path-dependent American-style options by backward induction. *SSRN eLibrary*, 2002.

V. V. Piterbarg. A Practitioner's guide to pricing and hedging callable LIBOR exotics in forward LIBOR models. *SSRN eLibrary*, 2003.

V. V. Piterbarg. Risk sensitivities of Bermuda swaptions. *International Journal of Theoretical and Applied Finance*, 7(4):465–510, 2004a.

V. V. Piterbarg. Computing deltas of callable Libor exotics in forward Libor models. *Journal of Computational Finance*, 7(3):107–144, 2004b.

V. V. Piterbarg. TARNs: Models, valuation, risk sensitivities. *Wilmott*, 14(11): 62–71, 2004c.

V. V. Piterbarg. Pricing and hedging callable Libor exotics in forward Libor models. *Journal of Computational Finance*, 8(2):65–119, 2005a.

V. V. Piterbarg. Stochastic volatility model with time-dependent skew. *Applied Mathematical Finance*, 12(2):147–185, 2005b.

V. V. Piterbarg. Time to smile. *Risk*, 18(5):71–75, 2005c.

V. V. Piterbarg. Smiling hybrids. *Risk*, 19(5):66–71, 2006.

V. V. Piterbarg. Practical multi-factor quadratic Gaussian models of interest rates. 5th WBS Fixed Income Conference, Budapest, 2008.

V. V. Piterbarg. Rates squared. *Risk*, 22(1):100–105, 2009a.

V. V. Piterbarg. Quadratic Gaussian models for CMS spread options. ICBI Conference, Rome, 2009b.

V. V. Piterbarg. Funding beyond discounting: Collateral agreements and derivatives pricing. *Risk*, 2:97–102, 2010.

V. V. Piterbarg and L. B. Andersen. Bermudan swaptions and callable Libor exotics. In R. Cont, editor, *Encyclopedia of Quantitative Finance*, pages 177–181. Wiley, 2010a.

V. V. Piterbarg and L. B. Andersen. Libor market model. In R. Cont, editor, *Encyclopedia of Quantitative Finance*, pages 1031–1036. Wiley, 2010b.

V. V. Piterbarg and M. A. Renedo. Eurodollar futures convexity adjustments in stochastic volatility models. *Journal of Computational Finance*, 9(3):71–94, 2006.

D. M. Pooley, K. R. Vetzal, and P. A. Forsyth. Convergence remedies for non-smooth payoffs in option pricing. *Journal of Computational Finance*, 6(4):25 – 40, 2003.

W. H. Press, B. P. Flannery, S. A. Teukolsky, and W. T. Vetterling. *Numerical Recipes in C: The Art of Scientific Computing*. Cambridge University Press, 1992.

P. E. Protter. *Stochastic Integration and Differential Equations*. Springer-Verlag, 2005.

S. Pruess. Properties of splines in tension. *Journal of Approximation Theory*, 17: 86–96, 1976.

R. Rannacher. Finite element solution of diffusion problems with irregular data. *Numerische Mathematik*, 43(2):309–327, 1984.

N. S. Rasmussen. Control variates for Monte Carlo valuation of American options. *Journal of Computational Finance*, 9(1):83–118, 2005.

R. Rebonato. *Interest-Rate Option Models*. Wiley, 1998.

R. Rebonato. *Modern pricing of interest rate derivatives: The Libor market model and beyond*. Princeton University Press, 2002.

Y. Ren, D. Madan, and M. Q. Qian. Calibrating and pricing with embedded local volatility models. *Risk*, 9:138–143, 2007.

R. Rendleman and B. Bartter. The pricing of options on debt securities. *Journal of Financial and Quantitative Analysis*, 15:11–24, 1980.

R. J. Renka. Interpolatory tension splines with automatic selection of tension factors. *SIAM Journal of Scientific and Statistical Computing*, 8(3):393–415, 1987.

P. Rentrop. An algorithm for the computation of exponential splines. *Numerische Matematik*, 35:81–93, 1980.

D. Revuz and M. Yor. *Continuous Martingales and Brownian Motion*. Springer-Verlag, 3d edition, 1999.

P. Ritchken and L. Sankarasubramanian. Volatility structure of forward rates and the dynamics of the term structure. *Mathematical Finance*, 5:55–72, 1995.

L. C. G. Rogers. Which model for term-structure of interest rates should one use? In *Proceedings of IMA Workshop on Mathematical Finance, IMA Vol 65*, pages 93–116, New York, 1995. Springer-Verlag.

L. C. G. Rogers. Gaussian errors. *Risk*, 9(1):42–45, 1996.

L. C. G. Rogers. Monte Carlo valuation of American options. University of Bath working paper, 2001.

L. C. G. Rogers and Z. Shi. The value of an Asian option. *Journal of Applied Probability*, 32:1077–1088, 1995.

M. Romano and N. Touzi. Contingent claims and market completeness in a stochastic volatility model. *Mathematical Finance*, 7(4):399–410, 1997.

C. Rouvinez. Going Greek with VaR. *Risk*, 10:57–65, 1997.

Y. Saad. *Iterative methods for sparse linear systems*. SIAM, 2003.

K. Sandmann and D. Sondermann. A note on the stability of lognormal interest rate models and the pricing of eurodollar futures. *Mathematical Finance*, 7(2): 119–125, 1997.

E. Schlögl. Arbitrage-free interpolation in models of market observable interest rates. In K. Sandmann and P. Schnbucher, editors, *Advances in Finance and Stochastics: Essays in Honour of Dieter Sondermann*, pages 197–218. Springer-Verlag, 2002.

J. Schoenmakers and B. Coffey. Stable implied calibration of a multi-factor Libor model via a semi-parametric correlation structure. WIAS Preprint No. 611, 2000.

J. Schoenmakers and A. Heemink. Fast valuation of financial derivatives. *Journal of Computational Finance*, 1:47–62, 1997.

M. Schroder. Computing the Constant Elasticity of Variance option pricing formula. *J. Finance*, 44:211–219, 1989.

D. G. Schweikert. An interpolating curve using a spline in tension. *Journal of Mathematics and Physics*, 45:312–317, 1966.

M. Selby and C. Strickland. Computing the Fong and Vasicek pure discount bond price formula. *Journal of Fixed Income*, 5:78–85, 1995.

C. E. Shannon. Communication in the presence of noise. *Proceedings of Institute of Radio Engineers*, 37:10–21, 1949.

W. F. Sharpe. Capital asset prices: A theory of market equilibrium under conditions of risk. *Journal of Finance*, 19:425–442, 1964.

G. S. Shea. Pitfalls in smoothing interest rate terms structure data: Equilibrium models and spline approximation. *Journal of Financial and Quantitative Analysis*, 19:253–269, 1984.

J. Sidenius. Libor market model in practice. *Journal of Computational Finance*, 3 (3):75–99, 2000.

C. Sin. Complications with stochastic volatility models. *Advances in Applied Probability*, 30:256–268, 1998.

R. Smith. An almost exact simulation method for the Heston model. *Journal of Computational Finance*, 11:115–125, 2007.

H. Soner, S. Shreve, and J. Cvitanic. There is no nontrivial hedging portfolio for option pricing with transaction costs. *Annals of Applied Probability*, 5:327–355, 1995.

M. N. Spijker and F. A. J. Straetemans. Error growth analysis via stability regions for discretizations of initial value problems. *BIT*, 37:442–464, 1997.

M. L. Stigum and F. L. Robinson. *Fixed Income Calculations: Money Market Paper and Bonds*. Irwin Professional Publishing, 1996.

G. Stoyan. Monotone difference schemes for diffusion-convection problems. *ZAMM*, 59:361–372, 1979.

Q. Su and C. Randall. General market greeks in practice. *Wilmott*, 8, 2008.

Y. Su and M. C. Fu. Optimal importance sampling in securities pricing. *Journal of Computational Finance*, 5:27–50, 2002.

H. Sutter and A. Alexandrescu. *C++ Coding Standards: 101 Rules, Guidelines, and Best Practices*. Addison-Wesley, 2004.

K. Sydsaeter and P. Hammond. *Essential Mathematics for Economic Analysis*. Prentice Hall, 2008.

D. Talay. Efficient numerical schemes for the approximation of expectations of functionals of an SDE, and applications. In *Lecture Notes in Control and Information Sciences, Vol. 61*, pages 294–313. Springer-Verlag, 1984.

D. Talay and L. Tubaro. Romberg extrapolations for numerical schemes solving stochastic differential equations. *Structural Safety*, 8, 1990.

N. Taleb. *Dynamic Hedging*. Wiley, 1997.

C. Tanggaard. Nonparametric smoothing of yield curves. *Review of Quantitative Finance and Accounting*, 9:251–267, 1997.

D. Tavella and C. Randall. *Pricing Financial Instruments – The Finite Difference Method*. Wiley, 2000.

C. Tezier. Short rate models. Linear and quadratic Gaussian models. Barclays Capital working paper, 2005.

H. Theil. *Principles of Econometrics*. Wiley, Amsterdam, 1971.

S. Traven. Pricing linear derivatives with a single discounting curve. Barclays Capital working paper, 2008.

J. N. Tsitsiklis and B. V. Roy. Regression methods for pricing complex American-style options. *IEEE Transactions on Neural Networks*, 12(4):694–703, 2001.

R. J. Van Steenkiste. Term structure model perturbation. Barclays Capital working paper, 2009.

O. Vasicek. An equilibrium characterization of the term structure. *Journal of Financial Economics*, 5:177–188, 1977.

M. Wichura. Algorithm AS 241: The percentage points of the normal distribution. *Applied Statistics*, 37:477–484, 1988.

P. Wilmott, J. Dewynne, and J. Howison. *Option Pricing: Mathematical Models and Computation*. Oxford Financial Press, 1993.

H. Windcliff, P. A. Forsyth, and K. R. Vetzal. Shout options: A framework for pricing contracts which can be modified by the investor. *Journal of Computational and Applied Mathematics*, 134:213–241, 2001.

L. Withington and L. Lucic. Noisy hedges and fuzzy payoffs: Using soft computing to improve risk stability. RBC Capital Markets working paper, 2009.

F. Wu, E. A. Valdez, and M. Sherris. Simulating exchangeable multivariate Archimedean copulas and its applications. UNSW working paper, 2006.

N. Yoshida. Asymptotic expansion for small diffusions via the theory of Malliavin-Watanabe. *Probability Theory and Related Fields*, 92:275–311, 1992.

L. Zadeh. Fuzzy sets. *Information and Control*, 8:338–353, 1965.

Z. Zhang and L. Wu. Optimal low-rank approximation to a correlation matrix. *Linear Algebra and its Applications*, 364:161–187, 2003.

R. Zvan, P. A. Forsyth, and K. R. Vetzal. Robust numerical methods for PDE models of Asian options. *Journal of Computational Finance*, 1(2):39–78, 1998.

R. Zvan, P. A. Forsyth, and K. R. Vetzal. Discrete Asian barrier options. *Journal of Computational Finance*, 3(1):41–67, 1999.

Index

absorbing boundary, *see* diffusion, absorbing barrier
accrual factor, *see* year fraction
ADI, *see* PDE, ADI scheme
adjusters method, *see* out-of-model adjustment, adjusters method
affine short rate model, 431–444, 512–520
  bond reconstruction formula, 433, 515–517
  calibration, 441–443
    multi-pass bootstrap, 442
  calibration to yield curve, 437–439
  characteristic function, 434
  European swaption, 439
    Fourier integration, 439
    Gram-Charlier expansion, 439
  extended transform, 433
    constant parameters, 434, 436
    piecewise constant parameters, 436
  Feller condition, 319, 432
  importance sampling, 1065
  moment-generating function, 434, 439
  Monte Carlo, 444
  multi-factor, 512–520
    bond dynamics, 516
    bond reconstruction formula, 515–517
    existence and uniqueness, 514
    exponential affine, 513
    Feller condition, 515
    forward rate correlation, 516
    forward rate dynamics, 516
    regularity issues, 514–515
    short rate state dynamics, 513
  one-factor, 431–444
    PDE, 444
    regularity issues, 432
    short rate domain, 432
    short rate dynamics, 431
    short rate state dynamics, 437
    swap rate volatility, 440
      affine approximation, 440
      time averaging, 440
    time-dependent, 433
    volatility skew range, 433
    volatility smile, 432
almost surely, 4
American capped straddle, 936
American swaption, 895–899
  accrued current coupon, 895
  approximating with Bermudan swaption, *see* Bermudan swaption, approximating American swaption
  discontinuity of exercise value in time, 895
  PDE, 897–899
    extra state variable, 898–899
    proxy Libor rate method, 897–898
American/Bermudan option, 30–41
  Bellman principle, 32, 33, 69
  Black-Scholes model, 839
  capped, 936
  conditional on no exercise, 31

continuation region, 33
discontinuity at expiry, 39
duality, 35
early exercise boundary, 37
early exercise premium, 36, 39, 41
exercise never optimal, 36
exercise policy, 30
exercise region, 33
exercise value, 30
high contact condition, 37
hold value, 32
integral representation, 39, 40
lower bound, see Monte Carlo, lower
    bound for American option
marginal exercise value decomposi-
    tion, 41
Monte Carlo, 158–165
    confidence interval for value, 164
    random tree, 164
    stochastic mesh, 165
PDE jump condition, 34
perfect foresight bias, 160
short-maturity asymptotics, 38
smooth pasting condition, 37
supermartingale, 31
upper bound, see Monte Carlo,
    upper bound for American option
annuity mapping function, see terminal
    swap rate model, annuity
    mapping function
annuity measure, see measure, annuity
arbitrage opportunity, 8
arbitrage pricing, 11
arithmetic put-call symmetry, 940
Arrow-Debreu security, 21, 76, 77, 79,
    458, 462, 1048
    backward Kolmogorov equation, 458
    forward Kolmogorov equation, 458
art of derivatives trading, 980
Asian option, 70
    Black model, 922
    Monte Carlo, see Monte Carlo, Asian
        option
    PDE, see PDE, Asian option
ATM backbone, see volatility smile,
    ATM backbone
autocorrelation, see inter-temporal
    correlation

averaging, see calibration, time
    averaging
averaging cash flow, 201, 722–723
    convexity adjustment, 722
averaging swap, see averaging cash flow

Bachelier model, see Normal model
backbone, see volatility smile, backbone
backward Kolmogorov equation, see
    Kolmogorov backward equation
balance-guarantee swap, 900
band swap, see flexi-swap
"bang-bang", 902
barrier option, 44
    Broadie adjustment for sampling
        frequency, see Monte Carlo,
        sampling extremes, adjusting
        barrier for sampling frequency
    continuous barrier, 64
    discrete barrier, 66
    importance sampling, 1074–1077
    Markovian projection, see Markovian
        projection, barrier option
    Monte Carlo, see Monte Carlo,
        barrier option
    on capped straddle, 937
    one-touch, 939
    pathwise differentiation method,
        1041–1044
        recursion, 1043
    payoff smoothing, see payoff
        smoothing, barrier option
    PDE jump condition, 66
    rebate, 64
    semi-static replication, 939
    step-down, 64
    step-up, 64
    tube Monte Carlo, 1024
    up-and-out, 44, 64, 66, 124, 126, 1134
basis point, 169
basis risk, see yield curve, basis risk
basket option, 205, 1146
    Black model, 924
    displaced log-normal approximation,
        1147
    local volatility model, 1145
    Monte Carlo, see Monte Carlo, Asian
        option on basket
    slope of volatility smile, 1148

stochastic volatility model, 1149
BDT model, *see* Black-Derman-Toy
    model
Bermudan cancelable swap, *see*
    Bermudan swaption; cancelable
    note
Bermudan option, *see* Ameri-
    can/Bermudan option
Bermudan swaption, 207, 875–920
    accreting, *see* Bermudan swaption,
        non-standard
    American, *see* American swaption
    amortizing, *see* Bermudan swaption,
        non-standard
    approximating American swaption,
        896
    bullet, *see* Bermudan swaption,
        vanilla
    carry, 908, 914
        impact on exercise decision, 914
    exercise fee, 899
    exercise value, XXXIV, 208, 875
    flexi-swap, *see* flexi-swap
    gamma-theta mismatch, 914
    hold value, XXXIV, 208
    lockout, 207, 875
    mid-coupon, 897, 899
    no-call, *see* Bermudan swaption,
        lockout
    non-standard, 880–899
        calibration by payoff matching,
            884–886
        calibration by PVBP matching,
            884, 886
        calibration by tenor matching, 883
        calibration to basket, 887–889
        calibration to representative
            swaption, 884
        calibration to row of European
            swaptions, 888
        Gaussian short rate model, 888
        global calibration, 881, 883
        Libor market model, 887
        local projection method, 881, 883
        lower bound, 893, 909
        Markov-functional model, 881
        quadratic Gaussian model, 888
        quasi-Gaussian model, 881, 888,
            891

    representative swaption for
        accreting Bermudan, 886
    representative swaption for
        amortizing Bermudan, 885, 886
    super-replication, 890–894
    upper bound, 891, 892, 909
    non-vanilla, *see* Bermudan swaption,
        non-standard
    PDE jump condition, *see* Ameri-
        can/Bermudan option, PDE jump
        condition
    strike, 875
    survival measure, 1047
    vanilla, 880
    zero-coupon, 894–895
Bermudan swaption calibration
    adjusters method, 955
    local projection method, 554,
        876–880
        Gaussian short rate model, 877
        non-standard Bermudan, *see*
            Bermudan swaption, non-
            standard
        quadratic Gaussian model, 877
        quasi-Gaussian model, 877
    smile calibration, 878–880
        at-the-money, 878
        exercise boundary, 879
        strike, 878
Bermudan swaption greeks
    pathwise differentiation method,
        1044–1050
        forward induction, 1049–1050
        performance, 1050
        survival density, 1048
        survival measure, 1047
    portfolio replication for hedging, 913
    Principal Components Analysis, 913
    robust hedging, 912–914
    static hedging, 912
Bermudan swaption valuation, 822–873
    control variate, 1086–1090
        non-linear, 1088
        sampled at exercise time, 1087
    fast pricing, 916
    impact of forward volatilities, 876
    impact of inter-temporal correlation,
        554, 877
    impact of mean reversion, 554, 876

impact of the number of factors, 877
Monte Carlo, 905–912
  exercise strategy, 906
  explanatory variables, 905
  parametric lower bound, 906–912
  regression lower bound, 905
Bermudanality, 879
Bessel function of the first kind, 282
Bessel process, 281, 282
best-of option, see MAX-option
best-of-calls option, 782
BGM model, see Libor market model
Black model, XXXIV, 21, 28, 202, 279, 283
  Asian option, see Asian option, Black model
  basket option, see basket option, Black model
  call option, 24
  CMS spread, 776
  delta, 350, 698
  effects of volatility mis-specification, 987
  Fourier integration, 329
  gamma-vega, 981
  log-likelihood ratio, 1060
  moment-generating function, 329
  PDE, 25
  stochastic interest rates, 28, 30
  strike-specific volatility, 698
  time-dependent parameters, 27, 983–985
  vega, 698
  use in calibration, 704
  with dividends, 27
Black shadow rate model, 452
Black-Derman-Toy model, 445–447
  mean-fleeting, 447
  short rate dynamics, 446
Black-Karasinski model, 447
Black-Scholes model, see Black model
Black-Scholes-Merton model, see Black model
BMA index, 192, 265
BMA rate, 192
Boltzman-Gibbs distribution, see out-of-model adjustment, path re-weighting method, Boltzman-Gibbs distribution

Bond Market Association, see BMA index
box smoothing method, see payoff smoothing, box smoothing
break-even rate, see forward swap rate
Broadie adjustment for sampling frequency of barriers, see Monte Carlo, sampling extremes, adjusting barrier for sampling frequency
Brownian bridge, 125, 647, 648
  conditional moments, 129
  Libor market model, see Libor market model valuation, Monte Carlo, Brownian bridge
  path construction, see Brownian motion, path construction by Brownian bridge
  sampling extremes, see Monte Carlo, sampling extremes, with Brownian bridge
Brownian motion, 4
  geometric, 16
  Haar function decomposition, see Brownian motion, path construction by Brownian bridge
  Ito integral, see Ito integral
  Karhunen-Loeve decomposition, see Brownian motion, path construction by Principal Components
  path construction, 106
  path construction by Brownian bridge, 128, 129
  path construction by Principal Components, 130
  Stratonovich integral, see Stratonovich integral
BSM model, see Black model

C^0, XXXIV
C^1, XXXIV
C^2, XXXIV
C^n, XXXIV
calibration, 299
  calibration norm, 630–633
    fit, 634
    regularity, 634
  cold start, 633

forward induction, 445, 458, 953
Levenberg-Marquardt, 633
local projection method, *see* local
    projection method
Markovian projection method, *see*
    Markovian projection
most likely path, 990
stochastic optimization method, 953
time averaging, 301, 307, 363,
    371–381, 550, 584, 667
  algorithm, 377–381
  non-zero correlation, 376
  skew, 373–374
  volatility, 371–373
  volatility of variance, 374–377
callable Libor exotic, *see* CLE
callable zero, *see* Bermudan swaption,
    zero-coupon
cancelable note, 214, 829, 830
  ATM, 859
  carry, 858, 914
cancelable swap, *see* cancelable note
cap, 186, 202
  caplet volatility from cap volatility,
    706
    interpolation, 707
    precision norm, 707
    relaxation, 708
    smoothness norm, 708
    splitting scheme, 708
  digital, 203, 209
  valuation formula, 202
Capital Asset Pricing Model, 357
capped floater, 209
Cauchy distribution, 98, 101
  Monte Carlo, 98
certificate of deposit, 194
CEV model, 280–286
  attainability of zero, 280
  displaced, 285
  European call option value, 282, 283
  explosion, 280
  regularization, 284
  relation to Bessel process, 281
  strict supermartingale, 280
  time-dependent, 304
    effective parameter, 305
  volatility skew, 283
characteristic function, 20

Cheyette model, *see* quasi-Gaussian
    model
chi-square distribution, 100
  Monte Carlo, 100, 102
  non-central, *see* non-central
    chi-square distribution
  PDF, 100
chooser cap, *see* flexi-cap
chooser swap, *see* flexi-swap
CIR model, *see* Cox-Ingersoll-Ross
    model
CLE, 213, 216, 628, 817–873, 875
  accreting at coupon rate, 216, 870
  carry, 858, 908, 915
    impact on exercise decision, 849,
    858
  definition, 822
  exercise value, XXXIV, 215, 822
  hold value, XXXIV, 215, 822
  lockout, 213
  marginal exercise value decomposi-
    tion, 824
  multi-tranche, 217
  no-call, *see* CLE, lockout
  optimal exercise, 823
  single-rate, 864
  smooth function of Monte Carlo
    path, 1029
  snowball, 216, 872
CLE calibration, 817–822
  local projection method, 864–869
    calibration targets, 865
    core swap rate analog, 867
    local models, 866–867
    quadratic Gaussian model, 866
    quasi-Gaussian model, 866
    two-factor Gaussian model, 866
    two-strike calibration, 867
    vega, 869
  low-dimensional models, 864–869
  model choice, 821
  single-rate, 864
  to forward volatility, 821
CLE greeks, 1036–1040
  as sum of coupon greeks, 1037
  discontinuity in Monte Carlo, 1041
  freezing exercise boundary, 835, 1039,
    1040
  freezing exercise time, 1038–1040

likelihood ratio method, *see*
  likelihood ratio method
pathwise differentiation method,
  1035–1040, 1058–1060
  computational complexity, 1052
  forward induction, 1049–1050
  survival density, 1048
  survival measure, 1047
perturbation method, 1040, 1059
  computational complexity, 1053
portfolio replication for hedging, 913
recursion, 1036
source of noise, 1040
tube Monte Carlo, 1029
CLE regression, 825–864
  automatic selection of regression
    variables, 857
  boundary optimization, 832
  cancelable note, 829–830
  choice of regression variables,
    850–856
  decision only, 830–832
  discrepancy principle, 861
  excluding suboptimal points, 858
  exercise value, 827–829
  explanatory variables, 851–856
    classification, 853
    CMS spread, 853
    core swap rate, 853
    stochastic volatility, 856
    with convexity, 854–856
  general-to-specific approach, 857
  generalized cross-validation, 861
  L-curve method, 861
  Libor market model, 851, 852
    state variables, 851
  lower bound, 833–835
  perfect foresight bias, 834
  pseudo-inverse method, 862
  quadratic Gaussian model, 851
  quasi-Gaussian model, 851
  regression operator, 826
  regression variables, 825
  rescaling, 863
  reuse exercise boundary, *see* CLE
    greeks, freezing exercise boundary
  ridge regression, *see* CLE regression,
    Tikhonov regularization
  robust implementation, 860–864

singular value decomposition, 104
  stabilization, 861
  state variables, 850–851
    Libor market model, 851
  SVD decomposition, 862, 863
    connection to Tikhonov regulariza-
      tion, 863
  Tikhonov regularization, 162, 255,
    861–863
    connection to TSVD, 863
  truncated SVD decomposition, 162,
    862, 863
  two-step, 859
  upper bound, 839–850
    alternative methods, 849
    computational cost, 843
    improvements to algorithm,
      847–849
    nested simulation algorithm,
      839–849
    non-analytic exercise values,
      845–847
    simulation within a simulation, *see*
      CLE regression, upper bound,
      nested simulation algorithm
CLE valuation, 215, 822–873
  as cancelable note, 829
  boundary optimization, 832
  confidence interval for value, 844
  control variate, *see* Bermudan
    swaption valuation, control
    variate
  discontinuous function of Monte
    Carlo path, 1041
  duality, 838, 1093
    multiplicative, 1093
  duality gap, 841, 844, 910
    in stochastic volatility models, 912
  exercise policy consistency conditions,
    835
  fast pricing, 918
  Hamilton-Jacobi-Bellman equation,
    823
  impact of forward volatility, 820
  impact of inter-temporal correlation,
    865
  impact of volatility smile dynamics,
    820, 821
  Libor market model, 826

lower bound, 836, 843, 847, 850
  by regression, see CLE regression,
    lower bound
  iterative improvement, 835
  iterative improvement by nested
    simulation, 837
  quality test, 1060
LS method, see CLE regression
Monte Carlo, 825–864, 905
optimal exercise policy, 835, 837,
    1039
PDE, 870–873
  accreting at coupon rate, 870
  path-dependent, 870–873
  similarity reduction, 871
  snowball, 872
perfect foresight bias, 834
policy fixing, 847
recursion, 823
regression method, see CLE
    regression
tube Monte Carlo, 1029
upper bound, 838–850
  cancelable note, 846
  nested simulation algorithm, 841,
    909
  non-analytic exercise values,
    845–847
weighted coupon decomposition, 918
CMS, 206
  annuity to forward measure change,
    737–739
  convexity adjustment, 723–745
    annuity mapping function, see
      terminal swap rate model, annuity
      mapping function
    correcting arbitrage, 733–735
    density integration method, 738
    impact of mean reversion, 735–736
    impact of volatility smile, 735
    impact on implied volatility, 776
    Libor market model, 731–733
    linear TSR model, 728–730
    out-of-model adjustment, 963, 964
    quasi-Gaussian model, 730–731
    replication method, 723–725
    stochastic volatility model, 739
    swap-yield TSR model, 727–728

vega hedging, see terminal swap
    rate model, linear TSR model,
    vega hedging
  hedging portfolio, 725
  quanto, see quanto CMS
CMS cap, 207, 697
  impact of CMS convexity on
    volatility smile, 741
  link to European swaptions, 741
CMS digital spread option, 791
  dimensionality reduction, 791
CMS floor, 207
CMS rate, 206
  distribution in forward measure,
    737–739
CMS spread option, 210, 211, 621, 690,
    765, 776
  by integration, 777
  copula method, 776–784
  dimensionality reduction, 789
  floating digital, 792
  Gaussian copula, 777
    correlation impact, 778
    vega to swaptions, 778
  implied copula, 781
  implied correlation, 778
  Libor market model, 619–621, 636,
    692, 808
    closed-form approximation, 810
  Libor market model calibration, 636
  local volatility model, 1145
  Margrabe formula, 812
  Markovian projection, 1145, 1149
  multi-stochastic volatility, see
    multi-stochastic volatility model
  non-standard gearing, 777, 792
    dimensionality reduction, 792
  Normal spread volatility, 776
  one-dimensional integration, 789
  out-of-model adjustment, 964, 966
  power Gaussian copula, 781
  quadratic Gaussian model, 810
    closed-form approximations, 810
  risk management with one-factor
    model, 971
  stochastic volatility
    correlation impact, 807
  stochastic volatility de-correlation,
    962

stochastic volatility model, 1149
  correlation impact, 805
vega in Libor market model, 1116
CMS swap, 206, 697
  valuation formula, 207
CMS-linked cash flow, 723–745
  direct integration method, 737
  replication method, 725
coherent risk measure, see risk measure,
  coherent
collateral, 192, 266
complementary Gamma function, 281
complete market, 11
compounded rate, 200
conditional expected value, 19
  iterated conditional expectations, see
    iterated conditional expectations
  projection approximation, see
    Markovian projection, conditional
    expected value by projection
constant elasticity of variance model,
  see CEV model
constant maturity swap, see CMS swap
contingent claim, see derivative security
continuity correction, see payoff
  smoothing, continuity correction
control variate, 145–148, 330, 654,
  1077–1094
  adjusters method, 955
  construction from MC upper bound,
    1093
  dynamic, 148, 654, 1090–1093
  regression-based, 1091
  efficiency, 147
  impact on risk stability, 1093
  instrument-based, 1086–1090
  model-based, 676, 1077–1085
  non-linear controls, 147–148
  path re-weighting method, 961
  proxy Markov LM model, 1078
  proxy model, see control variate,
    model-based
convexity adjustment
  averaging swap, see Libor-with-delay,
    convexity adjustment
  CMS, see CMS, convexity adjustment
  futures, see ED future, convexity
    adjustment

Libor-in-arrears, see Libor-in-arrears,
  convexity adjustment
Libor-with-delay, see Libor-with-
  delay, convexity adjustment
moment explosion, 760–763
  second moment, 760
copula, 770
  Archimedean, 772
    Monte Carlo, 800
  Clayton, 773
  conditional CDF, 792
  Frechet bounds, 771
  Gaussian, 768
    CMS spread option, see CMS
      spread option, Gaussian copula
    integration, 789
    joint CDF, 769
    joint PDF, 769, 777
    mixture, 774
    Monte Carlo, 799
  Gumbel, 773
  implied, 781
  independence, 770
  mixture, 774
    Monte Carlo, 800
  perfect anti-dependence, 771
  perfect dependence, 770
  power Gaussian, 775, 780
    parameter impact, 781
  product, 775
    Monte Carlo, 800
  reflection, 773
    Monte Carlo, 800
  Sklar's theorem, 771
copula density, 772
copula method, 768
  CMS spread option, see CMS spread
    option, copula method
  dimensionality reduction, 789–799
    by conditioning, 793–797
    by measure change, 797–799
  forward swaption straddle, 949
  integration, 786–799
    inverse CDF caching, 787
    singularities, 788
  limitations, 801–802
  mapping function, 795
  Monte Carlo, 799–801
  observation lag, 784

quanto options, 748
volatility swap, 934
core correlations, *see* inter-temporal correlation
core volatilities, 865, 876
correlation extractor, *see* Libor market model, correlation extractor
correlation risk sensitivity, 1119
correlation smile, 778
Cox-Ingersoll-Ross model, 432
   multi-factor, 519
   two-factor, 517
Crank-Nicolson scheme, *see* PDE, Crank-Nicolson scheme
credit risk, 260, 975
credit value adjustment, 266, 916
cross-currency basis swap, *see* floating-floating cross-currency basis swap
cross-currency basis swap spread, 261, 265
CRX basis swap, *see* floating-floating cross-currency basis swap
CRX spread, *see* cross-currency basis swap spread
cumulant-generating function, 154
curve cap, 211, 766
   range accrual, *see* range accrual, curve cap
CVA, *see* credit value adjustment

date rolling convention, 224
day count convention, 224–226
   30/360, 226
   Actual/360, 225
   Actual/365.25, 224
day count fraction, *see* year fraction
deflator, 9
delta, 18, 132, 355, 980
   bucketed interest rate deltas, 251, 1045
   forward rate, 253
   Jacobian method, *see* risk sensitivities, Jacobian method
   par-point, 251, 252, 256, 257, 993
   parallel, 257
   with backbone, 1120–1122
delta hedge, 18
density process, 9

derivative security, 11
   attainable, 11
   pricing, 11
diffusion, 4, 14
   absorbing barrier, 281, 289
   displaced, 285
   Feller boundary classification, 280
   Feller condition, 319
   Fubini's theorem, 409
   integration by parts, 120
   Ito integral, *see* Ito integral
   Ito process, 4
   local time, 25, 26, 294
   Ornstein-Uhlenbeck process, 413
   polynomial growth condition, 19
   predictable process, 7
   scale measure, 280
   SDE, 14
      generator, 19
      linear, 16
      locally deterministic, 172, 541
      strong Markov, 15
      strong solution, 15
      weak solution, 15
   speed measure, 280
diffusion invariance principle, 14
discount bond, XXXIV, 23, 167
   valuation formula, 172
discount curve, *see* yield curve
displaced CEV model, *see* CEV model, displaced
displaced log-normal model, 285
   basket option, 1147
   canonical form, 286
   explicit solution to SDE, 312
   Fourier integration, 328
   implied correlation, 811
   moment matching, 922
   moment-generating function, 329
   time-dependent, 304
      effective skew, 305
      explicit solution to SDE, 307
      range for process, 306
Dupire local volatility, 1131
   proof by Tanaka extension, 294, 1131
duration, 246
DVF model, *see* local volatility model
Dybvig parameterization, *see* short rate model, Dybvig parameterization

early exercise, 30
ED future, 168–170, 196–197, 697,
    750–760
  convexity adjustment, 187, 197,
      750–760
    from market inputs, 752
    Gaussian HJM model, 186
    impact of volatility smile, 752, 757
    Libor market model, 753, 757
    replication method, 753, 757
  delivery arbitrage, 170
  futures rate, 169
    definition, 196
    instantaneous, 170, 172, 173
    martingale in risk-neutral measure,
        173, 750
    martingale in spot Libor measure,
        751
    simple, 170
    to forward rate, 756, 760
  mark to market, 170
  yield curve construction, 231, 992
ED futures contract, see ED future
effective volatility
  local volatility model, see local
      volatility model, effective
      volatility
  stochastic volatility model, see
      stochastic volatility model,
      effective volatility
envelope theorem, 1038
Eonia, 193, 200
equivalent martingale measure, see
    measure, equivalent martingale
Esscher transform, see exponential
    twisting
Eurodollar futures contract, see ED
    future
European call option, 23
  at-the-money, 24
  Fourier integration, 324
  in-the-money, 24
  out-of-the-money, 24
  probability density from, see
      volatility smile, probability
      density from
European digital call option, 59
European option
  Fourier integration, 326

European put option, 24
  at-the-money, 24
  in-the-money, 24
  out-of-the-money, 24
European swaption, 203, 697–705
  cash-settled, 205, 744–745
    payoff, 744
    put-call parity, 745
    replication method, 744, 745
  core swaptions, 424, 819
  coterminal swaptions, see European
      swaption, core swaptions
  diagonal swaptions, see European
      swaption, core swaptions
  forward swaption straddle, see
      forward swaption straddle, 943
  midcurve, 223
  non-standard, see Bermudan
      swaption, non-standard
    Black formula, 889
  physically-settled, 205
  SV model calibration, 703–704
  swap-settled, 205, 744, 745
  swaption grid, 205, 703
  swaption strip, 423
  tenor, 204
  valuation formula, 204
  volatility cube, 697
European-style option, 95
  replication method, 337
  valuation by volatility mixing, 339
exchange market, 193
  Chicago Mercantile Exchange, 196
  London International Financial
      Futures and Options Exchange,
      196
  Marché à Terme International de
      France, 196
exotic swap, 205, 208, 209, 822, 951
  CMS spread, 766
  CMS-based, 210
  digital CMS spread, 766
  global cap, 219
  global floor, 219
  knock-out, 218
  Libor-based, 209
  multi-rate, 210, 766
  path-dependent, 212
  principal amount, 208

range accrual, *see* range accrual

snowball, 212

spread-based, 210

structured coupon, 208–211

expectations hypothesis, 173

expected hedging P&L, 988

exponential distribution, 98

Monte Carlo, 98

exponential integral, 334

exponential twisting, 154

extra state variable method, *see* PDE,
path-dependent options

"The Fed Experiment", 452

Federal funds future, 201

Federal funds rate, 192, 200, 201, 266

effective, 192

target, 192

Federal funds/Libor basis swap, 201,
266

Feller condition, *see* diffusion, Feller
condition

Feynman-Kac solution, 21

FFT, *see* stochastic volatility model,
Fourier integration

filtration, 3, 4

usual condition, 3

flexi-cap, 71

flexi-swap, 900–904

decomposition into Bermudan
swaptions, 901

local projection method, 901

marginal exercise value decomposition, 903

narrow band limit, 904

PDE, 900, 902

purely local bounds, 901

"flip-flop", 210

floating digital, 792, 794

dimensionality reduction, 792

floating digital spread option, 792

dimensionality reduction, 792

floating-floating cross-currency basis
swap, 262, 264, 265

floating-floating single-currency basis
swap, 201, 268

floor, *see* cap

Fokker-Plank equation, *see* Kolmogorov
forward equation

Fong-Vasicek model, 454–455, 517

bond reconstitution formula, 454

forward CMS straddle, 941, 944, 945

swaption, *see* forward swaption
straddle

volatility, *see* forward volatility

forward contract, 195

forward Kolmogorov equation, *see*
Kolmogorov forward equation

forward Libor model, *see* Libor market
model

forward Libor rate, XXXIV, 168, 191,
192, 196

accrual end date, 224

accrual period, 224

accrual start date, 224

martingale in forward measure, 174

tenor, 168

variance by replication method, 757

year fraction, *see* year fraction

forward par rate, *see* forward swap rate

forward price, 23, 168

forward rate, 167

continuously compounded, XXXIV,
168

instantaneous, XXXIV, 169

simple, 168

tenor, 168

volatility hump, 418, 494

forward rate agreement, *see* forward
contract

forward starting option, 222

forward swap rate, XXXIV, 171, 199

distribution in forward measure, *see*
CMS rate, distribution in forward
measure

expiry, 171

fixing date, 171

linking forward and annuity measure,
737

market-implied variance, 557

martingale in swap measure, 178

non-standard, 882

decomposition, 882

tenor, 171

weighted average of Libor rates, 171,
256

forward swaption straddle, 223,
945–950

copula method, 949
relation to CMS spread option, 948
triangulation, *see* forward volatility, triangulation
vanilla model, 946
vega exposure, 948
volatility, *see* forward volatility
forward volatility, 223
    connection to inter-temporal correlations, *see* inter-temporal correlation, connection to forward volatilities
    hedging, 914
    impact of rate correlation, 919
    impact of volatility smile, 945
    Libor rate, *see* volatility, forward volatility of Libor rate
    triangulation, 948
forward volatility derivative, 220, 222
    forward swaption straddle, *see* forward swaption straddle
    implied Normal volatility contract, 223
    midcurve swaption, *see* European swaption, midcurve
    volatility swap, *see* volatility swap
forward yield, *see* forward rate
Fourier transform, 325
    inverse, 325
FRA, *see* forward contract
Frobenius norm, *see* matrix, Frobenius norm
fundamental matrix, 486
fundamental theorem of arbitrage, 10
fundamental theorem of derivatives trading, 987
futures contract, *see* ED future
futures rate, *see* ED future, futures rate
fuzzy logic, *see* payoff smoothing, fuzzy logic
FX rate, 179, 746, 748
    dynamics in domestic risk-neutral measure, 180
    forward, 178
        martingale in domestic forward measure, 180

Gâteaux derivative, 253

gamma, 980
    pathwise differentiation method, *see* pathwise differentiation method, gamma
    payoff smoothing, 1019
    relationship to vega, 981
gamma distribution, 100
    Monte Carlo, 100, 102
    PDF, 100
Gamma function, XXXIII
    incomplete, *see* incomplete Gamma function
    quick approximation, 1153
Gauss-Hermite quadrature, *see* quadrature, Gauss-Hermite
Gaussian copula, *see* copula, Gaussian
Gaussian distribution, XXXIII
    conditional distribution, 648
    cumulant-generating function, 154
    imaginary mean, 798
    inverse CDF, 99, 165
    linear transform, 103
    measure change, 797
    multi-dimensional PDF, 103
    quadratic form, 523
    moment-generating function, 524, 535
    moments, 535
Gaussian HJM model, 184–187
    caplet, 186
    ED future convexity adjustment, *see* ED future, convexity adjustment, Gaussian HJM model
    time-stationary, 418
    zero-coupon bond option, 185
Gaussian multi-factor short rate model, *see* Gaussian short rate model, multi-factor
Gaussian one-factor short rate model, *see* Gaussian short rate model
Gaussian short rate model, 408, 415–431, 480–512
    as special case of affine model, 432
    Bermudan swaption, *see* Bermudan swaption calibration, local projection method, Gaussian short rate model
    bond dynamics, 417
    bond reconstruction formula, 416

efficient calculation, 417
calibration, 423
  bootstrap, 424
calibration to yield curve, 416
European swaption, 420, 423
  Jamshidian decomposition, 420
fast pricing of Bermudan swaptions,
  916
forward rate dynamics, 415
forward rate volatility, 415
  dynamics, 419
humped volatility structure, 418
in spot measure, 430
in terminal measure, 430
mean reversion, see mean reversion
mean reversion calibration, see mean
  reversion calibration
Monte Carlo, 427–431
  approximate, 429
  Euler scheme, 429
  exact, 427
  other measures, 430
multi-factor, 480–512
  benchmark rate parameterization,
    507–510
  benchmark rates, 508
  benchmark tenors, 508
  bond reconstitution formula, 480,
    483, 485
  bond volatility, 481
  calibration, 507
  classic development, 487–490
  correlated Brownian motions, 490
  correlation stationarity, 490
  European swaption, 501–507
  European swaption by Jamshidian
    decomposition, 505
  factors and loadings, see Gaussian
    short rate model, multi-factor,
    statistical approach
  forward rate correlation, 490–491
  forward rate volatility, 484
  Gaussian swap rate approximation,
    506–507
  loadings, 500
  mean reversion matrix diagonaliza-
    tion, 488–490
  Monte Carlo, 510–511
  PDE, 511–512

rotations, 486
separability, 480–487
short rate dynamics, 481
short rate state distribution, 487,
  510
short rate state dynamics, 481–487
short rate state dynamics,
  integrated, 487, 510
single Brownian motion, 498
statistical approach, 497–501
swap rate volatility, 506
PDE, 425–427
  boundary conditions from PDE,
    426
short rate distribution, 428
short rate dynamics, 415
short rate state dynamics, 416, 427
  integrated, 427
swap rate dynamics in annuity
  measure, 422
swap rate volatility, 422
time-stationary, 418
two-factor, 491–497
  bond reconstitution formula, 492,
    502
  CLE, see CLE calibration, local
    projection method, two-factor
    Gaussian model
  correlated Brownian motions, 492
  correlation stationarity, 493
  doubly mean-reverting form, 494
  European swaption by Jamshidian
    decomposition, 502–506
  forward rate correlation, 492–494
  forward rate dynamics, 492
  forward rate volatility, 492–496
  short rate state conditional
    distribution, 504
  short rate state correlation, 491
  short rate state dynamics, 491
  single Brownian motion, 497
  volatility hump, 494
Gaussian two-factor short rate model,
  see Gaussian short rate model,
  two-factor
generalized trigger product, 1074
  importance sampling, 1074–1077
  pathwise differentiation method,
    1041–1044

payoff smoothing, 1074–1077
trigger variable, 1074
tube Monte Carlo, *see* barrier option,
    tube Monte Carlo
Girsanov's theorem, 12, 13
Gaussian distribution, 797
Gram-Charlier expansion, 368, 439
greeks, *see* risk sensitivities
Green's function, 20
grid shifting, *see* payoff smoothing,
    grid shifting
GSR model, *see* Gaussian short rate
    model
Gyöngy theorem, *see* Markovian
    projection, Gyöngy theorem

H^2, 5
Hagan and Woodward parameteriza-
    tion, *see* short rate model, Hagan
    and Woodward parameterization
hat smoothing method, *see* payoff
    smoothing, hat smoothing
Heath-Jarrow-Morton model, *see* HJM
    model
hedge, 251
    best hedging strategy, 355
    beta, 357
    minimum variance, 355–357
    model-independent, 718
    semi-static, *see* replication method,
        semi-static
    shadow delta, *see* volatility smile,
        shadow delta hedging
    sub-replicate, 719
    super-replicate, 719, 979
    zero-beta, 357
Hermite matrix, 270
Heston model, *see* stochastic volatility
    model
HJM model, 181–190
    bond dynamics, 181
    forward bond dynamics, 182
    forward rate dynamics, 182
    Gaussian, *see* Gaussian HJM model
    Gaussian Markov, 187–189
        short rate dynamics, 188
    log-normal, 189–190
    Markovian, 407
    separable, 415

short rate dynamics, 183
stochastic basis, *see* HJM model,
    two-curve
two-curve, 680–683
    forward rate spread dynamics, 681
    Gaussian basis spread, 683
    index bond dynamics, 682
    index forward rate dynamics, 681
    index short rate dynamics, 681
    quanto correction, 682
Ho-Lee model, 408–412
    bond dynamics, 411
    bond reconstitution formula, 410
    calibration to yield curve, 409
    drawbacks, 412
    forward rate dynamics, 411
    short rate dynamics, 410
hybrid differentiation method, 1061

implied volatility, *see* volatility, implied
importance sampling, 146, 149–158,
    1063–1077
    application to payoff smoothing,
        1067
    barrier option, *see* barrier option,
        importance sampling
    density formulation, 149
    efficiency, 150
    generalized trigger product, *see*
        generalized trigger product,
        importance sampling
    least-squares, 153
    likelihood ratio, 150, 153, 155
    rare events, 154
        approximately optimal mean shift
            in multi-variate case, 157
        asymptotic optimality, 158
        efficiency, 156
        minimal variance, 155
        multi-variate, 156
        SDE, 151–154
    short rate model, *see* short rate
        model, importance sampling
    survival measure, 1067
    simulation under, 1072, 1074, 1076
    TARN, *see* TARN, importance
        sampling
incomplete Gamma function, XXXIII,
    281

index, 206

index option, *see* basket option

infinitesimal operator of SDE, *see* diffusion, SDE, generator

infinitesimal perturbation analysis, 136

information theory, 957

instantaneous futures rate, *see* ED future, futures rate, instantaneous

integration by parts for diffusion process, *see* diffusion, integration by parts

inter-temporal correlation, 424, 554, 820, 865, 876

  connection to forward volatilities, 820

  hedging, 914

  impact of mean reversion, 554

  impact of volatility smile, 945

  impact on Bermudan swaption, *see* Bermudan swaption valuation, impact of inter-temporal correlation

  impact on CLEs, *see* CLE valuation, impact of inter-temporal correlation

  impact on TARNs, 929

  mean reversion calibration to, *see* mean reversion calibration, to inter-temporal correlations

interbank money market, 192

International Swaps and Derivatives Association, 192, 266

intrinsic value, 26

inverse floater, 209

iterated conditional expectations, 176

Ito integral, 4, 5

Ito isometry, 5

Ito's lemma, 6

Ito-Taylor expansion, 118

Jacobian, *see* risk sensitivities, Jacobian method

Jamshidian decomposition

  American/Bermudan option, *see* American/Bermudan option, Jamshidian decomposition

  European swaption, *see* Gaussian short rate model, European swaption, Jamshidian decomposition

Kolmogorov backward equation, 19, 20

Kolmogorov forward equation, 20, 386, 459, 1048

  correct boundary conditions, 386

  discrete consistency with backward equation, 460

Kullback-Leibler relative entropy, 957

kurtosis, 375

L^1, XXXIV, 4

L^2, XXXIV, 4

ladder, 985

ladder swap, *see* ratchet swap

Lagrange basis functions, *see* PDE, Lagrange basis; payoff smoothing, Lagrange basis

Lagrange multiplier, 249, 958

least squares method, *see* CLE regression

LIA, *see* Libor-in-arrears

Libor curve, *see* yield curve

Libor market model, 451, 591–694, 731, 868, 911

  annuity mapping function, 732, 733

  asset-based adjustment, 963

  back stub, 657–662

    arbitrage-free, 659–661

    from Gaussian model, 661–662

    simple, 658–659

  choosing number of factors, 614

  CLE, 821

  CMS convexity adjustment, 964

  correlation extractor, 865

  deflated bond dynamics, 651

  delta with backbone, 1120–1122

  drift approximation, 646

    Brownian bridge, 1079

  drift freezing, 1052

  exercise boundary, 911

  exercise strategy, 908

  expected value of Libor rate in annuity measure, 671

  front stub, 662–667

    exogenous volatility, 663–665

    from Gaussian model, 666–667

    simple interpolation, 666

    zero volatility, 662–663

    in hybrid measure, 642

index function, *see* tenor structure,
    index function
Libor rate correlation, 603–614, 758
    correlation PCA, 611
    covariance PCA, 626
    historical estimation, 606
    majorization, 613
    parametric form, 608, 609
    PCA, 604–606
    poor man's correlation PCA, 614
    regularization, 610
Libor rate dynamics, 593–603
    annuity measure, 733
    in forward measures, 594–595
    in hybrid measure, 597
    in spot measure, 596
    in terminal measure, 595, 641
Libor rate inter-temporal correlation,
    758
Libor rate volatility
    from volatility norm, 625–627
    functional form, 622
    grid-based, 622–623
    interpolation, 624–625
Libor rate volatility link to HJM
    forward rate volatility, 598
link to HJM, 597
local volatility, 598–600
    CEV, 599
    displaced log-normal, 599
    existence and uniqueness, 599
    LCEV, 599
    log-normal, 599
Markov, 676, 1078–1085
    as control variate, 1084
    Brownian bridge, 1079
    calibration, 1082
    one-factor, 1079
    one-factor reconstitution formula,
        1080
    separable volatility, 1080
    two-factor, 1081
    two-factor reconstitution formula,
        1081
Markovian projection, 667, 669, 1139
model risk, 629
multi-stochastic volatility, 690–694,
    962
    caplet, 691

CMS spread option, 692
European swaption, 691
moment-generating function, 692
Musiela parameterization, 604
pathwise derivative
    forward Libor rate, 1051
    forward swap rate, 1055
    numeraire, 1054
    structured coupon, 1055
    stub bond, 1054
pathwise differentiation method,
    1051–1058
    computational complexity, 1052
PCA, *see* Principal Components
    Analysis
portfolio replication, 914
stochastic basis, *see* Libor market
    model, two-curve
stochastic variance dynamics, 689
stochastic volatility, 600–603
    moment-generating function, 689
    non-zero correlation, 687
stub volatility, 664, 667
swap rate correlation, 620–621
swap rate dynamics, 617, 669
    approximate, 618
time-stationary, 623
tool to extract forward volatility, 821
two-curve, 683–687
    deterministic spread, 687
    European swaption, 686
    Libor rate dynamics, 685
    Monte Carlo, 685
    swap rate dynamics, 686
vega, *see* vega, Libor market model
Libor market model calibration,
    622–637
    algorithm, 633, 636, 675
    bootstrap, 635
    for vega, 1111
    cascade, *see* Libor market model
        calibration, bootstrap
    choice of instruments, 627
    effective skew, 672
    effective volatility, 670
    global, 628
    grid-based, *see* Libor market model
        calibration, global
    local, 628

objective function, 630
PCA, 626
row-by-row, 633, 634
to spread options, 635, 808
volatility skew, 637
volatility smile, 674
Libor market model valuation
Bermudan swaption, see Bermudan
swaption valuation, Monte Carlo
caplet, 615
CLE, see CLE valuation, Libor
market model
CMS convexity adjustment, see
CMS, convexity adjustment,
Libor market model
CMS spread option, see CMS spread
option, Libor market model
curve interpolation, 657–667
European swaption, 616, 618, 667
Libor-with-delay, see Libor-with-
delay, Libor market model
Monte Carlo, 637
analysis of computational effort,
639
antithetic variates, 653
Brownian bridge, 647
choice of numeraire, 642
control variate, 654
discretization bias, 639
Euler scheme, 638
front stub, 664
high-order schemes, 650
importance sampling, 655
lagging predictor-corrector, 644
large time steps, 641, 646–649
log-Euler scheme, 638
martingale discretization, 650–653
Milstein scheme, 650
predictor-corrector, 643, 644, 647,
653
survival measure, 1072, 1075
two-curve, 685
variance reduction, 653–655
multi-rate vanilla derivative, 807
PDE, see Libor market model,
Markov
TARN, see TARN, Libor market
model

volatility swap, see volatility swap,
Libor market model
Libor rate, see forward Libor rate
Libor-in-arrears, 200, 716–719
convexity adjustment, 717
replication method, 718
sub-replicating portfolio, 719
super-replicating portfolio, 719
Libor-with-delay, 719–723
convexity adjustment, 719
Libor market model, 720, 722
quasi-Gaussian model, 720, 721
replication method, 720, 722
swap-yield TSR model, 720
likelihood ratio method, 139–142,
1060–1061
discontinuous payoff, 138
exploding variance, 1061
for Euler scheme, 140–141
for Milstein scheme, 141
log-likelihood ratio, 140
score function, 140
vega, 1124
linear regression, 146
Lipschitz function, 137
LM model, see Libor market model
local projection method, 560, 864, 953,
1097
Bermudan swaption, see Bermudan
swaption calibration, local
projection method
CLE, see CLE calibration, local
projection method
non-standard Bermudan swaption, see Bermudan swaption,
non-standard, local projection
method
TARN, see TARN, local projection
method
volatility swap, see volatility swap,
local projection method
local stochastic volatility model, 316,
1137–1145
calibration, see Markovian projection,
LSV calibration
Markovian projection, see Markovian
projection, LSV calibration
local time, see diffusion, local time
local volatility model, 277–312

approximation with displaced
log-normal model, 286
asymptotic expansion, 295–299
basket option, see Markovian
projection, basket option in LV
model
CEV, see CEV model
displaced log-normal, see displaced
log-normal model
effective convexity, 307–312
effective skew, 301–312
effective volatility, 301
expansion around displaced
log-normal model, 296
expansion around Gaussian model,
298
forward equation for call options, 293
PDE, 292–295
simultaneous for multiple
parameters, 293
space discretization, 292
transform to constant diffusion
coefficient, 87, 292
quadratic volatility, see quadratic
volatility model
range-bound, 287
small-noise expansion, see volatility,
small-noise expansion
smile dynamics, 279, 350, 352
time-dependent, 299–312
separable, 300
log-normal distribution, XXXIII, 16
moment matching, see moment
matching
moments, 16
Monte Carlo, 101
Longstaff-Schwartz method, see CLE
regression
Longstaff-Schwartz model, 517–519
bond reconstitution formula, 518
lookback option, 124
Monte Carlo, see Monte Carlo,
lookback option
LS method, see CLE regression
LSV model, see local stochastic
volatility model
LVF model, see local volatility model

Malliavin calculus, 142, 1042, 1060

Margrabe formula for spread option,
812
mark-to-model, 818
Markov process, 15
Feynman-Kac theorem, see
Feynman-Kac solution
strong, 15
transition density, 20
Markov-functional model, 472–478
calibration to yield curve, 475
criticism, 478
Libor parameterization, 473
log-normal, 474
no-arbitrage condition, 473
non-standard Bermudan swaption,
881
numeraire, 472
numeraire mapping, 472
Libor parameterization, 473
non-parametric, 476
swap parameterization, 475
PDE, 477
state process, 472
swap parameterization, 475
transition density, 472
Markovian projection, 805, 1129–1156
average option, 1133
barrier option, 1134
basket option in LV model,
1145–1148
basket option in SV model,
1149–1152
CMS spread option, 1145
conditional expected value by
Gaussian approximation,
1134–1135
conditional expected value by
projection, 726, 727, 1136–1137
displaced Heston model, 1149, 1151
non-perturbative approximation,
1151
displaced log-normal model, 1136,
1146
Gyöngy theorem, 1130
LSV calibration, 1139–1145
mapping function, 1142
proxy model, 1143–1145
quadratic volatility model, 1137,
1148

quasi-Gaussian model, *see* quasi-Gaussian model, Markovian projection
spread option, 1151
stochastic volatility model, 1138
martingale, 5
Doob-Meyer decomposition, 35
exponential, 12
Doleans exponential, XXXIII, 12
local, 5
bounded, 288
martingale representation theorem, 6
Novikov condition, 12
optional sampling theorem, 35
Snell envelope, 31, 823
square-integrable, 5
stopping time, *see* stopping time
submartingale, 5
supermartingale, 5, 360
CEV, *see* CEV model, strict supermartingale
quadratic volatility, *see* quadratic volatility model, strict super-martingale
SV model, *see* SV model with general variance process, strict supermartingale
matrix
exponential, 486
Frobenius norm, 105, 609–611, 626, 627, 851
infinity norm, 53
positive semi-definite, 103
Cholesky decomposition, 103
rank-deficient, 106
spectral norm, 53
stiffness, 1111
tri-diagonal, 47
MAX-option, 908
mean reversion, 316, 413, 552, 573
effects, 552–554
inter-temporal correlation, 554
swaption volatility ratio, 553
mean reversion calibration, 552–560, 573
to inter-temporal correlations, 557–559
to row of European swaptions, 555, 888

to volatility ratios, 554–557
mean-reverting square-root process, *see* square-root process
measure, XXXIII
absolutely continuous, 1067
annuity, 178, 204
change of numeraire, *see* numeraire, change of numeraire
domestic, 746
equivalent, 9, 1067
equivalent martingale, 8, 9, 14, 171
foreign, 746
hybrid, 176
local martingale, 10
risk-neutral, XXXIII, 22, 172
domestic and foreign, 179, 180
spot, XXXIII, 175
survival density, 1047
survival for Bermudan swaption, *see* Bermudan swaption, survival measure
survival in importance sampling, *see* importance sampling, survival measure
T-forward, XXXIII, 29, 174
domestic and foreign, 180
terminal, 176
min-max volatility swap, 222, 938
capped, 940
semi-static replication, 939
moment explosion, 323, 343, 344, 361, 760, 761
impact on convexity adjustment, *see* convexity adjustment, moment explosion
SABR model, *see* SABR model, moment explosion
stochastic volatility model, *see* stochastic volatility model, moment explosion
SV model with general variance process, *see* SV model with general variance process, moment explosion
moment matching, 889, 920–924
Asian option, 922
basket option, 924
moment-generating function, 13
Monte Carlo, 95–165

A-stable scheme, 110
Asian option, 107
Asian option on basket, 107
average rate option, *see* Monte Carlo,
    Asian option
barrier option, 124–128
  adjusting barrier for sampling
    frequency, 128
  double-barrier knock-out, 124
bias, 122
bias/standard error trade-off, 123
Brownian motion, *see* Brownian
    motion
calibration by stochastic optimization
    method, 953
central limit theorem, 96
convergence rate, 97
discretization bias, 428
efficiency, 143
Euler scheme, 110, 111
  linear SDE, 112
  region of stability, 111
  weak convergence order, 111
Euler-Maruyama scheme, *see* Monte
    Carlo, Euler scheme
Heun scheme, 116
higher-order schemes, 116
implicit Euler scheme, 113
  region of stability, 114
implicit Milstein scheme, 390
log-Euler scheme, 112, 113
lookback option, 125
low-discrepancy sequence, *see* Monte
    Carlo, random number generation,
    quasi-random
lower bound for American option, 34,
    35, 164
  parametric, 159, 160
  regression-based, 161
mean-square error, 123
Milstein scheme, 119, 121
  multi-dimensional, 121
modified trapezoidal scheme, *see*
    Monte Carlo, Heun scheme
optimal root-mean-square error, 123
perfect foresight bias, *see* Ameri-
    can/Bermudan option, perfect
    foresight bias
predictor-corrector, 115, 116

convergence order, 116
random number generation, 97
  acceptance-rejection method,
    99–101
  Box-Muller method for Gaussian
    distribution, 99
  composition method, 101–102
  conditional Gaussian, 1066
  correlated Gaussian, 103
  correlated Gaussian by Cholesky
    decomposition, 103
  correlated Gaussian by eigenvalue
    decomposition, 104
  inverse transform method, 98
  linear congruential generator, 97
  Marsaglia polar method for
    Gaussian distribution, 99
  Mersenne twister, 98
  period, 98
  pseudo-random, 97, 130
  quasi-random, 129
  Sobol, 129
region of stability, 110
Richardson extrapolation, 122, 470
sample mean, 96
sampling extremes, 124–128
  adjusting barrier for sampling
    frequency, 128, 937, 970
  with Brownian bridge, 125
SDE discretization, 108
second-order scheme, 119, 121
seed, 97
standard error, 97, 122
  for digital option, 133
  for greeks, 132, 135
strong convergence order, 111
strong law of large numbers, 96
strongly consistent, 109
third-order scheme, 470
upper bound for American option,
    34–36, 163, 164
variance reduction, *see* variance
    reduction
weak convergence, 109
weak convergence order, 110
weakly consistent, 109
most likely path, *see* volatility, implied,
    most likely path approximation
multi-rate vanilla derivative, 765–815

copula method, *see* copula method
Libor market model, 809
observation lag, 784
stochastic volatility, *see* multi-
stochastic volatility model
term structure models, 806
multi-stochastic volatility model,
803–808, 1149
correlation impact, 805
measure change by CMS caplet
calibration, 804
measure change by drift adjustment,
803
Monte Carlo
Quadratic-Exponential scheme, 805
multi-rate vanilla derivative, 803–808
multi-tranche, *see* CLE, multi-tranche

non-central chi-square distribution, 283
asymptotics, 392
CDF, 102, 319
in CEV model, 283
in delta-gamma VaR/cVaR, 998
in LS model, 519
two-dimensional, 519
Normal model, XXXIV, 283
CMS spread, 776
vega to swaptions, 777
numeraire, 10, 171
change of numeraire, 11
Girsanov's theorem, *see* Girsanov's
theorem
discrete money market account,
XXXIV, 175, 176
money market account, XXXIV, 22,
28, 172

OIS, *see* overnight index swap
one-dimensional integral for spread
option, 789
operator calculus, 998–999
OTC market, *see* over-the-counter
market
out-of-model adjustment, 951–971
adjusters method, 954–956
algorithm, 955
as control variate, 955
volatility adjustment, 956
asset-based adjustment, 963–964

CMS spread option, 964
coupon calibration, 952–954
delta-adjustment method, 956
extended calibration, 953
fee adjustment method, 967–969
additive, 967
blended, 968
impact on derivatives, 968
multiplicative, 968
issues, 961, 964
mapping function adjustment, 965
market adjustment, 965
path re-weighting method, 956–961
as control variate, 961
Boltzman-Gibbs distribution, 959
Boltzman-Gibbs weights, 959
dual, 961
inappropriate use, 958
partition function, 958
risk sensitivities, 961
PDE for coupon values, 953
proxy model method, 961
spread adjustment method, 966
strike adjustment method, 969–971
impact on derivatives, 970
over-the-counter market, 193
overhedge, 1023
overlay curve, *see* yield curve, overlay
curve
overnight index swap, 193, 200, 266

P&L, 698, 990–995
P&L analysis, 986
P&L attribution, *see* P&L explain
P&L explain, 993–995
bump-and-do-not-reset explain, *see*
P&L explain, waterfall explain
bump-and-reset explain, 994–995
waterfall explain, 993–994
P&L explanation, *see* P&L explain
P&L of hedged book, 987–990
P&L predict, *see* P&L prediction
analysis
P&L prediction analysis, 258, 991–993
first-order, 991
second-order, 991
unpredicted P&L, 991
par rate, *see* forward swap rate

parameter averaging, *see* calibration, time averaging
partial differential equation, *see* PDE
partition function, 958
pathwise delta approximation, *see* pathwise differentiation method, pathwise delta approximation
pathwise differentiation method, 135–138, 1035–1060
  adjoint method, 1056
  computational complexity, 1053, 1057
  barrier option, *see* barrier option, pathwise differentiation method
  Bermudan swaption, *see* Bermudan swaption greeks, pathwise differentiation method
  CLE, *see* CLE greeks, pathwise differentiation method
  computational complexity, 1052, 1053
  discontinuous payoff, 1042, 1061
  European option, 1054
  gamma, 1050, 1056
  generalized trigger product, *see* generalized trigger product, pathwise differentiation method
  Libor market model, *see* Libor market model, pathwise differentiation method
  money market account, 1046
  Monte Carlo models, 1051–1060
  pathwise delta approximation, 1059
  PDE models, 1044–1050
  sensitivity path generation, 138
  TARN, *see* TARN, pathwise differentiation method
  vega, 1050, 1056
payoff smoothing, 1001–1033
  adaptive integration, 1006
  adding singularity to grid, 78, 1007
  barrier option, 1074–1077
  benefits, 1012
  Bermudan swaption, *see* CLE greeks, tube Monte Carlo
  box smoothing, 1015–1018
    multiple dimensions, 1020
    on discrete grid, 1015
  by importance sampling, 1065–1077

  CLE, *see* CLE greeks, tube Monte Carlo
  continuity correction, 59, 1012
  fuzzy logic, 1028
  gamma, 1019
  grid shifting, 1007
  hat smoothing, 1019
  integration, 1012
  Lagrange basis, 59, 1018
  locality, 1019
  Monte Carlo, 1022–1030
  moving average, 1012
    choice of window, 1014
  multiple dimensions, 1019–1022
    box smoothing, 1020
    dominant dimension, 1022
  one dimension, 1014
  partial analytical integration, 76–78, 1010
  partial coupons, 1028
  PDE, 1012
  piecewise smooth function on a grid, 1016
  singularity removal, 1009
  TARN, *see* TARN, payoff smoothing; TARN, tube Monte Carlo
  tube Monte Carlo, *see* tube Monte Carlo
PCA, *see* Principal Components Analysis
PDE, 18, 43–93
  A-stable scheme, 55
  ADI scheme, 43, 82–85
    boundary conditions, 85
  Asian option, 70
  backward induction, 51
  Black-Scholes, *see* Black model, PDE
  boundary conditions
    for barrier options, 64
    from PDE itself, 385, 426
    linear at boundary, 48
    log-linear at boundary, 48
  Cauchy problem, 18, 44
  centering, 563
  conditional stability, 55
  consistent scheme, 56
  convection-dominated, 61–63
  convergent scheme, 56
  coupon-paying, 67

Craig-Sneyd scheme, *see* PDE,
    predictor-corrector scheme
Crank-Nicolson scheme, 50
    American options, 69
    not strongly A-stable, 55
    oscillations, 55, 58
Dirichlet problem, 44, 64
    space discretization, 46
dividends, 67, 68
domain truncation, 44
    stability of greeks, 1002
Douglas-Rachford scheme, 84, 91
    boundary conditions, 85
early exercise, 68
exponentially fitted schemes, 63
extra state variable method, *see*
    PDE, path-dependent options
for implied volatility, *see* volatility,
    implied, PDE for
forward equation, *see* Kolmogorov
    forward equation
fully implicit scheme, 50
greeks off grid, 1005
L-stable scheme, 55
Lagrange basis, 58, 59
Lax equivalence theorem, 56
local volatility model, *see* local
    volatility model, PDE
mesh refinement, 72, 79
    equidistant blocks, 73, 74
    non-equidistant, 74
multi-dimensional, 91
multi-exercise, 71
multi-level time-stepping, 58
non-equidistant discretization, 56
Nyquist frequency, 59
odd-even effect, 59
operator splitting, 82
orthogonalization, 86
    drawbacks, 88
partial analytical integration,
    see payoff smoothing, partial
    analytical integration
path-dependent options, 69, 71, 870,
    872, 898, 900, 932, 934
Peaceman-Rachford scheme, 84
    boundary conditions, 85
predictor-corrector scheme, 89–92
quantization error, 59

Rannacher stepping, 58–61, 67, 459
semi-Lagrangian methods, 63
Shannon Sampling Theorem, 59
similarity reduction, 71
sinh transform, 384
smoothing, 58–60
    continuity correction, 59
    grid dimensioning, 1002
    grid shifting, 59, 1002
    space discretization, 45
stable scheme, 53
strongly A-stable scheme, 55
time discretization, 49
    theta scheme, 50
two-dimensional, 80
two-dimensional with mixed
    derivatives, 85, 86, 89
upwinding, 62
variable transform, 44
von Neumann method, 53–56
    amplification factor, 54
    stability criterion, 54
    well-posed, 56
Poisson distribution, 102
portfolio replication, *see* Bermudan
    swaption greeks, portfolio
    replication for hedging
power Gaussian copula, *see* copula,
    power Gaussian
predictor-corrector, 89, 115, 382, 643
    Monte Carlo, *see* Monte Carlo,
        predictor-corrector
    PDE, *see* PDE, predictor-corrector
        scheme
present value of a basis point, *see* swap,
    annuity
principal component, 105
Principal Components Analysis, 105,
    106, 500, 604–606
principal factor, 105
product integral, 486
Profit-And-Loss, *see* P&L
pseudo-Gaussian model, *see* quasi-
    Gaussian model
pseudo-random number generator, *see*
    Monte Carlo, random number
    generation, pseudo-random
put-call parity, 24
PVBP, *see* swap, annuity

QG model, *see* quadratic Gaussian
    model
qG model, *see* quasi-Gaussian model
quadratic covariation, XXXIII, 7
quadratic Gaussian model, 443,
    520–534
  as affine model, 521
  benchmark rate parameterization,
    527
  Bermudan swaption, *see* Bermudan
    swaption calibration, local
    projection method, quadratic
    Gaussian model
  bond dynamics, 522
  bond reconstitution formula, 522
  calibration, 533–534
    multi-pass bootstrap, 533
  CLE, *see* CLE calibration, local
    projection method, quadratic
    Gaussian model
  CMS spread option, *see* CMS spread
    option, quadratic Gaussian model
  curve factor, 525
  European swaption, 528–533
    approximations, 530
    exact, 529
    Fourier integration, 531
    rank-2 approximation, 532
  Fourier integration, 531
  mean-reverting state variables, 521
  moment-generating function, 531
  Monte Carlo, 534
  one-factor, 443
  parameterization, 524–527
  PDE, 534
  quadratic approximation to swap
    rate, 530
  short rate, 520
  short rate in SV form, 527
  short rate state distribution
    in annuity measure, 528
    in forward measure, 523
  short rate state dynamics, 443, 520
    in forward measure, 522
    in annuity measure, 528
  smile generation, 524–525
  spanned stochastic volatility, 525,
    534

TARN, *see* TARN, local projection
    method, quadratic Gaussian
    model
  volatility factor, 525
  volatility smile, 533
  volatility swap, *see* volatility swap,
    quadratic Gaussian model
quadratic variation, XXXIII, 7
quadratic volatility model, 287–291
  European call option value, 290
  European put option value, 290, 291
  Markovian projection, 1137
  measure change, 289
  small-noise expansion, 308
  smile dynamics, 350
  strict supermartingale, 288
  time-dependent, 308
Quadratic-Exponential scheme, *see*
    square-root process, Monte Carlo,
    Quadratic-Exponential scheme
  multi-dimensional, *see* multi-
    stochastic volatility model, Monte
    Carlo, Quadratic-Exponential
    scheme
quadrature, 533, 788
  Gauss-Hermite, 533, 789
  Gauss-Legendre, 788
  Gauss-Lobatto, 788
quanto CMS, 745–750
  annuity mapping function, 749
  convexity adjustment, 749–750
  copula method, 748
  quanto adjustment, 747
  replication method, 747
quasi-Gaussian model, 539–590
  Bermudan swaption, *see* Bermudan
    swaption calibration, local pro-
    jection method, quasi-Gaussian
    model
  bond reconstitution formula, 540
  calibration, 584
  CEV local volatility, 547
  CLE, *see* CLE calibration, local pro-
    jection method, quasi-Gaussian
    model
  CMS convexity adjustment, *see*
    CMS, convexity adjustment,
    quasi-Gaussian model
  density approximation, 585

direct integration, 560, 585
Libor-with-delay, *see* Libor-
    with-delay, quasi-Gaussian
    model
linear local volatility, 547–550
    calibration, 550
    European swaption, 549
    for swaption strip, 549
    swap rate dynamics, 548
    swap rate inter-temporal
        correlation, 557
    swap rate variance ratio, 555
Markovian projection, 543, 566, 579,
    1139
mean reversion, *see* mean reversion
mean reversion calibration, *see* mean
    reversion calibration
Monte Carlo, 565
    Euler scheme, 565
multi-factor, 575–585
    benchmark rate correlations, 584
    benchmark rate parameterization,
        577
    bond reconstitution formula, 576
    calibration to spread options, 584
    correlation smile, 585
    loadings, 584
    local volatility, 576
    Monte Carlo, 585
    PDE, 585
    short rate state distribution in
        annuity measure, 580
    short rate state dynamics, 576
    stochastic volatility, 576–585
    swap rate dynamics, 579–583
    swap rate dynamics by Markovian
        projection, 579
one-factor local volatility, 541
    short rate state dynamics, 541
PDE, 562–565
    convection-dominated, 563
    domain truncation, 564
    space discretization, 563
short rate state distribution, 561
short rate state dynamics, 540
    in annuity measure, 544, 545
    in forward measure, 586
single-state approximation, 565–569
small-time asymptotics, 561

stochastic volatility, 569–574
    bond reconstitution formula, 570
    calibration, 572–573
    Monte Carlo, 574
    non-zero correlation, 574
    PDE, 574
    swap rate dynamics, 571–572
    unspanned, 570
swap rate dynamics, 542–547, 551
    approximate, 543–547
    approximate linear, 544
    approximate quadratic, 547
    swap rate variance, 546
    swap rate volatility, 542
TARN, *see* TARN, local projection
    method, quasi-Gaussian model
volatility swap, *see* volatility swap,
    quasi-Gaussian model

Radon-Nikodym derivative, 9, 1067
range accrual, 211
    CMS, 211
    CMS spread, 211, 766
    curve cap, 212, 766
    dual, 212, 766
    floating, 766
    product-of-ranges, 212
ratchet swap, 212
relative entropy, 957
replication method, 337, 724
    CMS, *see* CMS, convexity
        adjustment, replication method
    European option, *see* European-style
        option, replication method
    Libor-in-arrears, *see* Libor-in-arrears,
        replication method
    Libor-with-delay, *see* Libor-with-
        delay, replication method
    semi-static, 939
reserve, 986
rho, 980
Riccati, 364
Riemann zeta function, 128
risk limit, 985
risk measure, 996
    coherent, 996
risk sensitivities, 1093
    common definitions, 980
    delta, *see* delta

grid dimensioning for stability, 1002
grid shifting for stability, 1002
Jacobian method, 254–258, 985, 986,
    1105, 1106, 1111, 1118, 1119, 1121
off PDE grid, 1005
perturbation approach, 1050
vega, see vega
root search, 99
    Newton-Raphson method, 99, 116,
        235
    secant method, 235
Runge-Kutta method, 116, 365, 434,
    436, 516
running maximum, 124
running minimum, 124

SABR model, 343–345, 357, 951, 1121
    ad-hoc improvements, 705
    density tail, 761
    moment explosion, 344
    volatility smile expansion, 345
SALI tree, see tree, SALI
sausage Monte Carlo, see tube Monte
    Carlo
scripting language for trades, 220
SDE, see diffusion, SDE
SDE discretization, see Monte Carlo,
    SDE discretization
Sharpe ratio, 22
shifted log-normal model, see displaced
    log-normal model
short rate, 169
short rate model, 172
    affine, see affine short rate model
    affine one-factor, see affine short rate
        model, one-factor
    Black-Derman-Toy, see Black-
        Derman-Toy model
    calibration to yield curve, 457
        forward induction, 458
        forward-from-backward induction,
            460
    Cox-Ingersoll-Ross, see Cox-Ingersoll-
        Ross model
    Dybvig parameterization, 463–465,
        468
        HJM representation, 464
    econometric, 451
    empirical estimation, 451

forward volatility impact on
    Bermudan swaption, 878
Gaussian approximation, 1064
Gaussian model for basis spread, 683
Gaussian short rate, see Gaussian
    short rate model
Hagan and Woodward parameteriza-
    tion, 465–468
Ho-Lee, see Ho-Lee model
importance sampling, 1063–1065
log-normal, 445–451
    issues, 447
    Sandmann-Sondermann transform,
        448
Monte Carlo, 469–471
    Euler scheme, 469
    Milstein scheme, 469
    payoff construction issues, 470
    SDE discretization, 469
    variance reduction, 470
multi-factor, 479
path independence, 446
PDE, 456–457
    domain truncation, 456
power-type, 451
quadratic Gaussian, see quadratic
    Gaussian model
quasi-Gaussian, see quasi-Gaussian
    model
time-stationary, 418
volatility calibration, 461–463
    multi-pass bootstrap, 463
shout option, 935
    on capped coupon, 935
    optimal stopping time, 936
similarity reduction, 71, 871
    CLE, see CLE valuation, PDE,
        similarity reduction
    PDE, see PDE, similarity reduction
single-rate vanilla derivative, 697–763
    approximately single-rate, 709
    cap, see cap
    CMS cap, see CMS cap
    CMS floor, see CMS floor
    CMS swap, see CMS swap
    ED future, see ED future
    European swaption, see European
        swaption
    futures contract, see ED future

Libor-in-arrears, *see* Libor-in-arrears
Libor-with-delay, *see* Libor-with-delay
range accrual, *see* range accrual
singular value, 862
singular value decomposition, *see* CLE regression, SVD decomposition
truncated, *see* CLE regression, truncated SVD decomposition
singularity removal, *see* payoff smoothing, singularity removal
skew vega, *see* vega, skew vega
smile vega, *see* vega, smile vega
snowball, *see* CLE, snowball
snowbear, 213
snowrange, 213
snowstorm, 213
Sonia, 193, 200
spline, 229, 270–275
   Catmull-Rom, 238, 240, 271, 272
   cubic C^2, 273
   cubic smoothing, 248
   exponential tension spline, 243
   Hermite cubic, 238, 270–273
   interpolating, 248
   Kochanek-Bartels, 272
   least-squares regression, 248
   natural, 241
   natural cubic, 273
   shape preserving, 275
   smoothing, 234
   TCB, *see* spline, Kochanek-Bartels
   tension, 240, 243, 244, 246, 247, 250, 272, 274–275
      convergence to piecewise linear, 275
   tension factor, 243
spot Libor measure, *see* measure, spot
spot rate, *see* short rate
square-root process, 315
   $E(\sqrt{z})$, 1153, 1155
   basic properties, 319–320
   boundary behavior, 319
   conditional CDF, 319
   conditional moments, 319
   Feller condition, 319
   moment-generating function, 322, 342, 364, 372
      time-dependent parameters, 364
   moments, 375

Monte Carlo, 388–394
   Euler scheme, 389
   exact simulation, 388
   full truncation scheme, 390
   higher-order schemes, 390
   log-normal approximation, 390
   moment-matching schemes, 390
   Quadratic-Exponential scheme, 392, 394
   truncated Gaussian scheme, 391
multi-dimensional, 1152
PDF, 1153, 1155
stationary distribution, 320, 383
static replication, 210, 718
   CMS, *see* CMS, convexity adjustment, replication method
   European option, *see* European-style option, replication method
   Libor-in-arrears, *see* Libor-in-arrears, replication method
   Libor-with-delay, *see* Libor-with-delay, replication method
stochastic optimization method, 953
stochastic volatility model, 315–403, 571, 572, 1140
   as interpolation rule, 703
   ATM volatility, 348
   basket option, *see* Markovian projection, basket option in SV model
   calibration, 703–704
      calibration norm, 704
      normalization, 704
   caplet calibration, 707
   CEV type, *see* SABR model
   CMS convexity adjustment, 739
   correlation, 347
   dampening constant, 325
   delta, 699
   effective skew, 373
   effective volatility, 371, 372
   effective volatility of variance, 375
   European option, 327
      control variate, 328
      volatility mixing, 339
   explicit solution, 320
   for CMS rate, 739–744
      dynamics in forward measure, 741
   Fourier integration, 324–339

arbitrary European payoffs, 336,
    338
convolution, 325
direct integration, 330
discrete, 330
FFT, 330
for variance, 339–343
integration bounds, 330
strip of convergence, 329
with control variate, 328, 330
hedging, 353–358
level parameter, 318
link between forward and annuity
    measures, 740
LSV, see local stochastic volatility
    model
martingale property, 320
mean reversion speed, 316, 317, 348
half-life, 318
measure change, 322
moment explosion, 323
moment-generating function, 321,
    324, 327
branch cut, 330
singularities, 329
time-dependent parameters, 364
Monte Carlo, 387–398
Broadie-Kaya scheme, 395
Broadie-Kaya simplified scheme,
    396
exact scheme, 395
martingale correction, 398
Taylor-type schemes, 396
variance process, see square-root
    process, Monte Carlo
multi-dimensional, see multi-
    stochastic volatility model
PDE, 381–387
boundary conditions for stochastic
    variance, 385
boundary conditions from PDE
    itself, 385
discretizing spot, 387
discretizing stochastic variance, 383
for forward Kolmogorov equation,
    386
predictor-corrector, 382
quadratic discretization, 384
range for spot, 386

range for stochastic variance, 382
sinh transform, see PDE, sinh
    transform
sinh-quadratic discretization, 384
variable transform, 383, 385
process for variance, see square-root
    process
skew, 317, 346
smile dynamics, 347–349, 351, 353,
    354
SV volatility, 317
time-dependent, 363–403
asymptotic expansion, 366–370
averaging, see calibration, time
    averaging
Fourier integration, 363, 366
volatility of variance, 316, 317, 346
volatility of volatility, 318
stopping time, 6
straddle, 223
strategy, 7
doubling, 9
gains process, 8
permissible, 9
replicating, 11
self-financing, 8, 17
Stratonovich integral, 5
strike price, 23
structured note, see exotic swap
structured swap, see exotic swap
Student's t-distribution, 101
Monte Carlo, 101
survival measure
Bermudan swaption, see Bermudan
    swaption, survival measure
importance sampling, see importance
    sampling, survival measure
SV model, see stochastic volatility
    model
SV model with general variance process,
    359–361
martingale properties, 360
moment explosion, 361
properties, 359
stationary distribution, 360
strict supermartingale, 360
SVD, see CLE regression, SVD
    decomposition
SVI model, see volatility smile, SVI

swap, 197
  accreting, 199
  amortizing, 199
  annuity, XXXIV, 199
  annuity factor, 170
  averaging, see averaging cash flow
  cash-settled, 745
  CMS, see CMS swap
  effective date, 225
  fixed-floating, 198, 199, 230
    valuation formula, 199
  fixing dates, 198
  legs, 197
  Libor-in-arrears, see Libor-in-arrears
  Libor-with-delay, see Libor-with-
    delay
  par rate, see forward swap rate
  payer, 203
  payment dates, 198
  receiver, 203
  swap rate, see forward swap rate
swap market model, 619, 676–678
swap measure, see measure, annuity
swap rate, see forward swap rate
swaption grid, see European swaption,
    swaption grid

Tanaka extension of Ito's lemma, 7, 26,
    294, 1131
targeted redemption note, see TARN
TARN, 217, 218, 925–933
  cap at trigger, 219
  global model, 927
  impact of inter-temporal correlation,
    see inter-temporal correlation,
    impact on TARNs
  importance sampling, 1068–1077
    one-step survival conditioning, 1069
    removing first digital, 1068
  leverage, 927
  Libor market model, 927
  lifetime cap, see TARN, cap at
    trigger
  lifetime floor, see TARN, make whole
  local projection method, 928–931
    Gaussian short rate model, 929
    Markov-functional model, 931
    quadratic Gaussian model, 931
    quasi-Gaussian model, 931

make whole, 219
Markov-functional model, 475
multi-factor quasi-Gaussian model,
    927
partial analytical integration, 1011
pathwise differentiation method,
    1044
payoff smoothing, 1011, 1029,
    1068–1077
PDE, 931–933
  cap at trigger, 933
  make whole, 933
  Monte Carlo pre-simulation, 933
  upper bound for extra state
    variable, 932
tube Monte Carlo, 1029
valuation formula, 218
volatility smile, 927, 929–931
tenor structure, XXXIV, 170
  index function, 593
tension spline, see spline, tension
term parameters, 378
term structure model, 202, 277
terminal swap rate model, 709–716
  annuity mapping function, 710, 715,
    724, 726–727, 730, 731, 733
  as conditional expected value,
    726–727
  calibration to market, 730
  forward swap rate condition, 734
  forward value condition, 734
  in measure change, 737
  linear approximation, 729
  LM model, see Libor market model,
    annuity mapping function
  mean reversion, see CMS, convexity
    adjustment, impact of mean
    reversion
  multi-rate, 767
  swap rate squared condition, 735
  CMS convexity adjustment, see
    CMS, convexity adjustment,
    linear TSR model
  consistency condition, 710
  exponential TSR model, 714–715
  Libor-with-delay, see Libor-
    with-delay, swap-yield TSR
    model
  linear TSR model, 711

CMS convexity adjustment, *see*
CMS, convexity adjustment,
linear TSR model
forward CMS straddle, 941
mean reversion parameterization,
712
swap rate distribution in forward
measure, 738, 739
vega hedging, 714
loading from Gaussian model, 714
no-arbitrage condition, 710
PDF of swap rate in forward
measure, 739
from CMS caplets, 739
reasonableness, 710
swap rate distribution in forward
measure, 738
swap-yield TSR model, 715–716
CMS convexity adjustment, *see*
CMS, convexity adjustment,
swap-yield TSR model
theta, 980, 991
rolling yield curve, 992
Tikhonov regularization, *see* CLE regression, Tikhonov regularization
time decay, 52
time value, 26
"tip-top", *see* "flip-flop"
tower rule, *see* iterated conditional
expectations, 176
tree, 425
binomial, 446, 458
SALI, 78
trinomial, 51, 458
truncated Gaussian scheme, *see*
square-root process, Monte Carlo,
truncated Gaussian scheme
TSR model, *see* terminal swap rate
model
tube Monte Carlo, 1022–1030
barrier option, *see* barrier option,
tube Monte Carlo
Bermudan swaption, *see* CLE greeks,
tube Monte Carlo
CLE, *see* CLE greeks, tube Monte
Carlo
digital option, 1024
discrete knock-in barrier, 1028

generalized trigger product, *see*
barrier option, tube Monte Carlo
partial coupons, 1028
TARN, *see* TARN, tube Monte Carlo

underhedge, 1023
uniform distribution, XXXIII, 770
universal law of volatility, 1137
upwinding, *see* PDE, upwinding

value-at-risk, 501, 975, 995–998
conditional, 996
delta VaR, 998
delta-gamma VaR/cVaR, 998
Gaussian, 997
historical, 996
vanilla derivative, 697–815
multi-rate, *see* multi-rate vanilla
derivative
single-rate, *see* single-rate vanilla
derivative
vanilla model, 202, 277, 315, 1121, 1129
for multi-rate derivative, *see*
multi-rate vanilla derivative
for single-rate derivative, *see*
single-rate vanilla derivative
local volatility model, *see* local
volatility model
stochastic volatility model, *see*
stochastic volatility model
vanna, 980
VaR, *see* value-at-risk
variance reduction, 142–158
antithetic variates, 144
efficiency, 144
non-Gaussian, 145
common random number scheme,
132, 134
conditional Monte Carlo, 127
control variate, *see* control variate
from hedging strategy, *see* control
variate, dynamic
importance sampling, *see* importance
sampling
moment matching, 146
systematic sampling, 145
Vasicek model, 413–415
bond reconstitution formula, 414
bond volatility, 415

forward rate volatility, 415
short rate distribution, 413
short rate dynamics, 413
yield curve shapes, 414
vega, 355, 980, 1095–1125
    additivity, 1103
    Bermudan swaption, 1114
    bucketed shocks, 1099
    CMS spread option, 1116, 1120
        constant Libor correlations, 1120
    constant Libor correlations, 1115,
        1120
    constant term swap correlations,
        1116, 1118–1120
    cumulative shocks, 1099
    direct method, 1098–1102, 1110
        Bermudan swaption, 1103
        European swaption, 1102
        second-order effects, 1110
    European swaption, 1113
    flat shock, 1099
    forward swaption straddle, 948
    "good", 1102–1105
    hybrid method, 1111–1114
        algorithm, 1112
        Bermudan swaption, 1114
        CMS spread option, 1116
        European swaption, 1113
    in LM model
        coverage, 886
    indirect method, 1105–1110, 1121
        Bermudan swaption, 1109
        European swaption, 1108
        least-squares problem, 1106
        locality, 1107
        smoothing, 1107
    Jacobian method, see vega, indirect
        method; risk sensitivities,
        Jacobian method
    Libor market model, 1095–1125
        bootstrap calibration, 1111
        multi-factor, 1115
        projection, 1123
    local projection method, 869
    local vs. global, 1096
    locality, 1104
        benchmark set locality, 1104
        exotic locality, 1104
        full set locality, 1104

market vega, 984, 1096, 1110
model vega, 984, 1096, 1124–1125
pathwise differentiation method, see
    pathwise differentiation method,
    vega
projection, 1122–1124
relationship to gamma, 981
row shocks, 1099
running cumulative shocks, 1099
scaling, 1103
skew vega, 1114–1115
smile vega, 1114–1115
volatility, 27
    average convexity, 307
    Bachelier, see volatility, Normal
    basis point, see volatility, Normal
    Black, XXXIV, 204
    bp, see volatility, Normal
    CEV, 280, 625
    Dupire's, see Dupire local volatility
    factor volatility, 501
    forward volatility of Libor rate, 819
    Gaussian, see volatility, Normal
    implied, 278
        as average of realized, 989
        effects of mis-specification, 987
        most likely path approximation,
            990
        PDE for, 296
    local, see Dupire local volatility
    Normal, 204, 283, 625
    Normal for CMS spread option, 776
    separable, 300
    small-noise expansion, 307
    spanned stochastic volatility, 454
    spot volatility, 819
    spread, 776
        strike-dependent, 777
    stochastic, see stochastic volatility
        model
    unspanned stochastic volatility, 445
    "volatility squeeze", 424
volatility cube, see European swaption,
    volatility cube
volatility derivative, see forward
    volatility derivative
volatility skew, 279
volatility smile, 279, 315
    ATM backbone, 701, 702

backbone, 698
  adjustable, 699–702
curvature, 1138
dynamics, 279, 348, 698–702, 820
  sticky delta, 350, 352, 699
  sticky strike, 699
forward skew, 944
Gaussian backbone, 700
impact on forward volatilities, see
    forward volatility, impact of
    volatility smile
impact on inter-temporal cor-
    relations, see inter-temporal
    correlation, impact of volatility
    smile
probability density from, 278
SABR, see SABR model
shadow delta hedging, 699
skew vega, 1114
skew-dominated, 352
slope, 279
smile vega, 1114
SVI, 705, 951, 1121
upward sloping, 281
vega, 1114
volatility structure, 817
volatility swap, 220, 221, 933–945
capped, 937
CMS spread, 221
copula method, see copula method,
    volatility swap
fixed-expiry, 221, 940
fixed-tenor, 221, 940
impact of forward volatility, 944
impact of volatility smile dynamics,
    941
Libor market model, 933, 934
local projection method, 934
min-max, see min-max volatility
    swap
PDE, 934
quadratic Gaussian model, 941
quasi-Gaussian model, 941
with barrier, 222
with shout, 221, 935
volga, 980
Volterra integral equation, 438
vomma, 980

Wiener process, see Brownian motion

year fraction, 224
yield curve, 191, 230, 231, 233
  base index curve, 268
  basis risk, 270
  benchmark set, 230
  forecasting curve, see yield curve,
    index curve
  index curve, 261, 267, 679
  index-discounting basis, 197, 261
  instantaneous forward curve, 233
  joint evolution of discount and
    forward curves, 678
  multi-index curve group, 267–270
  overlay curve, 258
  perturbation locality, 230, 251–253,
    258
  Principal Components Analysis, see
    Principal Components Analysis
  ringing, 235, 242, 243, 252
  smooth, 258
  spread curve, 269, 886
  tenor basis, 230, 267
  TOY effect, 258
yield curve construction, 229–275
  benchmark set, 231
  bootstrapping, 234
    flat forward, 236
    linear yield, 235
  constrained optimization, 248
  cross-currency, 259
  cross-currency arbitrage, 260
  cubic spline C^2, 240–243
    problems, 242
  curve overlays, 258
  FX forwards, 259
  Hermite spline, 238–240
    iterative solution, 239
  Jacobian rebuild, 256
  multi-index curve group, 230, 265
  non-parametric fitting, 245–250
    norm specification, 245
    optimization algorithm, 245
  separate discount and forward curves,
    260
  spline, see spline
  spline fitting, 234–244
  tension spline, 243–244

yield curve risk, 250–258
    cumulative shifts, 256, 257
    forward rate approach, 252
    Jacobian method, *see* risk
        sensitivities, Jacobian method
    par-point approach, 251
    rolling for theta, 992

waterfall approach, *see* yield curve
    risk, cumulative shifts
yield curve spread option, *see* CMS
    spread option

zero-coupon bond, *see* discount bond
zero-coupon bond option, 185

Lightning Source UK Ltd.
Milton Keynes UK
UKHW051342080821
388278UK00007BA/438/J